CMOS Processors and Memories

ANALOG CIRCUITS AND SIGNAL PROCESSING

Consulting Editor: **Mohammed Ismail**, *Ohio State University*

For other titles published in this series, go to
www.springer.com/series/7381

Krzysztof (Kris) Iniewski
Editor

CMOS Processors and Memories

Springer

Editor
Krzysztof (Kris) Iniewski
CMOS Emerging Technologies, Inc.
Stanley Place 2865
V3B 7L7 Coquitlam British Columbia
Canada
iniewski@yahoo.ca

ISBN 978-90-481-9215-1 e-ISBN 978-90-481-9216-8
DOI 10.1007/978-90-481-9216-8
Springer Dordrecht Heidelberg London New York

Library of Congress Control Number: 2010933360

Printed on acid-free paper

Springer is part of Springer Science+Business Media (www.springer.com)

Contents

Part I Processors

1 **Design of High Performance Low Power Microprocessors** ... 3
Umesh Gajanan Nawathe

2 **Towards High-Performance and Energy-Efficient Multi-core Processors** ... 29
Zhiyi Yu

3 **Low Power Asynchronous Circuit Design: An FFT/IFFT Processor** ... 53
Bah-Hwee Gwee and Kwen-Siong Chong

4 **CMOL/CMOS Implementations of Bayesian Inference Engine: Digital and Mixed-Signal Architectures and Performance/Price – A Hardware Design Space Exploration** ... 97
Dan Hammerstrom and Mazad S. Zaveri

5 **A Hybrid CMOS-Nano FPGA Based on Majority Logic: From Devices to Architecture** ... 139
Garrett S. Rose and Harika Manem

Part II Memories

6 **Memory Systems for Nano-computer** ... 165
Yong Hoon Kang

7 **Flash Memory** ... 197
Taku Ogura

8 **CMOS-based Spin-Transfer Torque Magnetic Random Access Memory (ST–MRAM)** ... 233
B.C. Choi, Y.K. Hong, A. Lyle, and G.W. Donohoe

9 **Magnetization Switching in Spin Torque Random Access Memory: Challenges and Opportunities** 253
Xiaobin Wang, Yiran Chen, and Tong Zhang

10 **High Performance Embedded Dynamic Random Access Memory in Nano-Scale Technologies** 295
Toshiaki Kirihata

11 **Timing Circuit Design in High Performance DRAM** 337
Feng (Dan) Lin

12 **Overview and Scaling Prospect of Ferroelectric Memories** 361
Daisaburo Takashima

Part I
Processors

Chapter 1
Design of High Performance Low Power Microprocessors

Umesh Gajanan Nawathe

Abstract The field of Microprocessor design came into existence in the early 1970s with the first microprocessor from Intel (the 4004). Since then, the technology and complexity of Microprocessors has increased exponentially following Moore's Law, which implies that the complexity of integrated circuits doubles every 2 years. The Intel 4004 had a little more than 2,000 transistors. Some of today's microprocessors have more than two billion. Early on, microprocessor design philosophy focused on increasing performance without worrying about how it would affect power consumption. Today, power consumption has become a major design constraint. As a result, power-efficient design became a priority and 'power management' came into existence. Today, it is no longer enough to only reduce maximum power - it is equally important to reduce idle power and also have the ability to manage power and be able to operate at various performance-power points depending upon the customer's needs/choosing. Modern Semiconductor technologies have become very complex. Transistor Performance and Power is increasingly dependent upon its layout and also the layout that surrounds it. Variation of key Transistor parameters has increased and hence statistical analysis has become very important.

Keywords Microprocessor • Processor • CMT • Concurrent Multi-Threading • SPARC • Niagara • Power • Static Power • Leakage Power • Gate Leakage • Subthreshold leakage • Drain Leakage • Diode Leakage • Dynamic Power • Power-efficient design • Power Management • Dynamic Voltage and Frequency Scaling • Dynamic Frequency Scaling • DVFS • Leakage, Back-Bias • Clock Power • Clock Design • Clock Skew • Clock Uncertainty • Systematic Skew • Layout Dependent effects • SRAM design • Memory Design • SRAM redundancy • 6-T Memory Cell • Statistical Analysis

U.G. Nawathe (✉)
Oracle Corporation, 4110 Network Circle, Santa Clara, CA 95054, USA
e-mail: unawathe@yahoo.com

K. Iniewski (ed.), *CMOS Processors and Memories*,
Analog Circuits and Signal Processing, DOI 10.1007/978-90-481-9216-8_1,

1.1 Introduction

The field of Microprocessor design came into existence in the early 1970s with the first microprocessor from Intel (the 4004). Since then, the technology and complexity of Microprocessors has increased exponentially following Moore's Law, which implies that the complexity of integrated circuits doubles every 2 years. Apart from Intel, early processor designs came from Zilog(Z80), Motoroal (M6800/68000), and Texas Instruments (TMS1000). Over time, Intel's x86 based architecture emerged as the dominant architecture, especially for the Personal Computer (PC) industry. Advanced Micro Devices (AMD) was the other significant player designing processors on the same architecture, referred to as the x86 architecture. For more powerful higher end systems, there were computer architectures from other companies which gave Intel a run for their money. Chief among them were SUN Microsystems' SPARC architecture, MIPS Computer system's MIPS, and IBM's PowerPC and Power architectures. Early implementations of a lot of these architecture were based on RISC (Reduced Instruction Set Computing) as opposed to Intel's CISC (Complex Instruction Set computing). There were also a debate between Super-scalar vs. Super-pipelined approaches to designing processor. All these approaches were focused on improving performance at any cost without paying much attention to power dissipation. This mindset continued until a few years ago. The birth of the internet and its subsequent explosive growth caused the number of computer systems in a datacenter and the number of datacenters themselves to increase dramatically. Datacenters started running out of capacity because datacenter cooling limits were reached. In effect, datacenters started hitting the proverbial power wall. Thus, power-efficient design – not only at the processor/chip level, but also at the system level – became very important. The computer industry started looking into new ways of doing power efficient designs while at the same time maintaining/increasing performance and also ways to manage power consumption in real world scenarios. As a result, the 'Concurrent Multi-Threading' (CMT) approach of doing processor/system design came into existence [2, 3].

1.2 Concurrent Multi-threading (CMT)

The basic concept of CMT is to have the processor support the concurrent execution of multiple threads. A lot of high performance power-efficient processor designs today use the CMT architecture approach. CMT also helped reduce the impact of another bottleneck – memory latency. Figure 1.1 shows how memory speed (i.e. latency) has changed over time compared to processor frequency. Figure 1.2 illustrates how the CMT architecture attempts to reduce the impact of this issue. For a single thread, memory access is the single biggest bottleneck to improving performance. For programs that exhibit poor memory locality, only a modest throughput speedup is possible by reducing compute time. As a result, processors which are optimized for Instruction-Level-Parallelism have low utilization and wasted power. Having many threads makes it easier to find something

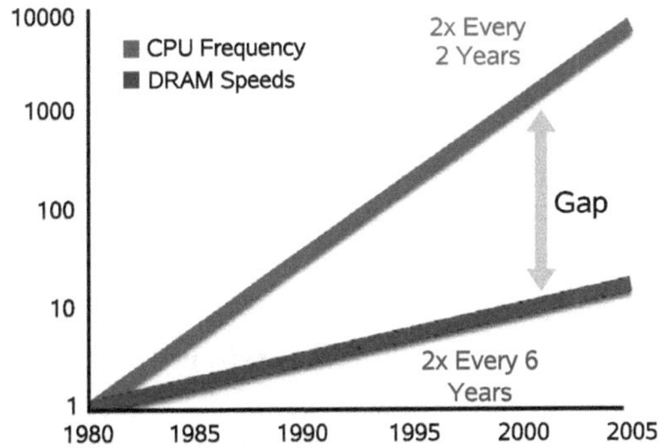

Fig. 1.1 Relative CPU frequency and memory speeds over time

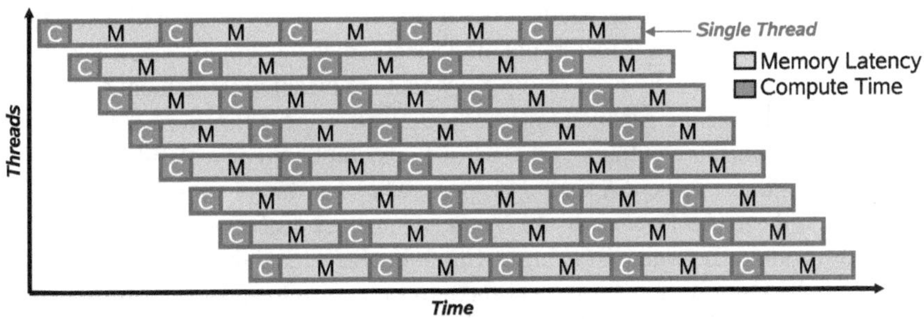

Fig. 1.2 Throughput computing using the CMT architecture (copyright © IEEE 2007)

useful to execute every cycle. As a result, processor utilization is higher, significant throughput speedups are achievable, and wasted power is reduced by reducing speculative execution (the result of which could get discarded).

1.3 Power and Power Management

As we discussed earlier, Processor/Chip and System Power consumption have become big design care-abouts. Hence it is important that we discuss Processor Power Consumption and Power Management in sufficient detail.

Total Chip power can be thought of as having two components: Dynamic Power and Static (Leakage) Power. Let's first focus on Dynamic Power.

1.3.1 Dynamic Power

Dynamic Power can be summarized using the following equation:

$$\text{Dynamic\_power} = \text{aCV}^2 f + \text{P\_crowbar},$$

where

a = Activity factor
C = Switching capacitance
V = Voltage (Vdd)
f = Frequency of operation.

1.3.1.1 Activity Factor and Switching Capacitance

Clock power is a big portion of the dynamic power. For one, the clock signal can switch twice every cycle (once low to high and once high to low). Also, the clock signal is integral to designing synchronous systems and is all-pervasive – every flip-flop or latch on the chip needs it. Reducing the activity factor will result in a linear reduction in power. Most, if not all, processors today employ clock-gating to achieve this. Conceptually a set of flip-flops that are meant to implement a specific architectural feature are grouped together and clock to these flip-flops is turned off during cycles when the outputs of these flip-flops don't hold any useful data and/or the flip-flops don't serve any useful purpose. The way this is implemented is that the clock to these flip-flops is driven by the same clock buffer or sets of clock buffers and its output is turned off using a 'Clock Enable' signal that is generated from logic signals. Even if one flip-flop in this group of flip-flops is performing useful work, the clock to all the flip-flops will have to toggle. Thus, the way flip-flops are grouped is important because it is important to find as may clock cycles as possible during which none of the flip-flops in the group is performing useful computation. Note that you are consuming extra power to generate the 'Clock Enable' signal. Obviously the grouping has to be done in such a way that the power saved by gating off clocks is more than the extra power needed to generate the 'Clock Enable' signals. Having an enable per flip-flop ends up costing too much power and having an enable per, say 1,000 flip-flops, generally ends up being not useful enough since the number of cycles when all the 1,000 flip-flops are not processing valid data is likely very small. Some studies have shown that having in the range of, say, 16 flip-flops per group is a reasonable number. In general, the above numbers are higher for data-paths than for control blocks containing random logic, since datapaths could have wide words (e.g. 128 bits wide) and all 128 flops in a row can generally be controlled by the same 'Clock Enable' signal. Note that the 'Enable' needs to satisfy timing constraints, i.e. it has to be evaluated every clock cycle and has to satisfy setup time constraints to the clock gate in the clock buffer.

Both Max Power and Idle Power are key distinct metrics that drive design decisions in modern processors. Idle power is especially important for processors used in laptops and other mobile devices to conserve battery charge. When the system is in the idle state, there are a lot more flip-flops and a lot more clock cycles when the flip-flops do not hold any useful data or don't do any useful work and hence lot more power savings can be achieved by clock gating when the system is idle.

There are several levels of clock gating implemented in modern processors. There is the 'flip-flop-group' level clock gating that we already discussed. There is also the 'functional-unit' level clock gating. For example, the clock to a 'floating point unit' can be turned off during cycles when the 'floating point unit' is not processing valid instructions. There is also 'chip-unit' level clock gating. For example, in today's multi-core processors, clocks to certain processor cores can be turned off when that processor core is not in use.

Other contributors to power are switching capacitances associated with routing wires, gates, and diffusions of transistors. Generally it is better to optimize wire classes (widths and spacings of wires) used for routing for minimizing 'power-delay product' as opposed to only 'delay' because if a designer does the latter, after a certain point he/she spends a lot on increased power consumption for a very small gain in performance – in effect the law of diminishing returns. Semiconductor technology companies are also making advances and making the transition to using 'lower-k' dielectrics which help reduce routing capacitance (in addition to reducing delay) and hence interconnect power.

Gate and diffusion capacitance of transistors also contribute significantly to power. It is always prudent to minimize sizes of gates in non-critical paths to minimize this power. As we will discuss later, this reduces leakage power as well, which is very important since, typically leakage power is a big (20–30%) portion of the total power in high performance processors.

Technology has a very important role to play in minimizing capacitance. Typically a linear shrink factor of 0.7–0.8 is achieved when a design in one technology node is ported to the next technology node (e.g. going from the 90 nm to the 65 nm technology node, or going from the 65 nm to the 45 nm technology node).

1.3.1.2 Voltage (VDD) and Frequency of Operation

Chip VDD has a big impact on power. As the equation at the beginning of Section 1.3.1 shows, dynamic power is directly proportional to the square of the voltage. As we will discuss later, Chip VDD also has a stronger impact on leakage power. In contrast, performance increases approximately linearly as VDD is increased. So, minimizing VDD is very important in order to reduce power consumption. Most processors today employ multiple voltages on chip. SRAM cell functionality is a big obstruction to reducing voltage because for voltages below what is called 'Vmin' of SRAM cells, the SRAM cells do not function reliably. As a result, quite often SRAM cells are put on a separate voltage supply (different from the rest of the chip) and the VDD to the rest of the chip is determined based on the chips performance/power requirements and tradeoffs. Very often, different parts of the logic on the chip is put on different supply voltages. For example, the high frequency processor cores are put on a higher supply voltage and the relatively lower frequency System on Chip (SOC) blocks are kept on a lower supply voltage. As we will discuss later, a lot of modern processors employ a technique called Dynamic Voltage and Frequency Scaling (DVFS). This means that when the processor is idle, its Frequency of operation and hence its VDD can be reduced to obtain a close to cubic power savings.

1.3.1.3 Crowbar Power

Crowbar Power is also a non-negligible component of dynamic power. This is really wasted power. When a logic gate switches, as the input transitions from 0V (VSS) to VDD or vice versa, there is a small amount of time when the N-transistors and P-transistors in a CMOS gate are conducting at the same time. During this time, some current flows directly from VDD to VSS. This current/power is wasted. The amount of this power strongly depends on the input and output slew rates. The larger the input slew rate, the longer the time when both N- and P-transistors are on and larger the crowbar current. The slew rate at the output of a gate also has an impact on the crowbar current because it determines the 'VDS' of the N- and P-transistors during the time when both of them are turned on. Processor physical designers always have limits on the maximum slew rates that signals on chips can have to limit crowbar power. Incidentally, other reasons why slew rate limits are in place are to limit age-related transistor degradation due to the Channel Hot Carrier (CHC) effect, improve signal noise immunity, and make sure the signals have a full transition between the power supply rails at the highest frequency of operation.

1.3.2 Static (Leakage) Power

As was previously stated, leakage power in processors today is typically 20–30% of the total power. Leakage power consists of sub-threshold leakage, gate leakage, and diode leakage.

1.3.2.1 Sub-Threshold Leakage

Typically, sub-threshold leakage is the largest component. As the name suggests this is the leakage power due to current flowing from the transistor's drain to source when the transistor is in the off state (VGS < Vth). For a CMOS inverter, when its input is at VSS, its output is at VDD and the leakage current is determined by the N-transistor leakage current. Similarly, when the inverter's input is at VDD, its output is at VSS and the leakage current is determined by the P-transistor leakage current. Leakage current of multi-stack gates, like nand2 and nor2, is lower than that of the inverter because there are two or more transistors in series.

Sub-threshold leakage is highly dependent upon the transistor channel length and threshold voltage of the transistor. Transistors with minimum channel length and lowest threshold voltage have the largest sub-threshold leakage. One of the techniques used to control sub-threshold leakage is to use longer channel length transistors. This is also very useful from a statistical point of view. The threshold voltage variation of longer channel length transistors is significantly smaller than minimum-channel-length transistors. Hence, statistically for a large group of transistors, leakage is significantly lower if the channel length is increased. It is especially important since leakage increases exponentially as threshold voltage reduces.

Typically, every process technology offers two to four kinds of core (i.e. Non-IO) transistors having different threshold voltages. The HVT (highest threshold voltage) transistors from this menu of transistors typically have 10–20% of the leakage of standard Vt transistors (at the cost of having 30–50% higher delay). These transistors can be used in logic paths where the number of levels of logic is small. Additionally, typically design teams have available to them a library of gates which use longer channel length transistors (sometimes they are referred to as GBIAS transistors since their gate lengths are 'biased' higher). These gates are typically designed to be footprint compatible with the corresponding cells from the library which use minimum channel length. Generally, the gate 'bias' is chosen so that the delay penalty is not as high as HVT devices. Consequently, the leakage reduction is not as much. Typically, 50% leakage reduction at a 15% delay penalty is targeted, though these numbers could vary based on design choices and semiconductor technology. Gates which are not in critical paths are substituted with these GBIAS gates to reduce leakage.

1.3.2.2 Gate Leakage

Several years back, gate leakage was fairly small compared to sub-threshold leakage. As semiconductor processing technology shrank from one generation to next, the gate oxide became thinner and thinner. Gate oxide thicknesses started approaching 10s of Angstroms, and gate leakage started increasing exponentially. Fortunately, the newest generations of process technology (45 nm and beyond) started employing metal gate technology, which effectively got rid of the gate leakage problem by reducing it by orders of magnitude. This can be seen from the four graphs in Figs. 1.3 and 1.4. These figures show sample results for a non-metal-gate and metal-gate process respectively. In these graphs, the magnitude of drain and gate leakage is shown relative to gate leakage at the lowest voltage/temperature point of the respective graph. As the graphs indicate, the ratio of gate to drain leakage current is lower in the metal-gate process (vs. a non-metal-gate process) by almost two orders of magnitude.

1.3.2.3 Diode Leakage

Diode leakage is the reverse biased leakage current flowing through the reverse biased diode formed by the source or drain of the transistor and the bulk or the well that the transistor resides in. Under normal bias circumstances, this current is very small. As we will see shortly, one of the techniques used for controlling sub-threshold leakage is to reverse bias the bulk (also called back bias) and this leads to an increase in diode leakage current.

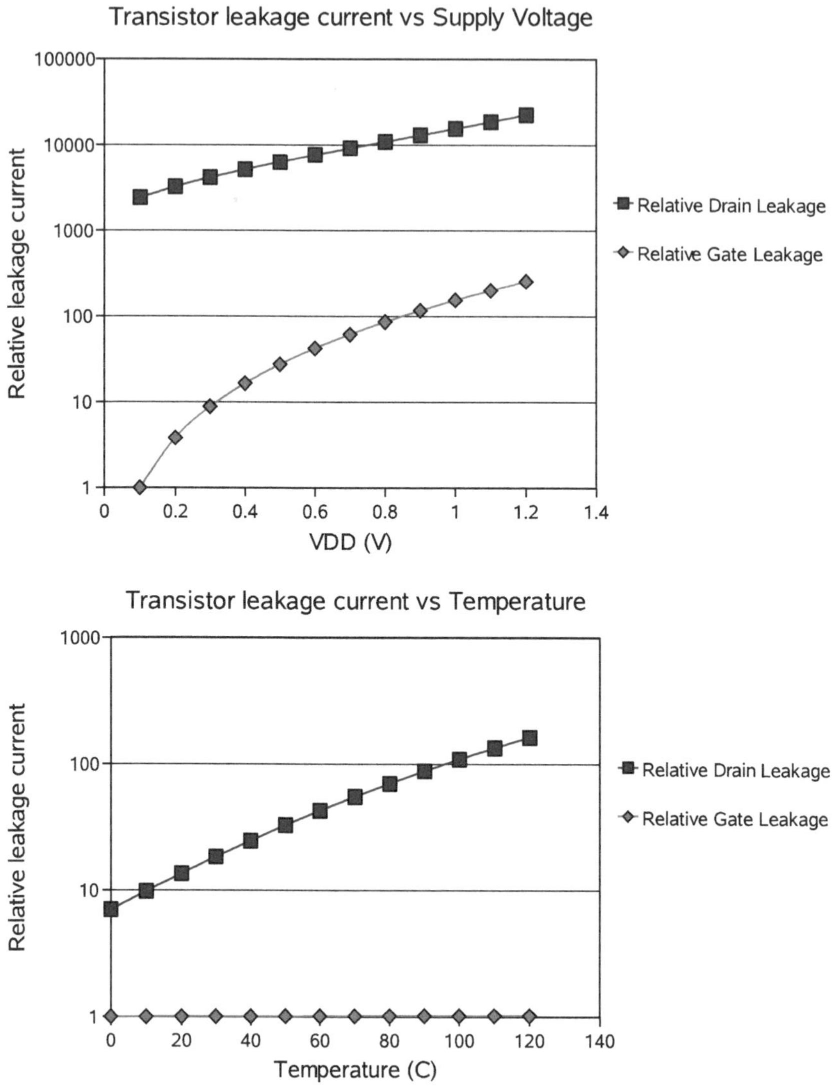

Fig. 1.3 Dependence of leakage on supply voltage and temperature for a non-metal-gate semiconductor process

1.3.2.4 Effect of VDD and Temperature on Leakage

One of the big overlaying factors that affects leakage is VDD. Leakage power has a larger than cubic relationship to VDD. Hence, in modern power-conscious processors, an attempt is made to put significant portions of the chip on lower supply voltages. That is also the reason why an attempt is made to turn off the power supply to cores or significant sections of the chip when they are not in use.

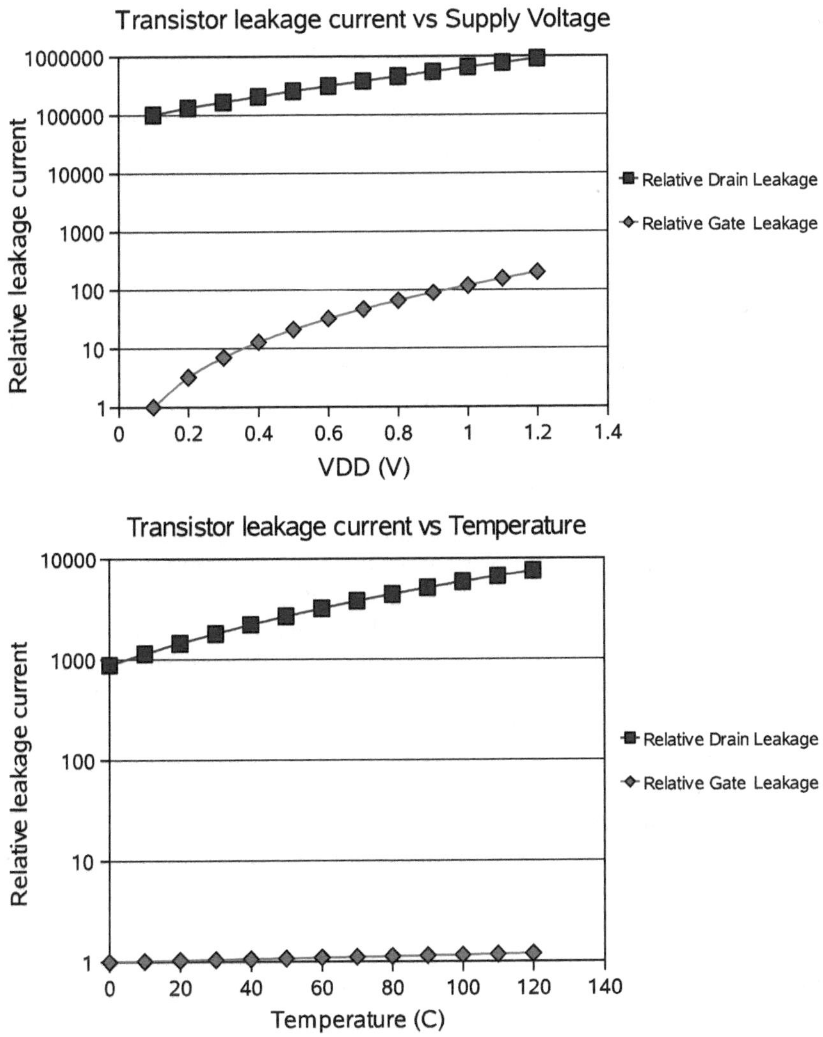

Fig. 1.4 Dependence of leakage on supply voltage and temperature for a metal-gate semiconductor process

Temperature also has a large effect on leakage. Gate leakage is relatively independent of temperature, but sub-threshold leakage and diode leakage increase as temperature increases. Hence, design of system cooling solutions is very important. Most systems are air cooled using a set of fans that blow air over chips in a system. Typical systems have heat sinks sitting on top of processors. These heat sinks are designed such that they have a large amount of surface area in contact with air and this helps heat escape faster and ultimately helps lower the maximum junction temperature. Under certain circumstances, it might be worth investing in a better cooling solution, e.g. liquid cooling. However the cost of the cooling solution has to be taken into account in the context of the type of system that is being designed. Typically, only the high end systems are able to afford advanced higher-cost cooling solutions.

Figures 1.3 and 1.4 show how much drain leakage (which is a big portion of the total leakage) changes with Temperature and Supply Voltage for a typical non-metal-gate and metal-gate process.

1.3.2.5 Back Bias

Back bias is a commonly used technique to control leakage power. The principle behind back bias is the effect that body bias has on the threshold voltage of the transistor. The equation of threshold voltage can be written as

$$\mathrm{Vth} = \mathrm{Vth0} +/- \gamma * (|\,\mathrm{VSB}\,|)^{1/2}$$

Where, the +ve sign is used for N-transistors and –ve sign is used for P-transistors. Vth0 is the threshold voltage of the transistor when the Source and the Bulk are at the same potential. As you can see, as the magnitude of VSB (voltage between the transistor Sources and the Bulk) increases, the diode formed by the source and the bulk becomes more reverse biased and the magnitude of Vth increases. This in turn causes sub-threshold leakage current to reduce. This is equally applicable to N- and P-transistors. In the case of P-transistors, as VSB (voltage between the transistor Source and the Nwell) becomes more negative, the Source-Nwell diode becomes more reverse biased and the magnitude of Vth increases (Vth becomes more negative). Consequently sub-threshold leakage current decreases. Note that an increase in 'back bias' reduces leakage current, but at the cost of increase in delay. After the 'back bias' is increased beyond a certain point, the reverse bias diode leakage current increases and becomes more dominant At that point, the effectiveness of back bias to reduce leakage current vanishes. Figure 1.5 illustrates the various leakage current components for a modern metal-gate process. In this particular example, you can see that beyond a back bias voltage of about 0.5 V, total leakage current is dominated by diode leakage current and back-bias as a technique to reduce leakage current is no longer useful. Note that, using the same principle, as you apply a 'forward bias' to the Source-Bulk or Source-Nwell junction, Vth of transistors reduces, the sub-threshold leakage current increases, and the transistor delay reduces helping improve performance.

1.3.3 Using VDD and Back-Bias to Optimize for Performance and Power

So, how is this used in the real world? Processors that have been shipping for a few years and that ship today have to meet frequency and power constraints. Due to process variation, a portion of the distribution of parts that comes from the fast

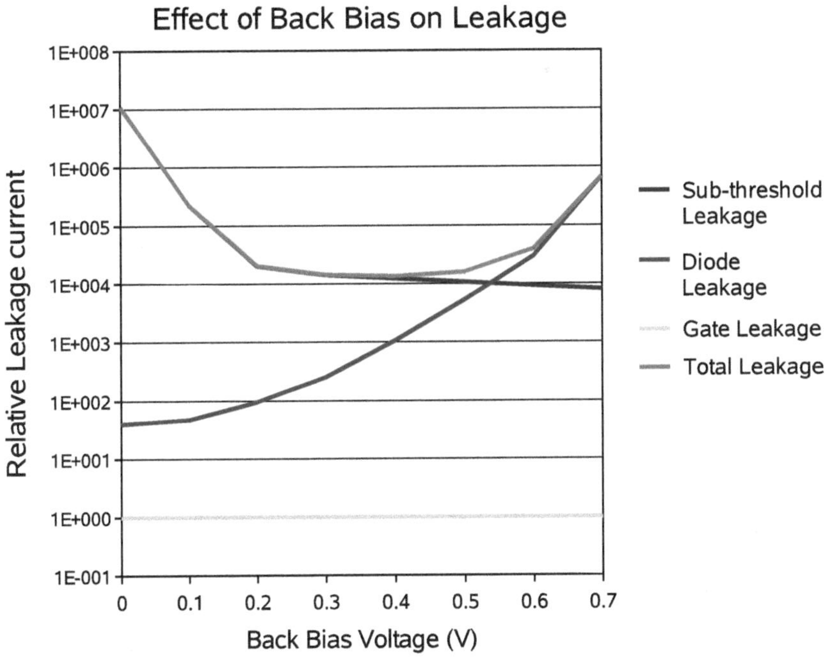

Fig. 1.5 Effect of back bias on leakage

corner of the transistor process could have more than enough margin in terms of meeting frequency requirements. However, some of these parts may be over the power constraint. For these parts, back bias is an effective technique to trade of frequency margin for a reduction in power. In most processors that are shipped today, the amount of N- and P-back bias is burned into the on-chip fuse lookup table at the time the chip is tested. This information is read by the system firmware and the value of the dc–dc converters on the board that drive the back-bias voltage is set accordingly.

Note that the back bias technique can also be used for the reverse reason for chips that come from that portion of the process distribution which has slower than typical transistors. These chips could be well below the power limit but are not fast enough to meet the frequency target. For these chips, a 'Forward bias' can be applied to trade off power for frequency.

Just as the back-bias voltages can be independently set per chip, the VDD can be independently set per chip to improve the frequency-power yield of a chip. For chips that are from the fast part of the distribution, the VDD is decreased to get their power below the power budget, as long as they still meet the target frequency. Similarly for chips that are from the slow part of the distribution, the VDD is increased to get their frequency above the minimum limit while keeping their power below the limit. As with back bias, the on-chip fuse array is used to implement this technique.

Note that there is a limitation on what the N and P back bias voltages can concurrently be set to. For example, if the N Back bias voltage is increased far more than the P back bias voltage, the P-transistor becomes stronger relative to the N-transistor and the effective 'Beta ratio' of a typical logic gate decreases by a large amount. Similarly, if the P Back bias voltage is increased far more than the N back bias voltage, the N-transistor becomes stronger relative to the P-transistor and the effective 'Beta ratio' increases by a large amount. Obviously, if the design of various circuits on the chip is not robust enough to take this into account, the circuits could fail. For this reason, the circuit design methodology and the design window has to be decided after taking into account the productization strategy of the chip.

1.3.4 Power Management: What and How?

Power Management in the context of a computer system/processor can be defined as the ability to manage the power consumption of the system/processor within a set power constraint/budget and/or opportunistically be able to reduce system/processor power consumption when the right set of circumstances present themselves. Having defined it, let's look at some commonly used techniques that are used to enable/implement Power Management.

1.3.4.1 Dynamic Voltage and Frequency Scaling

One very powerful Power Management technique is called Dynamic Voltage and Frequency Scaling (DVFS). The basic concept of DVFS is the ability to dynamically change the voltage and frequency at which the chip operates depending upon performance requirements and power constraints at any given point in time. For example, when the workload on a processing engine is low or when the processing engine is idle, the frequency and the voltage at which it operates can be lowered to reduce power consumption. Here, the word 'dynamically' means on the fly – i.e. without having to reset the system/processor. The processor and or processing engine will transition to a new frequency and/or voltage as it continues to run whatever code/workload it was running.

A common way of implementing DVFS is to have two different phase locked loops (PLLs) – say PLL1 and PLL2. Let us assume that the system is operating at a particular voltage and frequency using PLL1. Once the decision is made to change the voltage and frequency at which the processor is operating, the PLL2 is locked to the new operating frequency and the processor clock tree is switched to PLL2 using a multiplexer. Note that if the PLL2 frequency is lower than PLL1 frequency, the processor is switched to PLL2 first and then the voltage is lowered to the point at which it can support the new lower frequency. If PLL2 frequency is higher than PLL1 frequency, the voltage is increased to the point that is needed to support the higher frequency first before the clock tree is switched to PLL2.

Typically, DVFS is implemented through OS/software control. The software interacts with control registers on chip to implement the frequency change. In some implementations, it also interfaces with the DC–DC converter on chip to change the voltage at which the processor operates.

There are some processors in the market today, especially from Intel, where there is a dedicated Power Management unit inside the chip. In these chips, typically each processing core is powered by its own separate power domain and in some cases its own PLL, so that their frequency and voltage can be independently controlled. This unit has a lot of intelligence and can decide with little interaction with the OS what frequency and voltage each processing core should be running at.

Computer systems are typically designed for a specified maximum temperature that the transistor junction can be at. Having one or more temperature sensors on chip is vitally important to ensure that this specification is met. The thermal sensor continuously monitors the maximum chip junction temperature and provides this information to the on chip Power Management Unit or the OS/software. If the junction temperature exceeds the set limit, DVFS is used to reduce the voltage and frequency, which reduces power consumption and which results in the junction temperature reducing below the limit. Apart from DVFS, there are other ways of implementing this as well. For example, some systems throttle the chip (i.e. reduce instruction issue rates) to reduce power consumption and achieve the same result. Figure 1.6 illustrates this for Sun Microsystems' Niagara2 Processor [1–3].

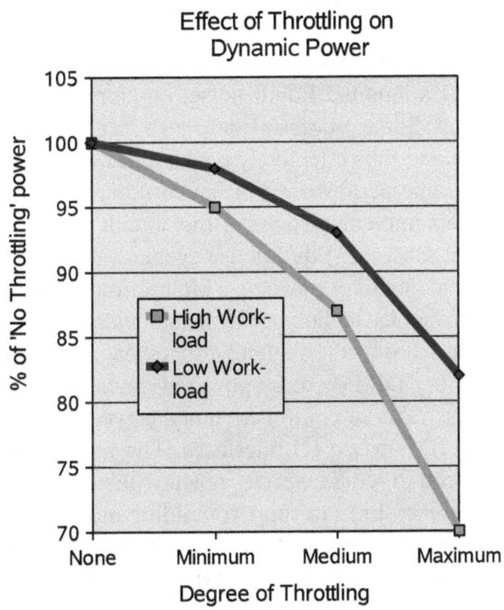

Fig. 1.6 Power reduction by throttling (copyright © IEEE 2007)

The temperature limit for which the system is designed is called the Thermal Design Point (TDP). Generally, the worst case average power consumed by real applications is about 10–20% lower than that consumed by a 'power virus'. Setting the TDP based on power consumed by a 'power virus' could lead to over-design and increased system cost. Generally, the design of the chip packaging solution, system cooling solution, etc is done in such a way that most (if not all) real applications operating at their worst case power points do not cause the junction temperature to exceed the TDP. In the pathological case when somebody is running a 'power virus', the system will get throttled back to make sure the maximum junction temperature (Tjmax) doesn't exceed the TDP. This way, over-design (and hence) system cost is reduced. Note that in general:

Worst case average power for power virus > TDP > Average CPU Power consumption.

For a multi-core processor, DVFS can also be used to get higher single thread performance than what you would have gotten under normal circumstances. Let's say for example that a processor has four processing cores and the power budget that each core is expected to operate under normal circumstances is 'P'. If there is a piece of code that requires higher single thread performance, three of the four cores can either be turned off or the frequencies that they operate at could be reduced and the power savings achieved by doing this could be transferred as extra power budget to one core which can be run at a higher than nominal frequency and voltage to get increased single thread performance.

1.3.4.2 Other Power Management Techniques

The on-chip circuit performance is dictated by the voltage that the transistors see, which is the Pin VDD minus the Ldi/dt noise, on-chip IR drops, and any other power supply tolerances. Some processors have also incorporated circuits to reduce the board Ldi/dt noise and thus effectively enable reduction of the Pin-VDD (and hence power) for the same performance or increase the processor performance at the same pin-Vdd. Most implementations of this circuit attempt to slow down the chip frequency when it senses an Ldi/dt noise event.

In a computer system, memory power is a big portion of the total power. So, lot of processors provide hooks to help control and manage memory power. For example, on Sun Microsystems' Niagara2 processor, the memory controller is design to enable Memory DIMMs to be powered down. It also has the ability to control Memory access rates to control Memory power consumption. Most processors today have SerDes-based IO interfaces. The number of SerDes lanes in operation can be reduced to reduce SerDes power consumption. Processors today also have several coherency links to support building multi-way glueless systems. For low end systems (e.g. 2-way system vs. an 8-way system), the number of coherency links required is less than the maximum. For these systems, the extra coherency links and the associated logic can be turned off to reduce processor and system power.

1.4 Clock Design

It is very important to design a low skew/uncertainty, low power clock distribution network for a high performance low power microprocessor. By its very nature, the quality of the clock distribution network has a truly chip-wide impact. For every peco-second (ps) of clock uncertainty that the clock distribution has (as compared to a reference design), the performance of the rest of the design has to be better by that amount to achieve the same chip frequency. Every addition ps of clock uncertainty will also mean there are some additional min-time paths (also called hold-time paths) that will need to get fixed to avoid functional failure. Also, the clock signal has, by far, the highest activity factor on chip and hence accounts for a large percentage of the total power (could be as much as 25–33% of the total power). Hence it is very important that the clock distribution architecture supports clock gating and other power saving techniques and the clock physical design is done to minimize power. The clock architecture must also be defined to support Design for Test (DFT) and help make chip debug easier.

The kind of clock distribution required is closely tied to the kind of synchronization element used on chip. For example, for a flip-flop based design a single-phase clock is required and for a latch-based design where a logic path can borrow time across the next cycle, a 2-phase clock is needed.

Typically the chip Global Clock Distribution starts at the on-chip PLL, goes to the center of the chip and is then distributed to different parts of the chip using a balanced distribution, like an H-tree distribution. This is illustrated in Fig. 1.7. The layout of the clock drivers is done with the best possible layout techniques to reduce variation due to processing differences. For example, all transistors in clock buffers are shielded with dummy transistors. The clock wires are shielded to prevent noise from getting injected into the distribution, which would increase clock jitter. The clock buffers are designed to support clock gating at various levels, e.g. processor core level, functional unit level, and flip-flop group level (group of flip-flops that are driven by the same clock driver/s).

Some processors also use clock grids at various levels to reduce skew. Typically, clock grids consume more power than clock trees but result in smaller clock skew. One of the requirements for designing a clock grid is that the difference between arrival time of the clock signal at the input of the grid drivers shouldn't be more than a certain limit. Otherwise, you will have multiple grid drivers driving the clock grid in different directions for a substantial amount of time. The result is a bad/unacceptable clock waveform and lot of wasted VDD to VSS crowbar current/power. However, note that one of the functions of the grid is indeed to reduce the amount of skew of the clock signal at the input of the grid-drivers that gets transferred to the output. The ratio by which the skew at the input of the grid gets reduced when it appears at the output of the grid is referred to as the skew reduction factor. Another advantage of using a clock grid is to help connect together different regions having different amounts of clock load and reduce the overall clock skew between these regions. There are some processors

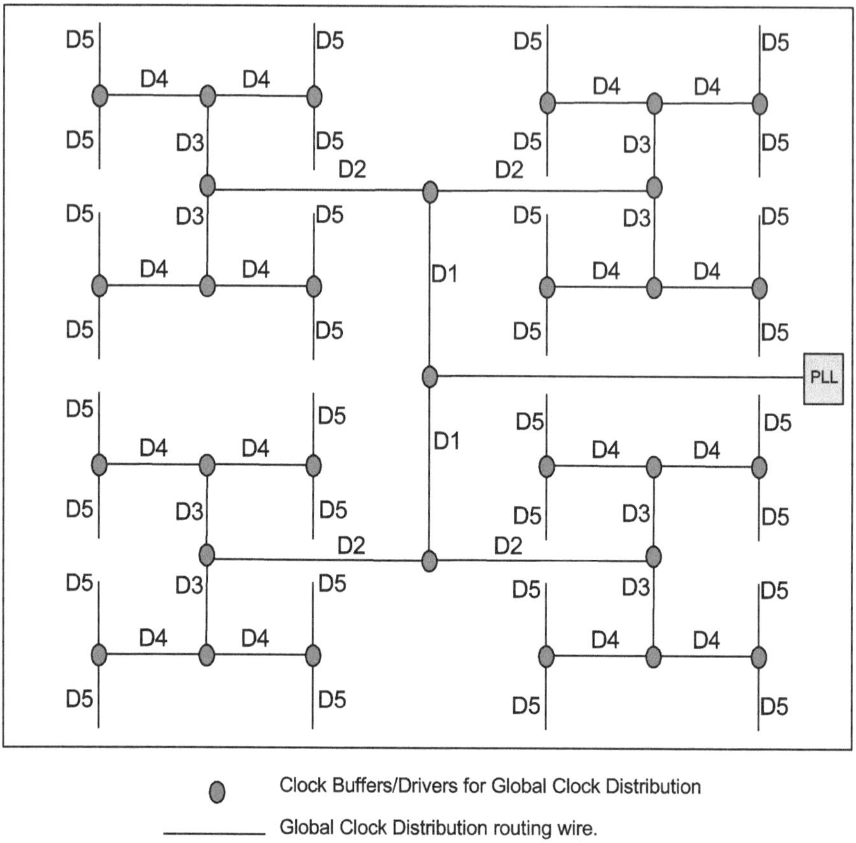

Fig. 1.7 Typical chip global clock distribution

which employ a single clock grid over the entire chip. However this is not very common since it consumes too much power. Other techniques are used to reduce clock skew as well. For example, the clocking scheme in Intel's Merom Processor [7] implements an active deskewing scheme using digital delay lines to minimize the skew across the die.

Modern processors have several asynchronous and/or mesochronous interfaces on chip. For example, let's take a look at AMD's Quad Core Optron Processor [5]. It is based on a flexible clocking architecture, designed to scale across designs containing different number of cores. Each core has its own PLL, clock distribution circuit, and power grid. The clock (and power) to each core is controlled independent of the others. This allows each of them to be operated at their own performance/power points. Communication between the cores and SOC (Northbridge) is asynchronous. A synchronous mode is provided for tester determinism.

Reference [6] talks about a low power SOC from PA Semiconductor (which was since acquired by Apple Computer). The clocking for this processor is built using a balanced H-tree distribution. The changing core VDD's poses a challenge to the clocking scheme. Due to the adjustable core VDD, the clock tree delay of the core is variable while the delay for the SoC is fixed. Keeping the core coherent with the coherent crossbar is vital for reducing Level 2 Cache (L2) and memory access latency and for simplicity of architecture and logic design. A phase detector is used to solve this problem. The phase detector detects the phase difference between the core and SOC clocks and the result is used to choose a specific source and destination clock pair to transfer data to maximize setup time and minimize hold time.

Modern processors have several hooks designed into the clocking schemes to support debug and also optimize frequency after testing silicon. Clock stop, clock stretch, and clock shrink have been used for a long time. Some processors employ clock drivers which support movement of clock edges in sections of the chip (w.r.t. the clock edges on the rest of the chip) as a debug aid. This is described in Ref. [7] and is illustrated in Fig. 1.8. Here, two sets of delay-tunable clock buffers 'A' and 'B' drive clocks to Clock domains 'A' and 'B' respectively. For all sets of frequency critical paths where the source flip flop and destination

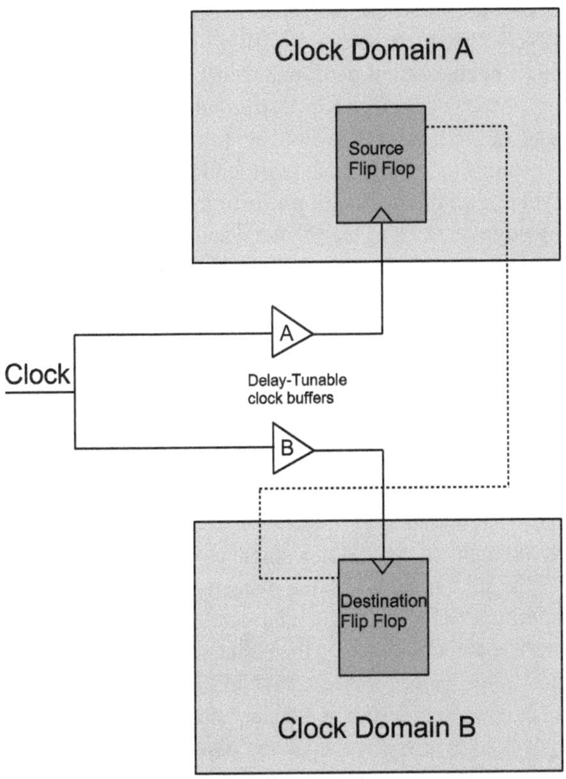

Fig. 1.8 Locating critical paths in silicon by moving clock edges

flip flop are in different clock domains, the frequency at which the critical paths operate can be changed by varying the delays of clock buffers 'A' and 'B'. This can help locate the exact critical path. This mechanism is referred to as the 'Locate Critical Path' (LCP) Mode in Ref. [7]. In LCP Mode, the delay of the local clock drivers is controlled by configuration registers programmed by the Test Access Port (TAP). Programming these enables after testing the chip can help us to get better performance for the same power point. IBM's Power6 processor [4, 8]) essentially uses the same concept to alter flip-flop launch and capture times using programmable delays in the local clock buffers to improve performance of chip frequency critical paths.

1.4.1 Clock Skew/Clock Uncertainty

Clock Skew or Clock Uncertainty is calculated differently when you are considering a frequency critical path vs. if you are considering a hold time path. Let's take an example of a flip-flop based design which is synchronized using a single phase clock. For a single clock cycle critical path, under ideal circumstances, the rising edge of clock cycle 'n + 1' at the destination flip-flop should occur later than the rising of clock cycle 'n' at the source flip-flop by exactly the clock cycle period. Similarly for a half cycle critical path, under ideal circumstances, the falling edge of clock cycle 'n' at the destination flip-flop should occur later than the rising edge of clock cycle 'n' at the source flip-flop by exactly half the clock cycle period. When you are considering a hold time path, under ideal circumstances, the rising edge of clock cycle 'n' at the destination flip-flop should occur exactly at the same time as the rising edge of clock cycle 'n' at the source flip-flop. The amount of time by which the clock edge at the destination flip-flop deviates from its ideal position is called clock uncertainty or clock skew.

1.4.1.1 Sources of Clock Skew/Clock Uncertainty

There are several sources of Clock Skew: Cycle-to-cycle PLL jitter, systematic skew, skew resulting from transistor and interconnect process variation, power-supply voltage noise induced skew, skew due to temperature gradient across chip. Assuming we are discussing Clock Skew in the context of a single clock cycle path, Cycle-to-cycle PLL jitter is defined as the amount by which the rising edge of clock n + 1 can deviate from its ideal position in time w.r.t. the rising edge of clock N at the output of the PLL. In other words, this is the amount by which the period of the cycle (at the source PLL) can differ from its ideal value. Hence it is also called PLL period jitter. Typically it is specified in terms of 'ps' per 'sigma' (i.e. a standard deviation). More often than not, the on-chip PLL is powered by a voltage regulated power supply to help minimize this.

Systematic skew is the skew induced due to the differences in the structure of the design. For example, ideally the clock designer attempts to match the length and metal type of every branch of the H-tree distribution. But due to some reason, e.g. chip floorplan restrictions, there could be a finite mismatch in the wire length between two branches. The skew due to this mismatch in length is classified as systematic skew. Referring to Fig. 1.7, all wire sections labeled 'D1' (or 'D2' … 'D6') ideally have same metal routing topology (metal layer, length, width, spacing, etc). For example, if all 'D1' segments do indeed have identical topologies but one of the D1 segments is longer than the remaining D1 segments, this length difference will introduce a skew which will be categorized as systematic skew. Also, if a clock buffer is driving a group of flip-flops, the difference in clock arrival times at the inputs of these flip-flops, say due to RC delay, is also classified as systematic skew. Obviously, systematic skew is under the control of the designer. Every attempt should be made to minimize it.

Process variation is another component of clock skew. Even though a set of clock buffers at different locations of the chip have identical designs (including identical layouts), the buffers will have some variations between them once they go through the various processing steps and actually get built on chip. The variations could result from, from example, differences in gate length, gate width, differences in transistor threshold voltage, etc. Wires with identical layouts will also vary w.r.t. one another once they get built on the chip – due to variation in wire thickness, wire width, dielectric thickness, etc. Hence the actual delay through buffers and wires with identical layouts will be different and will result in clock skew. This skew can be minimized by using highly controlled layouts. Basic statistics tells us that a buffer with large number of gate fingers will have a smaller variation than a buffer with smaller number of gate fingers. Now-a-days (increasingly so with 45 nm technologies and beyond), the context in which a gate resides (the transistors/channel area that surrounds the gate) also has an impact on its delay. Hence, the context should be as controlled as possible.

Power-supply noise induces changes in buffer delays and hence results in clock skew. For this reason (as well as for several other reasons), a good quiet power supply distribution is very important in order to design a high performance processor. Having a lot of on-chip de-coupling capacitors is very important to reduce power-supply noise. It is especially important to have a lot of de-coupling capacitors near drivers that switch at high frequencies and high activity factors – like clock drivers. Finally, depending upon the kind of workload that is being run on the processor, a temperature gradient can exist across the chip and since temperature affects transistor drive current, it will induces differences in delay, which translate into clock skew.

1.5 Memory Design

In this context, the word 'Memory' is a term loosely used to refer to all storage elements on the chip that are not flip-flops. In a typical microprocessor, multiple types of memory cells are used depending upon the design requirements. Typically, the semiconductor foundary provides one or more 6-transistor (6T) memory cells.

The smallest cell is generally used for the largest cache memory. This is also the memory that is farthest from the processor in terms of memory hierarchy and has the largest on-chip access time. Since this cell is small, it has the smallest read current. Generally, the foundary will provide a larger cell – which also has a larger read current and is used for memories that need higher performance, like the primary data/instruction caches. Cells with more transistors are generally used as well. The additional transistors in these cells are used to increase the number of read and write ports. These cells are generally used in structures like register files, queues, etc. However, the basic storage element is generally similar to the 6T cell. Hence it is instructive to study the 6T cell in more detail.

1.5.1 The 6-T Memory Cell

Figure 1.9 shows a basic representation of a 6T memory cell. The back-to-back inverters store the data. The cell is accesses using the Word Line (WL) and the Bit Lines (BL and BLb – collectively referred to as BLs) through the two N-pass transistors. Of the transistors in the back to back inverters, the P-transistors act as 'load' transistors and the N-transistors act as 'driver' transistors. The ratio of the drive strength of the 'driver' transistor to the pass-gate transistor is called the 'β' ratio. If we normalize all drive strength w.r.t. the pass-gate transistor, the drive strength of the 'driver' transistor will be 'β'. This ratio is an extremely important parameter in the design of a 6T cell. It affects the DC and AC performance of the cell and also affects the cell area. Ultimately, the 'β' ratio ends up being a tradeoff between write margin, memory cell stability, read current, and cell area. The cell is accessed by driving the WL high. To write the cell, the write drivers have to drive the BLs (which have several memory cells connected to them) and over-power the 'driver' transistor of the cell being written. To read data from the cell, the BLs are pre-charged high, WL of the accessed cell is asserted, and the 'driver' transistor of

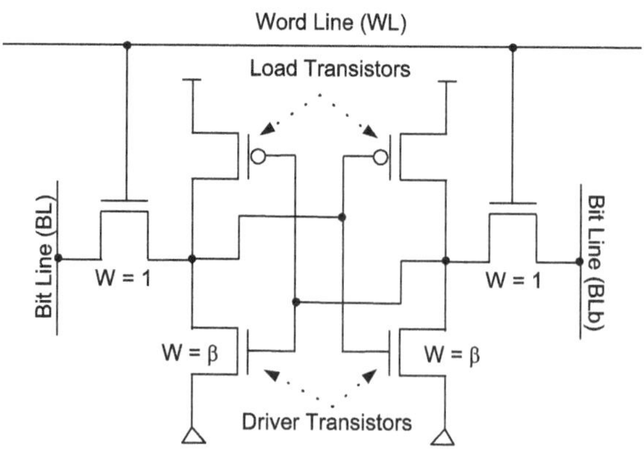

Fig. 1.9 6-T SRAM cell

the side that stores a low slowly starts pulling the BL low. Care has to be taken to make sure the accessed cell doesn't flip the data it has stored (read-disturb). If 'β' is too large, the cell area becomes large and it becomes difficult to write new data into the cell. Also, if we assume the area of the cell to be fixed, a large 'β' ratio will limit the drive strength of the 'pass-transistor' and thus limit read current and hence speed. If 'β' is too small, the cell becomes vulnerable to read-disturb due to power supply noise, channel length/width and threshold voltage imbalances between the two sets of transistors in the storage element, etc. The read current also suffers.

1.5.1.1 Important Metrics/Tests for Evaluating a 6-T Memory Cell

There are several kinds of tests that are performed on the memory cell itself to make sure it is robust enough. One of them is the Static Noise Margin (SNM) test. This tests the ability of the cell to store logic 0 or logic 1(VDD) in spite of external DC noise sources, e.g. variation in channel width/length and threshold voltages between like transistors on either half of the (symmetric) cell, unequal transistor degradation over the lifetime of the cell due to '1' being stored for a longer time than '0' or vice versa, etc. This is most often performed by plotting the widely used 'butterfly-plot' of the cell (Fig. 1.10). The diagonal of the largest square (squares in Fig. 1.10) that can fit inside the area enclosed by the transfer curves represents the SNM of the

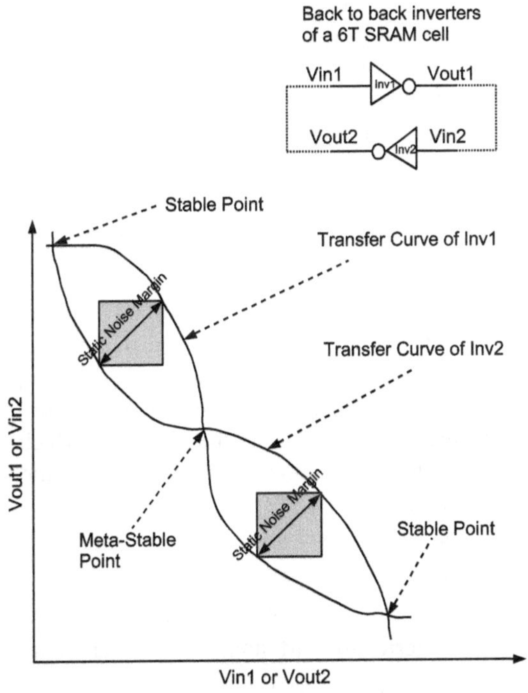

Fig. 1.10 Butterfly plot of a 6T SRAM cell

cell. A second test tests the writeability of the cell. This test find the difference in the voltages of the BLs required to flip the cell. A third test tests the stability of the cell at different values of VDD. This is performed by applying a very slow (to make it appear as a DC test) VDD ramp and checking the lowest VDD at which the cell is able to maintain its stored data (or the VDD at which the cell flips to the opposite state). This helps define the 'Vmin' of the cell (the minimum voltage at which the cell is functional) and also helps understand the robustness of the cell to Power Supply offsets. A fourth test tests the read current of the cell. This test is performed to help us measure the amount of current the cell can sink to pull one of the pre-charged bitlines below VDD. The larger the current, the faster the cell can pull the BL low and faster the cell data can be read.

1.5.2 Memory Redundancy

Redundancy strategy is a big part of memory design in microprocessors today. Most designs have built in redundant rows or columns of memory cells. The redundancy strategy has to be conceived after detailed discussions with the semiconductor foundary on yield, failure modes (rows/columns/random bit fails), etc and the processor architects on the desired memory architecture. The presence of redundant rows and/or columns of memory cells enables test engineers to recover chips that otherwise would have to be thrown away by replacing bad rows, columns, or bits of memory cell by spare functional rows or columns. This is implemented by simple multiplexers that are controlled by fuse bits that are burned into the fuse array at the time of testing the die. Needless to say, the redundancy strategy can have a big impact on the cost of the product as well.

1.5.3 The Importance of Statistical Analysis

Statistical analysis is becoming more and more important in the design of processors today. This is especially the case for memory design. There are several reasons as to why that is the case. A big reason is that the caches in modern processors generally are several MBytes in size. So, there are millions of 6T cells on the same chip, most (all if there is no redundancy built into the design) of which need to function for the die to yield. Also, transistor variability, especially for the small transistors used in the 6T cells, is increasing.

Variations can be broadly classified into:

1. Chip Mean (CM) variation
2. Across Chip variation, also called ACLV (Across Chip Line Variation)
3. Local Variation, also referred to as Mismatch (MM) variation

CM variation encompasses the entire spec range of the semiconductor manufacturing process. This includes the variation between different dies on the same wafer, wafer to wafer variation, and lot to lot variation. ACLV variation is the variation seen across the same die. In effect, this is the variation across the scanner field. MM variation is the variation between two identical transistors or wires (or any other identical component) that are placed next to each other (or within a local area) on a single die. The actual parameters that vary include (but are not limited to) transistor channel length, transistor channel width, transistor oxide thickness, transistor threshold voltage, interconnect thickness, interconnect width, inter/intra-layer dielectric thickness, etc.

Each metric of interest for a 6T cell has to be evaluated with statistical models that account for all these kinds of variation and the designer has to ensure that the appropriate operating margin exists after accounting for the statistics associated with the number of instances of each memory cell that the designer plans to have on the chip. Obviously, the desired yield and the amount of redundancy that is built into the design will have an impact on these calculations as well. The details of how these calculations are done are outside the scope of this chapter.

1.6 Process Technology and Impact of Layout on Performance and Power

As semiconductor manufacturing technology has continued to advance from one generation to next as per Moore's Law, transistor geometries have continued to shrink and the interaction between transistor layout and key transistor parameters has continued to increase. As a result, the way a transistor is laid out and the context in which the transistor exists on chip has come to have a significant impact on the performance of the transistor. So, layout and context-induced performance changes have to be taken into account when designing and characterizing circuits. The use of stress/strain to improve transistor performance has also contributed to this. For example, tensile stress along the channel direction of an NMOS transistor (and compressive stress along the channel direction of a PMOS transistor) improve the transistor's drive current. So, semiconductor manufacturers deposit stress liners on transistors to get the desired kind of stress to boost performance. If there are several transistors of the same width and length in a stack, each transistor will see different amounts of stress and will thus have different drive current. Shallow Trench Isolation (STI) technique is the technique of choice to isolate one transistor from another. The Oxide in the STI also generates stress in the adjoining diffusion thus affecting the transistor performance. The amount of stress will depend upon the distance of the STI from the diffusion of the transistor. Similarly, ions during well implants can scatter at the edge of the photoresist and can get embedded in transistor channel regions. These act as additional dopants and alter the threshold voltage of the transistor. The closer the transistor channel is to the edge of the well, the higher is the concentration of the scattered dopants. Photolithographic effects are also a big factor, especially since the geometries being drawn are smaller and smaller fractions of the wavelength of light being used. 90° angles on layout actually become rounded edges on silicon and this has to be taken into account when doing layouts and predicting performance of the transistors. In most modern semiconductor technologies, all transistors need to be shielded by dummy transistors or dummy field polys to ensure that the channel length variation is kept below a certain limit. For modern technologies, especially 45 nm and beyond, transistor spice models now have to take the layout type and the layout context into account, and hence the transistor models have become very complex.

1.7 Conclusion

Microprocessor design has dramatically evolved since the birth of the first Microprocessor (Intel's 4004) in the early 1970s, to today. The Intel 4004 had a little more than 2,000 transistors. Today's microprocessors can have more than two billion. As can be expected, design and manufacturing complexity has increased by several orders of magnitude. Computing power has found more and more applications. The hunger for computing power has increased and this has propelled the field of microprocessor design forward.

In the last 10 years, the growth of the internet caused a dramatic increase in the need for more computing power. Tens of thousands of data centers were built. Data centers started hitting the coolable power limits. Power-efficient design and Power-management rapidly gained in importance. CMT became popular.

What would the next 20 years bring? What would a microprocessor look like 20 years from now? How would the computing landscape change during the next 20 years? Only time will tell. It would be foolish to venture a guess but let me do so just for the sake of it. Microprocessors would likely have 100s of billions, of transistors. Computing threads will be available in plenty. These computing threads might be spread across multiple locations all around the globe separated by thousands of miles, but will likely be accessible enough through the internet. There would be great advances in software to make use of this computing power more efficiently. Lowering power consumption will continue to be extremely important. Network computing will truly come into its own. Whatever happens, one thing looks certain - the hunger for computing power will continue to increase and the field of microprocessor design will continue its march forward.

References

1. Umesh Nawathe et al., An 8-Core, 64-Thread, 64-Bit, Power Efficient SPARC System on a Chip, ISSCC, February 2007.
2. Greg Grohoski et al., Niagara2: A Highly Threaded Server-on-a-chip. Hot Chips Symposium, August 2006.
3. Robert Golla et al., Niagara2: A Highly Threaded Server-on-a-chip, Microprocessor Forum, October 2006.

4. Joshua Friedrich et al., Design of the Power6™ Microprocessor, ISSCC, February 2007.
5. J. Dorsey et al., An Integrated Quad-Core Optron™ Processor, ISSCC, February 2007.
6. Zongjian Chen et al., A 25W SoC with Dual 2GHz Power™ Cores and Integrated Memory and I/O Subsystems, ISSCC, February 2007.
7. Nabeel Sakran et al., The Implementation of the 65nm Dual-Core 64b Merom Processor, ISSCC, February 2007.
8. B. Curran et al., Power-constrained high-frequency circuits for the IBM POWER6 microprocessor, IBM J. RES. & DEV. VOL 51 NO 6, November 2007

Chapter 2
Towards High-Performance and Energy-Efficient Multi-core Processors

Zhiyi Yu

Abstract Traditional uni-core processors have met tremendous challenges to improve their performance and energy efficiency, and to adapt to the deep submicron fabrication technology. Meanwhile, traditional ASIC implementations are also widely prohibited due to their inherent inflexibility and high design cost. On the other hand, rapidly advancing fabrication technologies have enabled the integration of many processors into a single chip, called multi-core processors, and promise a platform with high performance, high energy efficiency, and high flexibility.

This chapter will discuss the motivations of shifting from traditional IC systems (including uni-core processors and ASIC implementations) to multi-core processors, investigate the design cases of multi-core processors and their key features, and look forward to the future work.

Keywords Multi-core • Processors • ASIC • Energy-efficient • Network-on-Chip (NoC)

2.1 Motivating Multi-core Processors

2.1.1 Challenges on Uni-core Processors

Previous IC designers have been mainly concerned with fabrication cost and performance. Minimizing the number of transistors to reduce the area is the main approach to reduce the cost, and increasing the clock frequency is the main approach to increase the performance. Currently, how to achieve energy efficiency and how to adapt to the advanced fabrication technologies also become important challenges.

Z. Yu (✉)
State-Key Laboratory of ASIC & System, Fudan University
No. 825 Zhangheng Rd., Shanghai, 201203, P.R. China
e-mail: zhiyiyu@fudan.edu.cn

K. Iniewski (ed.), *CMOS Processors and Memories*,
Analog Circuits and Signal Processing, DOI 10.1007/978-90-481-9216-8_2,

2.1.1.1 High Performance Innovations are Challenging

Increasing clock frequencies and using wide issue architectures has worked well to improve processor performance, but recently has become significantly more challenging.

Deeper pipelining is one of the key techniques to increase the clock frequency and performance, but the benefit of the deeper pipeline is eventually diminished when the inserted Flip-Flop's delay is comparable to the combinational logic delay. Moreover, deeper pipeline stages increase cycles-per-instruction (CPI) and negatively impacts the system performance. Research has found that the depth per pipeline is approximately 8 Fanout-4 (FO4) inverter delays to obtain the highest performance [1], which corresponds to 20–30 pipeline stages for a typical programmable processor. The pipeline delay of some modern processors is already close to ten FO4 [2] so the deeper pipelining technique for high performance is reaching its limit. Also, increasing pipeline stages necessitates more registers and control logic, thereby further increasing design difficulty as well as power consumption. As reported by A. Hartstein [3], the optimum pipeline depth for maximum energy efficiency is about 22.5 FO4 delay (about 7 stage pipeline), using $BIPS^3/W$ as the metric – BIPS are billions of instructions per second.

The other key technology – shrinking the size of transistors to increase the clock frequency and integration capability – has amazingly followed Moore's Law [4] for about 40 years. Although the pace of this innovation is still going on, the limitations are right ahead in another couple of generations: either because of the physical limit when the size of transistors approaches the size of atoms, or because of the fabrication cost prohibiting further progress.

Wide issue processor architectures such as VLIW and superscalar are other efficient approaches for high performance computation while their benefit is also quickly diminished when the issue width is more than 10; since most applications do not have many independently parallel executable instructions per fetch. For example, an eight-way VLIW TI high performance C6000 DSP [5] and an eight-wide superscalar RISC microprocessor [6] was reported in 2002, and there are no wider-issue examples reported since then.

As the result of all those challenges mentioned above, the nearly 50% performance improvement per year lasting about 20 years has comes to the end, as shown in Fig. 2.1. New computer architectures are urgently demanded to overcome those challenges.

2.1.1.2 Power Dissipation Becomes the Key Constraint

The high performance design of modern chips is also highly constrained by power dissipation as well as the circuit constraints. Power consumption is generally dominated by dynamic power with the trend that leakage power is playing a growing role. The dynamic power of a common static gate is described by Eq. 2.1

$$P = aCV^2f \tag{2.1}$$

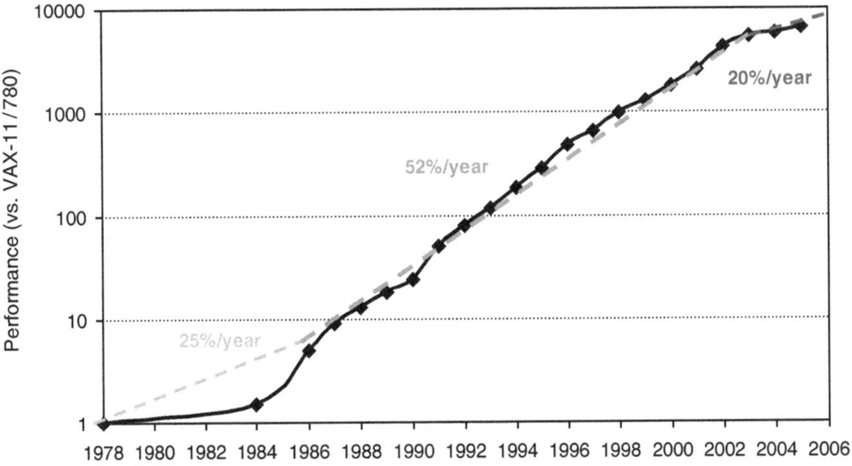

Fig. 2.1 Growth in processor performance since 1978 [7]. Further improving of the processor performance is challenging and the 16-year renaissance is over

where a is the circuit state transition probability, C is the switched circuit capacitance, V is the supply voltage, and f is the clock frequency. The leakage power mainly results from the reduction of the transistor threshold voltage [8] and is also highly dependent on the supply voltage. Most high performance techniques, such as increasing clock frequencies and increasing processor issue-widths (which means increasing number of circuits and increasing capacitance) result in higher power consumption. All these imply a new era of high-performance design that must now focus on energy-efficient implementations [9].

Portable devices powered by batteries are strongly affected by their power consumption since it determines their operational life time between each battery charging. Traditional non-portable systems such as PCs are also concerned with power consumption, since it highly determines the packaging costs, cooling system costs, and even limits the operation speeds and integration capacities of the systems. Figure 2.2 shows the power consumption of main Intel microprocessors from 1970 to 2006. The data between 1970 to 2000 is from S. Borkar [10] where he found that the changes of the power consumption follow the Moore's law increasing from 0.3 to 100 W, and estimated that the processor power consumption will go up to 1,000 W in a couple of years if this trend continues. Similarly to the power consumption, the power density increased significantly from a couple of W/cm² to about 100 W/cm² and becomes another key issue. This trend has been mostly halted recently thanks to low power techniques such as voltage and frequency control technologies. The power consumption of recently-reported microprocessors is still around 100 W. It also implies that 100 W is the power limit the current packaging technology and cooling technology can tolerate at reasonable cost. Power consumption has become the highest constraint for designers and limits the achievable clock frequency and processor performance [9].

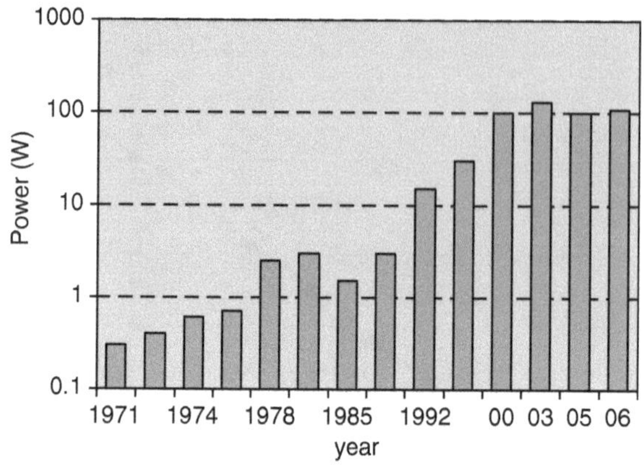

Fig. 2.2 Power consumption of Intel microprocessors from 1970 to 2006; data from 1970 to 2000 are from [10]; data in 03, 05, and 06 are from [11–13] respectively

2.1.1.3 The Gap Between Dream and Reality on Performance and Energy Efficiency

H. D. Man shows a future ambient intelligent computing example which illustrates the gap between the future requirement and the current reality in both performance and power [14]. In his opinion, there will be three major devices in the future intelligent computing system. One is the main powerful computation components, like today's PC; its target performance is 1 TOPS, with power consumption less than 5 W, corresponding to the energy efficiency 100–200 GOPS/W. This requirement is about 1,000 times higher than today's PC which has about 10 GOPS while consuming 100 W, corresponding to 0.1 GOPS/W. The second device is the handable devices powered by battery, like current mobile phone, targets to 10–100 GOPS with less than 1 W, corresponding to 10–100 GOPS/W. This requirement is about ten times higher than current solutions which use RISC processors and/or DSP processors. The third component is the sensor network to receive and transfer information, powered by energy harvesting methods such as mechanical vibration, with power consumption less than 100 μW; developing such low power components is another challenging topic.

2.1.1.4 Future Fabrication Technologies Imposing New Challenges

Future fabrication technologies are also imposing new challenges such as wiring and parameter variations.

In the early days of CMOS technology, wires could be treated as ideal. They transmit signals with infinite speed, without power consumption, and without coupling effect. This assumption is no longer true. For global wires, their length is nearly constant along with the technology scaling if the chip size stays the same; which makes their delay nearly constant. Compared to gate delay which scales

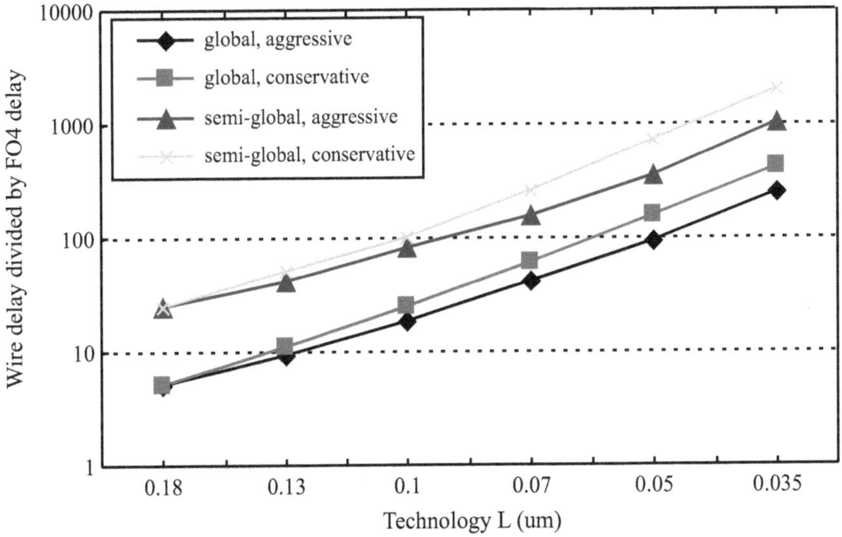

Fig. 2.3 Wire delays (in FO4s) for fixed-length (1 cm) wires [15]; global wires have larger width, height, and spacing which result in smaller resistance than semi-global wires; aggressive and conservative are two extreme scaling projections

down with the technology, the delay of the *global* and *long* wires scales up with the technology. As shown in Fig. 2.3 [15], a 1 cm long global wire delay in modern 65 nm technology is around 100 FO4 delay; which corresponds to more than one clock cycle period in modern processors and the global wires have to be pipelined or be eliminated through architecture level innovations. Similar with the delay, the power consumption of the long wires also scales up along with the technology compared to the gates. Besides the effect on delay and power consumption, the inductive and capacitive coupling between wires adds signal noise and impacts system reliability.

Furthermore, future fabrication technologies are expected to have tremendous variability compared to current technologies in both gates and wires. Fundamental issues of statistical fluctuations for submicron MOSFETs are not completely understood, but the variation increases leakage of transistor and causes a variation of the speed of individual transistors, which in turn leads to IC timing issues [16]. Borkar et al. reported chips fabricated in advanced nanometer technologies can easily have 30% variation in chip frequencies [17]. This variance will be present at time of fabrication and also have a time-varying component.

2.1.2 Challenges on ASIC Implementations

ASIC implementations have the key advantages of much higher performance and energy efficiency compared with other implementation techniques such as programmable processors and FPGAs, which makes ASICs widely used in applications

Table 2.1 The multiple HDTV transmission standards, traditionally implemented using ASICs

Standards	DVB-S2	DVB-C	DVB-T	ATSC	DTMB
Tran. media	Satellite	Cable	Wireless	Wireless	Wireless
Countries	Europe	Europe	Europe	USA	China

where high performance and high energy efficiency are the key design constraints. But unfortunately, ASIC implementations have inherent drawback of lacking flexibility, which make it less attractive along with the keep increasing design and fabrication cost, especially in application domains demanding multi-standard solutions.

Firstly, the design cost and fabrication cost of IC systems keep growing because of the increasing complexity of modern chips and the more advanced fabrication technologies. The total design and fabrication costs of a modern chip can easily run into the tens of millions of dollars. It is desirable to make the IC systems flexible enough so that the cost can be shared among a variety of applications.

In addition, nowadays many applications (such as communication, and multimedia) are implemented based on specific standards. Different application domains have different standards, and even in one application domain there can have multiple standards. As shown in Table 2.1, High-definition television (HDTV) transmission systems have at least five standards for different countries (areas) and different transmission media. Consumers certainly wish to have a product which can be used in different places and the product suppliers also wish to have ONE solution for all areas to simplify their management. In other words, people want to see flexible solutions which can support multiple standards of one application domain, or even different application domains.

Overall, it is highly desirable to have high flexible IC systems to lower the cost and to support multiple standards, but unfortunately ASICs lack flexibility inherently.

2.1.3 Solution: Multi-core Processors

In order to address the challenges the uni-core processors and ASIC implementations faced, innovations on computer architecture and design are required. Deep submicron fabrication technologies enable very high levels of integration such as a recent dual-core chip with 1.7 billion transistors [12], thus reaching a key milestone in the level of circuit complexity possible on a single chip. A highly promising approach to efficiently use these circuit resources and to address the current IC design challenges is the multi-core processors which integrate multiple processors onto a single chip. The multi-core processors will soon enter a new era called multi-core processors as the number of cores in a single chip keeps increasing.

Multi-core processors can easily achieve high performance through parallel processing. In addition, multi-core processors can provide high energy efficiency since they can allow the clock frequency and supply voltage to be reduced together to dramatically reduce power dissipation during periods when full rate computation is not needed. Giving a simple example, assuming one uni-core processor is capable of computing one application with clock rate f and voltage v, and consuming power p; now if using a dual core system and assuming the application can be partitioned into two cores without any overhead, then each core only needs a clock rate $f/2$ and the voltage can be reduced accordingly; assuming a linear relation between voltage and clock frequency and the voltage is reduced to $V/2$, then the power dissipation of the dual core system will only be about $p/4$ calculated by Eq. 2.1. Multi-core processors also provide the opportunity to independently control each processor's clock and voltage to achieve higher energy efficiency, if different processors are in separate clock and voltage domains.

Furthermore, multi-core processors are suitable for future technologies. The distributed architecture can potentially constrain the wires into one core and eliminate the global (long) wires. The multi-core processors also provide flexible approaches to treat each core differently by adjusting the mapped application, supply voltage, and clock rate; to utilize each core's specific features due to variations. For example, when one processor in the chip is much slower than the others, a low workload can be mapped on it without affecting system performance.

The multi-core processors have high scalability since a single processor can be designed and the system can be obtained by duplicating processors, thus the systems can be easily adjusted according to the required performance and cost constraints by changing the number of cores, which is much easier than changing the issue-width or the pipelining of uni-core processors.

Of course, multi-core processors have inherent flexibility through programming, and are applicable in much wider application domains than ASICs.

Multi-core processors started to emerge in 2003 [18] and became widely used in 2005 [12, 19, 20], and currently quad-core processors have become the main style. The trend to migrate from uni-core processors to multi-core processors is clear. The research community already presented some multi-core processors such as 16-core RAW [21] in 2003, 36-core AsAP [22] in 2006, and 80-core Intel chip [23] in 2007.

Although the trend is clear, there are many issues that remain unclear in designing efficient multi-core processors. For example, how to design cores or processing elements in the multi-core processors, how to efficiently connect these cores, and how to efficiently program multi-core processors, etc. As was already pointed out by the parallel computer group at the EECS department of UC Berkeley [24]: the "shift toward increasing parallelism is not a triumphant stride forward based on breakthroughs in novel software and architectures for parallelism; instead, this plunge into parallelism is actually a retreat from even greater challenges that thwart efficient silicon implementation of traditional uniprocessor architectures".

2.2 Pioneering Multiprocessor Systems and Multi-core Processors

2.2.1 Communication Model: Shared-Memory and Message Passing

The idea of multiprocessor systems can be traced back to early 1970s, when a *shared memory* multiprocessor architecture was proposed [25] in which processors exchange data through a global memory. In early days, the memory is normally located out of processors in a global fashion, called centralized (or symmetric) shared-memory architecture, as shown in Fig. 2.4a. In order to reduce the traffic between processors and global memory, each processor normally has some cache embedded; which raises the issue of multiple copies of a single memory word being manipulated by multiple processors. This so called *cache coherence* issue significantly increases hardware complexity. Another major issue of the shared memory system is its limited scalability due to its centralized structure. Some work such as DASH [26] tried to improve the scalability of shared-memory systems by distributing the global memory to each processor node, called distributed shared-memory architecture.

Later another multiprocessor system platform using *message passing* communication was proposed [27] where processors exchange information by sending/receiving data between processors in a point-to-point fashion. Message passing architectures simplify hardware and also increase system scalability, but normally increase programming complexity. Some work such as FLASH [28] built systems to support both share-memory and message passing communication protocols in one hardware platform efficiently.

Figure 2.4b can be used to demonstrate both the distributed shared-memory and message passing architectures. In shared-memory architectures the memories in different processors are physically local while logically global, and in message-passing architectures the memory in each processor is local in terms of both physically and logically.

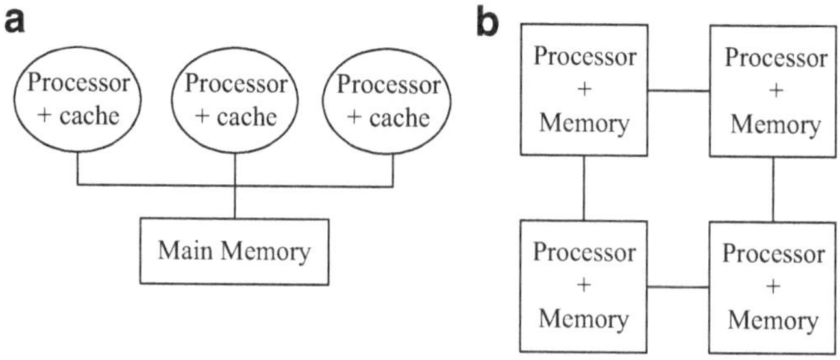

Fig. 2.4 (**a**) Basic structure of a centralized shared-memory multi-core processor; (**b**) basic structure of a distributed shared-memory or message-passing architecture

2.2.2 Interconnect Topology

Under the two communication models, there are many interconnect topologies. The most basic topology is probably the *global bus* where all elements communicate through the global wires. The bus topology is simple, but it doesn't scale well since increasing the number of nodes on the bus can quickly introduce congestion and reduce communication performance; furthermore, the increasing delay/power of long wires in modern fabrication technology also makes the bus unattractive.

Crossbar is a network which is organized by a grid of switches to provide the connection from one group processors (memories) to another group of processors (memories), which makes it suitable for the shared memory architecture. It can provide more powerful communication capability than the global bus; but its complexity grows exponentially along with the number of nodes. A hybrid topology between bus and crossbar called multistage topologies have been proposed where the network is organized by multiple stages and each stage provides part of the crossbar function; some examples includes *Omega* [29] and *perfect shuffle* [30].

As for the message passing communication, the most straightforward topology is probably the *completely-connected* network where each node has a direct link to every other node. It has the obvious disadvantage of the too large number of links. Other topologies suitable for the message passing communication include: 1-D linear array where each node is connected to its two neighbors, as shown in Fig. 2.5a; 2-D mesh array where each node is connected to its four neighbors, as shown in Fig. 2.5b; and 3-D mesh (cube) where each node is connected to its six neighbors, as shown in Fig. 2.5c. *2-D mesh* is especially attractive since it naturally fits into the 2-D chips.

2.2.3 Some Design Cases

The transputer [31] is a popular parallel processor originally developed in the 1980s. It pioneers the philosophy of using multiple relatively simple processors to achieve high performance. The transputer is designed for a multiple processor

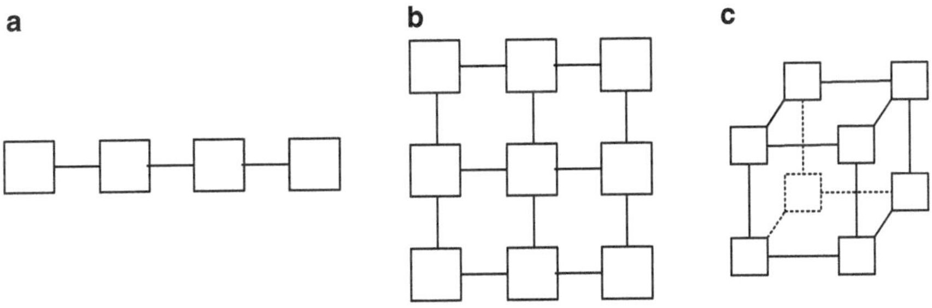

Fig. 2.5 Some basic interconnect topologies: (**a**) 1-D linear array; (**b**) 2-D mesh array; (**c**) 3-D cube

board, where each transputer processor is a complete standalone system. It uses a bit serial channel for inter-processor communication which can support communication of different word lengths to save hardware, but with dramatically reduced communication speeds.

Systolic processor [32, 33] is another parallel computing architecture proposed in the 1980s. It pioneers in showing that some applications can be partitioned and connected by a couple of subtasks, and then these sub-tasks can be mapped to an array processor to achieve higher performance. Systolic processors contain synchronously-operating elements which send and receive data in a highly regular manner through a processor array. Due to its strict timing requirement for the data flow, the suitable applications for Systolic processor are quite limited. Some projects such as ChiP [34] (Configurable Highly Parallel Computer) can be categorized into Systolic systems, but it provides more flexible programmable interconnect structure so that the communication is not limited to the neighboring elements.

Wavefront processors [35, 36] were proposed right after Systolic processors, mainly by S.Y. Kung. In order to release the strict data flow timing requirement of Systolic processors, the Wavefront processor permits a *data-driven, self-timed* approach to array processing, and substitutes the requirement of correct *timing* by correct *sequencing* and thus significantly broadens the number of suitable applications. Furthermore, S.Y. Kung studied the VLSI device scaling trend and predicted the importance of the global (long) wiring [36], and suggested each processing element could execute asynchronous with each other, which pioneered the concept of GALS although he did not implement it. The work finished by U. Schmidt et. al. [37] was an exciting experiment of using wavefront array processor for image processing.

2.3 Modern Multi-core Processors

Although multi-core research in the 1980s showed great potential, they unfortunately did not dominate the market. The main reason is that simply increasing the processor clock frequency is much easier and cheaper than designing a brand new multi-core architecture, so there was not enough motivation for the shifting from uni-core processors to multi-core processors.

There is a clear revival in the multi-core processor research starting from the middle 1990s, in both academic and industry, because of the great challenges faced by the uni-core processors. Compared to the previous multiprocessors which were normally distributed in multiple chips, the modern single chip multi-core systems have some unique concerns such as the power consumption, wiring issues, and the faster and higher intra-chip communication bandwidth. In addition, some researchers focus on the programming method to make the multiprocessor programming easier. Some modern multi-core processors will be discussed in this subsection, mainly from academic since they have more published information.

2.3.1 Design Cases of Modern Multi-core Processors

PADDI-2 [38] was developed at UC Berkeley in the middle 1990s for DSP applications. It is a MIMD multiprocessor architecture consisting of arrays of simple elements (PEs) connected by a reconfigurable 2-level hierarchical network. The level-1 communication network consists of 6 data buses and is responsible for communication within a cluster of 4 PEs; and the level-2 communication network consists of 16 buses and handles traffic among the 12 clusters. PEs communicate with each other exclusively using data streams and control streams in a data-driven manner. The chip was fabricated in a 1 μm technology; it occupies 12 × 12 mm, operates at 50 MHz and provides 2.4 GOPS peak performance.

RAW [21, 39] was developed at MIT starting from the late 1990s and is widely considered as the first tile-based multi-core processor. Their initial target applications were stream-based multimedia computations although later they believe RAW can be a universal solution for both general- and special-purpose applications. Motivated by the increasingly important wiring issues and high economic constraints of verifying new designs, they choose a tile-based architecture and use software to control inter-processor communication, wiring, as well as instructions. One of the key features of their design is the static and dynamic switching circuity which allows an extremely low communication latency about three clock cycles for inter-processor communication. To achieve this goal is not free though, about half of the processor area is devoted to communication circuity; whether this cost is worthwhile is not easy to conclude and depends on specific applications. A 4 × 4 RAW array was implemented and reported in 2003 [21] using IBM 0.18 μm CMOS technology, as shown in Fig. 2.6. Each core contains 32 KB instruction memory, 32 KB Dcache and 64 KB SMem, and occupies 4 × 4 mm, and the entire chip is 18.2 mm × 18.2 mm including PADs. The chip core averages 18.2 W at 425 MHz. The RAW processor was commercialized by tilera corporation in 2004.

Imagine [40] was developed at Stanford starting from the late 1990s, targeting stream style media processing. Motivated by the high parallelism and high computation locality of media applications, a streaming programming model and a corresponding Imagine architecture were developed. One of the main features of Imagine processor is that it provides a bandwidth hierarchy tailored to the demands of media applications which is organized by local register file, global register file and memory; and most of the computations can be finished in the registers thus improving performance as well as energy efficiency. The difference between global register file and cache is its high bandwidth which allows tens of words be fetched per cycle. A prototype Imagine processor was fabricated in a 0.15 μm CMOS technology [41]. It is an array containing 48 floating-point arithmetic units organized as eight SIMD clusters with a six-wide VLIW processor per cluster, with 9.7 KB of local register file and 128 KB of global register file. Each cluster occupies 5.1 × 0.8 mm and the die size is 16 × 16 mm. Imagine sustains 4.8 GFLOPS on QR decomposition while dissipating 7.42 W operating at 200 MHz. The Imagine processor

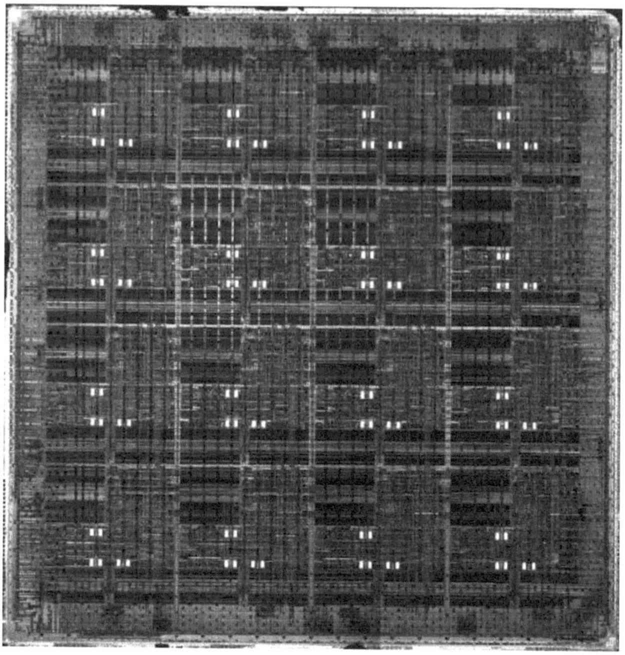

Fig. 2.6 Chip micrograph of the 16-core RAW [21]

was commercialized in 2007 in Stream Processors [42] in a 0.13 μm technology. The chip contains a 16-lane data-parallel unit with 5 ALUs per lane, 2 MIPS CPU cores, and I/Os. Each lane runs at 800 MHz at 1.0 V and occupies 3.2 mm². The peak performance is 128 GMACs with power consumption about 10 W.

Hydra [43] was proposed in Stanford in the late 1990s. From the high level view of its architecture, Hydra is similar to the traditional shared-memory multiprocessor systems: the chip connects a couple of RISC processors (such as MIPS) and a L2 Cache together by a global bus; this centralized feature might limit its scalability potential. The novel contribution of Hydra is mainly on the programming: it simplifies the parallel programming as well as improves the performance by hardware supported thread-level speculation and memory renaming. Some of the technologies of Hydra were successfully used in the Sun Niagara SPARC processor [44]. The chip contains eight symmetrical four-way multi-threaded cores sharing 3 MB of L2 cache. The die occupies 378 mm² in a 90 nm CMOS technology, operates at 1.2 GHz and consumes 63 W under 1.2 V. The second version of Niagara was reported in 2007 [45].

Pleiades Maia [46] was developed in UC Berkeley in the late 1990s; a reconfigurable DSP for wireless baseband digital signal processing. It combines an embedded microprocessor with an array of computational units of different granularities to achieve high energy efficiency; processing units are connected by a hierarchical reconfigurable interconnect network to provide flexibility; and handshake style GALS signaling was adopted to allow various modules to operate at different and dynamically varying rates. A prototype chip was implemented in a 0.25 μm CMOS process. The die size is 5.2 × 6.7 mm. The chip operates at 40 MHz on average and consumes 1.8 mW at a supply voltage 1 V; with energy efficiency between 10 and 100 MOPS/mW. The heterogeneous architecture is the main reason for Pleiades Maia' high energy efficiency; while it also limits its application domain.

Smart Memories [47] was proposed in Stanford in the early 2000s. It is a tile-based design made up of many processors, and has an ambitious goal to be a universal general design. In order to make it widely applicable, Smart Memories provides configurability to memories, wires and computational elements. This flexibility comes at a cost though. Compared to those programmable processors with specific domains such as Imagine processor, Smart Memories has about 50% performance degradation. The reconfigurable memory block of Smart Memories was fabricated in TSMC 0.18 μm technology [48] but the computation engine and the whole system has not been built. One reconfigurable memory with 512 × 36 b SRAM occupies 0.598 mm² (in which 61% of the area is SRAM and the others are the logic providing the configuration), operates up to 1.1 GHz and consumes 141.2 mW.

TRIPS [49] was developed at UT Austin in the early 2000s, with the ambitious goal to build a system which provides high performance across a wide range of application types. In order to provide configurability for both small and large-grain parallelism, TRIPS uses ultra-large cores – each core is a 16-wide-issue processor; the TRIPS can be configured to use ILP, TLP and DLP to be adapted to different applications and achieve high performance. It is a little questionable whether it is an achievable goal to build a universally energy efficient processor since the large core and additional configurability introduces non-negligible overhead. A prototype TRIPS chip was fabricated in a 0.13 μm technology [50]. The chip contains two cores and a separate on-chip network and occupies 336 mm². Its clock rate is 366 MHz and its power consumption is 36 W.

PipeRench [51] was developed at CMU in the early 2000s. The motivation of PipeRench is to efficiently execute numerically intensive applications, and its key feature is the dynamically reconfigurable datapath to match different applications. PipeRench is organized by 16 stripes and each stripe consists of 16 processing elements. A prototype PipeRench chip was fabricated in a 0.18 μm technology. Each processing element occupies 325 μm × 225 μm whose area is dominated by interconnect resources, and the whole die area is 7.3 × 7.6 mm. The chip can run at 120 MHz and executes a 40 tap 16-bit FIR filter at 41.8 million samples per second (MSPS). Operating at 33 MHz, the FIR filter consumes 518 mW without virtualization and 675 mW with virtualization.

WaveScalar [52] was proposed in University of Washington in 2003 targeting general purpose applications. A proposed architecture is organized as 16 clusters and each cluster contains tens of processing elements. Each processing element contains eight instruction buffers, and each four clusters contains a traditional L1 data cache. The processing elements communicate via a shared bus within a cluster, and the inter-cluster communication uses a dynamic routed on-chip network. More implementation prototypes were discussed in 2006 [53]. Furthermore, WaveScalar

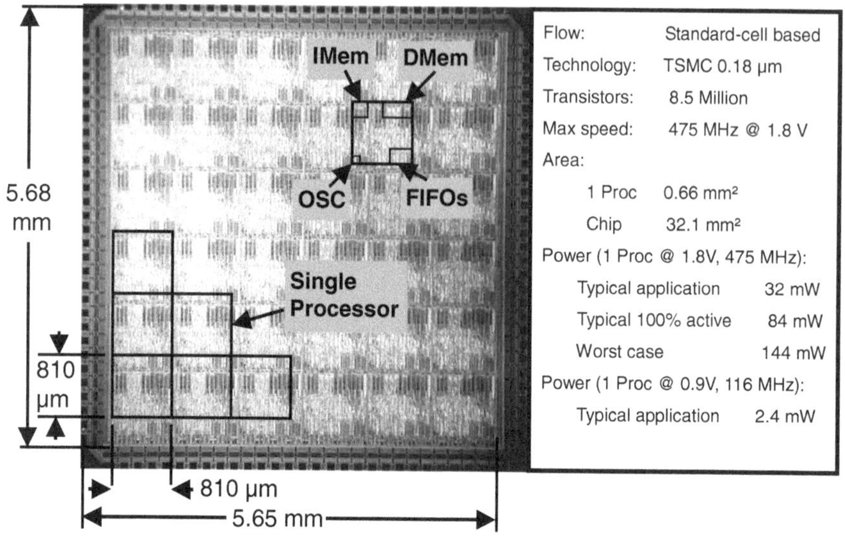

Fig. 2.7 Chip micrograph of the 36-core AsAP [22]

uses a dataflow programming model which eliminates the traditional program counter to increase the parallelism (Fig. 2.7).

AsAP [22] was developed in UC Davis in early 2000 and published in 2006. It is widely considered as the first tile-based multi-core processors with GALS clocking style. It contains 36 simple processors with independent clocks; is capable of computing DSP applications with high performance, high energy efficiency and is suitable for future fabrication technologies. The chip was fabricated during the summer of 2005 in TSMC 0.18 μm CMOS technology. Each processor occupies 0.66 mm^2; operates at 510 MHz, dissipates 34 mW while executing applications, and dissipates 90 mW while 100% active at 1.8 V. The second generation of AsAP chip was developed in 2007 which contains 164 simple processors as well as three special purpose modules [54].

Intel 80-core was published in 2007 [23]. It is a network-on-chip (NoC) architecture containing 80 tiles arranged as a 10 × 8 2D mesh network. The chip uses scalable global mesochronous clocking, which allows for clock-phase-insensitive communication across tiles and synchronous operation within each tile. The chip was fabricated in a 65 nm technology, each tile occupies 3 mm^2 and the whole chip is 275 mm^2. The chip operates at 3.13 GHz at supply voltage 1 V and 4 GHz at 1.2 V, and achieves 1.0 TFLOPS (trillion floating point operations per second) and 1.28 TFLOPS respectively.

There are some other chip multiprocessors such as Synchroscalar developed in UC Davis [55], RaPiD [56] developed in University of Washington, Picochip's PC102 [57], NEC's IMAP-CE [58], Xelerated AB's NPU [59], CISCO's Metro [60], IBM/sony/toshiba's CELL [61], Intellasys's SEAforth [62], Mathstar's Arrix FPOA [63],

RAPPORT's KC256 [64], Ambric's proc [65], CEA-LETI's FAUST [66], and Cavium networks' 16-core [67]. The following subsections will try to compare those parallel processors using tables instead of describing them separately.

2.3.2 *Distinguishing Multi-core Processors*

It is not easy to distinguish and categorize different multi-core processors by simply comparing their area, speed, power without looking into them deeply. This section tries to analyze different multi-core processors starting from their objective since that is the key to motivate their specific designs and features.

Table 2.2 summarizes the objectives of some multi-core processors and their one or two most important features. The first three processors including Transputer,

Table 2.2 Comparison of the objectives and distinguished key features of selected parallel processors

Processors	Targets/objectives	Distinguished features
Transputer [31]	High performance Multiprocessor	Bit serial channel
Systolic [32]	High performance array processing	Regular communication pattern
Wavefront [36]	High performance array processing	Data-driven, selftime execution
Hydra [43]	Program shared-mem systems	Thread speculation, memory renaming
WaveScalar [52]	Program dataflow systems	Memory ordering
RAW [39]	A universal system	Complex static/dynamic route
TRIPS [49]	A universal system	Large wide issue configurable core
Smart Memories [47]	A universal system	Configurable memory system
AsAP [22]	DSP applications	Fine grain processor array, GALS
PADDI-2 [38]	DSP applications	Data-driven control
RaPiD [56]	DSP applications	Reconfigurable pipelined datapath
PipeRench [51]	DSP applications	Dynamically configurable datapath
Ambric's proc [65]	DSP applications	Massively-parallel MIMD fabric
Synchroscalar [55]	Some of DSP applications	Rationally clocking and global comm.
CELL [61]	Multimedia applications	1 power proc and 8 synergistic procs
Imagine [40]	Stream applications	Hierarchical register file system
Pleiades Maia [46]	Wireless applications	Heterogeneous, reconfigurable connect
Picochip [57]	Wireless applications	Heterogeneous, reconfigurable

Systolic, and Wavefront are pioneering multiprocessor systems proposed and/or developed before 1990s and the others are modern multi-core processors. Some processors such as Hydra and WaveScalar target to simplify the programming. Some processors such as RAW, TRIPS, and Smart Memories try to be as flexible and reconfigurable as possible to make them suitable for wide applications; to achieve this goal, they introduce relatively large overhead such as complex interconnect, extremely large cores, and configurable memories. Some processors such as AsAP, PADDI, RaPiD, PipeRench, Synchroscalar, Imagine, Pleiades, Picochip, and Metro target one domain of applications; they use more specific architectures such as heterogeneous processing elements, simple processing cores, specific register file systems, and configurable datapaths to make the architecture efficient to the specific feature of those application domains.

Table 2.3 compares the listed parallel processors including the features of their processing elements, homogeneous/heterogeneous styles, chip sizes and power, etc. Most parallel processors can be easily differentiated by their processing element architectures which can be categorized into three broad types – heterogeneous, multiple execution units (often similar to classic SIMD), and multiple processors (MIMD).

A heterogeneous style such as the one used by Pleiades and Picochip makes the system efficient for specific applications. The disadvantage of heterogeneous style is its non-regular layout and potentially is difficult to scale.

Some projects such as Imagine, RaPiD, and PipeRench use multiple execution units to improve their performance. Strictly speaking, they can be categorized to parallel processing architectures but not multi-core systems since they have a centralized instruction controller. These architectures normally use hierarchical structures: combining some processing element into groups (called *cluster* in Imagine and *stripes* in PipeRench) and then those groups are organized into chip.

Quite a lot of systems can be categorized into MIMD processors, and normally they can be distinguished by the different granularities of their processing elements. The main reasons to determine the processor's granularity is the target applications: wider range of target applications normally will result larger processor granularity.

Table 2.4 compares the inter-processor communication network. Different processors choose their network for their target applications: from the simple bus, crossbar to 2-D mesh (including those called network-on-chip structure); and some processors use hierarchical network to treat local and long distance communication differently.

In terms of the clocking style, most other projects use a totally synchronous clocking style. Pleiades and FAUST use GALS clocking style and they use handshaking scheme to handle the asynchronous boundaries; AsAP uses the source-synchronous interprocessor communication style which is able to sustain a full-rate communication of one word per cycle; Ambric processor also uses the GALS clocking style but no details were reported. Intel 80-core employs mesochronous clocking where each processor has the same clock frequency while the clock phase can be different. Synchroscalar uses rationally related clocking style to achieve part of the clock/voltage scaling benefit.

Table 2.3 Comparison of the selected parallel processors; the data is scaled to 0.18 m technology and 1.8 V voltage supply; assuming a $1/s^2$ reduction in area, a s^2/v increase in clock rate, and a s/v^3 reduction in power consumption here s is the technology scaling factor and v is the voltage scaling factor; energy per operation is obtained from chip power divided by (issue width element number clock rate); the processors are sorted according to the area of each processing element

Proc.	Year	Homog./Heterog.	Proc.Elem.	Elem. size (mm²)	Mem. size (KB)	Elem. issue width	Elem. num.	Clock rate (MHz)	Chip power (W)	Energy per op (nJ/op)
Arrix	2006	Heterog.	ALU	<0.05	0.128	1	400	782	N/A	N/A
PADDI	1995	Homog.	ALU	0.09	0.052	1	48	277	N/A	N/A
SEAForth	2006	Homog.	Proc.	0.1	0.256	1	24	600	0.216	0.015
RaPiD	1999	Heterog.	ALU	0.65	N/A	1	16	277	N/A	N/A
AsAP	2006	Homog.	Proc.	0.66	0.64	1	36	530	1.28	0.07
Pleiades	2000	Heterog.	ASICs	0.8	N/A	N/A	22	138	0.014	N/A
Metro	2005	Homog.	Proc.	0.96	N/A	1	188	180	67	1.98
PipeRench	2002	Homog.	Cluster	1.17	0.256	16	16	33	0.675	0.08
Imagine	2002	Homog.	Cluster	5.9	9.7	8	8	200	7.4	0.58
IMAP-CE	2003	Homog.	Cluster	6	16	32	16	100	3	0.06
Stream	2007	Homog	Cluster	6.1	16	10	16	750	42	0.35
FAUST	2007	Heterog.	ASICs	7	N/A	N/A	20	126	2.68	N/A
RAW	2003	Homog.	Proc.	16	128	1	16	425	18.2	2.68
Intel 80-C	2007	Homog.	VLIW	23	5	8	80	735	206	0.44
Niagara1	2006	Homog.	Proc.	50	24	4	8	450	106	7.36
CELL	2005	Homog.	SIMD	58	256	4	8	1500	120	2.5
Niagara2	2007	Homog.	Proc.	75	24	8	8	300	133	6.93
TRIPS	2007	Homog.	Proc.	200	32	16	2	264	69	8.2
KC256	2006	Homog.	ALU.	N/A	N/A	1	256	100	0.5	0.02
NPU	2004	Homog.	Proc.	N/A	N/A	1	200	144	18	0.62
Cavium	2006	Homog.	Proc.	N/A	40	2	16	470	61	4.05
PicoChip	2003	Heterog.	Proc.	N/A	0.7–65	3	322	125	12	0.1

Table 2.4 Comparison of inter-element communication of selected parallel processors

Processor	Inter-connect	Details
Hydra [43]	Bus	Connect RISC processors and L2 Cache
Cavium 16-core [67]	Bus	Connect 16 processors and L2 Cache
CELL [61]	Bus	A bus composed of four 128b data rings
Niagara [44]	Crossbar	Connect between 8 cores and 4 L2 Caches
RaPiD [56]	1-D linear array	Linearly connect reconfigurable pipelined datapath
PipeRench [51]	1-D linear array	Linearly connect execution clusters
IMAP-CE [58]	1-D linear array	Linearly connect 128 processors
NPU [59]	1-D linear array	Linearly connect 200 processors
AsAP [22]	2-D mesh	Statically configurable
RAW [21]	2-D mesh	Including static and dynamic route
Ambric's proc [65]	2-D mesh	Configurable switches for distant comm.
TRIPS [49]	2-D mesh	Dynamically routed, with ALU operand network
Smart Memories [47]	2-D mesh	Packeted-based, dynamically-routed
FAUST [66]	2-D mesh	Packeted-based, dynamically-routed
Intel 80-core [23]	2-D mesh	Packeted-based, dynamically-routed
Pleiades [46]	Hierarchical	2-D mesh for both local and global
PADDI-2 [38]	Hierarchical	Bus for local and crossbar for global
Imagine [40]	Hierarchical	Switch for both local and global
WaveScalar [52]	Hierarchical	Bus for local and 2-D mesh for global
Picochip [57]	Hierarchical	Special picobus for both local and global
Metro [60]	Hierarchical	16 clusters each with 12 processing elements

2.4 Looking Forward to the Future of Multi-core Processors

2.4.1 There is no Universal Multi-core Processor

Some projects on multi-core processors have the ambitious goal to build a universal processor which can be used in anywhere and anyplace, and we believe this goal is un-achievable at least in the near future. As show in Fig. 2.8, there will be at least two types of multi-core processors: general-purpose multi-core processors and domain-specific multi-core processors.

The general-purpose multi-core processors should be developed based on current general-purpose uni-core processors, and their most important feature is to be suitable for variable application domains, and easy of programming. The domain-specific multi-core processors should push themselves as powerful and energy-efficient as ASICs, and to partly replace ASICs at some application domains. The very different requirement will make those two types of multi-core processors have different architectures and won't be merged in the near future.

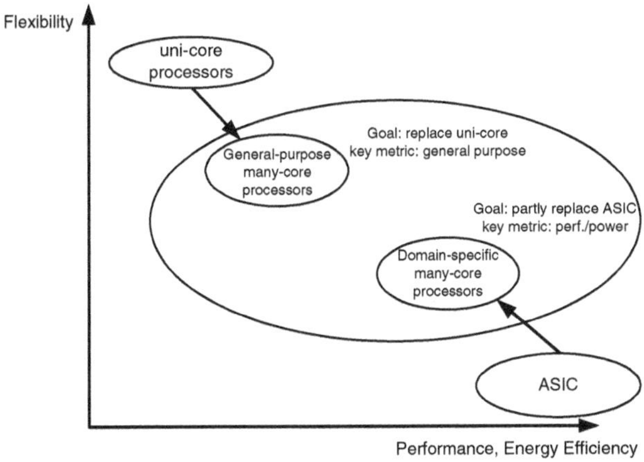

Fig. 2.8 A view of the future multi-core processors: uni-core processors will shift to general-purpose multi-core processors and part of ASICs will shift to domain-specific multi-core processors

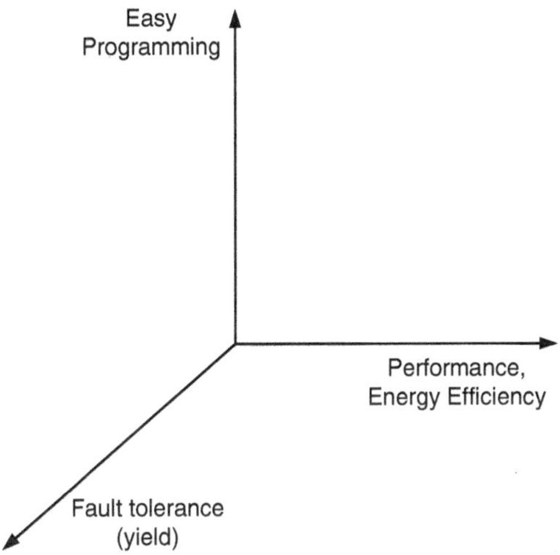

Fig. 2.9 Three dimension parameters for multi-core processors: performance (energy efficiency), easy programming, and fault tolerance

2.4.2 Fault-Tolerance Will Become a Key Issue in Multi-core Processors

Traditionally fault-tolerance computing is only for applications with very high demand on the reliability. The importance of the fault-tolerance computing will keep increasing along with the smaller fabrication technology nodes, larger chip size, and lower supply voltages. On the other hand, multi-core processors have the inherent advantage to handle the fault-tolerance issues due to their sufficient duplicated resources, and that duplicated resources can be used to detect and/or recover the faults. As shown in Fig. 2.9, performance (energy efficiency), easy programming, and fault tolerance are three of the key parameters for multi-core processors.

Reference

1. M.S. Hrishikesh, N.P. Jouppi, K.I. Farkas, D. Burger, S.W. Keckler, P. Shivakumar, The optimal logic depth per pipeline stage is 6 to 8 FO4 inverter delays, in *International Symposium on Computer Architecture (ISCA)*, May 2002, pp. 14–24
2. B. Flachs, S. Asano, S.H. Dhong, P. Hofstee, G. Gervais, R. Kim, T. Le, P. Liu, J. Liberty, B. Michael, H. Oh, S.M. Mueller, O. Takahashi, A. Hatakeyama, Y. Watanabe, N. Yano, A streaming processing unit for a CELL processor, in *IEEE International Solid-State Circuits Conference (ISSCC)*, Feb 2005, pp. 134–135
3. A. Harstein, T.R. Puzak, Optimum power/performance pipeline depth, in *IEEE International Symposium on Microarchitecture (MICRO)*, Dec 2003, pp. 117–125
4. G.E. Moore, Cramming more components onto integrated circuits. Electronics **38**(8), 114–117 (Apr 1965)
5. S. Agarwala, T. Anderson, A. Hill, M.D. Ales, R. Damodaran, P. Wiley, S. Mullinnix, J. Leach, A. Lell, M. Gill, A. Rajagopal, A. Chachad, M. Agarwala, J. Apostol, M. Krishnan, D. Bui, Q. An, N.S. Nagaraj, T. Wolf, T.T. Elappuparackal, A 600-MHz VLIW DSP. IEEE J. Solid State Circuits (JSSC) **37**(11), 1532–1544 (Nov 2002)
6. R.P. Preston, R.W. Badeau, D.W. Balley, S.L. Bell, L.L. Biro, W.J. Bowhill, D.E. Dever, S. Felix, R. Gammack, V. Germini, M.K. Gowan, P. Gronowshi, D.B. Jankson adn S. Mehta, S.V. Morton, J.D. Pickholtz, M.H. Reilly, M.J. Smith, Design of an 8-wide superscalar RISC microprocessor with simultaneous multithreading, in *IEEE International Solid-State Circuits Conference (ISSCC)*, Feb 2002, pp. 266–267
7. J.L. Hennessy, D. Patterson, *Computer Architecture – A Quantitative Approach,* 4th edn. (Morgan Kaufmann Publisher, 2007)
8. K. Roy, S. Mukhopadyay, H. Mahmoodi-meimand, Leakage current mechanisms and leakage reduction techniques in deep-submicrometer CMOS circuits. Proc. IEEE **91**(2), 305–327 (Feb 2003)
9. M. Horowitz, W. Dally, How scaling will change processor architecture, in *IEEE International Solid-State Circuits Conference (ISSCC)*, Feb 2004, pp. 132–133
10. S. Borkar, Low power design challenges for the decade, in *Asia and South Pacific Design Automatic Conference (ASP-DAC)*, 2001, pp. 293–296
11. J. Stinson, S. Rusu, A 1.5 GHz third generation Itanium processor, in *IEEE International Solid-State Circuits Conference (ISSCC)*, Feb 2003, pp. 252–253
12. S. Naffziger, T. Grutkowski, B. Stackhouse, The implementation of a 2-core multi-threaded Itanium family processor, in *IEEE International Solid-State Circuits Conference (ISSCC)*, Feb 2005, pp. 182–183, 592
13. S. Rusu, S. Tam, H. Muljono, D. Ayers, J. Chang, A dual-core multi-threaded Xeon processor with 16MB L3 cache, in *IEEE International Solid-State Circuits Conference (ISSCC)*, Feb 2006, pp. 102–103
14. H.D. Man, Ambient intelligence: Gigascale dreams and nanoscale realities, in *IEEE International Solid-State Circuits Conference (ISSCC)*, Feb 2004, pp. 29–35
15. R. Ho, K.W. Mai, M.A. Horowitz, The future of wires. Proc. IEEE **89**(4), 490–504 (Apr 2001)
16. International Roadmap Committee, International technology roadmap for semiconductors, 2005 edn. Technical report, ITRS, 2005. http://public.itrs.net/
17. S. Borkar, T. Karnik, S. Narendra, J. Tschanz, A. Keshavarzi, V. De, Parameter variations and impact on circuits and microarchitecture, in *IEEE International Conference on Design Automation (DAC)*, June 2003, pp. 338–342

18. S. Kaneko, K. Sawai, N. Masui, et al., A 600 MHz single-chip multiprocessor with 4.8GB/s internal shared pipelined bus and 512kB internal memory, in *IEEE International Solid-State Circuits Conference (ISSCC)*, Feb 2003, pp. 254–255
19. J. Hart, S. Choe, L. Cheng, C. Chou, A. Dixit, K. Ho, J. Hsu, K. Lee, J.Wu, Implementation of a 4th-generation 1.8GHz dual-core SPARC v9 microprocessor, in *IEEE International Solid-State Circuits Conference (ISSCC)*, Feb 2005, pp. 186–187
20. A. Bright, M. Ellavsky, A. Gara, R. Haring, G. Kopcsay, R. Lembach, J. Marcella, M. Ohmacht, V. Salapura, Greating the BlueGene/L supercomputer from lowpower SoC AISCs, in *IEEE International Solid-State Circuits Conference (ISSCC)*, Feb 2005, pp. 188–189
21. M.B. Taylor, J. Kim, J. Miller, D. Wentzlaff, F. Ghodrat, B. Greenwald, H. Hoffman, P. Johnson, W. Lee, A. Saraf, N. Shnidman, V. Strumpen, S. Amarasinghe, A. Agarwal, A 16-issue multiple-program-counter microprocessor with point-topoint scalar operand network, in *IEEE International Solid-State Circuits Conference (ISSCC)*, Feb 2003, pp. 170–171
22. Z. Yu, M. Meeuwsen, R. Apperson, O. Sattari, M. Lai, J. Webb, E. Work, T. Mohsenin, M. Singh, B. Baas, An asynchronous array of simple processors for DSP applications, in *IEEE International Solid-State Circuits Conference (ISSCC)*, Feb 2006, pp. 428–429
23. S. Vangal, J. Howard, G. Ruhl, S. Dighe, H. Wilson, J. Tschanz, D. Finan, P. Lyer, A. Singh, T. Jacb, S. Jain, S. Venkataraman, Y. Hoskote, N. Borkar, An 80-tile 1.28 TFLOPS network-on-chip in 65nm CMOS, in *IEEE International Solid-State Circuits Conference (ISSCC)*, Feb 2007, pp. 98–99
24. K. Asanovic, R. Bodik, B.C. Catanzaro, J.J. Gebis, P. Husbands, K. Keutzer, D.A. Patterson, W.L. Plishker, J. Shalf, S.W. Williams, K.A. Yelick, The landscape of parallel computing research: A view from berkeley, Technical Report UCB/EECS-2006-183, University of California, Berkeley, Dec 2006
25. W.A. Wulf, C.G. Bell, C.mmp – a multi-mini-processor, in *AFIPS Conference*, 1972, pp. 765–777
26. D. Lenoshi, J. Laudon, K. Gharachorloo, W.D. Weber, A. Gupta, J. Hennessy, M. Horowitz, M.S. Lam, The stanford DASH multiprocessor. IEEE Comp. **25**(3), 63–79 (Mar 1992)
27. C.L. Seitz, The cosmic cube. Commun. ACM **28**(1), 22–33 (Jan 1985)
28. J. Kuskin, D. Ofelt, M. Heinrich, J. Heinlein, R. Simoni, K. Gharachorloo, J. Chapin, D. Nakahira, J. Baxter, M. Horowitz, A. Gupta, M. Rosenblum, J. Hennessy, The Stanford FLASH multiprocessor, in *International Symposium on Computer Architecture (ISCA)*, Apr 1994, pp. 302–313
29. D.H. Lawrie, Access and alignment of data in an array processor. IEEE Trans. Comput. **24**(12), 1145–1155 (Dec 1975)
30. H.S. Stone, Parallel processing with the perfect shuffle. IEEE Trans. Comput. **2**, 153–161 (Feb 1971)
31. C. Whitby-Strevens, Transputers-past, present and future. IEEE Micro **10**(6), 16–19 (Dec 1990)
32. H. T. Kung. "Why systolic architectures?" *Computer Magazine*, 15(1), January 1982.
33. H.T. Kung, Systolic communication, in *International Conference on Systolic Arrays*, May 1988, pp. 695–703
34. L. Snyder, Introduction to the configurable, highly parallel computer. IEEE Comput. **15**(1), 47–56 (Jan 1982)
35. S.Y. Kung, K.S. Arun, R.J. Gal-Ezer, D.V. Bhaskar Rao, Wavefront array processor: Language, architecture, and applications. IEEE Trans. Comput. **31**(11), 1054–1066 (Nov 1982)
36. S.Y. Kung, VLSI array processors. IEEE ASSP Mag. **2**(3), 4–22 (July 1985)
37. U. Schmidt, S. Mehrgardt, Wavefront array processor for video applications, in *IEEE International Conference on Computer Design (ICCD)*, Sept 1990, pp. 307–310
38. A. Keung, J.M. Rabaey, A 2.4 GOPS data-driven reconfigurable multiprocessor IC for DSP, in *IEEE International Solid-State Circuits Conference (ISSCC)*, Feb 1995, pp. 108–110
39. E. Waingold, M. Taylor, D. Srikrishna, V. Sarkar, W. Lee, V. Lee, J. Kim, M. Frank, P. Finch, R. Barua, J. Babb, S. Amarasinghe, A. Agarwal, Baring it all to software: Raw machines. IEEE Comput. **30**(9), 86–93 (Sept 1997)

40. S. Rixner, W.J. Dally, U.J. Kapasi, B. Khailany, A. Lopez-Laguns, P. Mattson, J.D. Owens, A bandwidth-efficient architecture for media processing, in *IEEE international Symposium on Microarchitecture (MICRO)*, Nov 1998, pp. 3–13
41. B. Khailany, W.J. Dally, A. Chang, U.J. Kapasi, J. Namkoong, B. Towles, VLSI design and verification of the imagine processor, in *IEEE International Conference on Computer Design (ICCD)*, Sept 2002, pp. 289–294
42. B. Khailany, T. Williams, J. Lin, E. Long, M. Rygh, D. Tovey, W.J. Dally, A programmable 512 GOPS stream processor for signal, image, and video processing, in *IEEE International Solid-State Circuits Conference (ISSCC)*, Feb 2007, pp. 272–273
43. L. Hammond, B. Hubbert, M. Siu, M. Prabhu, M. Chen, K. Olukotun, The stanford Hydra CMP. IEEE Micro **20**(2), 71–84 (March 2000)
44. A. Leon, J. Shin, K. Tam, W. Bryg, F. Schumachier, P. Kongetira, D. Weisner, A. Strong, A power-efficienct high-throughput 32-thread SPARC processor, in *IEEE International Solid-State Circuits Conference (ISSCC)*, Feb 2006, pp. 98–99
45. U.G. Nawathe, N. Hassan, L. Warriner, K. Yen, B. Upputuri, D. Greenhill, A. Kumar, H. Park, An 8-core 64-thread 64b power-efficient SPARC SoC, in *IEEE International Solid-State Circuits Conference (ISSCC)*, Feb 2007, pp. 108–109
46. H. Zhang, V. Prabhu, V. George, M. Wan, M. Benes, A. Abnous, J.M. Rabaey, A 1-V heterogeneous reconfigurable DSP IC for wireless baseband digital signal processing. IEEE J. Solid State Circuits (JSSC) **35**(11), 1697–1704 (Nov 2000)
47. K. Mai, T. Paaske, N. Jayasena, R. Ho, W.J. Dally, M. Horowitz, Smart memories: A modular reconfigurable architecture, in *International Symposium on Computer Architecture (ISCA)*, June 2000, pp. 161–171
48. K. Mai, R. Ho, E. Alon, D. Liu, Y. Kim, D. Patil, M. Horowitz, Architecture and circuit techniques for a reconfigurable memory block, in *IEEE International Solid-State Circuits Conference (ISSCC)*, Feb 2004, pp. 500–501
49. K. Sankaralingam, R. Nagarajan, H. Liu, J. Huh, C.K. Kim, D. Burger, S.W. Keckler, C.R. Moore, Exploiting ILP, TLP, and DLP using polymorphism in the TRIPS architecture, in *International Symposium on Computer Architecture (ISCA)*, Feb 2003, pp. 422–433
50. M. Saravana, S. Govindan, D. Burger, S. Keckler, TRIPS: A distributed explicit data graph execution (EDGE) microprocessor, in *Hotchips*, August 2007
51. H. Schmit, D. Whelihan, M. Moe, B. Levine, R.R. Taylor, PipeRench: A virtualized programmable datapath in 0.18 micron technology, in *IEEE Custom Integrated Circuits Conference (CICC)*, May 2002, pp. 63–66
52. S. Swanson, K. Michelson, A. Schwerin, M. Oskin, Wavescalar, in *IEEE international Symposium on Microarchitecture (MICRO)*, Dec 2003, pp. 291–302
53. S. Swanson, A. Putnam, M. Mercaldi, K. Michelson, A. Petersen, A. Schwerin, M. Oskin, S.J. Eggers, Area-performance trade-offs in tiled dataflow architectures, in *International Symposium on Computer Architecture (ISCA)*, May 2006, pp. 314–326
54. D. Truong, W. Cheng, T. Mohsenin, Z. Yu, T. Jacobson, G. Landge, M. Meeuwsen, C. Watnik, A. Tran, Z. Xiao, E. Work, J. Webb, P. Mejia, B. Baas, A 167-processor Computational Platform in 65 nm CMOS. IEEE J. Solid State Circuits (JSSC) **44**(4), 1130–1144 (April 2009)
55. J. Oliver, R. Rao, P. Sultana, J. Crandall, E. Czernikowski, L.W. Jones, D. Franklin, V. Akella, F.T. Chong, Synchroscalar: A multiple clock domain, power-aware, tile-based embedded processor, in *International Symposium on Computer Architecture (ISCA)*, June 2004, pp. 150–161
56. D.C. Cronquist, P. Franklin, C. Fisher, M. Figueroa, C. Ebeling, Architecture design of reconfigurable pipelined datapaths, in *Conference on Advanced Research in VLSI*, March 1999, pp. 23–40
57. R. Baines, D. Pulley, A total cost approach to evaluating different reconfigurable architectures for baseband processing in wireless receivers. IEEE Commun. Mag. **41**(1), 105–113 (Jan 2003)
58. S. Kyo, T. Koga, S. Okazaki, R. Uchida, S. Yoshimoto, I. Kuroda, A 51.2GOPS scalable video recognition processor for intelligent cruise contol based on a linear array of 128 4-way VLIW

processing elements, in *IEEE International Solid-State Circuits Conference (ISSCC)*, Feb 2003, pp. 48–49
59. J. Carlstrom, G. Nordmark, J. Roos, T. Boden, L. Svensson, P. Westlund, A 40Gb/s network processor with PISC dataflow architecture, in *IEEE International Solid-State Circuits Conference (ISSCC)*, Feb 2004, pp. 60–61
60. W. Eatherton, The push of network processing to the top of the pyramid, in *Symposium on Architectures for Networking and communications systems*, Oct 2005
61. D. Pham, S. Asano, M. Bolliger, M.N. Day, H.P. Hofstee, C. Johns, J. Kahle, A. Kameyama, J. Keaty, Y. Masubuchi, M. Riley, D. Shippy, D. Stasiak, M. Suzuoki, M. Wang, J. Warnock, S. Weitzel, D. Wendel, T. Yamazaki, K. Yazawa, The design and implementation of a first-generation CELL processor, in *IEEE International Solid-State Circuits Conference (ISSCC)*, Feb 2005, pp. 184–185
62. Intellasys, SEAforth-24B, embedded array processor, Technical report. http: //www.intellasys.net/
63. Mathstar, Arrix family product brief, Technical report. http://www.mathstar.com/
64. Rapport, KC256 technical overview, Technical report. http://www.rapportincorporated.com/
65. A.M. Jones, M. Butts, TeraOPS hardware: A new massively-parallel MIMD computing fabric IC, in *Hotchips*, Aug 2006
66. D. Lattard, E. Beigne, C. Bernard, C. Bour, F. Clermidy, Y. Durand, J. Durupt, D. Varreau, P. Vivit, P. Penard, A. Bouttier, F. Berens, A telecom baseband circuit based on an asynchronous network-on-chip, in *IEEE International Solid-State Circuits Conference (ISSCC)*, Feb 2007, pp. 258–259
67. V. Yalala, D. Brasili, D. Carlson, A. Hughes, A. Jain, T. Kiszely, K. Kodandapani, A. Varadhrajan, T. Xanthopoulos, A 16-core RISC microprocessor with network extensions, in *IEEE International Solid-State Circuits Conference (ISSCC)*, Feb 2006, pp. 100–101

Chapter 3
Low Power Asynchronous Circuit Design: An FFT/IFFT Processor

Bah-Hwee Gwee and Kwen-Siong Chong

Abstract This book chapter pertains to circuit designs using the emerging asynchronous approach as opposed to the prevalent synchronous approach, with emphases on low voltage operation and low energy dissipation. The circuits designed herein span from microcells (adders and some handcrafted asynchronous basic cells) and macrocells (a multiplier and a memory) to a complete 128-point radix-2 decimation-in-time Fast Fourier Transform/Inverse Fast Fourier Transform (FFT/IFFT) processor for energy-critical audio applications, including hearing aids. The novelties of these designs, largely leveraged on the design philosophy of asynchronous approach, include the incorporation of different functional features and reduced spurious switching by timing, resulting in increased versatility, compactness, and lower energy dissipation. By means of appropriate asynchronous design techniques and the incorporations of the asynchronous microcells and macrocells, the asynchronous FFT/IFFT processor is demonstrated to feature superior energy attributes over its synchronous counterpart.

Keywords Asynchronous logic • Digital circuits • Fft processor • Low power/energy • Synchronous logic

3.1 Introduction

Asynchronous or 'clockless' approach [1–3] has recently become increasingly popular for low power energy efficient applications due to the potential advantages from its inherent operation properties – the operations are operated

B.-H. Gwee (✉)
School of Electrical and Electronic Engineering, Nanyang Technological University, 50 Nanyang Avenue, Singapore 639798, Singapore
e-mail: ebhgwee@ntu.edu.sg

K.-S. Chong
Temasek Laboratories @ Nanyang Technological University, Nanyang Technological University, 50 Nanyang Drive, Singapore 637553, Singapore
e-mail: kschong@ntu.edu.sg

K. Iniewski (ed.), *CMOS Processors and Memories*, Analog Circuits and Signal Processing, DOI 10.1007/978-90-481-9216-8_3,

asynchronously without using a global clock. The potential premises for low power/energy attributes in asynchronous circuits may include the simplified clocking infrastructure, reduced redundant operations, adaptive voltage scaling, reduced switchings, data-dependent operations, and many others [1–7]. Despite these, many researchers and designers, however, were frustrated by adopting asynchronous circuits to their designs. Most of the cases, 'simply applying asynchronous approach' always results in high power/energy circuits (compared to their synchronous counterparts) [1]; only a relatively small portion of practical asynchronous circuits were demonstrated to be power-/energy-efficient [4–7]. Put simply, the asynchronous circuits may not always be power-/energy-efficient; the power/energy efficiency of asynchronous circuits largely depends on their targeted applications.

Many reported low power/energy asynchronous circuits were designed for their respective applications [1–8], and their adopted techniques were different from one another due to their inevitably varied specifications. In essence, exploring the characteristics of the given applications (or specifications) that are favorable to the asynchronous approach is the key success to a truly low power/energy asynchronous circuit. Therefore, much detailed understanding and broader perspectives on asynchronous designs for low power/energy attributes are highly desirable by means of more engineering design examples. It is also important to fully quantify and qualify the asynchronous design (over its synchronous counterpart).

In this book chapter, a low energy asynchronous 128-point Fast Fourier Transform/Inverse Fast Fourier Transform (FFT/IFFT) processor, a core design embodied in the Digital Signal Processor (DSP) in hearing instruments, is presented. The underlying hypothesis to adopt asynchronous logic (for low energy dissipation) is because of the low speed requirement in hearing instruments (e.g. 1–2 MHz clock frequencies for synchronous DSPs), the FFT/IFFT processor can be controlled aptly (by the asynchronous approach) to reduce the redundant operations and switchings, at the cost of increased delay if necessary.

The organization of this book chapter is as follows. Section 3.2 reviews synchronization, both for the universal synchronous approach and the esoteric asynchronous approach. Section 3.3 presents several novel asynchronous microcells and macrocells that are employed in the asynchronous FFT/IFFT processor. Section 3.4 describes the FFT/IFFT processor and finally, Section 3.5 draws conclusions.

3.2 Synchronization: Synchronous and Asynchronous

Digital signals are generally assumed to be binary signals, and need to be synchronized for correct orders of events and sequence of operations, i.e. synchronization. This section first reviews the prevalent synchronous approach for a general understanding and then the asynchronous approach – a topic pertaining to this book chapter.

3.2.1 Synchronous Approach

The synchronous approach uses a global clock (global synchronization) to synchronize digital circuits [9]; an analogy to this global clock is akin to a baton in an orchestra to synchronize the notes and music. Figure 3.1 depicts a basic structure of a synchronous pipeline circuit and Table 3.1 tabulates timing notations.

Under the ideal condition when $t_{clk1} = t_{clk2}$, the proper operations of the synchronous circuits are constrained by the following expressions.

$$T \geq t_{reg} + t_{logic} + t_{setup} \tag{3.1}$$

$$t_{reg,cd} + t_{logic,cd} \geq t_{hold} \tag{3.2}$$

Equation (3.1) indicates that the clock period must be long enough for the data to propagate through the registers and logic (including all wire delays) and to be setup at the input registers before the next rising edge of the clock. Equation (3.2)

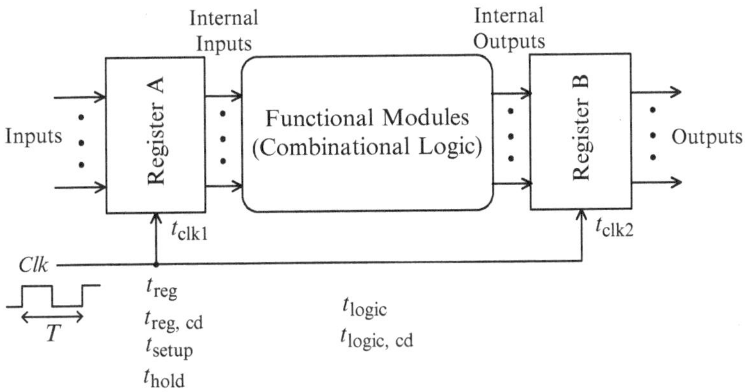

Fig. 3.1 Synchronous circuit

Table 3.1 Timing notations for the synchronous circuit

Symbol	Notation
t_{logic}	Maximum delay of the logic circuits
$t_{logic,cd}$	Contamination delay (minimum delay) of the logic circuits
t_{reg}	Maximum delay of the registers
$t_{reg,cd}$	Contamination delay (minimum delay) of the registers
t_{setup}	Setup time for the register
t_{hold}	Hold time for the register
t_{clk1}	Clock delay to the Register A
t_{clk2}	Clock delay to the Register B
T	Clock period

indicates that the hold time of the input registers must be shorter than the sum of the minimum delays of the logic circuits and registers. These restrictions make the synchronous circuits easy to understand and design. This, in part, explains why most digital circuits are synchronous based.

However, in real-life, the clock signal is never ideal. As a result, two additional clock-related parameters, namely clock skew and jitter, need to be considered in the synchronous system. Clock skew is referred to the spatial variation in arrival time of a clock transition on a circuit, and is caused by static mismatches in the clock paths and differences in the clock loads [9]. The clock skew makes the circuit more susceptible to race conditions, which may harm the correct operation of the system. The low skew clock network, however, usually requires a large number of clock buffers to balance the loads at different blocks and may result in higher power dissipation.

Clock jitter refers to the temporal variation of the clock period at a point on the circuit and the clock period may be reduced or expanded on a cycle-to-cycle basis [9] largely due to the process, temperature, and voltage (PVT) variations. The clock jitter degrades the circuit performance (speed), and to some extent, it makes the inputs (to the combinational circuits) be less synchronous, resulting in spurious switching – hence, increased power/energy dissipation.

Many sophisticated synchronous EDA tools have been successfully developed to resolve these clock-related issues. For a design operating at a lower frequency of up to few hundred Megahertz clock, conventional commercial synchronous EDA tools (such as tools from Synopsys and Cadence) can easily accommodate these issues. However, for a design operating at a very high frequency (e.g. >1 GHz), clock-related issues remain formidable.

To reduce the power/energy dissipation due to the clock in a synchronous design, clock gating is widely used. The clock gating approach reduces the power/energy dissipation of the circuit when computation is not required for an extended period of time (idle state). Although the clock gating approach to reduce power/energy dissipation is effective, the degree of reduction is somewhat constrained by the global clock infrastructure. Furthermore, to a large extent, it can only be used in a course-grain manner, that is, only a portion of the whole system can be gated. However, the clock gating approach tends to create more skew problems, and in turn complicates the design of the clock infrastructure.

3.2.2 Asynchronous Approach

This book chapter pertains to asynchronous designs with general objectives of low voltage operation and low energy dissipation. This section serves to review four relevant topics, namely (i) delay models, (ii) handshaking protocols and channels, (iii) data encoding, and (iv) asynchronous pipelines, and this review preambles the work presented in the subsequent sections in this book chapter.

3.2.2.1 Delay Models

Asynchronous circuits can be classified into four delay models: the delay-insensitive (DI), quasi-delay-insensitive (QDI), speed-independent (SI), and self-timed (ST) models.

A DI circuit is an asynchronous circuit whose functionality is unaffected by the gate delays and the interconnect wire delays. The assumptions here are that both the gate and wire delays are unbounded and arbitrary. The DI circuits are extremely robust for their delay insensitive property. However, these circuits can only be realized by C-Muller gates [10], hence impractical for many systems. A variation, called QDI, is also virtually 'delay-insensitive' if 'isochronic forks' is accounted for and is accommodated in the design, thereby more possible design implementations [10]. An isochronic fork is a forked wire where all the branches have the same delay.

An SI circuit is an asynchronous circuit whose functionality is unaffected by the delays in its components (e.g. gates). The assumptions here are that the gate delays are unbounded and arbitrary, and that the wire delays are zero (or negligible). It is clear that the assumption that the negligible wire delay is unrealistic (with respect to gate delays), particularly where the relative wire delay is significant in deep-submicron processes.

An ST circuit is an asynchronous circuit whose correct operation relies on engineering timing assumptions [1]. Of the engineering timing assumptions, the most prevalent assumption is arguably the bounded delay model. The bounded delay model is similar to the model used in synchronous circuits. In this model, the circuit assumes that both the gate and wire delays are known, or at least bounded, hence a delay line is used to match to the delay of the gates and wire. Such 'matched-delay' asynchronous circuits are usually much simpler than the QDI and SI circuits. However, the drawbacks are its worst-case operation and sensitivity to PVT variations.

3.2.2.2 Handshaking Protocols and Channels

Asynchronous circuits operate according to a set of handshaking signals (the request (*REQ*) and acknowledge (*ACK*) signals) [1, 11]. In general, the handshake signaling can be classified into two protocols (two-phase and four-phase) and four different channels (NONPUT, PUSH, PULL, and BIPUT) as depicted in Fig. 3.2a, b.

The two-phase handshaking protocol is an edge-sensitive communication – any transition (rising or falling) on *REQ* or *ACK* is considered a valid communication. One complete asynchronous communication requires one transition on *REQ* and one corresponding transition on *ACK*. The four-phase handshaking protocol, on the other hand, is a level-sensitive communication. One complete asynchronous communication requires the assertion and de-assertion on both the *REQ* and *ACK* signals.

In Fig. 3.2b there are two parties called the sender and receiver, and the black dot represents the active party that initiates the *REQ* signal. Either the sender or

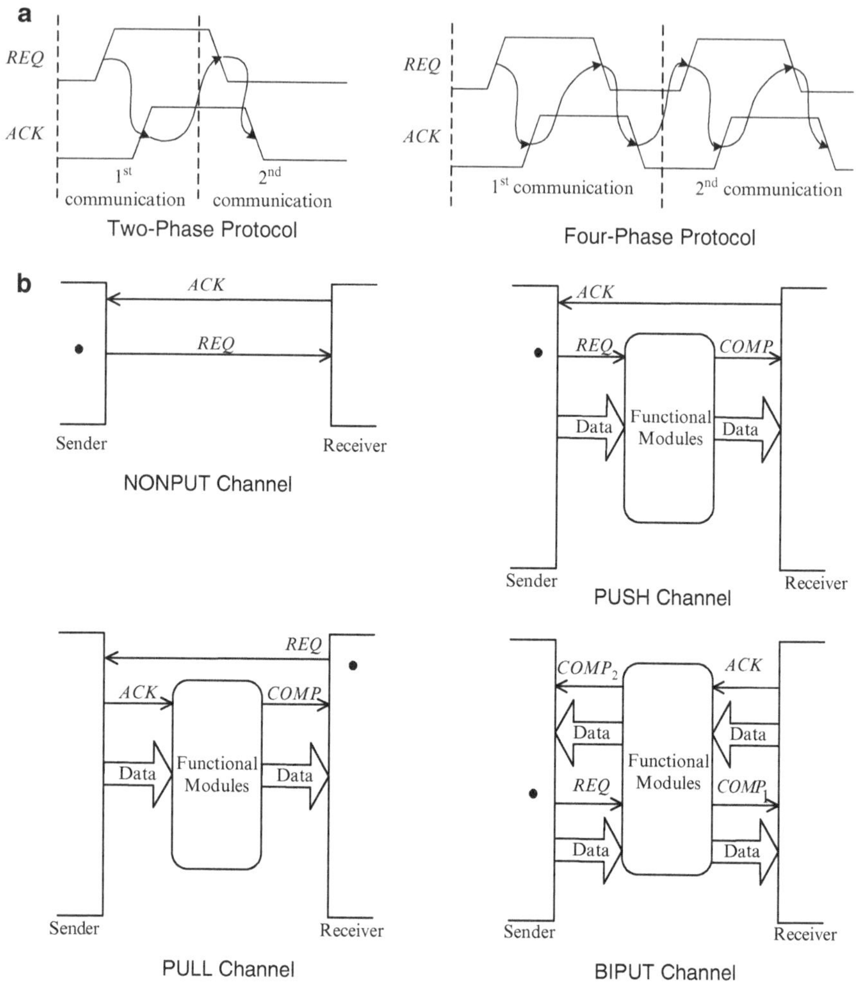

Fig. 3.2 (**a**) Handshake protocols: two-phase and four-phase, (**b**) channel types: NONPUT, PUSH, PULL and BIPUT

receiver can be the active party. For the NONPUT channel type, no data is exchanged between the sender and receiver. For the PUSH channel type, the sender pushes the data (by the *REQ* signal) to the receiver (via functional modules, if any), and the receiver sends the *ACK* signal once it receives the data. For the PULL channel type, the receiver pulls the data from the sender (via functional modules, if any), and the sender will acknowledge the data. For the BIPUT channel type, the sender sends the *REQ* signal together with the data (via functional modules, if any) to the receiver, and the receiver sends the *ACK* signal together with another data (via functional modules, if any) to the sender.

The handshaking protocols can be better interpreted depending on the assumptions of the data valid schemes (and the channel types). For instance, Fig. 3.3a

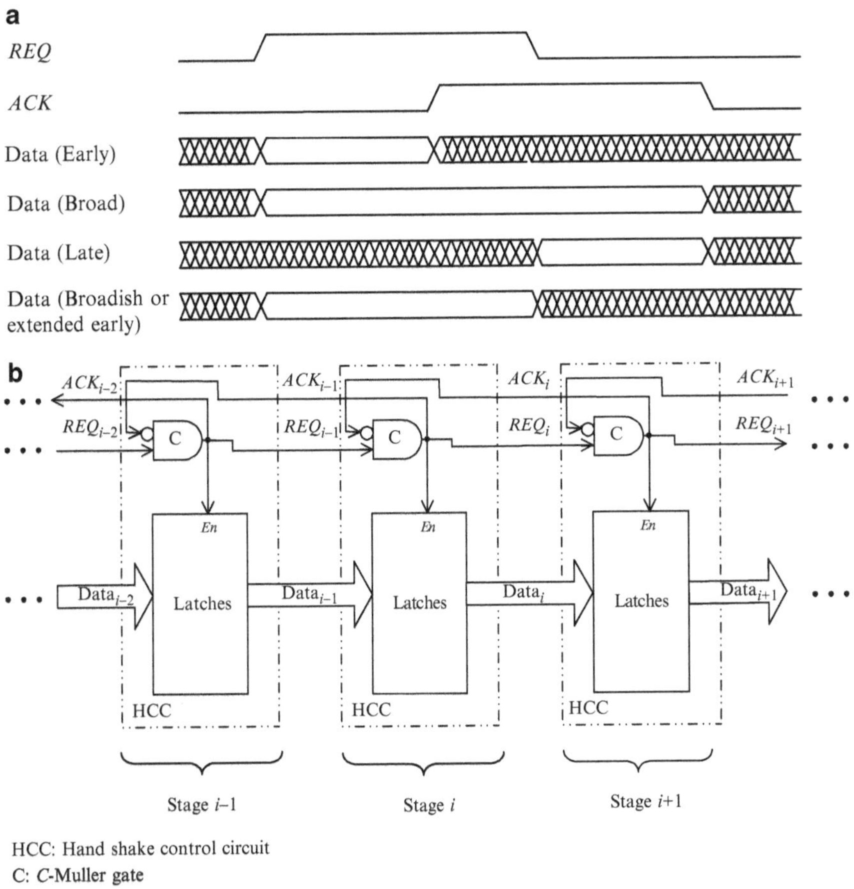

Fig. 3.3 (**a**) Push channel data valid schemes for the four-phase handshaking protocol, (**b**) a simple four-phase bundled data pipeline

depicts the data valid schemes for a PUSH channel in the four-phase protocol. In the four-phase handshaking protocol, there are four possible data valid schemes, namely early, broad, late, and broadish (extended early). In the early data valid scheme, the data is assumed to be valid when *REQ* is asserted until *ACK* is asserted. In the broad data valid scheme, the data is assumed to be valid when *REQ* is asserted until *ACK* is de-asserted. In the late data valid scheme, the data is assumed to be valid when *REQ* is de-asserted until *ACK* is de-asserted. In the broadish data valid scheme, the data is assumed valid when *REQ* is asserted until *REQ* is de-asserted. The data valid schemes for other channels (PILL and BIPUT) can be constructed accordingly and are well established [1, 11]. For brevity, Fig. 3.3b depicts a simple bundled data pipeline with a possibility of applying a broadish data valid scheme.

3.2.2.3 Data Encoding

Asynchronous circuits are designed to detect the arrival time of data, that is, the data must be encoded with some timing information. For data encoding in asynchronous circuits, bundled data and dual-rail data are commonly used.

The bundled data coding uses a bundle of data wires (each wire represents one-bit of information) with two control wires, *REQ* and *COMP*, to indicate the timing information; *COMP* serves as a completion detection signal. The bundled data coding is applicable to the bounded delay model discussed earlier.

The dual-rail coding uses two wires for one-bit information. One of the wires is dedicated as a 'TRUE' wire while the other one is dedicated as a 'FALSE' wire. Both wires cannot be '1' at the same time. Initially, both wires are '0' (no operation). Once an operation is asserted, only one of the wires can be '1', that is an event. The opposite logic states of the signals not only serve as logic information, but also detect the timing information simultaneously. The dual-rail coding technique is robust, and the circuits employing this technique can be QDI. However, this technique requires complementary signals which may increase the switching activity, and would require a larger IC area.

3.2.2.4 Asynchronous Pipelines

Figure 3.3b earlier depicts a conventional four-phase bundled data pipeline. For simplicity and as an illustration, the circuit in Fig. 3.3b is the pipeline without data processing, and functional modules (not shown) can be inserted between the latches in the consecutive stages for data processing.

In the four-phase handshaking protocol, either normally-closed or normally-opened latches [12, 13] can be used and the specific type affects the power/energy dissipation. Assuming that the initial handshaking signals are '0' and the normally-closed latches are used, when REQ_{i-2} is asserted, REQ_{i-1} (=ACK_{i-2}) will be asserted to open the latches in Stage $i-1$, and Data_{i-2} is passed to be Data_{i-1}. The ACK_{i-2} signal will acknowledge the receipt of the data to the preceding stage. The REQ_{i-1} signal will trigger the *C*-Muller gate in Stage i and REQ_i (= ACK_{i-1}) is asserted to open the latches in Stage i, and Data_{i-1} is passed to be Data_i. The ACK_{i-1} signal will acknowledge the receipt of the data for Stage $i-1$, and REQ_{i-2} can be de-asserted, and the latches in Stage $i-1$ are then closed. Data_{i-2} is now ready to be updated with new data. Thereafter, REQ_{i-1} (= ACK_{i-2}) is de-asserted to close the latches in Stage i, and Data_{i-1} is ready to be updated (if only when ACK_i is asserted). At this time, a new operation can begin by asserting REQ_{i-2}. This scenario applies to the subsequent stages. Pipeline circuits embodying normally-opened latches can similarly be analyzed. The normally-closed latch approach is usually applied for reducing spurious switching (for low power/energy reduction) while the normally-opened latch approach, on the other hand, is usually applied for high speed operations.

The two-phase bundled data pipeline is somewhat similar to the four-phase bundled data pipeline except without the need of 2 return-to-zero phases. More optimized circuit implementations have been reported in literature [1].

3.3 Low Power Asynchronous Micro/Macro Cells

In this section, several micro/macro cells, including a Latch Adder, two types of 2-bit carry completion sensing adders and three associated 16-bit adders, a 16 × 16-bit multiplier and a 128×16-bit memory, are described with specific emphases on low-voltage (1.1 V), low speed (<5 MHz) and power-critical operation. The intended application is for the design of DSPs for portable low-voltage power-critical bio-medical applications, including hearing instruments. For completeness, relevant literature reviews and comparisons will be provided.

3.3.1 Latch Adder

In most of the arithmetic operations (such as multiplication, division, etc.), full adders are the most rudimentary cells. The designs of full adders are generally mature in literature [14–18]. A full adder sums three inputs, *A*, *B* and carry-in C_{in}, and produces two outputs, sum *S* and carry-out C_{out}. The prevalent adder designs for low voltage operations that provide rail-to-rail output include the T28 adder, T18 adder, T16 adder, *N-P* dynamic adder (DA) and Domino DA (see Fig. 3.4a–e). Other adder designs with a small transistor count (e.g. <15 transistors per full adder), for example designs operate unreliably under low voltage conditions ($V_{DD} \approx |V_{Tp}| + V_{Tn}$), and are hence unsuitable for the low voltage applications including the intended hearing instrument. Pass-logic adders [9] are sensitive to voltage scaling, and hence also unsuitable for the low voltage applications.

A simple way to reduce spurious switching is to place separate latches in front of the full adder to synchronize the inputs to the adder. However, the cost in terms of IC area (three latches are required per full adder) and the added power dissipated by these latches are high. To circumvent this cost and yet obtain a latch function at the input to the adder to reduce the spurious switching, a latch is integrated into an adder, termed as Latch Adder (LA) [14, 15]. By integrating a latch (that latches all the three inputs simultaneously) within an adder instead of being separate and external, the overhead in LA is low (see later). The circuit schematic of the LA is depicted in Fig. 3.4f.

The latch function within the LA is controlled by means of the enable signals, E_i and $\overline{E_i}$, and serves to synchronize the three input signals (*A*, *B* and C_{in}) to the LA. In Fig. 3.4f, when E_i = ‘1’ ($\overline{E_i}$ = ‘0’) or asserted, the LA functions as a full-adder. When E_i = ‘0’ ($\overline{E_i}$ = ‘1’) or negated, the LA will hold its output signals (*S* and C_{out}) by means of weak (small *W/L* ratio) feedback PMOS transistors (*P*1–*P*4), and the outputs of the adder are hence latched. Note that although the LA design depicted in Fig. 3.4f is a semi-static CMOS design, its operation is robust. This is shown by considering the following two scenarios. First, assume that *S* is initially ‘1’. If E_i = ‘1’, there is no ambiguity. For the same initial condition, if E_i = ‘0’, $\overline{S}$ is at a high impedance state, hence an undesirable condition. This is because charge leaking

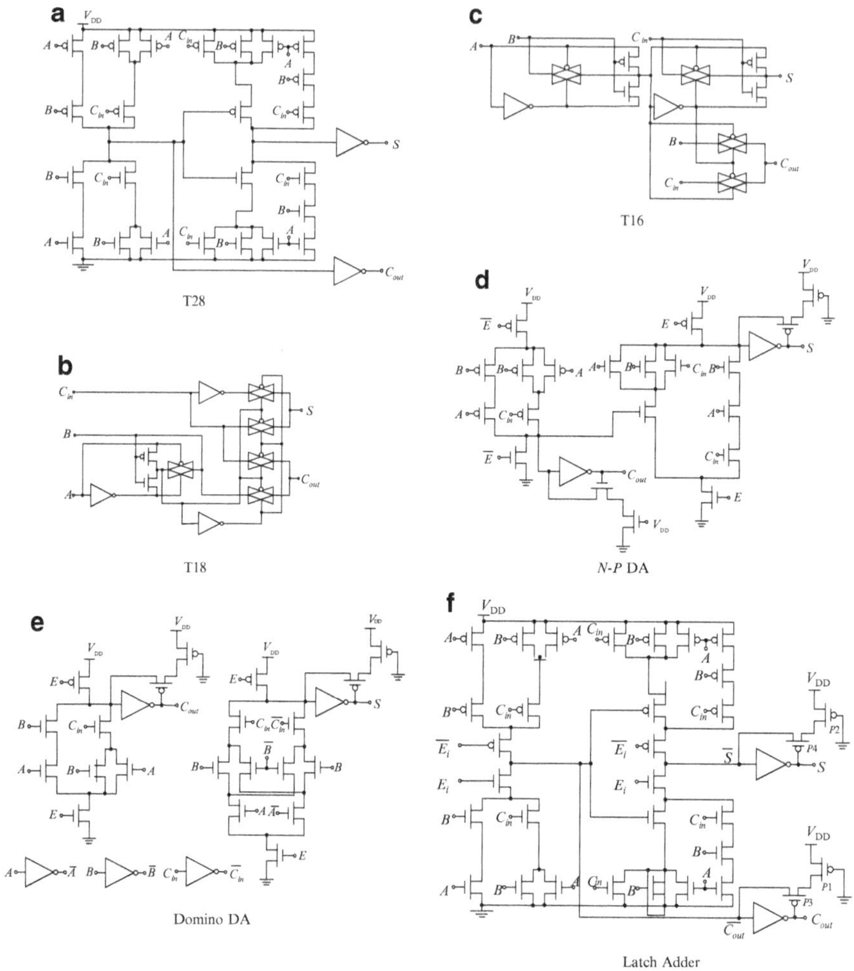

Fig. 3.4 Full adders that provide rail-to-rail signal swing at low voltage operation: (**a**) T28 adder, (**b**) T18 adder, (**c**) T16 adder, (**d**) *N-P* DA, (**e**) Domino DA, and (**f**) Latch Adder (LA)

into $\overline{S}$ may result in an undesirable change in state. This is, however, not an issue because for E_i = '0', the output is not used. Put simply, for E_i = '0', the output is inconsequential because the LA is in idle state. Second, assume that *S* is initially '0'. If E_i = '1', there is again no ambiguity. For the same initial condition, if E_i = '0', $\overline{S}$ is a well-defined '1' due to feedback inverter. The same scenarios apply to signal C_{out}. In short, the operation of the LA is robust.

To retain the low power/energy advantage in the LA, the weak feedback inverters are designed by cascading *P*1 and *P*2 to *P*3 and *P*4 respectively. The *P*3 and *P*4 transistors are sized to minimum-size (constrained by the design rules), and the *P*1 and *P*2 transistors are sized to have an effective low transistor *W/L* ratio. In this manner, these transistors present a low capacitive loading at outputs *S* and C_{out}. This is as opposed to the conventional method where only *P*3 and *P*4 (both with much longer *L*) are used, thereby presenting a larger capacitive load at outputs *S* and C_{out}. Also note that the short-circuit current is negligible when the targeted operating voltage is as low as 1.1 V (close to $|V_{Tp}| + V_{Tn}$ in the 0.35 μm CMOS process).

It would be of interest to note that if the supply voltage is relatively low ($V_{DD} < |V_{Tp}| + V_{Tn}$), there is no power dissipation due to short-circuit current, and the dynamic CMOS approach can be adopted (without weak feedback PMOS transistors, *P*1–*P*4). If the supply voltage is relatively high ($V_{DD} > |V_{Tp}| + V_{Tn}$), for the robustness, for reducing the short-circuit current and for enhancing its logic state, weak NMOS transistors can be included (fully static CMOS approach) at the expense of relatively slower speed, larger area and higher switching power dissipation (compared to dynamic CMOS approach). The semi-static CMOS approach provides a good compromise between the dynamic CMOS and fully static CMOS approaches. Compared to the former approach, the semi-static CMOS approach results in some short-circuit current reduction. Compared to the latter approach, the semi-static CMOS approach results in a lower switching power dissipation and higher speed due to relatively smaller output loads.

The LAs are particularly useful to reduce the spurious switching in multiplier designs, and more illustrations will describe later on how this low spurious switching can be achieved.

Two benchmarked comparisons are made on the basis of computer simulations. First, the LA is compared against the reported full adders (see Fig. 3.4a–e). Second, the LA is further compared against the reported full adders with separate latches. The separate latch is based on the low power semi-static non-inverting C^2MOS (with two series weak PMOS transistors implementation). In all the designs, the transistors are sized to have ×1 driving ability (the same as the standard library cells), specifically (2/0.3 μm) and (1/0.3 μm) for PMOS and NMOS respectively (except for the weak PMOS and NMOS transistors). Note that all designs are based on the same 0.35 μm CMOS dual-poly three-metal fabrication process. The low transistor-count adders and pass-logic adders that do not operate reliably at 1.1 V are not considered here.

On the basis of 500 random input signals and on simulations, Table 3.2 summarizes the energy dissipation, delay, EDP, and IC area of the different adders.

Table 3.2 Energy, delay, energy-delay-product (EDP) and IC area of the Latch Adder against that of other full adders @ 1.1 V, 1 MHz

Adder	Energy (10^{-8} W/MHz)	Delay (ns)		EDP (10^{-23} J s)	Core area (μm^2)
		C_{out}	*S*		
T28	6.0	3.4	4.8	28.8	389
T18	7.3	4.1	4.3	31.4	374
T16	7.2	5.0	5.9	42.5	387
N-P DA	11.7	7.0	8.3	97.1	486
Domino DA	11.8	1.9	2.7	31.9	660
Latch Adder	10.1	4.4	7.0	70.7	501

Table 3.3 Energy, delay, energy-delay-product (EDP) and IC area of the Latch Adder against that of other adders with latches @ 1.1 V, 1 MHz

Adder	Energy (10^{-8} W/MHz)	Delay (ns) C_{out}	Delay (ns) S	EDP (10^{-23} J s)	Core area (μm^2)
T28 with latches	15.4	5.6	7.3	112.4	1,180
T18 with latches	15.8	8.3	8.3	131.1	970
T16 with latches	15.8	9.4	11.0	173.8	1,000
Latch Adder	10.1	4.4	7.0	70.7	501

From Table 3.2, it is noted that among the conventional adders (T28, T18, T16) and the LA, the LA dissipates the highest energy, is the slowest speed (for *S* signals) and occupies the largest IC area. This is expected due to the added integrated latch function. On the other hand, among all adders compared, the Domino DA is the fastest adder but dissipates the highest energy due to its dynamic operation. Despite featuring seven less transistors than the Domino DA, the *N-P* DA features high energy dissipation, comparable to the Domino DA. This is because of its dynamic operations, additional control line $\overline{E}$ and the PMOS tree having a larger capacitance. The low mobility of the PMOS transistor (lower than that of NMOS transistor) makes the PMOS tree (and its associated NMOS weak transistor for charge sharing prevention) less attractive in the *N-P* DA. As a result, the *N-P* DA features a relatively poor speed performance compared to the Domino DA.

The advantage of the LA design becomes more apparent in the second benchmarked comparison as tabulated in Table 3.3. In this table, the circuits now have the same circuit functionality as the LA described above. When compared against the T28, T18 and T16 adders with separate latches, the LA features the lowest energy dissipation, the least delay (for *S* signal), and occupies the smallest area. As described previously, these desirable parameters are attributed to the integrated latch in the LA.

3.3.2 Asynchronous Carry Completion Sensing Adders

Apart from the full adder, an asynchronous adder is also one of the most rudimentary arithmetic cells [7, 19]. The asynchronous adder has the inherent ability to determine when the computation commences and when the computation finishes.

Two type of 2-bit asynchronous adders (based on the domino logic), termed as *Type-*α adders and *Type-*β adders [7], are depicted in Fig. 3.5a, b respectively. The Carry Out signals, C_{i+1} and $\overline{C_{i+1}}$, are dual-rail signals and for hardware simplicity, the Sum signals, S_i and S_{i+1}, are single-rail signals. This is because the C_{i+1} and $\overline{C_{i+1}}$ signals are used to encode dual-rail signals to generate the Completion signal, *COMP*, whereas the Sum signals are not required in the subsequent adder stage.

The *Type-*α design is based on the reported 2-bit carry look ahead (CLA) adder [19], but with an improvement by removing four transistors and obtain lower power dissipation and smaller IC area attributes without compromising speed.

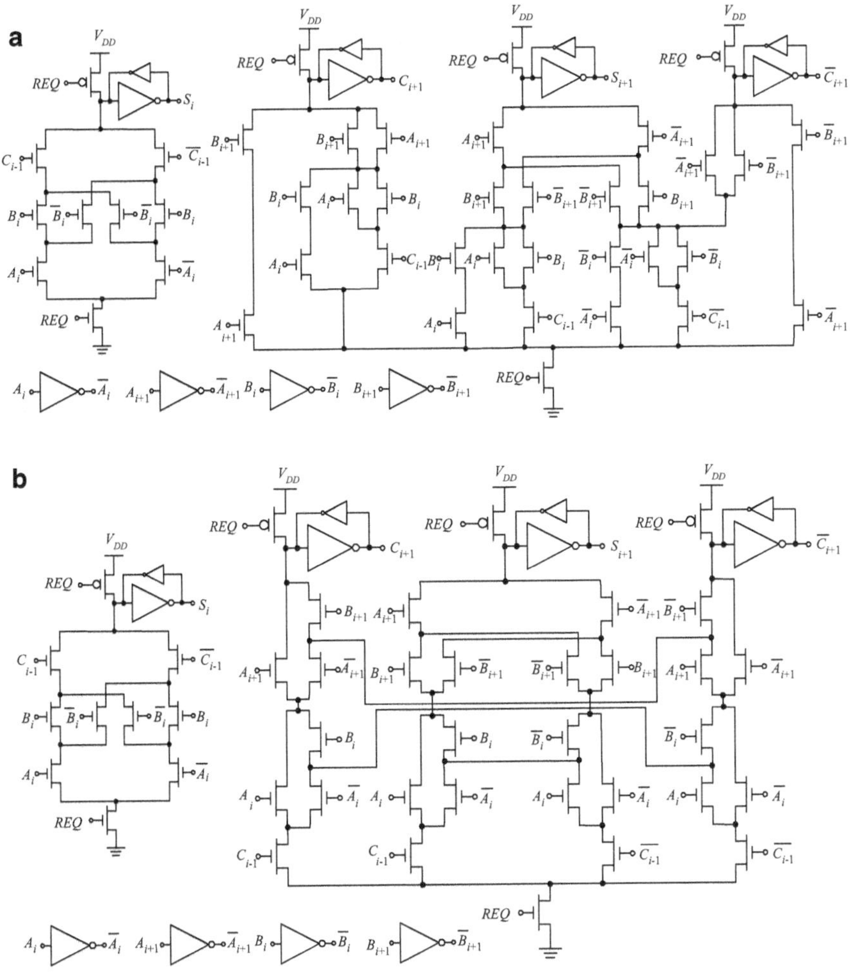

Fig. 3.5 Circuit schematic of (**a**) *Type-α* 2-bit adder and (**b**) *Type-β* 2-bit adder. The inputs are Carry In (C_{i-1} and $\overline{C_{i-1}}$), A_i, B_i, A_{i+1} and B_{i+1} . The outputs are Sum (S_i and S_{i+1}) and Carry Out (C_{i+1} and $\overline{C_{i+1}}$)

It was shown in Ref. [19] that the most efficient way of constructing a long word-length adder (e.g. ≥16-bit) is to cascade a number of 2-bit rudimentary adders. In the case of the *Type-β* adder, two 1-bit asynchronous full adders (AFAs) [4] are integrated to form an adder. The *Type-α* adder generates C_{i+1} and $\overline{C_{i+1}}$ faster than the *Type-β* adder. This is because the C_{i+1} and $\overline{C_{i+1}}$ signals of the latter need to wait for the previous C_{i-1} and $\overline{C_{i-1}}$ to be available before the addition process commences. Put differently, the C_{i-1} and $\overline{C_{i-1}}$ signals in the *Type-β* adder simultaneously function as the input and control signals, whereas the C_{i-1} and $\overline{C_{i-1}}$ in the *Type−α* adders serve only as input signals. In other words, the carry block in the *Type-α* adder is in weakly indicating [1], and the carry block in the *Type-β* adder is in strongly indicating [1].

It is worthwhile to note that the sum block in the *Type-α* and *Type-β* adders is based on single-rail logic – the *Type-α* and *Type-β* adders are not purely speed-independent. As a result, delay assumptions may be required – all sum outputs are assumed to be ready by the time the Carry Out signals (or preferable the completion signal) become valid. The well-established methods to address this include transistor sizing, dummy transistor insertion [4], and delay-line insertion. In this case, transistor sizing is adopted in the completion detector circuits for the delay assumption (see later).

Three 16-bit asynchronous adders are presented by cascading eight *Type-α* adders or eight *Type-β* adders or a combination of *Type-α* and *Type-β* adders. The choice for employing *Type-α* or *Type-β* adders depends on the power and speed considerations. Figure 3.6a depicts the block diagram of the 16-bit asynchronous adder, *Adder-A*, that is essentially a cascade of eight *Type-α* adders. When *REQ* is asserted, *Adder-A* commences the addition process and generates *COMP* when the addition is completed. The average speed of *Adder-A* is fast because the addition is independent of C_{i-1} and $\overline{C_{i-1}}$ when inputs, A_{i+1} and B_{i+1}, are both 0 or both 1. In these specific instances, C_{i+1} and $\overline{C_{i+1}}$ will be computed earlier (than when the A_{i+1} and B_{i+1} inputs are of opposite logic states) and the addition result is obtained in a shorter time. The delay assumption is realised by sizing the completion detector circuit (see the rectangular box in Fig. 3.6a) to match at least 1× maximum skew delay between the slowest delay of S_{i+1} and the fastest delay of Carry Out signals in the *Type-α* adder. The 1× maximum skew delay is sufficient because the completion detector circuit will only be asserted after all the *Type-α* adders have been asserted, hence all sum signals will be ready before *COMP* is generated.

Figure 3.6b depicts the second 16-bit asynchronous design, *Adder-B*, that is essentially a cascade of the *Type-β* adders. Of particular interest, this design features a very simple completion detector circuit – an OR gate – and is independent of the wordlength of the adder. As the completion signal of the *Type-β* adders propagates from one stage to another from least significant bit (LSB) to most significant bit (MSB), this adder is relatively slow. It is, however, very energy efficient. The delay assumption is realised by sizing the OR gate to match at least 1× maximum skew delay between the slowest delay of S_{i+1} and the fastest delay of Carry Out signals in the *Type-β* adder. This is because the delay of the Carry Out signals is longer than the skew delay (from the preceding *Type-β* adder). Hence, only the delay margin of last *Type-β* adder (i.e. 1× maximum skew delay) needs to be considered.

By appreciating the merits of the respective 16-bit *Adder-A* and *Adder-B* designs, a hybrid 16-bit asynchronous adder design is presented by combining the *Type-α* adders and *Type-β* adders and call this the 16-bit *Adder-C*, see Fig. 3.6c. In this design, the first four *Type-β* adders in *Adder-B* are replaced with the *Type-α* adders. The delay of this adder depends on the speed of the generation of C_7 and $\overline{C_7}$ signals. Consequently, this adder features a higher speed than, comparable hardware

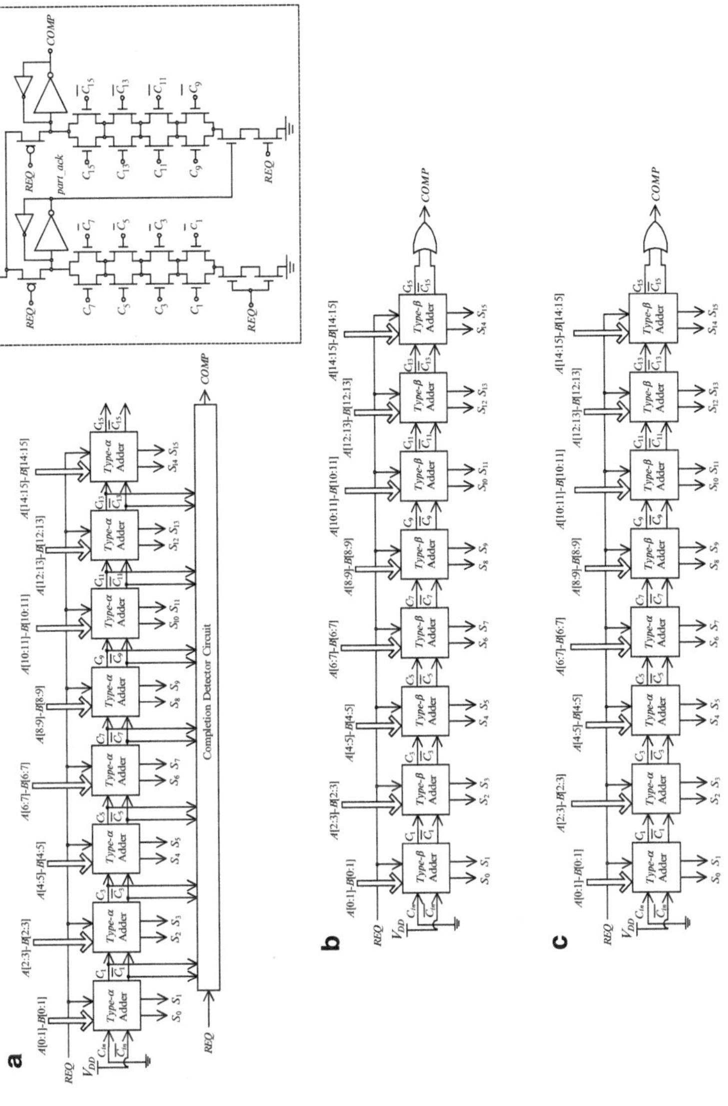

Fig. 3.6 The asynchronous 16-bit adders: **(a)** high-speed adder, *Adder-A*, **(b)** low power adder, *Adder-B*, and **(c)** medium-speed low power adder, *Adder-C*

and comparable power/MHz with *Adder-B*. When compared to *Adder-A*, this hybrid design features simpler hardware but slower speed. This hybrid design is recommended in place of *Adder-B* for applications where the speed demand is moderate and where power dissipation is critical.

Also note that for *Adder-C*, the delay of the four *Type-*β adders (in MSBs) is faster than that of the four *Type-*α adders (in LSBs), resulting in some inherent delay margin for the sum signals in the *Type-*α adders. Hence, the delay assumption required is similar to *Adder-B*, by sizing the OR gate to match at least 1× maximum skew delay between the slowest delay of S_{i+1} and the fastest delay of Carry Out signals in the *Type-*β adder.

The respective asynchronous adders were also fabricated based on a 0.35 μm dual-poly three-metal CMOS process ($V_{Tn} \approx 0.47$ V, $V_{Tp} \approx 0.62$ V) @ 1.1 V, 1 MHz.

On the basis of simulations, Table 3.4 summarizes a comparison of the *Type-*α and *Type-*β adders with the reported CLA and 2 × 1-bit adder. From Table 3.4, the *Type-*α and CLA feature the highest speed, and of the two, *Type-*α dissipates lower power/MHz (hence lower EDP) and requires a smaller IC area. The reason for the former attribute was delineated earlier. The *Type-*β adder dissipates the lowest power/MHz (marginally less than the *Type-*α adder), is relatively slow and requires a relatively small IC area. It is probably worthwhile reiterating that the hardware of the completion detector circuit is independent of the length of the adder embodying *Type-*β adders. In other words, the hardware of long wordlength adders embodying the *Type-*β adders is likely to be simple. Finally, as expected, using the two 1-bit AFAs to construct a 2-bit adder will result in the poorest performance in terms of power/MHz and speed. This design, however, features the smallest area largely due to the simpler interconnects and routing when compared to 2-bit asynchronous adders. Note that the IC area of *Type-*α and *Type-*β adders are comparable.

Table 3.5 summarizes a comparison between the 16-bit asynchronous *Adder-A*, *Adder-B* and *Adder-C*, two reported asynchronous adders (the CLA and the carry ripple adder [CRA]) and one synchronous adder (constructed by 1-bit T-28 full adders in a ripple structure with flip-flop outputs).

From Table 3.5, note that that although *Adder-A* has the same speed as the relatively high-speed 16-bit asynchronous CLA adder, the former dissipates on average 4% less power/MHz. Put simply, *Adder-A* features the lowest EDP among all adders. *Adder-B* and *Adder-C* exhibit the lowest power/MHz (*Adder-B* being marginally lower) and *Adder-C* is the faster (by ~33%) of the two. All three adders, *Adder-A*, *Adder-B* and *Adder-C*, dissipate lower power/megahertz than the reported 16-bit asynchronous and synchronous adders. The designs of *Adder-A*, *Adder-B* and *Adder-C* are verified by means of measurements on prototype ICs and the measurement results generally agree with the post-layout simulations. In terms of IC area, *Adder-B* is the second smallest (~7% bigger than the CRA adder). Among the designs, *Adder-B* is the smallest, *Adder-A* is largest and *Adder-C* lies in between. In summary, in the perspective view of power-critical applications with moderate speed and area requirements, *Adder-C* adder is recommended.

Table 3.4 Average delay, power dissipation, energy-delay-product (EDP) and IC area of various 2-bit adders @ *VDD* = 1.1 V, 1 MHz, obtained from simulations

	Average delay (ns)		Power (nW/MHz)				
2-bit Asynchronous Adder	Pre-charge	Evaluation	Driver	Adder	Total	EDP (10^{-21} J s)	Area (μm^2)
CLA	1.9	4.6	78.9	152.0	230.9	1.1	1350
2×1-bit	1.7	7.4	74.4	175.3	249.7	1.8	1232
Type-α	1.9	4.6	78.5	143.6	222.1	1.0	1332
Type-β	1.9	6.0	74.4	141.1	215.5	1.3	1275

Table 3.5 Delay, power dissipation, energy-delay-product (EDP) and IC area of various 16-bit adders @ *VDD* = 1.1 V, 1 MHz based on post-layout simulations and measurements on prototype ICs

	Simulations			Measurements			
16-bit Adder	Average delay (ns)	Power (μW/MHz)	EDP (10^{-21} J s)	Average delay (ns)	Power (μW/MHz)	EDP (10^{-21} J s)	Area (μm^2)
CLA Adder	27.6	2.22	61.3	–	–	–	15,002
CRA Adder	51.2	2.45	125.4	–	–	–	9,956
Synchronous Adder	52.2	3.52	183.7	–	–	–	17,325
Adder-A	27.6	2.14	59.1	30.9	2.09	64.6	14,387
Adder-B	49.9	1.77	88.3	51.4	1.72	88.4	10,702
Adder-C	33.2	1.80	59.8	33.9	1.74	59.0	11,520

3.3.3 Multiplier

A parallel multiplier [20] generally comprises a partial product generator (PPG) and an Adder Block. The majority of parallel multipliers employ the modified Booth's algorithm [21] for the PPG. This is because the use of the Booth algorithm results in a small number of partial products and this in turn reduces the number of rows (and hence the number of adders) in the Adder Block. In the case of the 2's complement format, a correction term is usually added so that the hardware required for the sign bit extensions (that would otherwise incur substantial hardware overhead) is small. Figure 3.7 depicts the partial products of a 16 × 16-bit 2's complement Booth radix-4 multiplier generated by the Booth PPG. Rows *a* to *h* contain the partial products and the LSB and the MSB are respectively at the extreme right and at the extreme left sides. The partial products in row *r* correspond to modified Booth encoded carry-in signals.

In a parallel array multiplier, the multiplication process simply involves the summation (by means of full adders) of the partial products row by row in a regular fashion in the Adder Block. Although the regularity of the array structure is changed slightly to accommodate the sign bit in 2's complement format, the majority of the full adders are still placed in a regular fashion. The array structure has the advantage (over a tree-based structure) of simple interconnects between the full adders.

Figure 3.8 depicts the block diagram of the 16 × 16-bit LA-based Booth array multiplier core [14] embodying the Booth PPG, delay circuits (D_1–D_7) and Adder Block. Note that LA means latch adder as mentioned in Section 3.3.1 earlier. The Adder Block comprises seven rows (Row 1–Row 7) of LAs and a final row of carry ripple adders (CRAs) that serves as a carry propagation adder.

The operation of the multiplier is as follows. When a multiplication is initiated with the multiplicand *X* and multiplier *Y* inputs, the Start signal asserts D_1. After a predetermined delay in D_1 that is at least equal to the duration required of the PPG to generate the partial products, E_1 and $\overline{E_1}$ will assert the latches in the LAs of Row 1 and assert D_2. As the inputs to the adders in the LAs in Row 1 are synchronized, there is little, if any, spurious switching in this row. In the remaining rows of the Adder Block, there is no switching as the LAs there remain negated. Subsequently, after a predetermined delay in D_2 that is at least equal to the duration of the delay in the adders in Row 1, E_2 and $\overline{E_2}$ will assert the latches in the LAs in Row 2 and assert D_3. Similarly, as the inputs to the adders in the LAs in Row 2 are synchronized, there

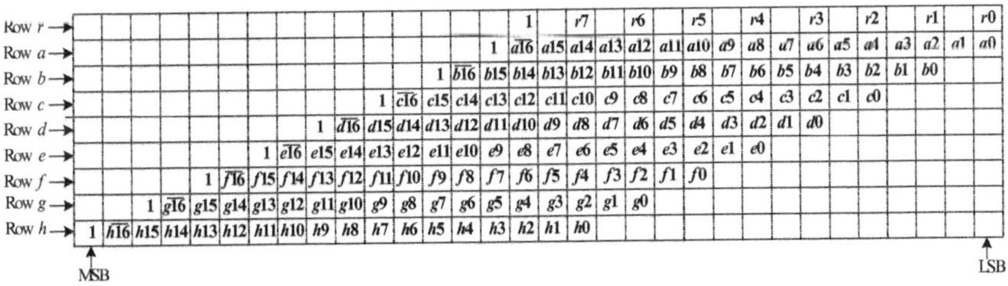

Fig. 3.7 The partial products in a 2's complement 16 × 16-bit Booth radix-4 multiplier

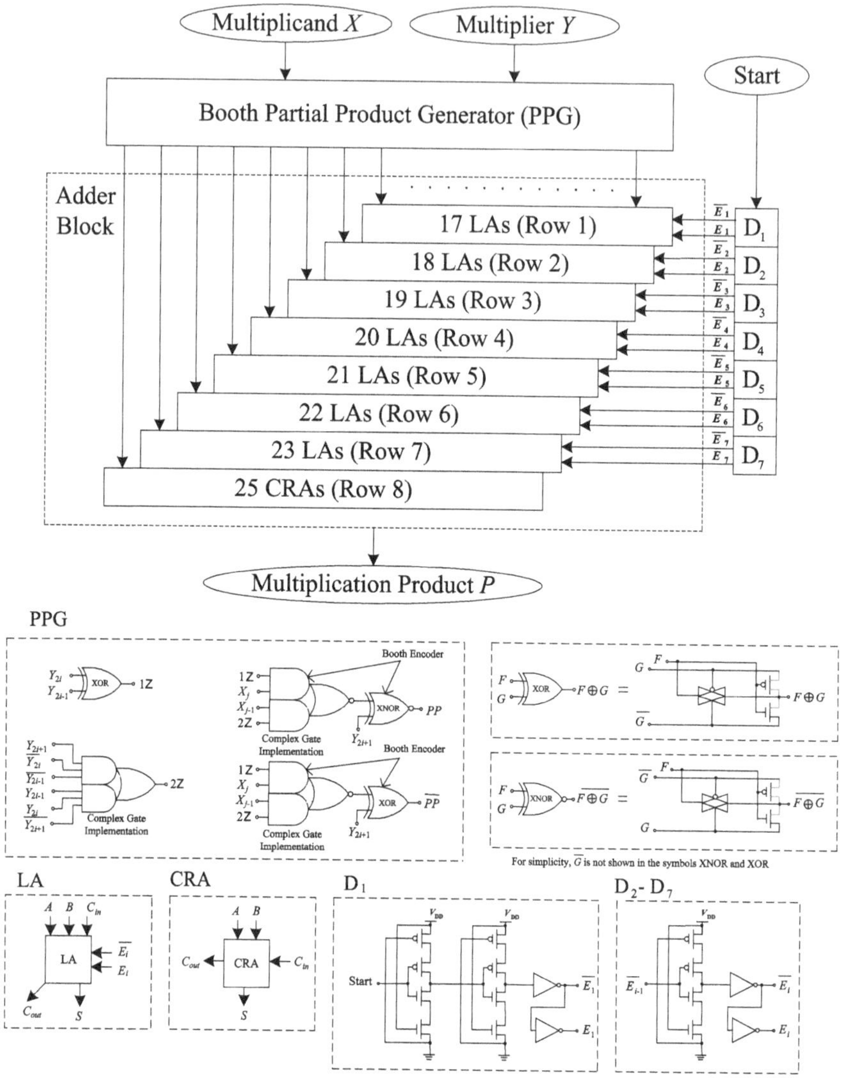

Fig. 3.8 Block diagram of the 16 × 16-bit LA-based Booth array multiplier

is little, if any, spurious switching in this row. Similarly, there is no switching in the remaining rows as the LAs there remain negated. The process repeats until Row 7, and the spurious switching in Rows 1–7 of the Adder Block is virtually eliminated. The last row of CRAs computes the final multiplication product P.

To delineate the efficacy of the LA-based multiplier approach to reduce spurious switching, Fig. 3.9 depicts, from simulations with 10,000 random inputs, the detailed breakdown of the power dissipation of the different blocks of different 16 × 16-bit and 32 × 32-bit array multipliers operating at 1 MHz @ 1.1 V; their energy dissipation can be found in μW/MHz accordingly. The different blocks include the Booth PPG,

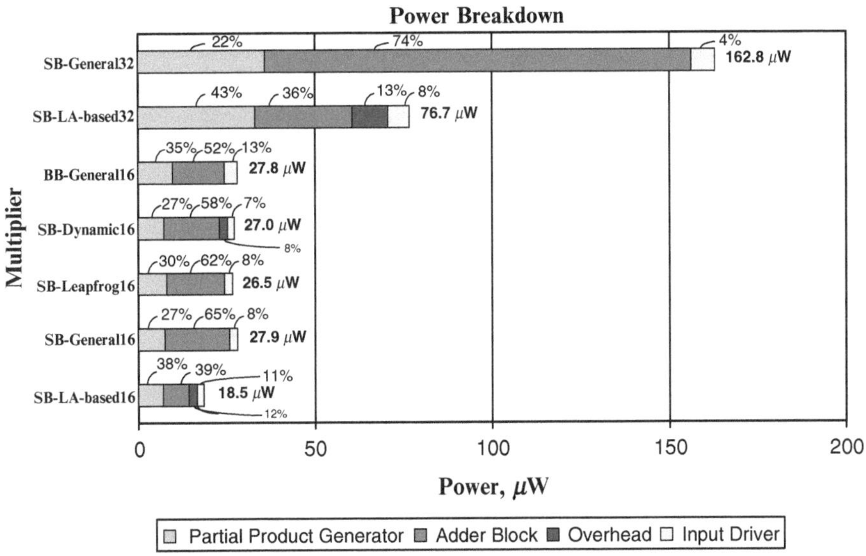

Fig. 3.9 Power breakdown (% and μW) of different array multipliers based on simulations @ 1.1 V, 1 MHz

Adder Block and Input Drivers (buffers for inputs). For the SB-Dynamic [22] and the LA-based multipliers, the 'Overhead' block refers to the power dissipated by the delay circuits to control the Domino DAs and LAs respectively. The T28 adders (see Fig. 3.4a earlier) are used in the Leapfrog [23] and standard (General) designs, and are used as the CRA in the LA-based multiplier design. Note that the total power dissipation of the multipliers is based on the post-layout simulations but the power breakdown distribution is estimated from pre-layout simulations.

In Fig. 3.9, two different PPGs are considered in the General array multipliers. The first PPG (embodying 12-transitor Booth encoders, see PPG in Fig. 3.8) is the standard Booth PPG and this standard Booth PPG is denoted as 'SB' PGG. The second PPG (embodying 18-transistor Booth encoders [24]) is used to balance (equalize) the arrival times of partial products to the Adder Block, resulting in potentially reduced spurious switching in the Adder Block and this Balanced Booth PPG is denoted as the 'BB' PPG.

From Fig. 3.9, it is noted that although the BB-General16 embodying the BB PPG reduces the power dissipation in the Adder Block (compared to the SB-General16 employing the SB PPG from 65% (18.1 μW) to 52% (14.4 μW)), the overall power dissipation of the two multipliers is approximately the same. This is because the reduced power in the Adder Block embodied in the BB-General16 is offset by the higher power dissipation in the BB PPG and the input drivers. Furthermore, the BB PPG is more sensitive to *W/L* ratio sizing, floorplan in layout, and PVT variations. Consequently, design considerations to minimize these undesirable susceptible effects in the BB PPG are critical – the BB PPG may otherwise generate partial products with very different times, much less synchronous than expected. These different arrival times of the inputs consequently result in larger amount of spurious switching in the Adder Block, hence higher power dissipation in the multiplier. Furthermore, the IC area required for the BB PPG is larger (compared to the SB PPG). In view of the potentially higher power dissipation, undesirable added design considerations and larger IC area required of the BB PPG, the BB PPG is not considered and instead only the SB PPG is considered in reported multiplier designs (except BB-General16).

From Fig. 3.9, it is also noted that for the reported SB-Dynamic, SB-Leapfrog, SB-General, and BB-General, 16 × 16-bit multipliers, an average of 59% (16.1 μW) of the total power is dissipated in the Adder Block. Put simply, the Adder Block dissipates the largest power in typical 16 × 16-bit multipliers. By means of the LA-based multiplier approach to significantly reduce the spurious switching and with little overhead, the Adder Block now dissipates 39% (7.2 μW) of the total power. In other words, the LA-based multiplier design on average dissipates ~32% less power than the reported designs compared herein, and ~30% less power than the lowest power reported design, the SB-Leapfrog multiplier.

To quantify the degree of reduced spurious switching, Fig. 3.10 depicts the number of switchings per adder in the different rows of the Adder Block based on information obtained from pre-layout simulations by the transistor-level Nanosim simulations. The total number of switching (i.e. changed voltage levels from V_{DD} to ground or vice versa) is first recorded for all the adders in the specific row, and is then averaged by the total number of adders in that row. The average number of switchings per adder for the SB-Leapfrog16, SB-General16, and BB-General16 multipliers is 5.3, 6, and 5.3 respectively, and the amount of switching increases as

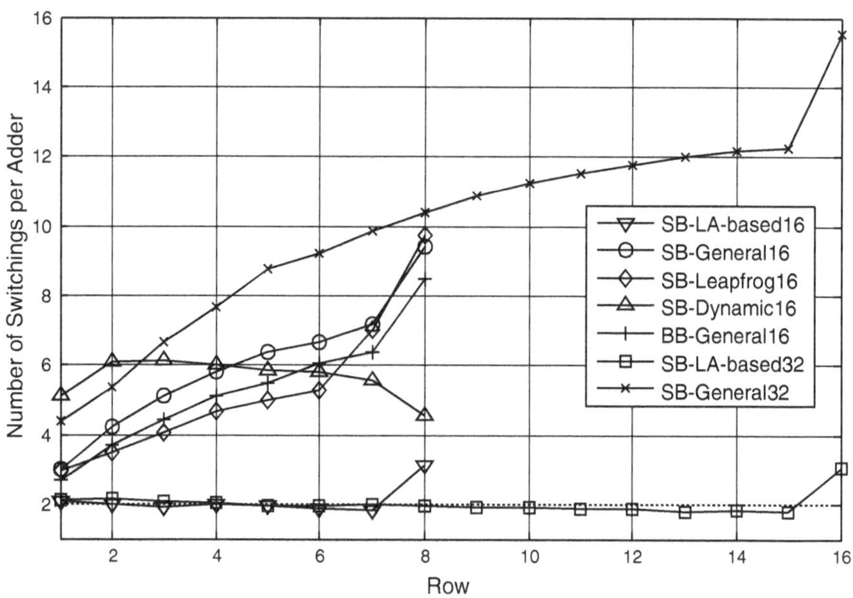

Fig. 3.10 Switching analysis in the Adder Block of various multipliers

the row number increases. This is not unexpected as the inputs to the adders become increasingly less synchronous in the latter rows. In the case of the multiplier with Domino DAs (termed SB-Dynamic16), the number of switchings per adder remains largely unchanged, an average of 5.6 switchings per adder. This is because the Domino DAs always first pre-charge and then evaluate when the inputs are ready. The high number switching in the Domino DAs is due to its dynamic operation (constant precharge and evaluation). By comparison, the number of switchings per adder in the LA-based multiplier design is substantially low at 2 switchings per adder, leading to the much desired lower power attribute. The average 3 switchings per adder (instead of 2) in the last row of the SB-LA-based16 and SB- LA-based32 multipliers is largely due to the carry ripple effect of the multiplication product in the CRAs in the last row.

The efficacy of the LA-based multiplier approach is more pronounced in longer wordlength multipliers, for example a 32 × 32-bit multiplier. This is because there is a larger number of rows and because the input signals to the adders in the latter rows are even less synchronous. Figure 3.10 also depicts the number of switchings per adder of a General 32 × 32-bit array multiplier. On average, there are ~10 switchings per adder @ 1.1 V, 1 MHz. By means of the LA-based multiplier approach, the number of switchings per adder remains approximately the same as the LA-based 16 × 16-bit multiplier design – 2.1 switchings per adder. In power terms @ 1.1 V, 1 MHz from Fig. 3.9, the Adder Block dissipates 74% (120.5 μW) and 36% (27.6 μW) of the total power of the SB-General multiplier and the LA-based multiplier design respectively. In other words, the LA-based multiplier design dissipates ~53% less power than the SB-General multiplier.

Two 16 × 16-bit multiplier designs, the SB-LA-based16 and the conventional SB-General16, have been realized in monolithic form using a 0.35 μm dual-poly three-metal CMOS process. On the basis of measurements on prototype ICs and on simulations, Table 3.6 summarizes the parameters of the different multipliers. The measurement results agree well with the simulations.

Of the 16 × 16-bit designs in Table 3.6, the LA-based multiplier design features the lowest energy dissipation, one of the slowest designs but the lowest EDP. In some applications, such as that for the intended hearing instrument application where the embodied DSP does not require high speed operations, the delay is inconsequential. The speed of the LA-based multiplier design can be increased by adopting a more aggressive timing for the delay controlled by the delay elements D_1–D_7 in Fig. 3.8. However, this may compromise the degree of spurious switching reduction, and hence the energy dissipation. The speed can also be increased by replacing the CRAs in the final row of the adder to a faster adder such as a carry look-ahead adder but at the slight cost of increased energy dissipation. For even higher speed operation, V_{DD} would need to be increased, for example from 1.1 to 3.3 V – in this case, for the same degree of spurious switching reduction, the delay of the SB-LA-based16 design reduces from 131 ns (7.6 MHz) to 26 ns (38 MHz) and the power dissipation remains low at ~273 μW/MHz. Note that at 3.3 V operation, the short-circuit current is no longer negligible, dissipating ~30% of the total power. For the other multipliers, the short-circuit

Table 3.6 Average number of switchings, energy, delay, area and energy-delay-product (EDP) of various array multipliers @ 1.1 V, 1 MHz based on post-layout simulations and measurements on prototype ICs

	Simulations based on post-layouts					Measurements based on prototype ICs		
Multiplier	Average switching	Energy (µW/MHz)	Delay (ns)	Area (mm^2)	EDP (a J s)	Energy (µW/MHz)	Delay (ns)	EDP (a J s)
SB-General16	1,412	27.9	104	0.18	2.9	28.3	122	3.5
SB-Leapfrog16	1,320	26.5	103	0.17	2.7	–	–	–
SB-Dynamic16	1,375	27.0	134	0.24	3.6	–	–	–
BB-General16	1,291	27.8	96	0.20	2.7	–	–	–
SB-LA-based16	657	18.5	131	0.20	2.4	18.8	145	2.7
SB-General32	8,350	162.8	201	0.45	32.7	–	–	–
SB-LA-based32	2,752	76.7	260	0.51	19.9	–	–	–

current is similarly not negligible. To reduce the short-circuit power dissipation, the adders and delay lines would need to be resized.

For the 32 × 32-bit multipliers, the comparison had been discussed earlier in Fig. 3.9 and the comments had also been made therein.

In terms of IC area, both the 16 × 16-bit and 32 × 32-bit LA-based multiplier designs are slightly larger than the 16 × 16-bit SB-General and SB-Leapfrog multipliers, and the 32 × 32-bit SB-General multiplier. This is due to the overhead of the LAs and the delay circuits. However, the LA-based multiplier design is smaller than the 16 × 16-bit SB-Dynamic multiplier and is comparable to the 16 × 16-bit BB-General multiplier.

Furthermore, specifically in the computation of the FFT/IFFT algorithm, a large number of multiplications are trivial due to the coefficients of ×1, ×–1, and ×0 [25]. It is noted that for a 128-point FFT, about 42% of the multiplications are trivial. The computation of trivial multiplications can be simply obtained by passing the inputs directly to be the outputs (for ×1), by setting the outputs to be zeros (for ×0), or by negating the inputs (for ×–1). In these instances, the effective energy dissipation due to the multiplier can be reduced by having the appropriate control circuits output these trivial multiplication products accordingly. Figure 3.11 depicts the asynchronous 16 × 16-bit LA-based multiplier with the added control circuits (comprising a control logic, a converting circuit, and multiplexers) and the multiplier is termed as Control-Multiplier [5]. The multiplier core design, is based on the earlier multiplier design where the spurious switching/glitch therein is largely eliminated by means of LAs timed by the delay lines, D_1–D_7.

The inputs are the multiplication *X* (input signals) and the multiplier *Y* (coefficient of the twiddle factor). In Fig. 3.11, the shaded inverting latches block the inputs to the multiplier core in cases of trivial multiplications. For non-trivial multiplications, the shaded inverting latches will be updated with the new data to the multiplier core for multiplications. The control circuits determine if the multiplications are trivial or otherwise, initiate multiplications when *REQ* is asserted and generate *COMP* upon the completion of a multiplication. The *COMP* signal is matched (by means of delay lines) to the worst-case delay for either one of two situations – either for the trivial multiplications or the non-trivial multiplications.

Table 3.7 tabulates the simulated energy dissipation and IC area for the Control-Multiplier and the LA-based Multiplier core. Based on simulations with 40% (estimated % of the actual multiplications) of the multiplications being trivial, the Control-Multiplier features ~22% lower energy dissipation. The Control-Multiplier, however, requires about 7% larger IC area than the multiplier core. For completeness, the delay of the Control-Multiplier is ~60 ns @ 1.1 V for trivial multiplications and ~160 ns @ 1.1 V for non-trivial multiplications respectively.

In short, for FFT/IFFT algorithms where the *N* is small (e.g. <512-point), the realization of the Control-Multiplier is worthwhile due to its low energy dissipation attribute.

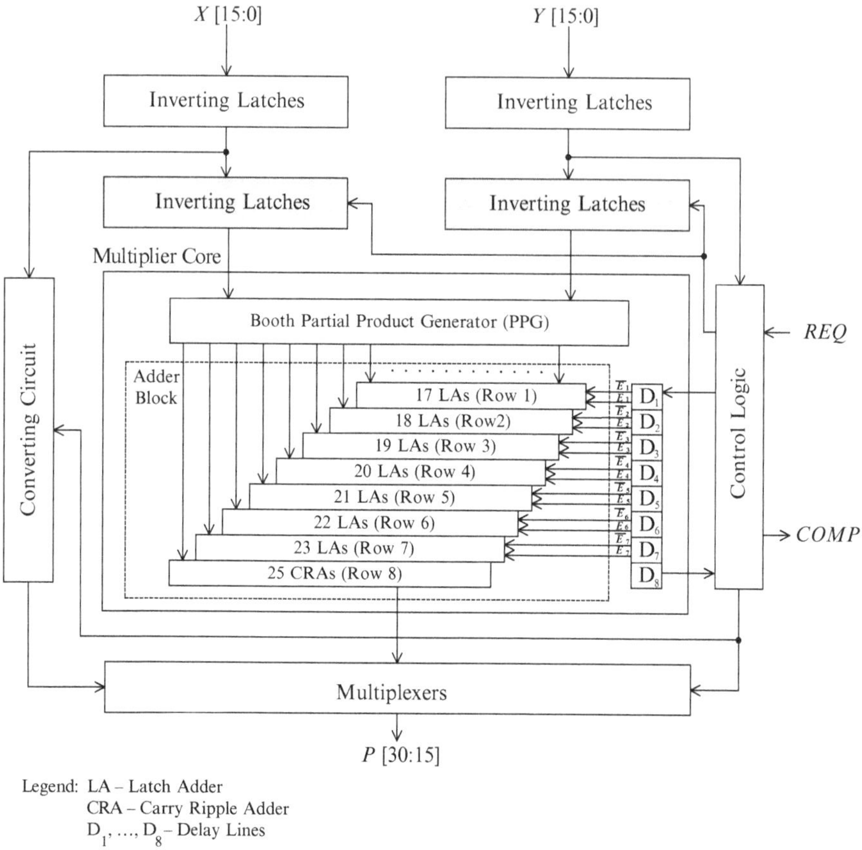

Fig. 3.11 Block diagram of the Control-Multiplier

Table 3.7 The energy and IC area for the Control-Multiplier and the multiplier cores @ 1.1 V, 1 MHz

	Energy (μW/MHz)	Area (mm²)
Control-Multiplier	14.1	0.134
Multiplier core	18.0	0.125

3.3.4 Memory

Figure 3.12a depicts the block diagram of the asynchronous 128 × 16-bit single-port memory macrocell. For low energy dissipation, the well-established partitioning approach is adopted to reduce the capacitance. In this approach, the memory is sub-divided into four 32 × 16-bit sub-blocks (Blocks A–D) and these sub-blocks are controlled by their respective Word Line Controllers. The Row Decoder, Column Decoder and Control Circuitry control the memory for the write and read accesses.

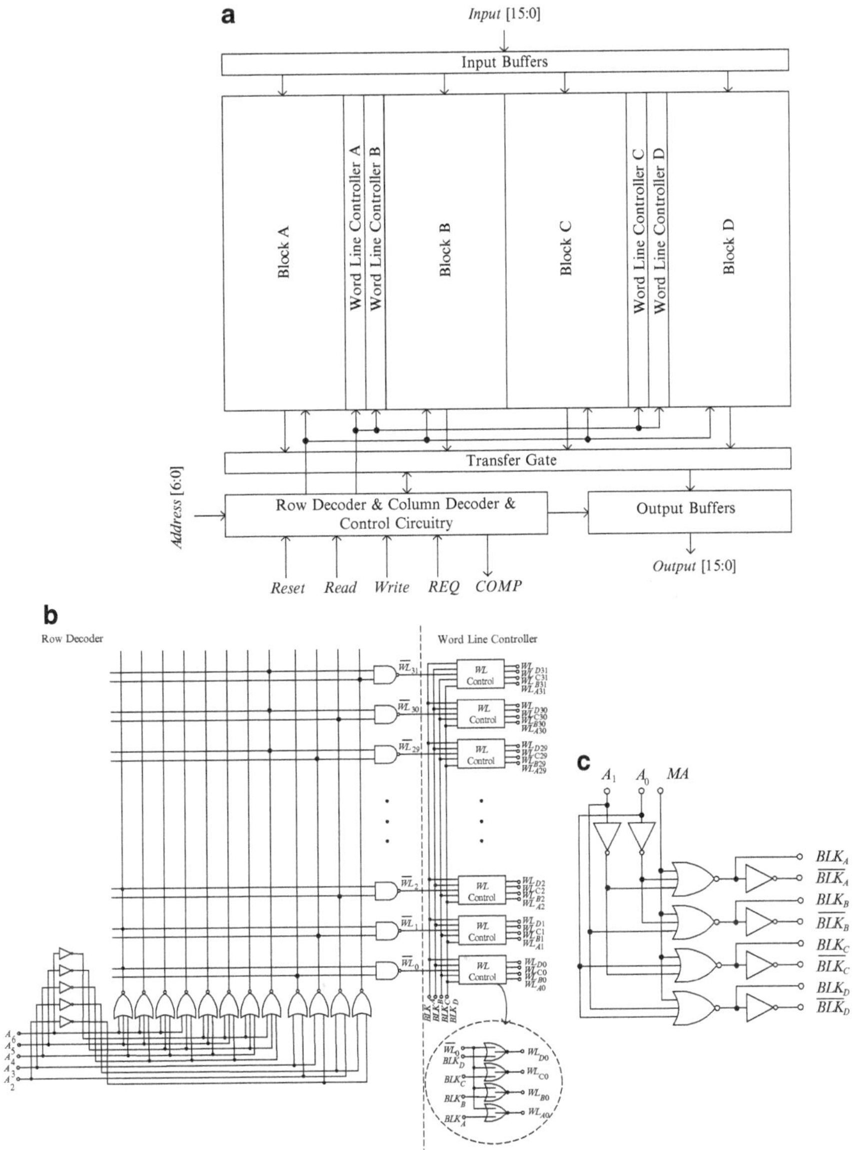

Fig. 3.12 Memory: (a) block diagram, (b) row decoder and word line controller, and (c) column decoder

For the read access, the stored data are retrieved (via the Transfer Gate from one of the four sub-memory blocks) to the output buffers, according to the address bus, $A[6{:}0]$. For the write access, the new input is written (from the input buffers) into one of the four sub-memory blocks, according to $A[6{:}0]$. The *REQ* and *COMP* signals respectively indicate the request and completion signals of the memory.

Table 3.8 The characteristics of the asynchronous 128x16-bit memory macrocell @ 1.1 V, 1 MHz

	Characteristics
Supply voltage	1.1 V
Energy	6 μW/MHz
Read access time	60 ns
Write access time	40 ns
Area	0.142 mm^2

The standard six-transistor single-port SRAM is employed as the basic memory storage. For low energy reasons [9], the two-stage decoding technique in a NOR-NAND structure is adopted to construct the Row Decoder (see left-hand side in Fig. 3.12b) according to the address bus $A[6:2]$. The Word Line Controllers (see right-hand side in Fig. 3.12b) activate the selected sub-memory block and prevent multiple assertions of unselected sub-memory blocks, resulting in reduced energy dissipation. The Word Line Controller is controlled by the Column Decoder (see Fig. 3.12c) which selects the specific column of the memory according to the address bus $A[1:0]$. The Memory Access (*MA*) signal is a buffered signal from *REQ* and the memory will only be selected when *REQ* (and *MA*) is asserted.

Table 3.8 tabulates the characteristics of the asynchronous 128 × 16-bit memory macrocell based on the post-layout computer simulations @ 1.1 V, 1 MHz. The read or write access time is defined as the time from the *REQ* signal being asserted until the *COMP* signal is asserted. The read and write access times of the memory macrocell are 60 and 40 ns, respectively, and the energy dissipation of the memory macrocell is 6 μW/MHz. The entire asynchronous memory macrocell occupies the area of 410 × 346.4 μm (or 0.142 mm^2) @ dual-poly four-metal CMOS process.

3.4 Low Power Asynchronous FFT/IFFT Processor

In this section, the design of an energy-efficient asynchronous 128-point 16-bit FFT/IFFT processor based on the asynchronous approach for energy-critical applications is demonstrated. The asynchronous design is benchmarked against its synchronous counterpart in terms of energy, delay, and IC area. Both designs are realized with the same functionality and similar architecture, and fabricated using the same 0.35 μm CMOS process. In the case of the synchronous design, the well established clock gating approach is adopted (idle state) to reduce energy dissipation when computation is not required.

This section is organized as follows. Section 3.4.1 presents an overview of the FFT/IFFT algorithm that provides a preamble to the architecture adopted for the synchronous and asynchronous approaches. Sections 3.4.2 and 3.4.3 present the benchmark synchronous and the asynchronous FFT/IFFT Processors. Finally, Section 3.4.4 gives a comparison of the synchronous and asynchronous designs.

3.4.1 FFT/IFFT Algorithm

In this section, an overview of the FFT/ Discrete Fourier Transform (DFT) algorithm is briefly described followed by a description of the synchronous and asynchronous FFT/IFFT processor designs.

Given a sequence of time samples $x(n)$ or a sequence of frequency samples $X(k)$, an N-point DFT and its inverse are respectively:

$$X(k) = \sum_{n=0}^{N-1} x(n)\cdot W_N^{nk} \quad k = 0,1,.,N-1 \tag{3.3}$$

$$x(n) = \frac{1}{N}\sum_{k=0}^{N-1} X(k)\cdot W_N^{-nk} \quad n = 0,1,.,N-1 \tag{3.4}$$

where the twiddle factor $W_N^{nk} = e^{-j2nk\pi/N} = \cos(2nk\pi/N) - j\sin(2nk\pi/N)$.

The FFT is essentially derived from the DFT and the former reduces the computational complexity to $O(\log rN)$ (as compared to $O(N^2)$ in the latter) where r is the radix number.

In this section, both the synchronous and asynchronous FFT/IFFT processors [5] are based on a 128-point radix-2 decimation-in-time FFT algorithm. The prevalent radix-2 is adopted for its operational regularity and for its hardware simplicity. The inverse (IFFT) is obtained simply by negating the sine coefficients and multiplying the outputs by a scaling factor of 1/128 and using the same FFT algorithm. Equation (3.5a–d) describe the operations for one radix-2 butterfly.

$$R'_a = R_a + [(C)\cdot R_b - (-S)\cdot I_b] \tag{3.5a}$$

$$R'_b = R_a - [(C)\cdot R_b - (-S)\cdot I_b] \tag{3.5b}$$

$$I'_a = I_a + [(C)\cdot I_b + (-S)\cdot R_b] \tag{3.5c}$$

$$I'_b = I_a - [(C)\cdot I_b + (-S)\cdot R_b] \tag{3.5d}$$

where $R_a + jI_a$ and $R_b + jI_b$ are the two inputs, and $R'_a + jI'_a$ and $R'_b + jI'_b$ are the corresponding two outputs. C and S are the cosine and sine coefficients of the twiddle factor.

To compute (3.5a–d), two multipliers and three adders perform one butterfly in two data flows to compute the real and the imaginary outputs. Each data flow requires five pipeline stages: Memory Read, Scaling, Multiplications, Additions/Subtractions, and Write Back. In Memory Read, the signals are retrieved from the memory and in Scaling, the signals are scaled (where necessary) to accommodate the signal levels and hence avoid overflow. In Multiplications, two multiplications are performed and in Additions/Subtractions, three additions/subtractions are performed. Finally, in Write Back, the outputs are stored.

3.4.2 Benchmarked Synchronous FFT/IFFT Processor

The block diagram of synchronous FFT/IFFT processor is depicted in Fig. 3.13 where two clock signals, Clk1 and Clk2, are used. The frequency of Clk2 is 2× Clk1 and Clk2 is used to trigger the memory to obtain two sets of 32-bit memory data for every Clk1 cycle (one butterfly operation). Clk1 is used as the system clock for the remaining modules. The synchronous FFT/IFFT processor requires 2,430 Clk2 cycles and 959 Clk1 cycles for one FFT/IFFT computation (including loading new samples and with result outputs). The well-established coarse-grain and fine-grain

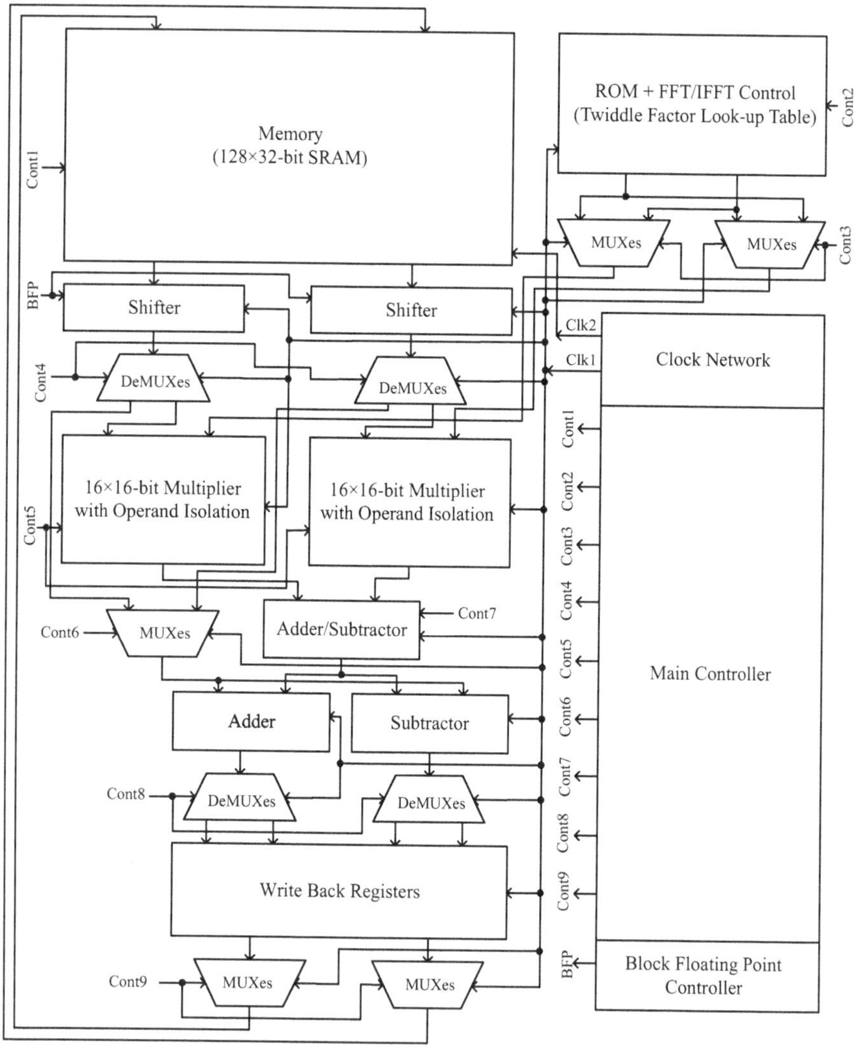

Fig. 3.13 Block diagram of the synchronous FFT/IFFT processor

clock gating approaches were considered and as both approaches yield somewhat similar energy savings, the former is adopted for its simplicity. Specifically, in the coarse-grain clock gating approach, Clk2 is gated after 2,430 Clk2 cycles and Clk1 gated after 959 Clk1 cycles. For a typical hearing instrument where Clk1 is 1 MHz and for a time duration of 4 ms (64 samples @ 16 KHz) for one FFT/IFFT computation, the coarse-grain clock gating blocks ~76% (3,041 of 4,000) and ~70% (5,570 of 8,000) of Clk1 and Clk2 respectively; note that the datapath modules in the synchronous design circuits are highly pipelined, thereby resulting in useful computation in most of the active clock cycles. If fine-grain clock gating is otherwise employed, only small percent of clock cycles (i.e. a further 63 out of 4,000 Clk1 cycles and a further 126 out of 8,000 Clk2) may be further gated, and collectively this translates to a small 6% improvement in energy dissipation. However, this small improvement comes at a cost including a design of higher complexity (including consideration for synchronization issues, skew problems and race conditions) and this would largely defeat the small energy advantage.

3.4.3 Asynchronous FFT/IFFT Processor

The asynchronous FFT/IFFT processor is based on the four-phase protocol (instead of the two-phase protocol [8] for the advantages already established in literature) and is partitioned into the control portion (Asynchronous FFT/IFFT Controller and Pulse Circuit, see Fig. 3.14a) and the datapath portion (Memory, Shifter, Multiplier, Adder, and Write Back Datapaths, see Fig. 3.14b). The former portion pipelines the entire operation asynchronously within the processor and the latter portion executes the butterfly operations.

In Fig. 3.14a, the Main Sequence Controller, Sample Loading Controller, Data Loading Controller, Butterfly Unit Operation Controller, Write Back Controller and Block Floating Point Controller are collectively grouped as the Asynchronous FFT/IFFT Controller. This Asynchronous FFT/IFFT Controller controls the respective datapath circuits according to the control and handshake signals. The Pulse Circuit functions as a local 'clock' to synchronize the sequences of each operation in the Asynchronous FFT/IFFT Controller and the local 'clock' is generated asynchronously.

Figure 3.14b depicts five datapath modules and their associated handshake and control signals. The Memory Datapath reads and writes data during the FFT/IFFT operations. The Shifter Datapath is for conditional block floating point scaling to retain the dynamic range of the signals, avoiding overflow – the signals therein are scaled when overflow is detected in the Write Back Datapath. The Multiplier Datapath is used to multiply the coefficients (from ROMs) with the signals from the Shifter Datapath. The Adder Datapath performs the additions and subtractions, and finally the Write Back Datapath stores and writes the outputs back into the Memory Datapath. For sake of illustration, only the main modules are shown in each datapath. These datapaths are controlled by their respective handshake control circuits (HCCs).

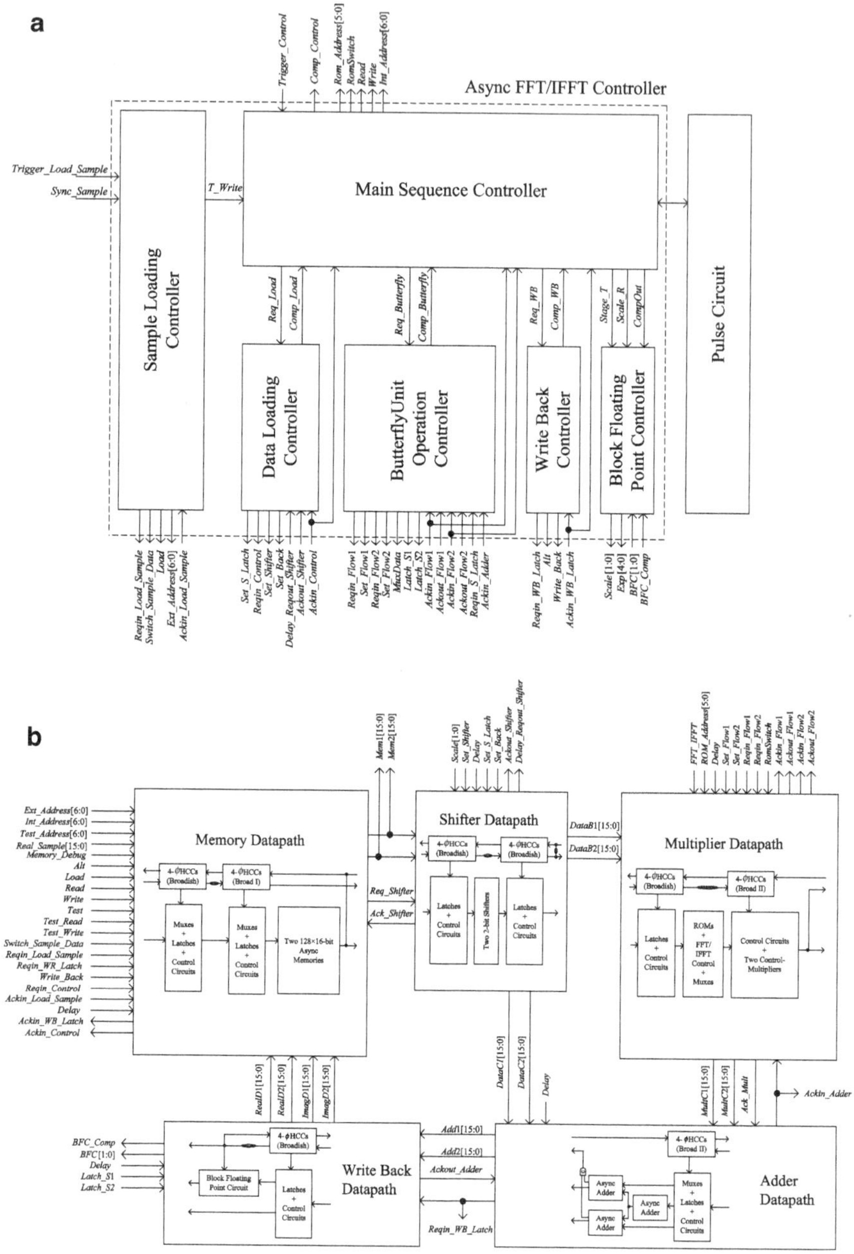

Fig. 3.14 (a) Control portion and (b) datapath portion in the asynchronous FFT/IFFT processor

Figure 3.15 depicts the simplified signal transition graph for one butterfly operation in the asynchronous FFT/IFFT processor, with the initialization of *Req_Butterfly*. The solid lines indicate the direct interface relationship between the signals and the dashed lines indicate the indirect interface relationship between the signals (e.g. via

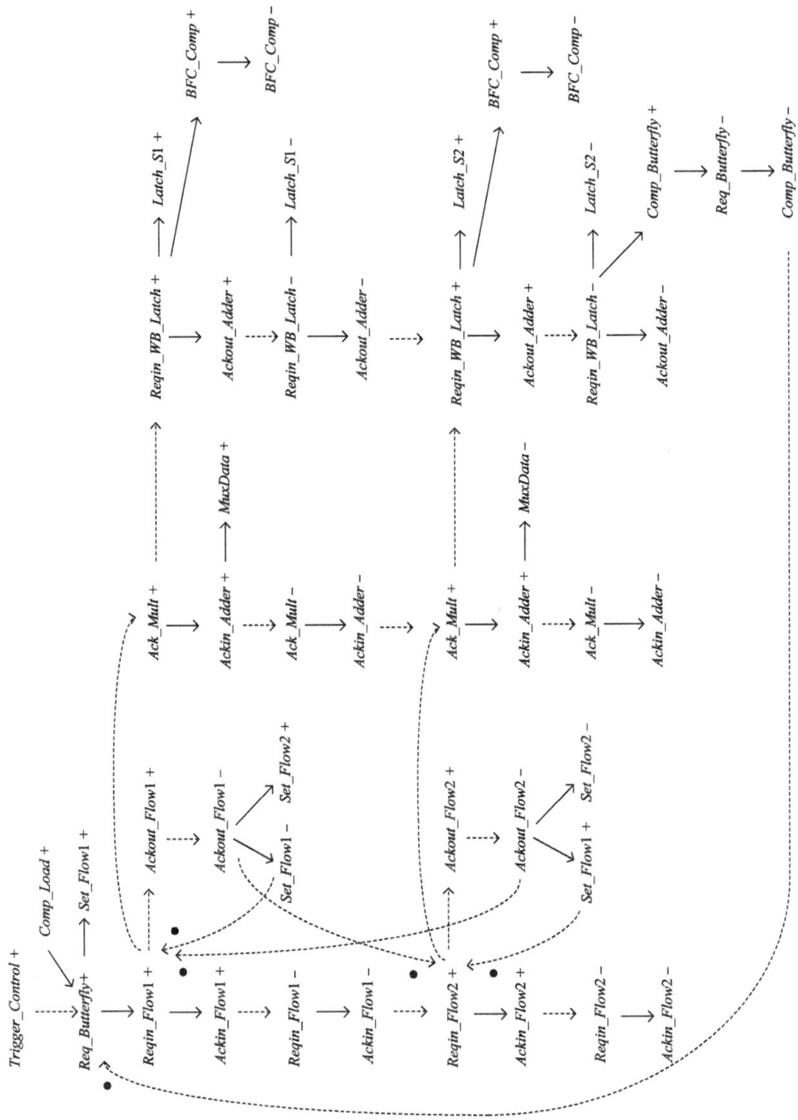

Fig. 3.15 Simplified signal transition graph for one butterfly operation

some circuits that depend on other control signals). Two data flows (*xxx_Flow*1 and *xxx_Flow*2, where *xxx* indicates the prefix labels for various signals) compute the multiplications and additions. After a butterfly operation is completed, *Comp_Butterfly* will be triggered. In total, there are 7 × 64 butterfly operations.

Three types of four-phase asynchronous latch controllers are employed and are depicted in Fig. 3.16 where the top figures are the schematic diagrams and the lower figures are their corresponding signal transition graphs. The first controller [13] in Fig. 3.16a is in Broadish data valid when the *En* signal is ready to be asserted after R_{out} is low ('–'). Similarly, the second latch controller [13] in Fig. 3.16b is in Broad data valid when the *En* signal is asserted after A_{out} is low ('–'). A potential problem with these two latch controllers is that R_{out} can be asserted even though *En* has not been reset to low. This results in the possibility of an incomplete data transfer and the subsequent macrocells may undesirably perform redundant operations when their inputs are not ready. To circumvent this potential problem, a different Broad controller is employed and this is depicted in Fig. 3.16c. In this case, R_{out} will only be asserted when *En* is low.

One assumption made in the operation of these three latch controllers is that the predetermined delay pulse (from *En*– to *En*+ to *En*–) has to be sufficiently long to allow the complete data transfer to the latches.

Figure 3.17a–d depict four other handcrafted circuits used widely in the asynchronous FFT/IFFT processor. Figure 3.17a, b respectively depict an inverting latch [26] and a non-inverting latch [11], and these latches are more energy-efficient than the usual flip-flops in synchronous designs. Figure 3.17c depicts a pulse

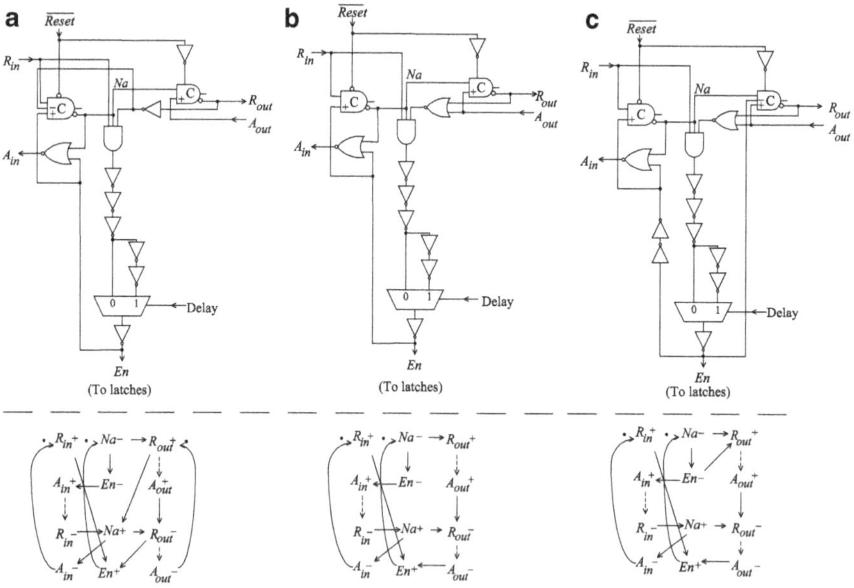

Fig. 3.16 Latch Controllers: (**a**) Broadish, (**b**) Broad I, and (**c**) Broad II

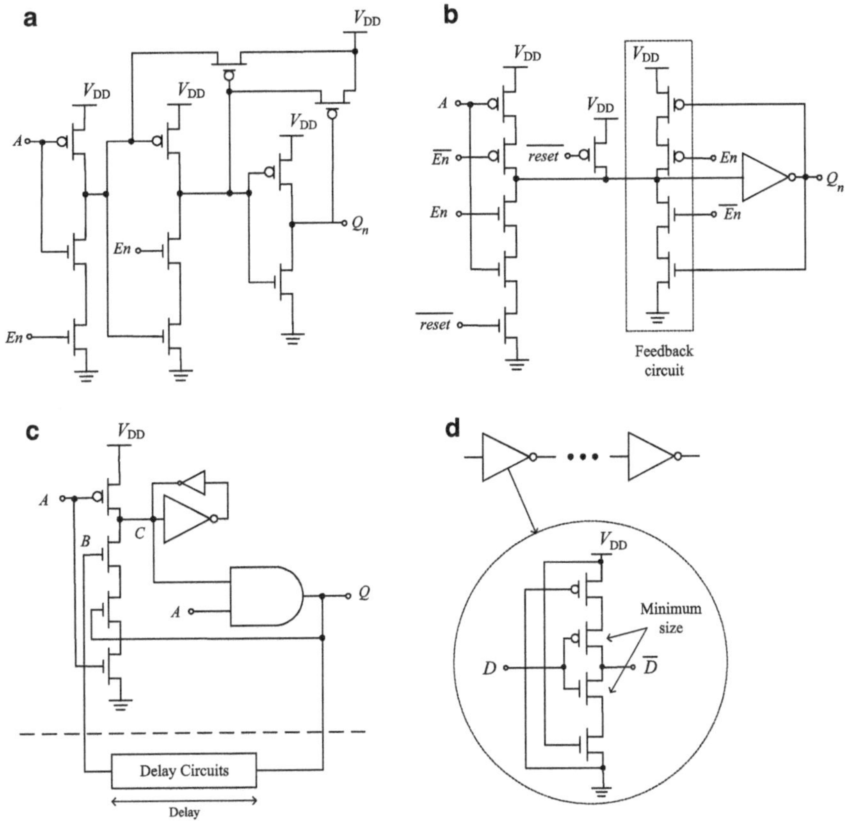

Fig. 3.17 (**a**) Inverting latch, (**b**) non-inverting latch, (**c**) pulse generator, and (**d**) delay line

generator circuit where the generated pulse width is determined by the delay lines (Fig. 3.17d). The self-reset property (low to high to low) of the pulse generator circuit makes it very easy to control the asynchronous controllers. Figure 3.17d depicts the delay line where the transistors connected to V_{DD} and ground are sized to obtain the desired amount of delay with little penalty in energy dissipation.

The Shifter Datapath employs conventional static CMOS multiplexers. In a radix-2 FFT algorithm, the maximum overflow that can occur per butterfly operation is 2 bits. Hence, the shifters only need to be implemented using simple 3-to-1 multiplexers. The Adder Datapath comprises three carry-completion sensing adders [7] where only the carry blocks are implemented using dual-rail logic for completion detection. The Write Back Datapath primarily comprises latches and the Block Floating Point circuit for temporary storage of the intermediate outputs and for the overflow detection respectively.

The Asynchronous FFT/IFFT Controller is modeled using Verilog and its interfaces are modified for asynchronous operations. The simple linear pipeline for the

Controller is adopted as it does not require high speed operation in view of the intended hearing instrument application. In other words, although the processing within one butterfly is operated at different pipeline stages and at different rates, the butterflies are in fact realized sequentially.

3.4.4 Comparison of the Synchronous and Asynchronous Designs

Tables 3.9 and 3.10 summarize the parameters and the operating conditions of the synchronous and asynchronous FFT/IFFT processors. The synchronous design is based on standard digital cells and a 128 × 32-bit memory obtained from the foundry. The asynchronous design is based on several fully and partially hand-crafted asynchronous cells (including two blocks of asynchronous 128 × 16-bit memories, two asynchronous multipliers, etc.) and where pertinent, on the same cells used in the synchronous design. The basic philosophy here is an attempt to make the comparison of the synchronous design and the asynchronous design as fair as possible.

Figure 3.18a, b respectively depict the microphotographs of the prototype synchronous and asynchronous FFT/IFFT processor ICs based on a 0.35 μm CMOS process with $|V_{Tp}| = 0.69$ V and $V_{Tn} = 0.5$ V. The asynchronous FFT/IFFT processor occupies 1.6 mm^2 IC area, 10% larger than its synchronous counterpart.

Figure 3.19a depicts the energy dissipation from measurements on prototype ICs for the asynchronous and synchronous designs. The asynchronous design features

Table 3.9 Synchronous (benchmark) and asynchronous FFT/IFFT processors

	Synchronous design	Asynchronous design
Algorithm	128-point Radix-2 Decimation-in-time	
Wordlength	16-bit	
Computation effort	7 × 64 butterfly operations	
Process technology	0.35 μm dual-poly four-metal CMOS process	
Datapath circuits	2 shifters, 2 multipliers, 3 adders	
Memory	1 memory bank (128 × 32-bit)	2 memory banks (128 × 16-bit each)
Cell library	Standard cells	Standard cells + asynchronous cells

Table 3.10 Operating conditions for the synchronous and asynchronous FFT/IFFT processors

	Condition
Voltage	1.1–1.4 V
Computation Time	4 ms (64 samples @ 16 kHz sampling frequency)
Operation	128 new samples loaded + one FFT/IFFT transform + 128 outputs

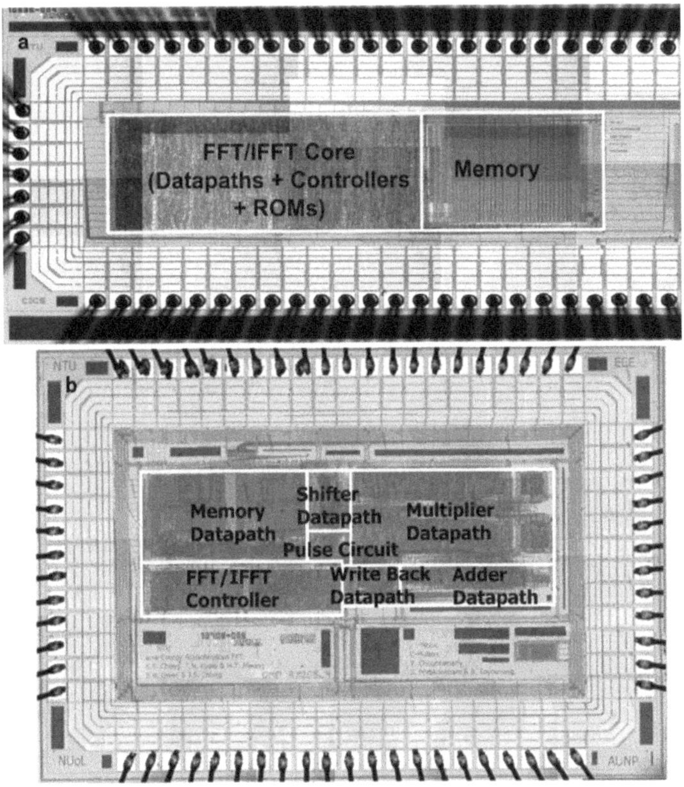

Fig. 3.18 Microphotographs of the FFT/IFFT processors: (**a**) synchronous design and (**b**) asynchronous design

lower energy dissipation where on average from 1.1 to 1.4 V (typical voltage range for hearing instruments), the asynchronous design dissipates ~37% lower energy per FFT/IFFT computation than the synchronous design and this is attributed to three reasons.

First, in the synchronous design, despite the coarse-grain clock gating described earlier, a small number of datapath circuits (e.g. during the initialization phase where signals are not ready for the particular pipeline stage) remain asserted by the clock signal(s), resulting in ~6% redundant operations in some pipeline stages. To design the synchronous circuits to eliminate the 6% redundant operations by means of fine-grain clock gating would require a more complex synchronous controller with multi-phasic clocks and complex logic circuits, hence largely defeating the 6% saving and this would likely incur further costs due to the associated overheads. This redundant switching does not occur in the asynchronous design. Further, the 96 flip-flops that serve as delay registers to pipeline the data sequence to the datapath circuits in the synchronous design are unnecessary in the

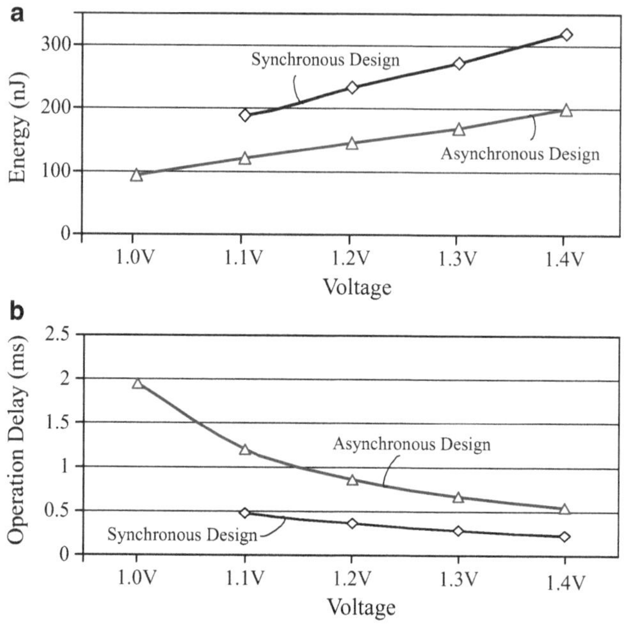

Fig. 3.19 (**a**) Energy dissipation and (**b**) delays for the asynchronous and synchronous FFT/IFFT processors

asynchronous design. Overall, this first reason accounts for an overall reduced energy of ~9%.

Second, the spurious switchings in the asynchronous datapath circuits are reduced by synchronizing the arrival time of the inputs to the custom-designed datapath circuits (e.g. memories, multipliers, adders, etc.) with better control by means of the latches and delay lines, by gating the redundant hazards and transitions with asynchronous handshake circuits, and by disabling the unused blocks (particularly for latches whose data are not updated for computations). For example, the asynchronous multiplier is ~30% lower energy than the synchronous multiplier (with operand isolation). In the synchronous design, the inputs of the signals can also be synchronized at added cost and increased design complexity, and this is applicable only to the circuits where input flip-flops are present. Overall, this second reason resulting in reduced switchings accounts for ~18% energy saving.

Third, a latch is typically dissipated 50% lower energy than a flip-flop and by comparison, there are a total of 783 flip-flops in the synchronous design against the 122 flip-flops and 459 latches in the asynchronous design. The synchronous design can also be designed based on a less prevalent latch-based approach, a somewhat unconventional approach that is more complex including the need to consider the signal synchronization more carefully. Overall, this third reason accounts for ~10% energy saving.

It is worthwhile to note that the clocking energy in the synchronous design is in fact small, ~5% of the total energy. This is largely because of the low clock rate, the FFT/IFFT architecture and the simplicity of a simple buffer network for the clock infrastructure. In other words, in this specific low clock frequency example, the absence of a global clock infrastructure in the asynchronous realization does not contribute significantly to the energy savings. The energy dissipation due to the clock infrastructure is likely to be more significant if the clock rate is high and/or if the IC is large.

Figure 3.19b depicts the measured delays for the asynchronous and synchronous designs, where the delay is defined as the minimum time for one complete FFT/IFFT computation. By means of increasing the clock frequency until the synchronous design fails, the delay of the asynchronous design is deemed to be ~1.4 times worse. The modality of increasing the clock frequency in the synchronous design is somewhat equivalent to reducing the delay of the delay elements in the matched delay asynchronous design – the former can be adjusted externally but the latter is an on-chip adjustment. On the basis of simulations, if the delay of the matched delay elements is more aggressively tuned (instead of simply designing to meet the 1.215 ms per computation requirement), the delay of the asynchronous design can be reduced to ~0.6 times worse than the synchronous design. Further, if speed was the primary objective, the asynchronous design can be designed to be faster by means of adopting high speed latch controllers [1], and with largely the same energy dissipation – specifically, the slow Broad controllers adopted in the asynchronous design can be replaced by the higher speed Broadish controllers (see Fig. 3.16a). Overall, by means of the aggressively decreasing the delay of the matched delay elements and adopting the Broadish controllers, the delay of the asynchronous design can in fact be made faster than the synchronous design by 0.24 times and largely without compromising energy. For completeness, the Broad controllers used were chosen and designed largely for convenience as the speed specification was relatively relaxed and not the primary concern.

An added observation from Fig. 3.19a, b is the lower functional operating voltage for the asynchronous design albeit the longer delay. It fails <1 V largely due to the delay mismatch in some matched delay components. A dual-rail delay-insensitive approach [1] can resolve the delay mismatch problem, thereby theoretically allowing the asynchronous design to operate as long as $V_{DD} > V_T$, but at the expense of higher energy dissipation and area overhead. In the case of the synchronous FFT/IFFT design, the prototype IC fails @ $V_{DD} < 1.1$ V. This is largely attributed to clock skew because clock skew is aggravated when gating is applied. For example from simulations, at $V_{DD} = 1$ V, the clock skew with clock gating ≈ 8.2 ns while the clock skew without clock gating ≈ 6.0 ns.

Table 3.11 tabulates a comparison of several low energy synchronous FFT designs [27–30], including the benchmarked synchronous FFT/IFFT and the novel asynchronous FFT/IFFT design herein. The entries in the last column, normalized FFTs per energy, have been scaled to a 0.35 μm CMOS, 1.1 V, 16-bit wordlength, and 128-point FFT algorithm based on (6) [29].

Table 3.11 Characteristics of reported FFT processors and the asynchronous and the benchmarking synchronous FFT/IFFT processors

	V_{DD} (V)	L_{eff} (μm)	N	W_{bit} (bit)	α	Algorithm	Energy @ computation (μJ)	Normalized FFTs (million) per energy
Baas, 1999	1.10	0.60	1,024	20	0.27	Radix-2	3.14	7.6
Wang, 2005	0.35	0.18	1,024	16	0	Radix-2	0.16	3.7
Ding, 1999	3.30	0.35	64	24	0	Radix-4	1.95	3.5
Yeh, 2003	3.30	0.35	64	12	0	Split-radix	0.33	8.0
Sync, Fig. 3.18a	1.10	0.35	128	16	0	Radix-2	0.19	5.3
Async, Fig. 3.18b	1.10	0.35	128	16	0	Radix-2	0.12	8.3

$$\text{Normalized FFTs per Energy} = \frac{\left(\frac{L_{eff}}{0.35}\right) \times \left(\frac{V_{DD}}{1.1}\right)^2 \times \left[\begin{matrix} \frac{2}{3} \times \frac{W_{bit}}{16} + (1-\alpha) \\ \times \frac{1}{3} \times \left(\frac{W_{bit}}{16}\right)^2 \end{matrix}\right] \times \left(\frac{(N/2)\log_2 N}{448}\right)}{\text{Energy Dissipation}} \tag{3.6}$$

where L_{eff} is the effective transistor length of the CMOS process, W_{bit} is the word-length, α is the truncated portion for multipliers therein, and N is the number of points of the FFT algorithm.

Although a comparison between the various designs is contentious due to large variations of the designs and parameters therein, it is nonetheless worthwhile to note that the synchronous and asynchronous designs described in this section are indeed energy-efficient and the latter being the more efficient.

3.5 Conclusions

This chapter explores the use of asynchronous circuit techniques for low energy dissipation in a VLSI digital processing circuit and evaluates the asynchronous design against its synchronous counterpart qualitatively and quantitatively. The demonstrated circuit therein is a 128-point radix-2 decimation-in-time FFT/IFFT processor (~74 K transistors) and this circuit pertains to a low voltage (1.1–1.4 V) energy-critical hearing instrument. The prototype asynchronous FFT/IFFT processor dissipates 120.7–200.7 nJ @ 1.1–1.4 V, features <1.2 ms @ voltage ≥1.1 V, and occupies 1.60 mm^2 @ 0.35 μm dual-poly four-metal CMOS process.

With a benchmark circuit (a synchronous FFT/IFFT processor), the energy efficacy in the asynchronous FFT/IFFT processor is shown, by as much as ~37% compared to its synchronous counterpart (@ 1 MHz) with the clock gating approach. The clock gating approach therein is that the system clock of the synchronous design is gated upon the completion of one FFT or IFFT transform. The computation times of the asynchronous FFT/IFFT processor are comparable (or even better for some operating conditions, i.e. at high voltage) to its synchronous counterpart. The IC area of the asynchronous design, however, is ~10% larger than the synchronous design.

As a basic principle, the low energy asynchronous FFT/IFFT processor comes at the collective efforts from all levels and aspects. In this chapter, a bottom-up approach has been adopted, by building the basic library microcells, designing the arithmetic macrocells (adders, multipliers, and memories), and finally constructing the asynchronous FFT/IFFT processor. These microcells and macrocells form an asynchronous cell library, with the primary design techniques of low energy dissipation (e.g. reduced spurious switchings). The development of the asynchronous library cells is particularly important for asynchronous designs because there are no commercial asynchronous library cells accessible to the public to date; despite some universities, research institutes and company have developed their owns.

Finally, it must be said that the asynchronous design and synchronous design are not opposites. Most of the circuits today are neither purely synchronous nor purely asynchronous; they involve synchronous and asynchronous parts. Both designs have their advantages (although the choice in literature is favorable to the synchronous design). Nevertheless, the inherent properties of asynchronous approaches to digital circuits are unique, divergent, and inspiring. In essence, the asynchronous approaches not only include the basic framework of clock-domain (pulse-based) designs, but also have many varieties and flexibilities to perform the necessary synchronization, communication, and sequencing of operations. The crux of asynchronous approaches is to fully exploit their inherent properties and potential advantages to a particular objective (e.g. low energy consumption) according to the targeted applications and design constraints. The asynchronous FFT/IFFT processor demonstrated herein for hearing instrument applications is, of course, another successful example.

References

1. J. Sparsø, S. Furber, *A Systems Perspective. Principles of Asynchronous Circuit Design* (Kluwer, 2001)
2. S. Hauck, Asynchronous design methodologies: an overview. Proc. IEEE **83**(1), 69–93 (Jan 1995)
3. A.J. Martin, M. Nyström, Asynchronous techniques for system-on-chip design. Proc. IEEE **94**(6), 1089–1120 (Jun 2006)
4. L.S. Nielsen, J. Sparsø, Designing asynchronous circuits for low power: an IFIR filter bank for a digital hearing aid. Proc. IEEE **87**(2), 268–281 (Feb 1999)
5. K.-S. Chong, B.-H. Gwee, J.S. Chang, Energy-efficient synchronous-logic and asynchronous-loigc FFT/IFFT processors. Proc. IEEE **87**(2), 268–281 (Sept 2007)
6. K.S. Stevens et al., An asynchronous instruction length decoder. IEEE J. Solid State Circuits (JSSC). **36**(2), 217–227 (Feb 2001)
7. K.-S. Chong, B.-H. Gwee, J.S. Chang, Design of several asynchronous-logic macrocells for a low-voltage micropower cell library. IET Circuits Dev Syst **1**(2), 161–169 (Apr 2007)
8. Y.W. Li, G. Patounakis, K.L. Shepard, S.M. Nowick, High-throughput asynchronous datapath with software-controlled voltage scaling. IEEE J. Solid State Circuits (JSSC) **39**(4), 704–708 (Apr 2004)
9. J.M. Rabaey, A. Chandrakasan, B. Nikoli , *Digital Integrated Circuits: A Design Perspective*, 2nd edn. (Prentice Hall, Upper Saddle River, NJ, 2002)
10. A.J. Martin, The limitations to delay-insensitivity in asynchronous circuits, in *Proceedings of Sixth IEEE/MIT Conference on Advanced Research in VLSI*, 1990, pp. 263–278
11. A. Peeters, K. van Berkel, Single-rail handshake circuits, in *Proceedings of IEEE Conference Asynchronous Design Methodologies*, 1995, pp. 53–63
12. K.T. Christensen, P. Jensen, P. Korger, J. Sparsø, The design of an asynchronous TinyRISC™ TR4101 microprocessor core, in *Proceedings of the IEEE International Symposium on Advanced Research in Asynchronous Circuits & Systems* (*ASYNC*), 1998, pp.108–119
13. M. Lewis, J. Garside, L. Brackenbury, Reconfigurable latch controllers for low power asynchronous circuits, in *Proceedings of the IEEE International Symposium on Advanced Research in Asynchronous Circuits & Systems* (*ASYNC*), 1999, pp. 27–35

14. K.-S. Chong, B.-H. Gwee, J.S. Chang, A micropower low-voltage multiplier with reduced spurious switching. IEEE Trans. VLSI Syst. **13**(2), 255–265 (Feb 2005)
15. J.S. Chang, B.-H. Gwee, K.-S. Chong, A digital multiplier with reduced switching by means of latch adders, U.S. Patent No 7206801, 2007
16. A.M. Shams, T.K. Darwish, M.A. Bayoumi, Performance analysis of low power 1-bit CMOS full adder cells. IEEE Trans. VLSI Syst. **10**(1), 20–29 (Feb 2002)
17. M. Sayed, W. Badawy, Performance analysis of single-bit full adder cells using 0.18, 0.25, and 0.35 μm CMOS technologies. Proc. IEEE Int. Symp. Circuits Syst. ISCAS **3**, 559–562 (2002)
18. C.H. Chang, J. Gu, M. Zhang, A review of 0.18 μm full adder performances for tree structured arithmetic circuits. IEEE Trans. VLSI Syst. **13**(6), 686–695 (June 2005)
19. D. Johnson, V. Akella, Design and analysis of asynchronous adders. IEE Proc. Comput. Digital Tech. **145**(1), 1–8 (Jan 1998)
20. B. Parhami, *Computer Arithmetic: Algorithms and Hardware Designs* (Oxford University Press, New York, 2000)
21. A.D. Booth, A signed binary multiplication technique. Quart. J. Mech. Appl. Math **4**(2), 236–240 (Jun 1951)
22. G.E. Sobelman, D.L. Raatz, Low-power multiplier design using delay evaluation. Proc. IEEE Int. Symp. Circuits Syst. (ISCAS) **3**, 1564–1567 (1995)
23. S.S. Mahant-Shetti, P.T. Balsara, C. Lemonds, High performance low power array multiplier using temporal tiling. IEEE Trans. VLSI Syst. **7**(1), 121–124 (Mar. 1999)
24. W.-C. Yeh, C.-W. Jen, High-speed Booth encoded parallel multiplier design. IEEE Trans. Comput. **49**(7), 692–701 (Jul. 2000)
25. S.K. Mitra, *Digital Signal Processing: A Computer-Based Approach*, 2nd edn. (McGraw-Hill, New York, 2002)
26. J. Yuan, C. Svensson, High-speed CMOS circuit technique. IEEE J. Solid State Circuits (JSSC) **24**(1), 62–70 (Feb 1989)
27. W.-C. Yeh, C.-W. Jen, High-speed and low-power split-radix FFT. IEEE Trans. Signal Process. **51**(3), 864–874 (Mar. 2003)
28. T.J. Ding, J.V. McCanny, Y. Hu, Rapid design of application specific FFT cores. IEEE Trans. Signal Process. **47**(5), 1371–1381 (May 1999)
29. B.M. Baas, A low-power, high-performance, 1024-point FFT processor. IEEE J. Solid State Circuits (JSSC) **34**(3), 380–387 (Mar. 1999)
30. A. Wang, A. Chandrakasan, A 180-mV subthreshold FFT processor using a minimum energy design methodology. IEEE J. Solid State Circuits (JSSC) **40**(1), 310–319 (Jan. 2005)

Chapter 4
CMOL/CMOS Implementations of Bayesian Inference Engine: Digital and Mixed-Signal Architectures and Performance/Price – A Hardware Design Space Exploration

Dan Hammerstrom and Mazad S. Zaveri

Abstract In this chapter, we focus on aspects of the hardware implementation of the Bayesian inference framework within the George and Hawkins' computational model of the visual cortex. This framework is based on Judea Pearl's Belief Propagation. We then present a "hardware design space exploration" methodology for implementing and analyzing the (digital and mixed-signal) hardware for the Bayesian (polytree) inference framework. This particular methodology involves: analyzing the computational/operational cost and the related micro-architecture, exploring candidate hardware components, proposing various custom architectures using both traditional CMOS and hybrid nanotechnology CMOL, and investigating the baseline performance/price of these hardware architectures. The results suggest that hybrid nanotechnology is a promising candidate to implement Bayesian inference. Such implementations utilize the very high density storage/computation benefits of these new nano-scale technologies much more efficiently; for example, the throughput per 858 mm2 (TPM) obtained for CMOL based architectures is 32–40 times better than the TPM for a CMOS based multiprocessor/multi-FPGA system, and almost 2000 times better than the TPM for a single PC implementation. The assessment of such hypothetical hardware architectures provides a baseline for large-scale implementations of Bayesian inference, and in general, will help guide research trends in intelligent computing (including neuro/cognitive Bayesian systems), and the use of radical new device and circuit technology in these systems.

D. Hammerstrom (✉)
Professor — Department of Electrical and Computer Engineering,
Associate Dean – Maseeh College of Engineering and Computer Science,
Portland State University, Portland, OR 97207, USA
e-mail: strom@cecs.pdx.edu

M.S. Zaveri
Assistant Professor, Dhirubhai Ambani Institute of Information and Communication Technology, Gandhinagar, Gujarat 382007, India
e-mail: mazad_zaveri@daiict.ac.in

K. Iniewski (ed.), *CMOS Processors and Memories*,
Analog Circuits and Signal Processing, DOI 10.1007/978-90-481-9216-8_4,

Keywords Bayesian Inference • Pearl - belief propagation • Cortex • CMOS • CMOL • Nanotechnology • Nanogrid • Digital • Mixed-signal • Hardware • Nanoarchitectures • Methodology • Performance • Price

4.1 Introduction

The semiconductor/VLSI industry has been following Moore's law since the past several decades, and has made tremendous progress, but as it approaches the lower nano-scale regime, it faces several challenges, including power density, interconnect reverse scaling, device defects and variability, memory bandwidth limitations, performance overkill[1], density overkill[2], and increasing design complexity [1–6].

The two most important aspects of technology scaling are speed and density. The transistor density available with modern state of the art μ-processors has almost reached one billion per chip [7]. Transistor switching delays are only a few picoseconds, and now these processors are moving to multiple core architectures, which continue to improve the performance [7]. Emerging nanotechnologies will further increase the densities by two to three orders of magnitude [5]. However, it has become increasingly difficult to efficiently use all these transistors [6]. In addition, volume markets, for the most part, are no longer performance driven [6]. If we exclude general-purpose processors/architectures, an important challenge then is: Where else can we efficient utilize silicon technology, and its capabilities: speed, density, and multicore concepts?

One potential "architectural solution" to some of the above challenges includes hardware architectures for applications/models in the neuro-inspired and intelligent computing domain [8]. Most of these applications are inherently compute and memory intensive, and also massively parallel; consequently, it is possible that such applications would be able to fully utilize nano-scale silicon technology much more efficiently.

In recent years, nanoelectronics has made tremendous progress, but Borkar (from Intel®) [4] indicates that, as yet, there is no emerging nanoelectronic candidate with the potential for replacing, within the next 10–15 years, Complementary Metal-Oxide-Semiconductor (CMOS) as it is being used today in state of the art μ-processors. However, recent research has shown that certain kinds of nano-scale electronics, when used in hybrid configurations with existing CMOS, could be useful in creating new computing structures, as well as high density memory, and hybrid and neuromorphic architectures [3, 5, 6, 9–13]. One of the most promising of these hybrid technologies is CMOL, developed by Likharev [11], which is used in this work.

Another challenge is that of intelligent computing [6]. The term "Intelligent Computing" encompasses all the difficult problems that require the computer to

[1] "Performance overkill" is where the highest-volume segments of the market are no longer performance/clock frequency driven.

[2] "Density overkill" is where it is difficult for a design team to effectively design and verify all the transistors available to them on a typical design schedule.

find complex structures and relationships through space and time in massive quantities of low precision, ambiguous, noisy data [2]. The term *Intelligent Signal Processing* (ISP), has been used to describe algorithms and techniques that involve the creation, efficient representation, and effective utilization of large complex models of semantic and syntactic relationships [2, 6]. ISP augments and enhances existing Digital Signal Processing (DSP) by incorporating such contextual and higher level knowledge of the application domain into the data transformation process [2, 6]. Several areas in intelligent computing, such as AI, machine learning, fuzzy logic, etc. have made significant progress; however, "general-purpose artificial intelligence has remained elusive" [14], and we still do not have robust solutions to even remotely approach the capabilities of biological/cortical systems [6]. Consequently, a number of researchers are returning to neuro and cognitive sciences to search for new kinds of Biologically Inspired Computational Models (BICMs) and "computing paradigms" [2, 15]. Recently, the neuroscience community has begun developing scalable Bayesian computational models (based on Bayesian inference framework [16–18]) inspired from the "systems" level structural and functional properties of the cortex, including the feedforward and feedback interactions observed in the "subsystem" layers of the visual cortex, and integrating them with some higher-level cognitive phenomena [17, 19, 20]. These computational models, also referred to as "Biologically Inspired Computational Models" (BICMs) [2], have the potential of being applied to large-scale applications, such as, speech recognition, computer vision, image content recognition, robotic control, and making sense of massive quantities of data [6, 21]. Some of these new algorithms are ideal candidates for large-scale hardware investigation (and future implementation), especially if they can leverage the high density processing/storage advantages of hybrid nanoelectronics [3, 5].

The effective design and use of these nanoelectronic structures requires an application-driven total systems solution [5, 6]. Consequently, in this chapter we propose a new family of application-specific hardware architectures, which implement Bayesian (polytree) inference algorithms (inspired from computational neuroscience), using traditional CMOS and hybrid CMOS + nanogrid (CMOL[3]) technologies.

We began this work by choosing the George and Hawkins' Model (GHM) [16] of the visual cortex, as a useful family of Bayesian BICMs. Recently, George and Hawkins gave a prognosis about the hardware platforms for their model, and indicated that, learning algorithms will be initially implemented in parallel software, partly because they are still evolving, while the inference framework could be implemented in embedded hardware [22]. One of the biggest problems with Bayesian inference in general is its compute intensiveness (it has been shown to be NP-Hard [6]).

Consequently, in this chapter, we exclusively focus on the Bayesian inference functionality of the GHM, and then develop a set of baseline hardware architectures and their relative performance/price measures using both traditional CMOS and a hypothetical CMOL technology. In the larger perspective, we perform a

[3] We use the term CMOL to describe a specific family of nanogrid structures as developed by Likharev et al.

"hardware design space exploration" (in a limited subset of the space), which is a first step in studying radical new computing models and implementation technologies. The architecture evaluation methodology used here (and elsewhere [3, 23]) has the potential of being applied to a broader range of models and implementation technologies.

This chapter is organized as follows: Section 4.2 provides a brief introduction to the more generic concept of hardware virtualization spectrum, and provides the methodology of "architecting" the hardware [24]. Section 4.3 describes the Bayesian Memory (BM) module, and its Bayesian inference framework. Section 4.4 defines various hardware architectures that support BM. Sections 4.5 and 4.6 discusses CMOS/CMOL digital and mixed-signal hardware architectures for BM. Finally, we discuss the hardware performance/price results in Section 4.7.

4.2 Hardware for Computational Models

4.2.1 Hardware Virtualization Spectrum

When creating a baseline architecture for implementing any computational model or algorithm (from several domains, such as neural networks, intelligent computing, etc.), the one single decision that has the greatest impact on the performance/price of the implementation is the degree of "virtualization" of the computation. This term, incidentally should not be confused with virtual machines. We use *virtualization* to mean "the degree of time-multiplexing of the 'components' of computation via hardware resources" [3]. Since some of the computational models consist of fine grained networks, very fine-grained parallel implementation is generally possible, but varying the degree of virtualization allows us to make a wider range of trade-offs for a more efficient implementation.

A typical "hardware virtualization spectrum" is shown in Fig. 4.1. It essentially describes a software and hardware design space for implementing massively parallel algorithms [3]. As we move from left to right, the time-multiplexing of hardware resources decreases and parallelism increases. Likewise, the general flexibility/programmability decreases.

The right bottom corner represents a design in which the algorithm is fully hardwired into the silicon (for example, analog architectures in region 8), it achieves maximum parallelism, but has little or no virtualization, and is the least flexible [3]. Such designs are generally very fast and compact [25]. However, in general, it is difficult to multiplex analog components [25, 26], so as we move to the left, towards more virtualization, digital design tends to dominate. However, even in the analog domain, analog communication is not feasible, so digital multiplexed communication is used [25–28]. For some particular computational models, depending on their dynamics, such designs could be an inefficient implementation [3].

Previous work (other than GHM) has shown that, with possible future hardware technologies, semi-virtualized hardware designs (regions 5–7) will scale well,

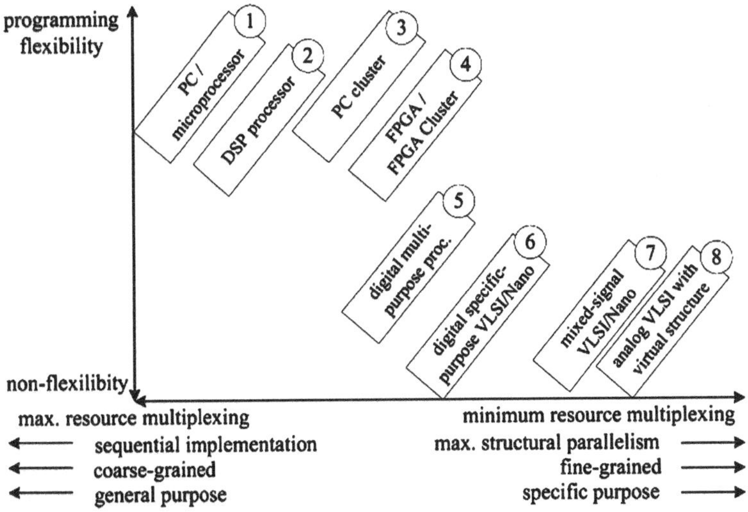

Fig. 4.1 Hardware virtualization spectrum (Adapted from [3], © 2007 IEEE). *Numbered boxes* correspond to various hardware design options, and are referred to as "region 1" through "region 8"

because they allow large-scale integration and component level multiplexing, and are suitable for multiplexed communication schemes [3, 10, 23].

For several application domains, designs from region 5–7, have not generally been explored, and hence, warrant research and investigation.

4.2.2 *Existing Hardware Implementations of George and Hawkins' Model*

The work by George and Hawkins [16, 29] (and the related work at Numenta Inc.) mostly falls into region 1, and only a little bit into region 3. The recent work by Taha [21, 30, 31] explores the combined implementations in regions 3 and 4, for the George and Hawkins Model (GHM) [16]. At this time, we are not aware of any work on custom hardware implementation of the GHM [16]. Hence, we can conclude that most of the work on the hardware implementation of GHM [16], is almost exclusively concentrated in region 1, 3 and 4, and hence, custom hardware implementations in regions 5–7 are still unexplored and present an opportunity for a different kind of hardware/computer architecture research. Consequently, the work presented here explores application-specific hardware designs/architectures[4] for the Bayesian inference framework (i.e. Pearl's Belief Propagation Algorithm for polytree) within

[4]From a traditional computer engineering perspective, the detailed implementation of designs from regions 5–8 could be referred to as micro-architectures, but for simplicity, we refer to them as architectures.

the GHM [16], in regions 5–7, and investigates the hardware performance/price of these architectures.

Note: The analog community has not yet explored Region 8 for GHM, but in the future, with appropriate research, interesting fully analog designs are no doubt possible. It is not in the scope of this chapter to cover fully analog designs, because the focus of this work is to explore designs which can transition to, and utilize CMOL nanogrid structures. In addition, we do not claim that the designs proposed in this work are better than potential fully analog designs.

It should be realized that there are no universal benchmarks or *baseline configurations* available to allow a comparison of the hardware for GHM or similar models [3]. In addition, factors that vary over the various models, such as data precision, vector/matrix sizes, algorithmic configurations, and hardware/circuit options, make these comparisons even more difficult. The *baseline configuration* for any hardware takes into account conservative assumptions & worst case design considerations, the design tends to be more virtualized and perhaps less optimized; hence, the price (area & power) and performance (speed) are on the lower-end, or at the baseline. Consequently, in the future, sophisticated designs would probably have better performance, but at a higher price.

4.2.3 Hardware Design Space Exploration: An Architecture Assessment Methodology

Figure 4.2 summarizes the "hardware design space exploration" methodology[5] used here for implementing the Bayesian inference framework, and investigating their performance/price. Our previous work on associative memory (digital and mixed-signal) implementations [3, 23], did not formally summarize/discuss this methodology, but was based on it; a brief example showing the use of this methodology for an associative memory model is given in the Appendix (Section 4.9.5). In this chapter, with respect to the steps in Fig. 4.2, Section 4.3 covers steps 1–4; Section 4.4 covers steps 6–8; and Sections 4.5 to 4.7 cover steps 5 and 9–12.

4.3 A Bayesian Memory (BM) Module

This work is based on the George and Hawkins Model (GHM) [16]. GHM has several desirable characteristics which include: the underlying fundamental concepts have a high degree of "interpretability" [32], the perception of time, and a basis in some

[5]This methodology has been used to study digital and mixed-signal designs/architectures in the past; in order to use it for other designs (including analog designs) one may need to make some adjustments.

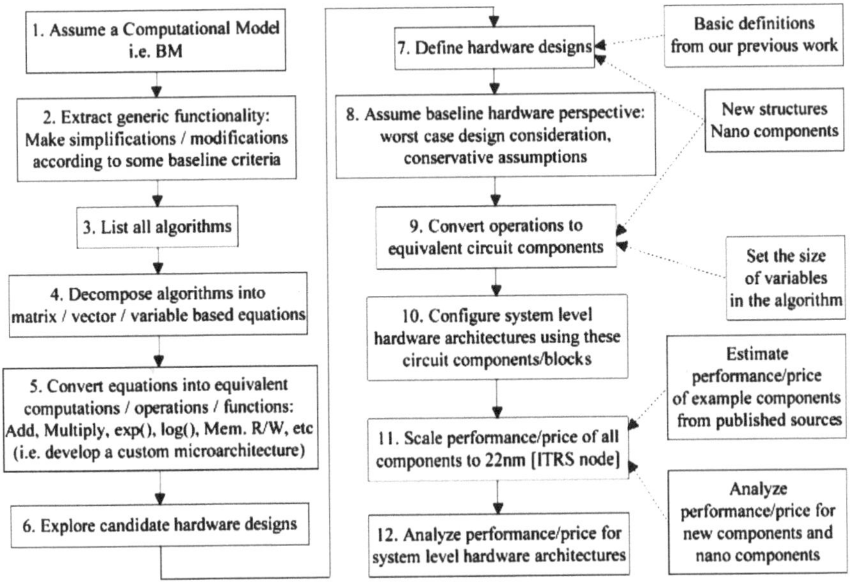

Fig. 4.2 Hardware design space exploration – methodology

proposed cognitive mechanisms [33, 34]. It has a modular and hierarchical structure, which allows shared and distributed storage; and is scalable [32, 33]. It also has a simpler non-"loopy" probabilistic framework as compared to [18]. All these characteristics potentially make the GHM [16] the simplest and most generic Bayesian model currently available in the neuroscience community. For all these reasons, it is a good candidate Bayesian model for investigating the related baseline hardware.

In this chapter, each individual module of the GHM [16], is referred to as a "Bayesian Memory" (BM) module [35], as shown in Fig. 4.3a. The BM module has two conceptual/functional parts (Part-A and Part-B), as shown in Fig. 4.3a. Part-A is largely involved in learning, consists of a set of quantized vectors, and is referred to as the Code Book (CB) of vectors [35]. During learning, the CB captures and accumulates the data on the inputs that the network sees [16, 35]. The conditional probability relations between these CB vectors are also captured during learning, which is used later for inference [16, 35]. Part-B is the Bayesian inference framework, it performs inference, and is based on the bidirectional Belief Propagation Algorithm (BPA), pioneered by Pearl [36]. Figure 4.3b shows the details of Part-B, with a single parent BM and multiple child BMs [35]. A hierarchical system for recognition and related tasks can be built using BM modules, as shown in Fig. 4.3c.

Pearl's BPA [36] for a polytree is summarized by Eqs. 4.1–4.6, which use a matrix/vector notation. As shown in Fig. 4.3b, the particular module of interest is called BM-y. BM-y has n_c child modules, each denoted by BM-x_i, where $i = 1, \ldots, n_c$. BM-y has a single parent module denoted by BM-z. BM-y exchanges probabilistic messages/vectors with its parent and children. In 4.1, notation (x_i,y) denotes that a message goes from BM-x_i to BM-y, and notation (y) denotes an internal

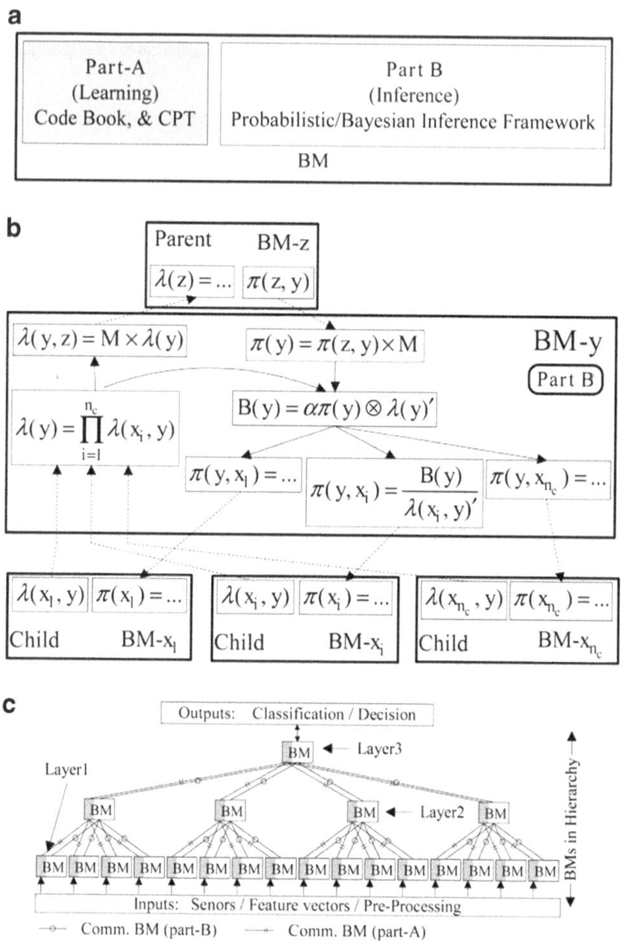

Fig. 4.3 Bayesian Memory (BM) module. (**a**) Basic parts of a BM. (**b**) Detailed Part-B of a BM. (**c**) Hierarchical system of BMs

message. The same style applies for 4.2–4.6. Equations 4.5 and 4.6 are equivalent, but 4.6 is easier to implement. Notation ⊗ denotes a $\prod$ operation, but for only two vectors. If there are n_c child BMs, then 4.6 has to be evaluated n_c times, corresponding to each child BM-x_i.

This chapter focuses almost exclusively on the implementation of the Bayesian inference framework (Part-B) of the GHM [16], and henceforth, the term BM denotes only Part-B, unless explicitly stated otherwise.

$$\lambda(y) = \prod_{i=1}^{n_c} \lambda(x_i, y) \tag{4.1}$$

$$\lambda(y,z) = M \times \lambda(y) \tag{4.2}$$

$$\pi(y) = \pi(z,y) \times M \tag{4.3}$$

$$B(y) = \alpha\pi(y) \otimes \lambda(y)' \tag{4.4}$$

$$\pi(y,x_i) = \alpha B(y) / \lambda(x_i,y)' \tag{4.5}$$

$$\pi(y,x_i) = \alpha\pi(y) \otimes \left(\prod_{k=1,k\neq i}^{n_c} \lambda(x_i,y)' \right) \tag{4.6}$$

The following discussion briefly summarizes the significance of the variables and operations in the equations of the BPA.

- $\lambda(x_i,y)$ is the input diagnostic message[6] coming from child BM-x_i, $\lambda(y)$ is the internal diagnostic message of BM-y, and $\lambda(y,z)$ is the output diagnostic message going to the parent BM-z.
- $\pi(z,y)$ is the input causal message coming from parent BM-z, $\pi(y)$ is the internal causal message of BM-y, and $\pi(y,x_i)$ is the output causal message going to the child BM-x_i.
- $\lambda(x_i,y)$ represents the belief[7] that BM-x_i has in the CB of BM-y. $\lambda(y,z)$ represents the belief that BM-y has in the CB of parent BM-z. $\pi(z,y)$ represents the belief that parent BM-z has in its own CB, while ignoring the diagnostic message $\lambda(y,z)$. $\pi(y,x_i)$ represents the belief that BM-y has in its own CB, while ignoring the diagnostic message $\lambda(x_i,y)$.
- M acts as a bidirectional probabilistic associative memory, which stores the probabilistic relationships between the CB of BM-y and BM-z. M is also called the Conditional Probability Table (CPT). $B(y)$ represents the final belief that BM-y has in its own CB, considering all possible diagnostic and causal messages.
- In 4.1, $\lambda(y)$ is formed by combining all the input diagnostic messages using the $\prod$ operation. In 4.2, $\lambda(y,z)$ is formed using a [matrix × vector] operation that converts the internal diagnostic message into the equivalent output diagnostic message. In 4.3, $\pi(y)$ is formed using a [vector × matrix] operation that converts the input causal message to the equivalent internal causal message. In 4.4, $B(y)$ is formed by combining internal diagnostic and causal messages using $\otimes$ operation. In 4.6, $\pi(y,x_i)$ is formed by combining internal causal messages and all input diagnostic messages, except $\lambda(x_i,y)$, using a $\prod$ operation.

In short, the BPA can be summarized as: each BM in a hierarchical system (like the one shown in Fig. 4.3c) first updates its internal belief using the incoming belief messages from its parent and children, and later, sends the updated version of its belief back to its parent and children.

---

[6] The "diagnostic" message is also called bottom-up evidence, and the "causal" message is also called top-down evidence.

[7] Probabilistic belief or Bayesian confidence is simply referred to as "belief".

4.4 Hardware Architectures for Bayesian Memory

4.4.1 Definition of Hardware Architectures for BM

The hardware architectures investigated in this chapter are categorized in Fig. 4.4. (The categorization and hardware definitions presented here are partly based on the hardware configurations proposed in [3, 6].) There are two major groups of architectures: digital and mixed-signal. The digital architectures are categorized into two groups: digital CMOS and digital CMOL. Each is further categorized into three basic "computational" groups: Fixed-Point (FXP), Logarithmic Number System (LNS), and Floating-Point (FLP) architectures. FXP architectures are further categorized based on the data precision, i.e. 4 bits, 8 bits, 12 bits, …, 32 bits, and are referred to as FXP4, FXP8, FXP12, …, FXP32 respectively. The mixed-signal architectures are categorized into two groups: mixed-signal CMOS and mixed-signal CMOL architectures.

All of the architectures studied here implement Eqs. 4.1–4.4, and 4.6 of the BPA. These equations can be categorized based on the type of matrix/vector operations: "Vector-Matrix Multiplication" (VMM) [37], Product (Π Off Vectors (POV), and Sum-Normalization (SNL). VMM has two subtypes: (Matrix × Column-Vector) (MCV) and (Row-Vector × Matrix) (RVM). Finally, Eqs. 4.1–4.4, and 4.6 correspond to operations POV, MCV, RVM (POV & SNL), and (POV & SNL) respectively. To implement these matrix/vector operations in hardware requires several different circuit components. Hence, the exact definition of any architecture depends on the particular circuit/hardware components that it uses to implement the equations of the BPA.

Table 4.1 summarizes the definitions of the (limited) architecture space explored here; it provides an overview of how the arithmetic operations, storage, and communication for each equation are implemented for those architectures. The digital CMOS and digital CMOL architectures are almost identical, except for the memory components; digital CMOS architectures use SRAM[8], while digital CMOL architectures use CMOL memory [3].

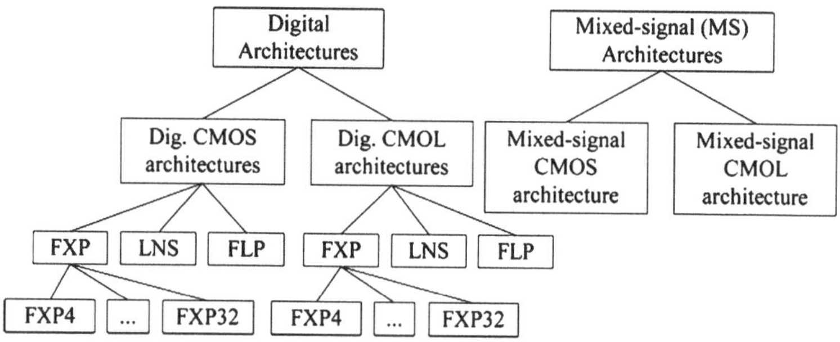

Fig. 4.4 Categorization of hardware architectures for Bayesian Memory

[8]A viable alternative (not explored here) to SRAM is eDRAM, which is denser, but is significantly slower.

Table 4.1 Definition of hardware architectures for Bayesian memory (BM)

Component		Equation 4.1			Equation 4.3			Equation 4.4			Equation 4.6			Equation 4.2		
		S	O	C	S	O	C	S	O	C	S	O	C	S	O	C
Memory (MEM)	SRAM	♣♥			♣♥			♣♥			♣♥			♣♥		
	CMOL MEM	♦♠			♦♠			♦♠			♦♠			♦♠		
Conventional arithmetic/ logic components	Digital CMOS		♣♦♥♠			♣♦♥♠			♣♦♥♠			♣♦♥♠			♣♦♥♠	
	Mixed-signal CMOS					♥♠									♥♠	
Structure based components (storage & computation)	Mixed-signal CMOS				♥	♥								♥	♥	
	Mixed-signal CMOL nanogrid				♠	♠								♠	♠	
MEM access based communication	SRAM			♣♥			♣♥			♣♥			♣♥			♣♥
	CMOL MEM			♦♠			♦♠			♦♠			♦♠			♦♠

♣ = Digital CMOS architecture; ♦ = Digital CMOL architecture; ♥ = Mixed-signal CMOS architecture; ♠ = Mixed-signal CMOL architecture S = storage; O = arithmetic Operations; C = communication

The mixed-signal architectures are based on the "Internally Analog, Externally Digital" (IAED) Structure for VMM (SVMM), proposed in [37]. This is a mixed-signal array based SVMM, which combines storage, signal-communication, and computation; storage and input/output are digital, and the analog computations occur along the wires as small increments in charge/current [3, 37]. The difference between the digital and mixed-signal architectures is that Eqs. 4.2 and 4.3 are partially implemented using mixed-signal SVMM or structure based components, because these equations are computationally the most intensive; the implementation of the remaining equations is the same as in digital architectures. The mixed-signal architectures also require conventional digital CMOS arithmetic/logic components and conventional mixed-signal CMOS components for post processing in 4.2 and 4.3.

When selecting a mixed-signal architecture, the single most important criterion was to select a mixed-signal CMOS design, which could later be easily mapped to CMOL nanogrid structures; the SVMM proposed in [37] was such a match, and in addition, it has its advantages and is close to the future realm of "digitally assisted analog circuits" [38]. Consequently, our mixed-signal architecture judiciously replaces some of the important operations using analog/mixed-signal components, according to the CMOS/CMOL design considerations suggested in [3].

The storage in all architectures is digital/binary-encoded. Consequently, it is easier to implement a memory access (read & write) based communication for all architectures.

4.4.2 General Issues

The general issues and assumptions related to our analysis of the hardware implementation of the BM are now briefly discussed.

4.4.2.1 Precision/Bits

The data precision (n_{bit} = bits representing a single element) required to implement any computational model is dependent on the algorithm, the structural configurations, the application domain, etc. Studies [39, 40], suggest that neuromorphic hardware generally requires a precision of 4–8 bits, but Bayesian methods, which are higher level, more abstract models, when applied to real-world problems generally require 8–16 bits for sufficient accuracy [21, 41]. In this work we have varied the precision over a range of values to understand its contribution to the overall performance/price trade-off. The precisions used here were: 4–32 bit FXP, 32 bit FLP, and 32 bit LNS. All architectures, even mixed-signal, store/communicate data in digital/binary-encoded format.

4.4.2.2 Communication

Communication is generally sequential and often involves memory access. Internal communication (for internal variables) is as simple as accessing an internal memory. External communication (between parent/child) requires inter-module buses with data, address, and control signals. Depending on the particular variable being read by BM-y, some of the communication is assumed to be virtualized. To evaluate 4.1, BM-y needs to read $\lambda(x_i,y)$ from each of its child BM-x_i; for this variable, one communication bus is multiplexed across all n_c child BMs. To evaluate 4.3, BM-y needs to read $\pi(z,y)$ from its parent BM-z; for this variable a separate communication bus is assumed. It should be noted that $\lambda(x_i,y)$ is required to evaluate 4.1 and 4.6; also, 4.6 is evaluated n_c times. Hence, each $\lambda(x_i,y)$ value is used in multiple equations. So instead of reading from the child BM-x_i multiple times through the virtualized communication bus, we assume that it is read once to evaluate 4.1, and a local copy[9] is stored in BM-y that can be used later for 4.6. In fact, we assume this for all variables that are read from parent or child BMs.

4.4.2.3 Number of Parent and Child BMs, and Code Book (CB) Size

The BPA used in the BM is for singly connected polytree structures, which means that each BM has a single parent BM, and multiple child BMs. Polytrees allow exact inference to be performed in one iteration of messages [36]. We assume four child BMs ($n_c = 4$) as input to each BM, which will allow the receptive field-sizes to increase by 4× in each layer of a typical hierarchy [29]. For the analysis done here, we assume that Part-A of any BM can learn a maximum of 4,096 (or 4K) CB vectors. Hence, the CB size (n_{CBy}) of BM-y and the CB size (n_{CBz}) of parent BM-z are both assumed to be 4 K. These CB sizes determine the size of all the variables in Eqs. 4.1–4.6. Table 4.2 summarizes the specifications (variables and their sizes)

Table 4.2 Specifications of a Bayesian memory (BM) module

Variable	Size
n_{bit}	4–32 bit FXP, 32 bit FLP, and 32 bit LNS
n_c	4
n_{CBy}	4,096
n_{CBz}	4,096
$\lambda(x_i,y)$	[4,096 × 1]
$\lambda(y)$	[4,096 × 1]
$\lambda(y,z)$	[4,096 × 1]
$\pi(y,x_i)$	[1 × 4,096]
$\pi(z,y)$	[1 × 4,096]
$\pi(y)$	[1 × 4,096]
$B(y)$	[1 × 4,096]
M	[4,096 × 4,096]

[9] Here we are trading-off local computation and storage for longer range communication.

of a BM; the hardware analysis is based on these specifications. From Table 4.2 it can be seen that the largest storage required is for the *M* matrix. Depending on a particular application, *M* could be sparse, but here we assume the worst case full density *M*.

4.4.2.4 Virtualization

In general, we have the following two ways to virtualize the arithmetic/logic components: virtualize (share resources) these components for similar operations over several BMs, or virtualize different operations/equations within each BM. We chose the latter because it keeps the hardware analysis restricted to one BM, and is more suitable for systems with distributed memory. Most of the designs considered here tend to be more virtualized, which provides a more favorable performance/price. However, as we shall see later, the cost of most BM architectures (except the mixed-signal CMOL architecture) is dominated by memory. Consequently, extreme virtualization of the arithmetic/logic computations may not be a significant advantage; but for the mixed-signal CMOL architecture, virtualization was the only way to make the architecture feasible and scalable.

4.4.2.5 Hybrid Nanotechnology – CMOL

The nanotechnology components used in this chapter are based on CMOL, which is a hybrid CMOS + nanogrid architecture developed by Likharev [11]. CMOL is mostly used here as digital memory and as an application specific, mixed-signal computational structure for the VMM operation. We use the CMOL/nanogrid modeling described in [3], and follow a similar strategy for the performance/price analysis of the CMOL components. Some of the assumptions for the CMOL analysis are:

- For CMOL memory, read and write can be performed using a common interface/path (row and column decoders, etc.). Only for the variables communicated between parent/child BMs, will we need a simultaneous read and write capability. According to [42], all these assumptions are plausible.
- The area analysis does not consider the CMOS to nanogrid routing (similar to [3]), i.e., the CMOS layer and nanogrid layer are analyzed separately. The total area is taken as the sum of CMOS and nanogrid layers; which is a worst case consideration.
- Our analysis uses a future value of $F_{nano} = 3$ nm (projected in 2028 [43]). CMOL technology should be mature by then, hence plausible defect rates would be around 2% to 5% [42]; and a recent roadmap suggests defect rates as low as 0.03% [43]. So in the worst case, the hardware overhead for defect-tolerance through redundancy/reconfiguration would be around 5% to 10% of the results shown in this chapter.

4.5 Digital CMOS and CMOL Hardware Architectures for Bayesian Memory (BM)

The all digital architectures considered here assume traditional CMOS arithmetic/logic components. The only difference between digital CMOS and digital CMOL architecture is the storage medium. The digital CMOS architectures use SRAM, while digital CMOL architectures use CMOL memory [3].

In our analysis, we assumed a custom [4K × 4K × 32 bit] CMOL memory, instead of using the results from [42]. In the BM application assumed here, all matrix/vector elements are accessed sequentially. Hence, our memory does not require random access decoders. So we were able to use shift register based decoders for sequential access [44], which reduces the total silicon area for the memory. This also decreases memory access time due to the decreased logic-depth of the decoder circuitry. Moreover, defect-tolerance can be achieved simply through reconfiguration to redundant/new nanogrid locations, not requiring special error-correction circuitry. A block diagram of the CMOL memory is shown in Fig. 4.5b. Each block consists of a [256 × 256] CMOL nanogrid, along with the necessary decoders, as shown in Fig. 4.5a. All row decoders connect to horizontal nanowires; all column decoders are assumed to connect to pass transistors/gates. We analyzed the performance/price of this particular CMOL memory, generalizing its performance/price to a single bit, and used this result to evaluate memories of varying sizes. The performance/price analysis assumes that 10% of

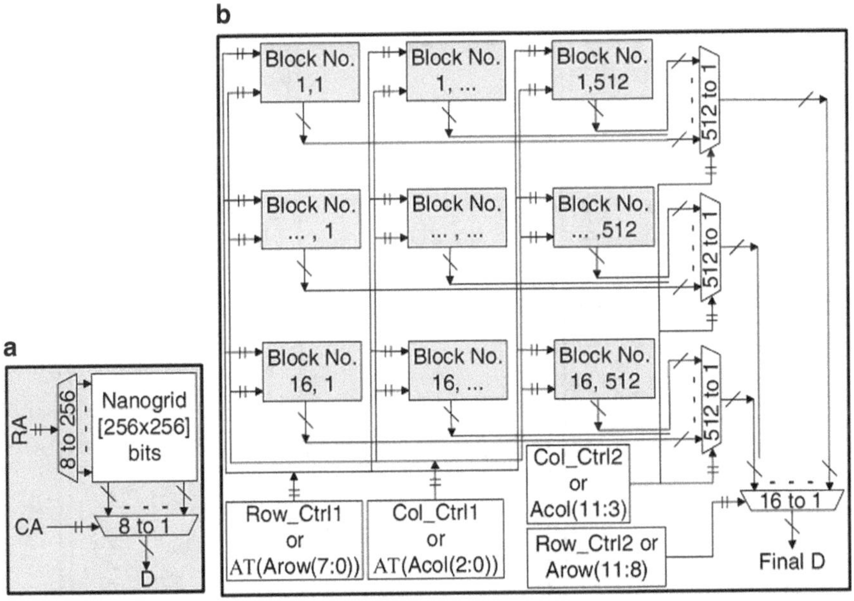

Fig. 4.5 CMOL memory: (**a**) One block of CMOL memory with a [256 × 256] nanogrid and decoders, (**b**) general block diagram of CMOL memory [4K × 4K × 32 Bits]

the total memory dissipates power during memory access [45]; this is a worst case consideration, because ideally, only one block (out of 8,192) would be drawing current.

4.5.1 Floating-Point (FLP) Architecture

The FLP architecture uses a 32 bit single precision floating-point data format [46]. The block diagram of the FLP architecture is show in Fig. 4.6. Blocks B1 to B5 perform the operations for Eqs. 4.1–4.4 and 4.6, respectively. Block B6 performs the Sum-Normalization (SNL) operation for 4.4. A similar additional block (not shown) is also needed for the SNL operation in Eq. 4.6. Digital counters (CT) and logical Address Transformations (AT) are used to generate the proper addresses for the SRAMs corresponding to all variables. SRAM is used to store the *M* matrix and has a dual-ported read capability. We require only two Floating-Point arithmetic Units (FPUs), which are shared by (virtualized over) all blocks in Fig. 4.6. Block B5 is virtualized for computing variables $\pi(y,x_1)$ to $\pi(y,x_{n_c})$ (i.e. 4.6 is computed n_c times, corresponding to each child BM). Table 4.3 shows the arithmetic and memory computations required for each equation in the BM. It also shows the FPU used by each equation.

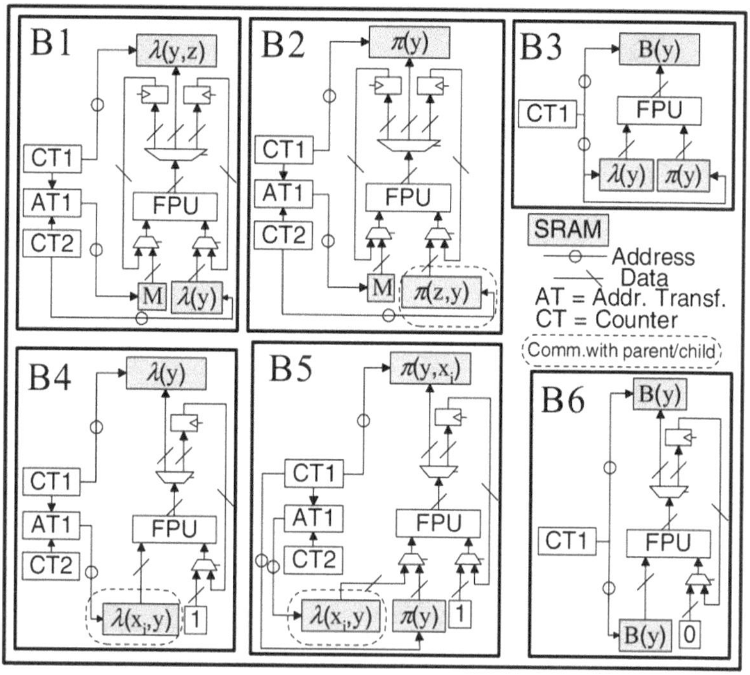

Fig. 4.6 Block diagram of digital floating-point (FLP) architecture

Table 4.3 Computations for a Bayesian memory (BM) module

Equation	Matrix/vector operations	FLP multiplication	FLP addition	FLP division	Memory read	Memory write	FPU used
4.1	POV	$n_c n_{CBy}$	0		$n_c n_{CBy}$	n_{CBy}	FPU1
4.2	MCV	$n_{CBz} n_{CBy}$	$n_{CBz} n_{CBy}$	0	$n_{CBz} n_{CBy}$	n_{CBz}	FPU1
4.3	RVM	$n_{CBy} n_{CBz}$	$n_{CBy} n_{CBz}$	0	$n_{CBy} n_{CBz}$	n_{CBy}	FPU2
4.4	POV	$2\,n_{CBy}$	0	0	$2\,n_{CBy}$	n_{CBy}	FPU1
	SNL	n_{CBy}	n_{CBy}	1	$2\,n_{CBy}$	n_{CBy}	FPU1
4.6	POV[a]	$n_c(n_c n_{CBy})$	0	0	$n_c(n_c n_{CBy})$	$n_c(n_{CBy})$	FPU2
	SNL[a]	$n_c(n_{CBy})$	$n_c(n_{CBy})$	$n_c(1)$	$n_c(2\,n_{CBy})$	$n_c(n_{CBy})$	FPU2

[a]Computations shown are the total for all n_c child BMs

Even though Table 4.3 specifically shows the FLP operations, it still provides a general overview of the computations within the BM. From Table 4.3, we can infer that 4.2 and 4.3 have a computational complexity of $O(N_1N_2)$, where $N_1 \propto n_{CBy}$ and $N_2 \propto n_{CBz}$; while 4.1, 4.4 and 4.6 have a computational complexity of $O(N)$, where $N \propto n_{CBy}$ and n_c is negligible relative to n_{CBy}.

4.5.2 *Logarithmic Number System (LNS) Architecture*

The LNS architecture is based on a 32 bit single precision LNS data format [46]. The LNS architecture is similar to that shown in Fig. 4.6, except that all FPUs are now replaced by Logarithmic Number system arithmetic Units (LNUs). The earlier discussion on FLP architecture also applies here. Table 4.3 can be used by replacing the FLP operations with LNS operations. Both the FLP and LNS architectures use a 32 bit data format; hence, the datapath structure remains the same, only the requirements with respect to arithmetic operations differ.

4.5.3 *Fixed-Point (FXP) Architecture*

The FXP architecture also has blocks B1 to B5, which are similar to those in Fig. 4.6. The FXP architecture does not require block B6. In the FXP architecture, each block in Fig. 4.6 is replaced with two blocks (the a & b type blocks, as shown in Fig. 4.7). For example, block B1 in Fig. 4.6 has to be replaced with block B1a and block B1b shown in Fig. 4.7. In the FLP architecture, the FPU in block B1 was able to perform both the add and multiply operations required for 4.2. But in the FXP architecture (as shown in block B1a) we need a separate digital adder and digital multiplier (of particular bit sizes) for 4.2. Using the same idea, each block in Fig. 4.6 has to be replaced by an equivalent block (with separate adders/multipliers) to form the FXP architecture in Fig. 4.7.

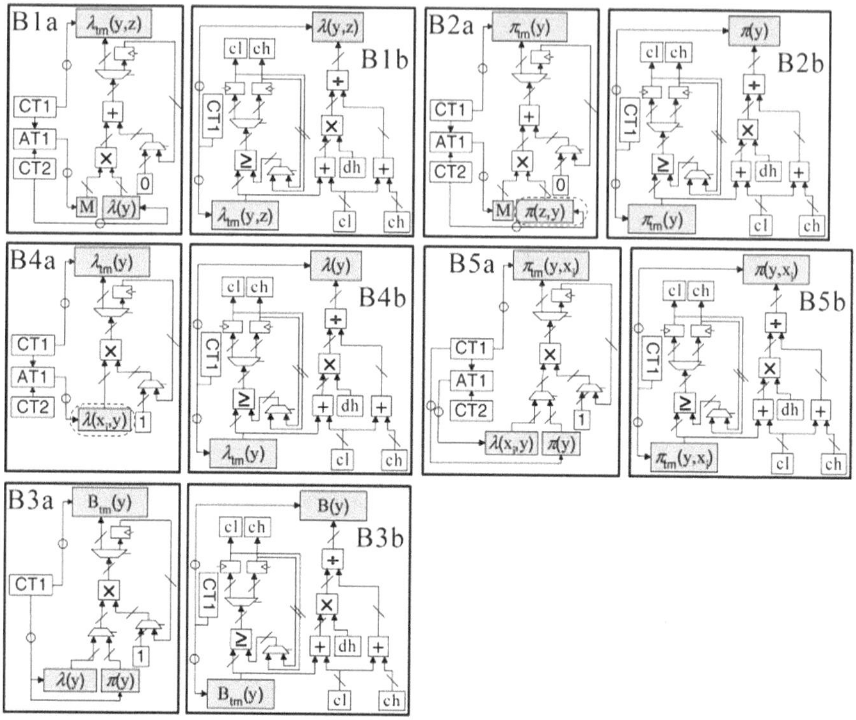

Fig. 4.7 Block diagram of digital fixed-point (FXP) architecture

The FXP architecture has to consider the bit size or precision of the intermediate results of all arithmetic operations. For example, multiplying two n bit numbers will lead to a $2n$ bit result. When m numbers, each of size n bits, are added, the result could grow to $n + ceil(\log_2(m))$ bits. If these results are simply truncated back to n bits, then error/noise will enter the system [47]. In order to maintain accuracy with the FXP architecture (to compare to the FLP and LNS architectures), and for the purposes of the worst case hardware analysis being done here, no intermediate results are truncated. Hence, to maintain intermediate results, we require digital components of different bit sizes, depending on the number and type of the operation. To maintain accuracy and reasonable functional equivalence to the FLP version, the FXP implementations should utilize the full dynamic range of the data [39] as much as possible. One way to accomplish this is to use range normalization. This is a conservative assumption, since it is likely that real implementations can get by with significant precision reduction even with fixed-point representations. However, to maintain reasonable functional equivalency, we have added such normalization to the FXP implementations. Range normalization typically has a hardware/operation overhead consisting of a digital adder/comparator, a multiplier and a divider, as shown in Fig. 4.7 (b type blocks).

In Fig. 4.7, Block B1a computes an intermediate version $\lambda_{tm}(y,z)$, which is then range normalized by block B1b to obtain $\lambda(y,z)$. Similarly blocks B2b to B5b, which are attached to the corresponding blocks B2a to B5a, perform range normalization. In general, we cannot[10] virtualize a single range normalization block across all other blocks, because the sizes of the digital components in the range normalization blocks are dependent on the size of the intermediate results. In blocks B1b to B5b, *cl*, *ch*, *dh* denote current minimum, current maximum, and desired maximum respectively. These correspond to the minimum and maximum values/elements in a vector that is to be range normalized. As in the FLP architecture, blocks B5a and B5b are virtualized for computing variables $\pi(y,x_1)$ to $\pi(y,x_{n_c})$. The overhead for the range normalization pushes the FXP architecture analysis towards an extreme worst case; i.e. most other simpler techniques for rounding/truncation will have a smaller overhead.

4.6 Mixed-Signal (MS) CMOS and CMOL Hardware Architectures for Bayesian Memory (BM)

Consider the Vector Matrix Multiplication (VMM) operation given by 4.7. Its computational complexity is $O(pq)$. To compute one element (i.e. Y_j) of the output vector Y roughly requires p (add & multiply) operations, as shown in 4.8. To compute all elements of Y, we need to repeat this process q times, hence the total time[11] required is $q\,p\,(t_{add} + t_{mult})$, when implemented sequentially. To reduce this time, many of the operations in 4.7 can be performed in parallel. The following discussion investigates the use of hardware components for parallel VMM operation of 4.2 and 4.3 in the BM.

$$\begin{bmatrix} Y_1 & Y_2 & \dots & Y_q \end{bmatrix} = \begin{bmatrix} X_1 & X_2 & \dots & X_p \end{bmatrix} \times \begin{bmatrix} M_{1,1} & M_{1,2} & \cdots & M_{1,q} \\ M_{2,1} & M_{2,2} & \cdots & M_{2,q} \\ \vdots & \vdots & \ddots & \vdots \\ M_{p,1} & M_{p,2} & \cdots & M_{p,q} \end{bmatrix} \tag{4.7}$$

$$Y_j = \sum_{i=1}^{p} X_i M_{i,j} \tag{4.8}$$

$$\begin{aligned} Y_1 &= X_1 M_{1,1} + X_2 M_{2,1} + X_3 M_{3,1} = \dots \\ &\dots[011 \times 100] + [011 \times 111] + [101 \times 110] \end{aligned} \tag{4.9}$$

[10] Considering that $n_{CBy} \neq n_{CBz}$, we cannot virtualize the blocks across 4.2 and 4.3. We could virtualize the range normalization blocks for 4.4 and 4.6, but have not done so here for simplicity's sake.

[11] Where, t_{op} denotes the time required to complete 'op' type operation.

Cauwenberghs [37] proposed an "Internally Analog, Externally Digital" (IAED) mixed-signal array based structure for parallel VMM, for massive (100–10,000) matrix dimensions. The IAED-Structure for VMM (SVMM) effectively combines storage and analog computation, and is better than other analog-only techniques for the VMM computation [37]. Internally, the storage is digital. The array structure allows inherent analog processing [37]. Externally, it provides the convenience and precision of pre/post digital processing [37]. Hence, we use the SVMM as the fundamental operation implemented in our mixed-signal architectures. Our mixed-signal CMOS architecture partly virtualizes SVMM, and our mixed-signal CMOL architecture replaces some of the traditional components of the SVMM with nano components, providing a novel, mixed-signal CMOL nanogrid based SVMM.

In the traditional SVMM [37], all elements of Y are computed in parallel. Each element is internally computed in a semi-parallel fashion, requiring only a few iterations through the SVMM. The mixed-signal CMOS SVMM is shown in Fig. 4.8b. The number of iterations required is n_{bitx}, which denotes the number of bits/precision used to represent each element X_i of vector X. Let t_{SVMM} denote the time for a single pass through the SVMM. Then the total time required to complete the VMM operation is $n_{bitx}\, t_{\text{SVMM}}$, which is much less than $q\,p\,(t_{\text{add}} + t_{\text{mult}})$ (for sequential operations), because $n_{bitx} << q$, and $t_{\text{SVMM}} <$[12] $p\,(t_{\text{add}} + t_{\text{mult}})$. Since all the elements of vector Y are being computed in parallel, multiple ADCs and other arithmetic/logic components are required. For a very large M (such as in our case where M is [4K x 4K]), we would require thousands of ADCs and adders, which would be costly in terms of hardware area and power, compromising feasibility and significantly limiting scalability. For this reason, we virtualized the ADCs and arithmetic/logic components over all the elements of Y; so now each Y_j is computed sequentially. In this case, the total time required to complete the VMM operation is $q\,n_{bitx}\,t_{\text{SVMM}}$. For a virtualized SVMM, we only achieve better performance than the sequential implementation if $n_{bitx}\,t_{\text{SVMM}} < p\,(t_{\text{add}} + t_{\text{mult}})$; our hardware analysis will address this issue. Such a method of computing each element of Y separately fits well to our BM model, because communication and memory access are sequential in the BM.

The formal equations behind the SVMM operation are presented in more detail in [37], and so are not repeated here. Instead, the operation of these, "semi-virtualized," mixed-signal CMOS and CMOL SVMM circuits are presented via a simple example.

4.6.1 Mixed-Signal CMOS Architecture

Consider a VMM operation where $q = p = 3$ and $n_{bitx} = 3$. The computation required for the first element Y_1 is given in 4.9, along with some example numbers for X_i and $M_{i,1}$. Each number is represented in "little endian" binary format (the

[12] This may not be always true, because it depends on the size of M.

leftmost bit is the MSB and the rightmost bit the LSB), for example $X_1 = 011$ and $M_{3,1} = 110$. Figure 4.8a shows a decomposition of the VMM operation into its constituent sub-operations. To illustrate the SVMM function, we first concentrate

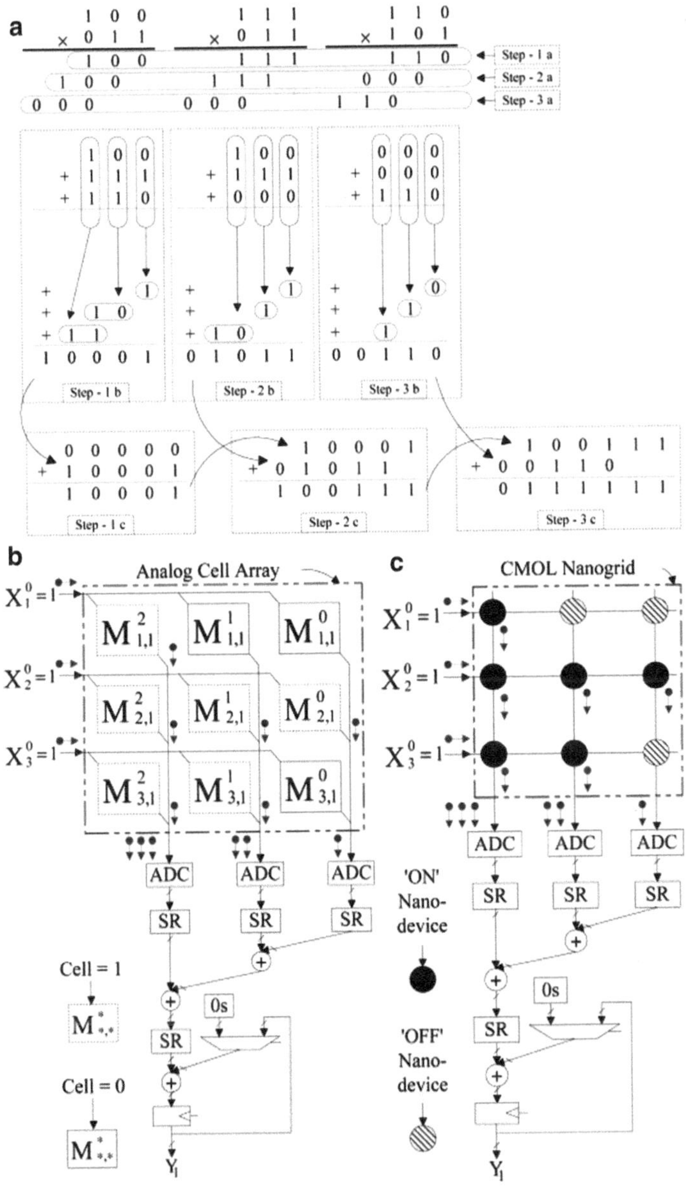

Fig. 4.8 Mixed-signal computations and structures: (**a**) Example of a VMM computation decomposed into sub-operations, (**b**) mixed-signal CMOS SVMM, and (**c**) Mixed-signal CMOL nanogrid SVMM

on step-1 (a, b, and c). In Fig. 4.8a, step-1a shows the partial products resulting from the multiplication of the first bit (LSB) of each X_i with the corresponding bits of each $M_{i,1}$. The equivalent operation in the mixed-signal CMOS SVMM is shown in Fig. 4.8b. The first bit of each X_i, i.e. X_i^0 (denoted by X_i^b, where i = element index, b = bit index, and $b = 0$ to $n_{bitx} - 1$) is presented from the left along the horizontal wires. Each cell (denoted by $M_{i,1}^b$) in the analog cell array, in Fig. 4.8b, is a "CID computational cell with integrated DRAM" [37]. The analog cell array corresponds to the first column of M in 4.7. Each cell can store one binary value, and can compute one binary multiplication. In the compute mode, it can contribute some charge/voltage on the vertical wire if its internal (stored) value is 1 and the external input is 1. For example, the cell $M_{1,1}^0$ has internal value 1, and its external input X_1^0 is also 1; hence it contributes some charge/voltage (shown by a vertical arrow ⟷) to the first vertical wire. Those cells that have an internal value of 0, do not contribute charge/voltage to the vertical wires. The analog cell array in Fig. 4.8b performs the equivalent of step-1a and the initial part of step-1b in Fig. 4.8a. The total charge/voltage (shown by multiple arrows) on each vertical wire is converted to digital outputs by the ADCs, and these outputs correspond to the partial sums shown in step-1b. The Shift Registers (SRs) and adders following the ADCs then complete step-1b. Step-1c is accomplished by the remaining SR and adder. This concludes step-1 or a single pass through the SVMM. In the next iteration we present the second bit of each X_i, i.e. X_i^1, on the horizontal wires of the SVMM, to accomplish step-2 (a, b, and c). Step-3 is completed in a similar manner. In summary, to obtain (one element) Y_j we have to iterate n_{bitx} times through the SVMM.

For computing the remaining elements of vector Y, we require similar steps; but for each Y_j we require a distinct analog cell array that corresponds to the jth column of M in 4.7. The input X remains the same, and is still presented bit-serially as described earlier. The ADCs and arithmetic/logic units are virtualized across all elements of Y.

In our hardware analysis, we assume a precision of $n_{bitx} = 8$ for both X and M. This makes this architecture comparable to the digital FXP8 architecture. (The block diagram of mixed-signal CMOS/CMOL will be similar to Fig. 4.7, but blocks B1a and B2a are replaced with equivalent SVMM – Fig. 4.8b/c.) The required ADC resolution is 12 bits, corresponding to the actual size of M (specified in Table 4.2). We virtualize all the addition operations in Fig. 4.8b by using a single physical adder.

4.6.2 Mixed-Signal CMOL Architecture

The mixed-signal CMOL nanogrid SVMM is shown in Fig. 4.8c. This is similar to the mixed-signal CMOS SVMM shown in Fig. 4.8b, except that the CMOS analog cell array is replaced with a CMOL nanogrid, and each cell is replaced with an equivalent nanodevice, which functions as a binary switch. If the nanodevice is ON, and the input X_i^b on that particular horizontal nanowire is 1, then an "on" current will flow in the corresponding vertical nanowire (shown by a vertical arrow ⟷). Each vertical nanowire accumulates some units of "on" current as shown in Fig. 4.8c. The ADC then converts the current (or equivalent voltage) to digital. The specific functionality is the same as discussed earlier for the mixed-signal CMOS SVMM and is not repeated here.

The storage for X requires special consideration; elements of X are written by sequential memory access, but we need to read one bit from each X_i in parallel. We do not go into details of how that can be done, but instead use a worst case consideration that storage for X has extra routing/circuit overhead of around 50% for this increased functionality.

4.7 Performance/Price Analysis and Results

4.7.1 Performance/Price Analysis

The circuit components used by the various architectures analyzed here are summarized in Table 4.4. For each architecture, a basic dataflow block diagram (Figs. 4.6 and 4.7) is generated from these components. The performance[13] (speed) and price (area and power) are derived using the computational requirements of

Table 4.4 Major circuit components used in implementing a Bayesian memory (BM) module

Components	Digital CMOS architecture			Digital CMOL architecture			Mixed-signal CMOS architecture	Mixed-signal CMOL architecture
	FXP	FLP	LNS	FXP	FLP	LNS		
Digital Adder	Y Y_R			Y Y_R			Y Y_R	Y Y_R
Digital Multiplier	Y Y_R			Y Y_R			Y Y_R	Y_R
Digital Divider	Y_R			Y_R			Y_R	Y_R
Floating-Point Unit[a]		Y			Y			
Log Num. sys. Unit[a]			Y			Y		
ADC							Y	Y
Memory – SRAM	Y	Y	Y				Y	
Memory – CMOL				Y	Y	Y		Y
Analog CID/DRAM Cell							Y	
CMOL nanogrid								Y

[a]FPU and LNU internally comprise of traditional digital arithmetic/logic and memory components
Y = component utilized for major operations
Y_R = component utilized for range normalization operations

[13] In accordance with our baseline assumption, the timing analysis for the digital architectures assumes that most of the operations within each block are executed in a sequential manner without any pipelining; however, some of the blocks are executing in parallel.

the BM (Table 4.3), and the performance/price measures corresponding to the circuit components that accomplish these computations. The detailed equations for deriving the performance/price of all architectures are given in the Appendix (Sections 4.9.1 to 4.9.4).

The performance/price values for traditional circuit components are adapted from Table 4.3 in [3]. The performance/price measures for the remaining components were obtained from the literature: FPU [48], LNU [49], ADC [50], divider [51], and analog CID/DRAM cell [37]. All digital components are scaled to a hypothetical 22 nm technology (as done in [3]), using first-order constant field scaling rules [45]. To make the comparisons as realistic as possible, we chose the 22 nm as a likely process technology when nanogrids become commercially available. Design of analog circuits below 90 nm is challenging [52], hence we conservatively scale analog circuits to 90 nm. The performance/price for CMOL components is based on the nanogrid analysis presented in [3]; the CMOL is scaled to $F_{nano} = 3$ nm (with underlying CMOS at 22 nm). Performance/price measures of various components are summarized in Table 4.5.

Table 4.5 Performance/price measures for various circuit components

Component	Area (= A) (mm²)	Power (W)	Time (s)
SRAM	$6\times10^{-7}N_b/2.85$	$\alpha 0.64A$ where, $\alpha = 0.1$	0.025×10^{-9}
CMOL MEM	$1.803\times10^{-9}N_b$	$8.974\times10^{-12}N_b$	1.72×10^{-9}
Dig. Adder	$\frac{1.2\times10^{-2}a_t^2 N\log_2(N)}{32\log_2(32)}$	$0.04A$ where, $a_t = 22/180$	$\frac{0.75\times10^{-9}a_t\log_2(N)}{\log_2(32)}$
Dig. Multiplier	$\frac{1.2\times10^{-4}N(M+1)}{16(16+1)}$	$\frac{1.7\times10^{-5}N(M+1)}{16(16+1)}$	$\frac{0.2\times10^{-9}(N+M)}{(16+16)}$
Dig. Divider	$\frac{3.09a_t^2 N\log_2(N)}{55\log_2(55)}$	$0.04A$ where, $a_t = 22/1200$	$160\times10^{-9}a_t N/55$
Analog CID/ DRAM Cell	$3.24\times10^{-5}a_t^2$	$5\times10^{-8}a_t^2$ where, $a_t = 90/500$	$1\times10^{-5}a_t$
ADC – 12b	$17.22a_t^2$	$0.033a_t^2$ where, $a_t = 90/1200$	$2\times10^{-7}a_t$
FPU	$0.258a_t^2$	$1.032\times10^{-2}a_t^2$ where, $a_t = 22/350$	
–Adder			$3a_t 7.692\times10^{-9}$
–Multiplier			$3a_t 7.692\times10^{-9}$
–Divider			$15a_t 7.692\times10^{-9}$
LNU	$16.6a_t^2$	$0.355a_t^2$ where, $a_t = 22/1200$	
–Adder			$a_t 2.631\times10^{-8}$
–Multiplier			$a_t 2.631\times10^{-8}$
–Divider			$a_t 2.631\times10^{-8}$

N = number of bits in operand-1; M = number of bits in operand-2, and $N \geq M$; N_b = total number of bits in memory; a_t denotes technology scaling$_{IC}$ factor

4.7.2 Performance/Price Results and Discussion

The hardware analysis results for the digital architectures are summarized in Figs. 4.9 to 4.13, and for the mixed-signal architectures in Table 4.6.

In Fig. 4.9, we see that the "Memory components" (MEM) dominate the total area of the BM. This is a major advantage, especially when using CMOL memory, which is 100 times denser than CMOS memory (SRAM). For the "Arithmetic/logic components" (ARL) we see that the FLP and LNS architectures consume less area, as compared to the FXP8–FXP32 architectures, since the FXP architectures are not virtualized to the same extent. In addition, the FXP architectures also had overhead for several range normalization circuits, which were also not virtualized. But because memory dominates the area of a BM processor, this overhead does not add much to the total cost. The ARL in the FLP architecture occupies less area than the ARL in LNS architecture.

In Fig. 4.10, we again see that MEM dominates the total power consumption of the BM. CMOL MEM consumes 1,000 times less power than CMOS MEM, but this is somewhat misleading, since the power density of the hotspots in CMOL nanogrids could be as high as 200 W cm^{-2} (the maximum allowed by ITRS [43]). The SRAM has much lower power density for obvious reasons. The ARL in the FXP architectures consumes more power when compared to the ARL in the FLP and the LNS architectures, for reasons stated earlier, but again this is not a big issue. The ARL in the FLP architecture consumes less power when compared to the ARL in LNS architecture.

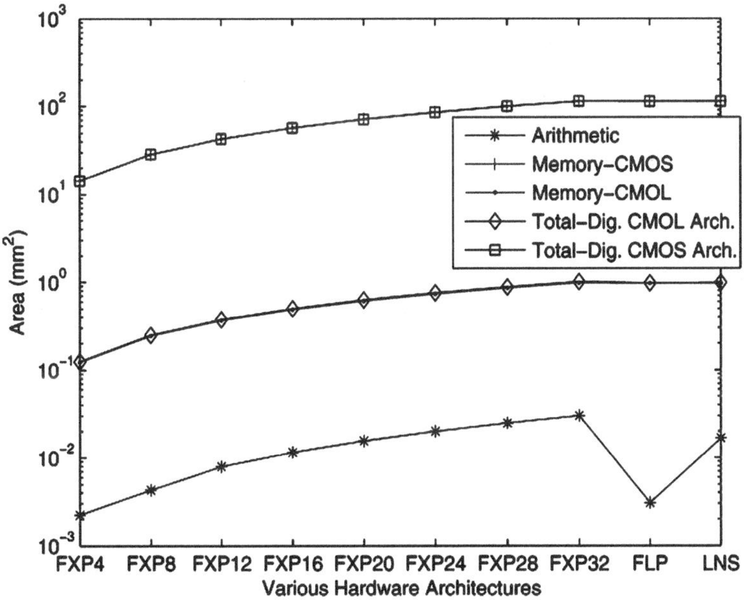

Fig. 4.9 Area occupied by single BM: digital architectures

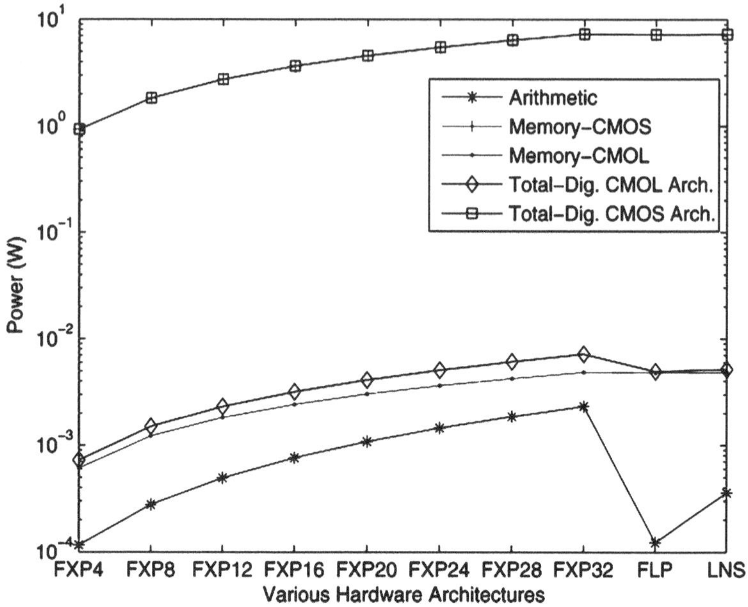

Fig. 4.10 Power consumed by single BM: digital architectures

Figure 4.11 shows the time required to update (execute the BPA) of a single BM. As compared to the FLP and LNS architectures, the FXP architectures have better performance. We assumed the range normalization overhead to obtain the maximum accuracy in a system with FXP precision (as compared to a system with FLP or LNS precision); our assumption did not affect the performance of the system. The LNS architecture is faster than the FLP architecture, because the specific LNU used here is three times faster than the FPU; in general, depending on the ratio of multiplication/addition operations, LNU based architectures can perform 1.5–2.6 times faster than FPU based architectures [49, 53]. All CMOL architectures are slower than the corresponding CMOS architectures by around 30 ms; this was expected because CMOL nanogrids are slow [3]. For digital architectures in general, timing can be easily reduced by adding more ARL components for semi-parallel functioning, and using multi-ported memories.

Figure 4.12 shows how many BMs we can fit in a maximum size chip of 858 mm^2 (i.e. a max-chip), which corresponds to the maximum reticle lithographic field size at 22 nm [3]. Because CMOL architectures are denser than CMOS architectures, we can fit around 100 times more BMs in the same area.

Figure 4.13 corresponds to Fig. 4.12, and shows the power consumed by all BMs residing in a max-chip; it assumes that all of these are concurrently active. The CMOL chip consumes about one-tenth the power of the CMOS chip; though the CMOL nanogrids have a higher power density.

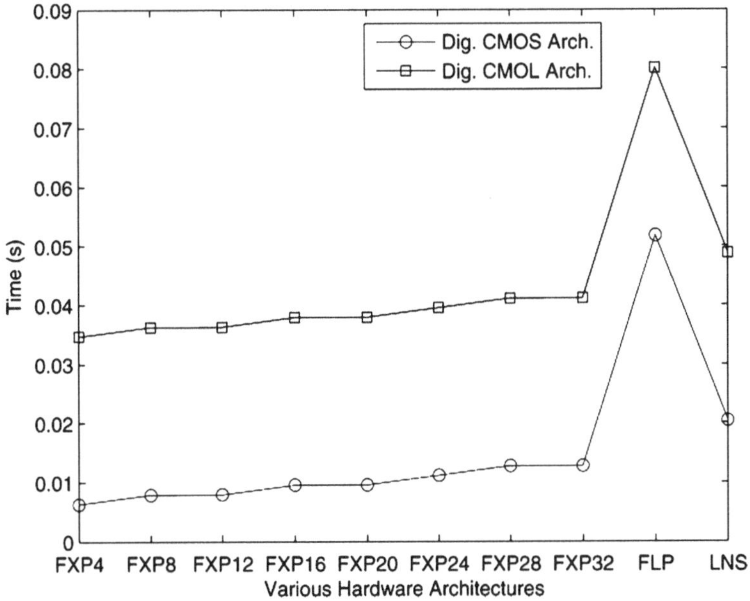

Fig. 4.11 Time to update single BM: digital architectures

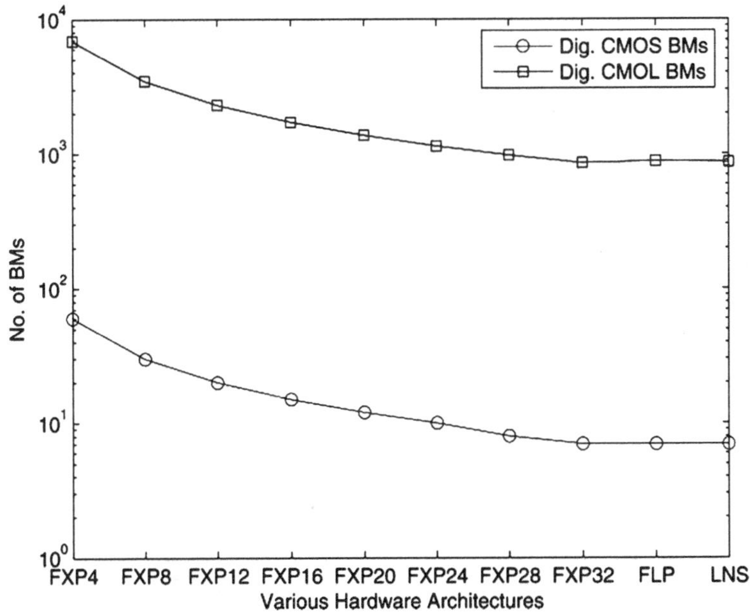

Fig. 4.12 BMs implemented on one max-chip: digital architectures

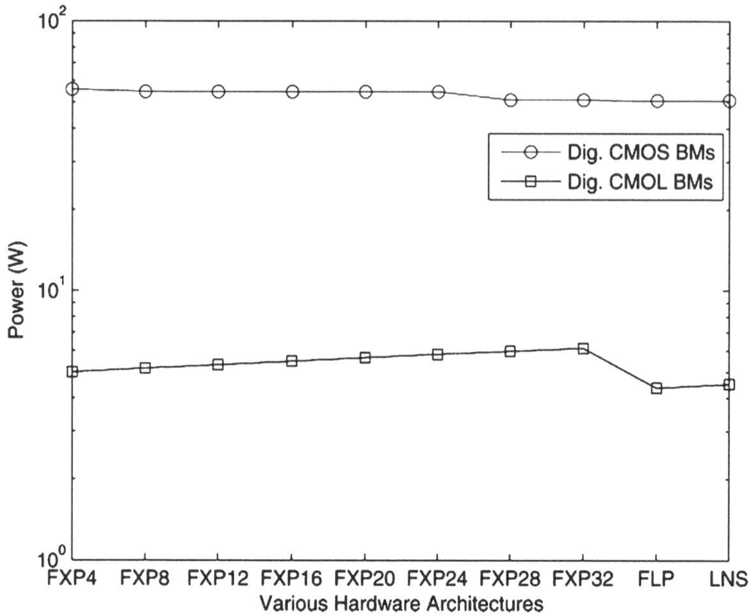

Fig. 4.13 Total power consumed by BMs on one max-chip: digital architectures

From Table 4.6, when comparing digital CMOL to Mixed-Signal (MS) CMOL, we were able to achieve the speed-up (i.e. $n_{bitx}\, t_{\text{SVMM}} < p\,(t_{\text{add}} + t_{\text{mult}})$) using our virtualized SVMM technique. But the same is not true for the digital CMOS and MS CMOS architectures. The MS CMOS architecture provides the worst performance, with the highest price; only 3 BMs can fit one max-chip, which suggests it is not as scalable of an architecture. The digital CMOS and MS CMOL architectures have approximately the same performance, but MS CMOL is less expensive, allowing us to fit 18 times more BMs in one max-chip, with one-fourth the total power consumption. Notice that the power density of a max-chip for all architectures is well below the allowed 200 W cm^{-2} (for 22 nm CMOS/3 nm CMOL, according to ITRS [43]).

In summary, the MS CMOL architecture, which consists of CMOL memory and a MS nanogrid implementation of the VMM operation is clearly the most cost-effective of the architecture options examined here, providing the best performance at a reasonable price.

An important criterion for comparing various hardware architectures is the "performance/price ratio", which measures the usefulness/efficiency of silicon area in solving the problem (or set of problems) at hand. For the hardware implementations of computational models, this criterion is defined as a "Module update rate per 858 mm^2" or "Throughput Per Max-chip" (TPM) [3, 54]. Table 4.6 shows that the TPM for digital CMOL architecture is 25 times the TPM of digital CMOS architecture; the TPM for MS CMOL architecture is 20 times the TPM for digital CMOS architecture. The TPM for a PC MATLAB implementation of BM is 33, and the TPM

Table 4.6 Comparison of digital and mixed-signal architectures

	Single BM							Max-chip[c]			
	Area (mm^2)			Power (W)							
Architecture	MEM[a]	ARL[b]	Total	MEM	ARL	Total	Time (s)	No. of BMs	Total Power (W)	Throughput Per Max-chip [f] (TPM)	Normalized TPM
Digital CMOS[d]	28.525	0.0048	28.529	1.82562	0.00028	1.82590	0.00790	30	54.78	3797	76
Mixed-Signal CMOS[e]	282.06	1.5546	283.62	0.46599	0.00324	0.46923	0.05980	3	1.41	50	1
Digital CMOL[d]	0.2443	0.0048	0.2491	0.00122	0.00028	0.00149	0.03627	3443	5.13	94848	1897
Mixed-Signal CMOL[e]	0.0121	1.5581	1.5703	0.02127	0.00324	0.02450	0.00727	546	13.38	75103	1505

[a]MEM denotes memory components
[b]ARL denotes Arithmetic/logic components
[c]Max-chip denotes a maximum reticle field chip size of 858 mm^2
[d]Performance/price corresponds to the digital CMOS/CMOL FXP8 architecture in Figs. 4.9 to 4.13
[e]MS architectures externally represent data digitally, with 8 bit FXP precision
[f]Throughput Per Max-chip (TPM) is measured as (modules updated per second) per 858 mm^2 area. In this table, the TPM is derived by dividing the "No. of BMs" column by the "Time" column

for BM implementation on Cray XD1 (supercomputer accelerated with FPGAs) is 2,326 (derived[14] from the throughputs reported in [21]). Hence, we see that the TPM of CMOL based custom architectures is 32–40 times better than Cray XD1 multi-processor/multi-FPGA system.

The nanogrids in all CMOL architectures studied here were designed to have a worst case power density of ~200 W cm^{-2} (allowed by ITRS [43]) at hotspots. Hence, all CMOL performance/price numbers correspond to that density. If the power density budget were increased, then performance could be improved. For example, the time to update a MS CMOL architecture based BM reduces to 0.00421 s, if the power density budget were doubled. So there is a clear trade-off between power density and performance for CMOL nanogrids; the same was also suggested in [3].

4.7.3 Scaling Estimates for BM Based Cortex-Scale System

This chapter has exclusively focused on the hardware implementations of Part-B of the BM. The hardware assessment methodology (given in Fig. 4.2) has also been used to investigate the hardware implementations of Part-A of the BM, which deals with learning/training. The detailed discussion on hardware architectures for Part-A of the BM (with a simplified algorithm for learning spatial CB vectors) is given in [20], and not repeated here. The results therein conclude that Part-B of the BM dominates the overall hardware requirements and performance/price of the complete BM (including Part-A and Part-B). The results in [20] also indicate that it is not feasible to map analog/non-linear functions such as Euclidean and Gaussian functions onto CMOL-like nanogrid/nanodevices [55]. Consequently, the MS architectures (for Part-A of the BM) have to depend on traditional analog CMOS circuit components to implement such functions [20, 55], and this limits the usefulness of CMOL structures for implementing the operations within Part-A of the BM [2]. This is contrary to the hardware analysis results for Part-B of the BM, which shows that MS CMOL structures were particularly more cost-effective.

Table 4.7 presents four prospective hardware designs/architectures for the BM, and the scaling estimates for BM based (human) Cortex-Scale System (BMCSS). A single BM specified by Table 4.2 emulates approximately 32.8×10^6 synapses, according to the artificial neural network equivalent for BPA proposed in [56]. Consequently, 4.87×10^6 BMs operating in parallel are required to scale[15] up to 1.6×10^{14} synapses in BMCSS. Table 4.7 proposes four different designs by selecting

[14] This system has 864 AMD processors, and 150 FPGAs; the reported Nodes/s is 2.25×10^6 for large network of BMs. The TPM is derived by using (22/90) technology-scaling factor for 22 nm, and ($1{,}014 \times 200/858$ mm^2) area factor; we assume each processor-chipset or FPGA (with SRAM banks) will have an area of 200 mm^2.

[15] Such a crude scaling estimate is intended only for guidance purposes. No structural / functional equivalence to the actual cortex and/or human intelligence is intended or claimed.

Table 4.7 BM based cortex-scale system: final performance/price and scaling estimates

Proposed design	Part-A of BM	Part-B of BM	Total time[b] (s)	Total area[a] (mm^2)	Norm. area[c]	No. of (30 cm) wafers	Perf./Price ($s^{-1}mm^{-2}$)	Norm. perf./price	Total power (W)	Net power density ($W\ mm^{-2}$)
1	Dig. CMOS	Dig. CMOL	3.63×10^{-2}	1.67×10^{6}	6.96	27	1.65×10^{-5}	1.00	3.73×10^{4}	2.23×10^{-2}
2	Dig. CMOS	MS CMOL	7.28×10^{-3}	8.11×10^{6}	33.79	128	1.69×10^{-5}	1.03	1.49×10^{5}	1.84×10^{-2}
3	Dig. CMOL	Dig. CMOL	3.64×10^{-2}	1.22×10^{6}	5.08	20	2.25×10^{-5}	1.36	8.37×10^{3}	6.85×10^{-3}
4	Dig. CMOL	MS CMOL	7.45×10^{-3}	7.66×10^{6}	31.92	121	1.75×10^{-5}	1.06	1.21×10^{5}	1.57×10^{-2}

[a]Approximately 4.876×10^6 BMs are required to reach cortex-scale
[b]Assumes that all BMs are operating in parallel
[c]Total Area normalized by the actual area of cortex ($2.4 \times 10^5\ mm^2$)

the type of designs for Part-A and Part-B of the BM. We used two designs (Digital CMOL and MS CMOL) with the best performance/price ratio (≅ TPM) for Part-B of the BM, and two designs (Digital CMOS and Digital CMOL) with the best performance/price ratio (≅ TPM) for Part-A of the BM.

The normalized performance/price ratio for all architectures for the BMCSS is approximately equal, except the all-digital CMOL design (proposed design no. 3), which has the highest performance/price ratio. However, in terms of only silicon area, or the number of 30 cm wafers required to implement the BMCSS, the digital architectures are 5× to 7× more compact compared to the architectures with MS parts in the BM. For the purpose of a crude scaling estimate (in terms of density), the total area for the BMCSS can be compared to the actual area (2.4×10^5 mm^2 [39]) of cortex, as shown in the column labeled "Norm. Area"; the artificial BMCSS is approximately 5× to 33× larger than the actual cortex. Hence, we can optimistically conclude that with the possible advances in nanoelectronics, we are approaching biological densities, with manageable power (i.e. power density is within the ITRS allowed limits 64–200 W cm^{-2} [57]); this is similar to the claims in [43, 54, 57].

4.8 Conclusion, Contribution and Future Work

The results in this chapter suggest that implementation of Bayesian inference engines and/or Bayesian BICMs is going to be significantly memory dominated. We conclude then that any enabling technology for large-scale Bayesian inference engines will need very high density storage (with potential for inherent computations), and that this storage needs to be accessed via high bandwidth, which limits the use of off-chip storage. Hybrid nanotechnologies such as CMOL have emerged as a very useful candidate for building such large-scale Bayesian inference engines.

We have shown how to effective use the CMOL hybrid nanotechnology as memory, and as a mixed-signal computational structure, for implementing Bayesian inference engines, which forms the core of many AI and machine learning techniques.

We have also proposed a novel use of a mixed-signal CMOL structure for Vector-Matrix Multiplication (VMM), VMM being one of the most fundamental operations used in many computational algorithms. This mixed-signal CMOL structure provides speeds comparable to digital CMOS, which is due to the efficient integration of storage and computation at the nanodevice/grid level.

For our particular application, we have shown that silicon real-estate is utilized much more efficiently (at least 32–40 times better TPM ≅ speed per unit-area) in the case of CMOL, as compared to, for example, a multi-processor/multi-FPGA system, that uses CMOS technology. When compared to a single general-purpose processor/PC, the TPM of CMOL based architectures is 2,200–2,800 times better. It is obvious that CMOL is denser (and slow when used as memory), but what is not as obvious is that CMOL when used in mixed-signal mode can provide reasonable speed-up. However, while using mixed-signal CMOL, external post-processing

CMOS components dominate the area, hence, their virtualization is a crucial constraint for design and scaling.

In addition to the results of the architecture study, another important contribution of this chapter is the development and use of a "hardware design space exploration" methodology for architecting hardware and analyzing its performance/price; this methodology has the potential of being used as an "investigation tool" for various computational models and their implementations. This is particularly true for when chip designers start transitioning their existing CMOS solutions to nanotechnology based solutions.

As future work, one needs to explore the remaining design space using both traditional and CMOL technology, and also search for new nanoelectronics circuit candidates that may be better suitable for implementing Bayesian inference engines. Another important approach is to investigate the possibility of more approximate inference techniques that trade-off accuracy for performance.

4.9 Appendix

Note: The values of the variables (related to Pearl's algorithm) used in the following equations are given in Table 4.2. The performance/price measures for the various circuit components used in the following equations are given in Table 4.5. Most of the subscripts are self-explanatory, and denote the particular component from Table 4.5. The superscripts denote the number of bits for that particular circuit component from Table 4.5. All timing numbers for digital components first need to be normalized as a multiple of $t_{\text{clk}} = 9.32 \times 10^{-11}$ s, before being used in these equations. The subscript 'BPA' refers to Pearl's belief propagation Eqs. 4.1–4.4 and 4.6; subscript 'ARL' denotes arithmetic/logic components; subscript 'MEM' denotes memory.

4.9.1 Digital FLP or LNS Architecture

The following equations are for FLP architecture; for LNS architecture, replace subscript 'FPU' with 'LNU'. Also, the following equations are for digital CMOS architectures; for digital CMOL architectures replace subscript 'SRAM' with 'CMOL-MEM'. Here, $n_{bit} = 32$, corresponding to single-precision FLP/LNS data.

4.9.1.1 Time

$$t_{\text{BPA1}} = \left(n_c \cdot n_{CBy} \cdot (t_{\text{FPU-mul}} + t_{\text{SRAM}})\right) + (n_{CBy} \cdot t_{\text{SRAM}})$$

$$t_{\text{BPA2}} = \left(n_{CBz} \cdot n_{CBy} \cdot (t_{\text{FPU-add}} + t_{\text{FPU-mul}} + t_{\text{SRAM}})\right) + (n_{CBz} \cdot t_{\text{SRAM}})$$

$$t_{\mathrm{BPA3}} = \left(n_{CBy} \cdot n_{CBz} \cdot (t_{\mathrm{FPU\text{-}add}} + t_{\mathrm{FPU\text{-}mul}} + t_{\mathrm{SRAM}})\right) + (n_{CBy} \cdot t_{\mathrm{SRAM}})$$
$$t_{\mathrm{BPA4}} = \left(2n_{CBy} \cdot (t_{\mathrm{FPU\text{-}mul}} + t_{\mathrm{SRAM}}) + (n_{CBy} \cdot t_{\mathrm{SRAM}})\right) + \ldots$$
$$\left(n_{CBy} \cdot (t_{\mathrm{FPU\text{-}mul}} + t_{\mathrm{FPU\text{-}add}}) + (3n_{CBy} \cdot t_{\mathrm{SRAM}})\right) + t_{\mathrm{FPU\text{-}d}}$$
$$t_{\mathrm{BPA6}} = \left(n_c \cdot n_c \cdot n_{CBy} \cdot (t_{\mathrm{FPU\text{-}mul}} + t_{\mathrm{SRAM}}) + (n_c \cdot n_{CBy} \cdot t_{\mathrm{SRAM}})\right) + \ldots$$
$$\left(n_c \cdot n_{CBy} \cdot (t_{\mathrm{FPU\text{-}mul}} + t_{\mathrm{FPU\text{-}add}}) + (3n_c \cdot n_{CBy} \cdot t_{\mathrm{SRAM}})\right) + (n_c \cdot t_{\mathrm{FPU\text{-}div}})$$
$$t_{\mathrm{final}} = \max\left((t_{\mathrm{BPA1}} + t_{\mathrm{BPA2}} + t_{\mathrm{BPA4}}), (t_{\mathrm{BPA3}} + t_{\mathrm{BPA6}})\right)$$

4.9.1.2 Area

$$n_{membits} = n_{bit} \cdot (4n_{CBy} + n_c \cdot n_{CBy} + n_{CBz} + n_{CBz} \cdot n_{CBy})$$
$$A_{\mathrm{MEM}} = A_{SRAM}^{(n_{membits})}$$
$$A_{\mathrm{ARL}} = 2A_{\mathrm{FPU}}$$

4.9.1.3 Power

$$P_{\mathrm{MEM}} = P_{SRAM}^{(n_{membits})}$$
$$P_{\mathrm{ARL}} = 2P_{\mathrm{FPU}}$$

4.9.2 Digital FXP Architecture

Here, n_{bit} is varied in the range [4, 8, …, 32], to obtain the performance/price of FXP4, FXP8, …, FXP32, architectures respectively.

$$n_{b1} = (n_c - 1) \cdot n_{bit}$$
$$n_{b3} = ceil(\log_2(n_{CBy})) + 2n_{bit}$$
$$n_{b4} = ceil(\log_2(n_{CBz})) + 2n_{bit}$$

4.9.2.1 Time

$$t_{\mathrm{BPA1}} = \left(n_c \cdot n_{CBy} \cdot (t_{\mathrm{FXP\text{-}mul}}^{(n_{b1}, n_{bit})} + t_{\mathrm{SRAM}})\right) + (n_{CBy} \cdot t_{\mathrm{SRAM}}) + \ldots$$
$$\left(n_{CBy} \cdot (t_{\mathrm{FXP\text{-}add}}^{(n_{b1}+n_{bit})} + t_{\mathrm{FXP\text{-}mul}}^{(n_{b1}+n_{bit}, n_{bit})} + t_{\mathrm{FXP\text{-}div}}^{(n_{b1}+2n_{bit})} + t_{\mathrm{SRAM}})\right)$$

$$t_{\mathrm{BPA2}} = \left(n_{CBz} \cdot n_{CBy} \cdot (t_{\mathrm{FXP\text{-}add}}^{(n_{b3})} + t_{\mathrm{FXP\text{-}mul}}^{(n_{bit}, n_{bit})} + t_{\mathrm{SRAM}})\right) + (n_{CBz} \cdot t_{\mathrm{SRAM}}) + \ldots$$
$$\left(n_{CBz} \cdot (t_{\mathrm{FXP\text{-}add}}^{(n_{b3})} + t_{\mathrm{FXP\text{-}mul}}^{(n_{b3}, n_{bit})} + t_{\mathrm{FXP\text{-}div}}^{(n_{b3}+n_{bit})} + t_{\mathrm{SRAM}})\right)$$

$$t_{\text{BPA3}} = \left(n_{CBy} \cdot n_{CBz} \cdot (t_{\text{FXP-add}}^{(n_{b4})} + t_{\text{FXP-mul}}^{(n_{bit},n_{bit})} + t_{\text{SRAM}})\right) + (n_{CBy} \cdot t_{\text{SRAM}}) + \ldots$$
$$\left(n_{CBy} \cdot (t_{\text{FXP-add}}^{(n_{b4})} + t_{\text{FXP-mul}}^{(n_{b4},n_{bit})} + t_{\text{FXP-div}}^{(n_{b4}+n_{bit})} + t_{\text{SRAM}})\right)$$
$$t_{\text{BPA4}} = \left(2n_{CBy} \cdot (t_{\text{FXP-mul}}^{(n_{bit},n_{bit})} + t_{\text{SRAM}}) + (n_{CBy} \cdot t_{\text{SRAM}})\right) + \ldots$$
$$\left(n_{CBy} \cdot (t_{\text{FXP-add}}^{(2n_{bit})} + t_{\text{FXP-mul}}^{(2n_{bit},n_{bit})} + t_{\text{FXP-div}}^{(3n_{bit})} + t_{\text{SRAM}})\right)$$
$$t_{\text{BPA6}} = n_c \cdot \left(n_{CBy} \cdot (t_{\text{FXP-mul}}^{(n_{b1},n_{bit})} + t_{\text{SRAM}}) + (n_{CBy} \cdot t_{\text{SRAM}})\right) + \ldots$$
$$n_c \cdot \left(n_{CBy} \cdot (t_{\text{FXP-add}}^{(n_{b1}+n_{bit})} + t_{\text{FXP-mul}}^{(n_{b1}+n_{bit},n_{bit})} + t_{\text{FXP-div}}^{(n_{b1}+2n_{bit})} + t_{\text{SRAM}})\right)$$
$$t_{\text{final}} = \max\left((t_{\text{BPA1}} + t_{\text{BPA2}} + t_{\text{BPA4}}), (t_{\text{BPA3}} + t_{\text{BPA6}})\right)$$

4.9.2.2 Area

$$A_{\text{BPA1-ARL}} = A_{\text{FXP-mul}}^{(n_{b1},n_{bit})} + \left(A_{\text{FXP-add}}^{(n_{b1}+n_{bit})} + A_{\text{FXP-mul}}^{(n_{b1}+n_{bit},n_{bit})} + A_{\text{FXP-div}}^{(n_{b1}+2n_{bit})}\right)$$
$$A_{\text{BPA2-ARL}} = \left(A_{\text{FXP-add}}^{(n_{b3})} + A_{\text{FXP-mul}}^{(n_{bit},n_{bit})}\right) + \left(A_{\text{FXP-add}}^{(n_{b3})} + A_{\text{FXP-mul}}^{(n_{b3},n_{bit})} + A_{\text{FXP-div}}^{(n_{b3}+n_{bit})}\right)$$
$$A_{\text{BPA3-ARL}} = \left(A_{\text{FXP-add}}^{(n_{b4})} + A_{\text{FXP-mul}}^{(n_{bit},n_{bit})}\right) + \left(A_{\text{FXP-add}}^{(n_{b4})} + A_{\text{FXP-mul}}^{(n_{b4},n_{bit})} + A_{\text{FXP-div}}^{(n_{b4}+n_{bit})}\right)$$
$$A_{\text{BPA4-ARL}} = A_{\text{FXP-mul}}^{(n_{bit},n_{bit})} + \left(A_{\text{FXP-add}}^{(2n_{bit})} + A_{\text{FXP-mul}}^{(2n_{bit},n_{bit})} + A_{\text{FXP-div}}^{(3n_{bit})}\right)$$
$$A_{\text{BPA6-ARL}} = A_{\text{FXP-mul}}^{(n_{b1},n_{bit})} + \left(A_{\text{FXP-add}}^{(n_{b1}+n_{bit})} + A_{\text{FXP-mul}}^{(n_{b1}+n_{bit},n_{bit})} + A_{\text{FXP-div}}^{(n_{b1}+2n_{bit})}\right)$$
$$A_{\text{ARL}} = A_{\text{BPA1-ARL}} + A_{\text{BPA2-ARL}} + A_{\text{BPA3-ARL}} + A_{\text{BPA4-ARL}} + A_{\text{BPA6-ARL}}$$
$$n_{membits} = n_{bit} \cdot (4n_{CBy} + n_c \cdot n_{CBy} + n_{CBz} + n_{CBz} \cdot n_{CBy}) + \ldots$$
$$\left((n_c + 1) \cdot n_c \cdot n_{bit} \cdot n_{CBy} + n_{b3} \cdot n_{CBz} + n_{b4} \cdot n_{CBy} + 2n_{bit} \cdot n_{CBy}\right)$$
$$A_{\text{MEM}} = A_{\text{SRAM}}^{(n_{membits})}$$

4.9.2.3 Power

$$P_{\text{BPA1-ARL}} = P_{\text{FXP-mul}}^{(n_{b1},n_{bit})} + \left(P_{\text{FXP-add}}^{(n_{b1}+n_{bit})} + P_{\text{FXP-mul}}^{(n_{b1}+n_{bit},n_{bit})} + P_{\text{FXP-div}}^{(n_{b1}+2n_{bit})}\right)$$
$$P_{\text{BPA2-ARL}} = \left(P_{\text{FXP-add}}^{(n_{b3})} + P_{\text{FXP-mul}}^{(n_{bit},n_{bit})}\right) + \left(P_{\text{FXP-add}}^{(n_{b3})} + P_{\text{FXP-mul}}^{(n_{b3},n_{bit})} + P_{\text{FXP-div}}^{(n_{b3}+n_{bit})}\right)$$
$$P_{\text{BPA3-ARL}} = \left(P_{\text{FXP-add}}^{(n_{b4})} + P_{\text{FXP-mul}}^{(n_{bit},n_{bit})}\right) + \left(P_{\text{FXP-add}}^{(n_{b4})} + P_{\text{FXP-mul}}^{(n_{b4},n_{bit})} + P_{\text{FXP-div}}^{(n_{b4}+n_{bit})}\right)$$
$$P_{\text{BPA4-ARL}} = P_{\text{FXP-mul}}^{(n_{bit},n_{bit})} + \left(P_{\text{FXP-add}}^{(2n_{bit})} + P_{\text{FXP-mul}}^{(2n_{bit},n_{bit})} + P_{\text{FXP-div}}^{(3n_{bit})}\right)$$
$$P_{\text{BPA6-ARL}} = P_{\text{FXP-mul}}^{(n_{b1},n_{bit})} + \left(P_{\text{FXP-add}}^{(n_{b1}+n_{bit})} + P_{\text{FXP-mul}}^{(n_{b1}+n_{bit},n_{bit})} + P_{\text{FXP-div}}^{(n_{b1}+2n_{bit})}\right)$$
$$P_{\text{ARL}} = P_{\text{BPA1-ARL}} + P_{\text{BPA2-ARL}} + P_{\text{BPA3-ARL}} + P_{\text{BPA4-ARL}} + P_{\text{BPA6-ARL}}$$
$$P_{\text{MEM}} = P_{\text{SRAM}}^{(n_{membits})}$$

4.9.3 Mixed-Signal CMOS Architecture

For mixed-signal CMOS architectures, each element of the CPT, and external data have (n_{bit} =) 8 bit representation. The mixed-signal CMOS SVMM implements only 4.2 and 4.3 of the BPA; for 4.1, 4.4, and 4.6, use the corresponding performance/price equations from FXP architecture.

4.9.3.1 Time

$$t_{\text{BPA2}} = \left(n_{CBz} \cdot n_{bit} \cdot (t_{\text{CID}} + n_{bit} \cdot t_{\text{FXP-add}}^{(n_{b3})} + t_{\text{ADC}})\right) + \ldots$$
$$\left(n_{CBz} \cdot (t_{\text{FXP-add}}^{(n_{b3})} + t_{\text{FXP-mul}}^{(n_{b3},n_{bit})} + t_{\text{FXP-div}}^{(n_{b3}+n_{bit})} + t_{\text{SRAM}})\right)$$
$$t_{\text{BPA3}} = \left(n_{CBy} \cdot n_{bit} \cdot (t_{\text{CID}} + n_{bit} \cdot t_{\text{FXP-add}}^{(n_{b4})} + t_{\text{ADC}})\right) + \ldots$$
$$\left(n_{CBy} \cdot (t_{\text{FXP-add}}^{(n_{b4})} + t_{\text{FXP-mul}}^{(n_{b4},n_{bit})} + t_{\text{FXP-div}}^{(n_{b4}+n_{bit})} + t_{\text{SRAM}})\right)$$
$$t_{\text{final}} = \max\left((t_{\text{BPA1}} + t_{\text{BPA2}} + t_{\text{BPA4}}), (t_{\text{BPA3}} + t_{\text{BPA6}})\right)$$

4.9.3.2 Area

$$A_{\text{BPA2-ARL}} = \left(A_{\text{FXP-add}}^{(n_{b3})} + n_{bit} \cdot A_{\text{ADC}}\right) + \left(A_{\text{FXP-add}}^{(n_{b3})} + A_{\text{FXP-mul}}^{(n_{b3},n_{bit})} + A_{\text{FXP-div}}^{(n_{b3}+n_{bit})}\right)$$
$$A_{\text{BPA2-MEM}} = n_{CBz} \cdot n_{CBy} \cdot n_{bit} \cdot A_{\text{CID}}$$
$$A_{\text{BPA3-ARL}} = \left(A_{\text{FXP-add}}^{(n_{b4})} + n_{bit} \cdot A_{\text{ADC}}\right) + \left(A_{\text{FXP-add}}^{(n_{b4})} + A_{\text{FXP-mul}}^{(n_{b4},n_{bit})} + A_{\text{FXP-div}}^{(n_{b4}+n_{bit})}\right)$$
$$A_{\text{BPA3-MEM}} = n_{CBy} \cdot n_{CBz} \cdot n_{bit} \cdot A_{\text{CID}}$$
$$n_{membits} = n_{bit} \cdot (4.5 n_{CBy} + n_c \cdot n_{CBy} + 1.5 n_{CBz}) + \ldots$$
$$\left((n_c + 1) \cdot n_c \cdot n_{bit} \cdot n_{CBy} + n_{b3} \cdot n_{CBz} + n_{b4} \cdot n_{CBy} + 2 n_{bit} \cdot n_{CBy}\right)$$
$$A_{\text{ARL}} = A_{\text{BPA1-ARL}} + A_{\text{BPA2-ARL}} + A_{\text{BPA3-ARL}} + A_{\text{BPA4-ARL}} + A_{\text{BPA6-ARL}}$$
$$A_{\text{MEM}} = A_{\text{SRAM}}^{(n_{membits})} + A_{\text{BPA2-MEM}} + A_{\text{BPA3-MEM}}$$

4.9.3.3 Power

$$P_{\text{BPA2-ARL}} = \left(P_{\text{FXP-add}}^{(n_{b3})} + n_{bit} \cdot P_{\text{ADC}}\right) + \left(P_{\text{FXP-add}}^{(n_{b3})} + P_{\text{FXP-mul}}^{(n_{b3},n_{bit})} + P_{\text{FXP-div}}^{(n_{b3}+n_{bit})}\right)$$
$$P_{\text{BPA2-MEM}} = n_{CBz} \cdot n_{CBy} \cdot n_{bit} \cdot P_{\text{CID}}$$
$$P_{\text{BPA3-ARL}} = \left(P_{\text{FXP-add}}^{(n_{b4})} + n_{bit} \cdot P_{\text{ADC}}\right) + \left(P_{\text{FXP-add}}^{(n_{b4})} + P_{\text{FXP-mul}}^{(n_{b4},n_{bit})} + P_{\text{FXP-div}}^{(n_{b4}+n_{bit})}\right)$$
$$P_{\text{BPA3-MEM}} = n_{CBy} \cdot n_{CBz} \cdot n_{bit} \cdot P_{\text{CID}}$$

$$P_{\text{ARL}} = P_{\text{BPA1-ARL}} + P_{\text{BPA2-ARL}} + P_{\text{BPA3-ARL}} + P_{\text{BPA4-ARL}} + P_{\text{BPA6-ARL}}$$

$$P_{\text{MEM}} = P_{\text{SRAM}}^{(n_{membits})} + P_{\text{BPA2-MEM}} + P_{\text{BPA3-MEM}}$$

4.9.4 Mixed-Signal CMOL Architecture

For mixed-signal CMOL architecture, each element of the CPT, and external data have (n_{bit} =) 8-bit representation. The mixed-signal CMOL-based SVMM implements only 4.2 and 4.3 of the BPA; for 4.1, 4.4, and 4.6, use the corresponding performance/price equations from FXP architecture.

4.9.4.1 MS CMOL Nanogrid for the SVMM

The following equations for CMOL nanogrid modeling are adapted from [3] and [58] (more details can be found there).

$$D = 18/(8\ \text{nm}/F_{\text{nano}})^2$$

$$R_{off} = 4000R_{on}$$

$$R_{con} = \rho / F_{\text{nano}}^2$$

$$L = 2(N+M)F_{\text{nano}}$$

$$R_{wire} = L\rho_0(1+(\ell / F_{\text{nano}}))/F_{\text{nano}}^2$$

$$\tau = (2R_{con} + 1.5R_{wire} + (R_{on}/D)) \bullet C_{wire}$$

$$P_{on} = \alpha\beta NMV^2 / (2R_{con} + R_{wire} + (R_{on}/D))$$

$$P_{leak} = \alpha(1-\beta)NMV^2 / (2R_{con} + R_{wire} + (R_{off}/D))$$

$$P_{dyn} = \alpha\gamma(N+M)C_{wire}V^2/\tau$$

$$P_{grid} = P_{on} + P_{leak} + P_{dyn}$$

$$A_{grid} = 4NMF_{nano}^2$$

Where, $\rho_0 = 2\mu$ cm, $l = 10$ nm, $C_{wire}/L = 0.2\text{fF}\mu\text{m}^{-1}$ [9]; $\rho = 10^{-8}\ \Omega\,\text{cm}^2$ [3]; $R_{on} = 100\text{M}\Omega$, $V = 0.3$ V, $\alpha = 0.5(1/M)$, $\beta = \gamma = 0.5$ [58]; and for our implementation $N = n_{CBy}$ for 4.2, $N = n_{CBz}$ for 4.3, and $M = n_{bit}$.

4.9.4.2 Time

$$t_{\text{BPA2}} = \left(n_{CBz} \cdot n_{bit} \cdot (\tau + n_{bit} \cdot t_{\text{FXP-add}}^{(n_{b3})} + t_{\text{ADC}})\right) + \ldots$$
$$\left(n_{CBz} \cdot (t_{\text{FXP-add}}^{(n_{b3})} + t_{\text{FXP-mul}}^{(n_{b3}, n_{bit})} + t_{\text{FXP-div}}^{(n_{b3}+n_{bit})} + t_{\text{CMOL-MEM}})\right)$$

$$t_{\mathrm{BPA3}} = \left(n_{CBy} \cdot n_{bit} \cdot (\tau + n_{bit} \cdot t_{\mathrm{FXP\text{-}add}}^{(n_{b4})} + t_{\mathrm{ADC}})\right) + \ldots$$
$$\left(n_{CBy} \cdot (t_{\mathrm{FXP\text{-}add}}^{(n_{b4})} + t_{\mathrm{FXP\text{-}mul}}^{(n_{b4}, n_{bit})} + t_{\mathrm{FXP\text{-}div}}^{(n_{b4}+n_{bit})} + t_{\mathrm{CMOL\text{-}MEM}})\right)$$

$$t_{\mathrm{final}} = \max\left((t_{\mathrm{BPA1}} + t_{\mathrm{BPA2}} + t_{\mathrm{BPA4}}), (t_{\mathrm{BPA3}} + t_{\mathrm{BPA6}})\right)$$

4.9.4.3 Area

$$A_{\mathrm{BPA2\text{-}ARL}} = \left(A_{\mathrm{FXP\text{-}add}}^{(n_{b3})} + n_{bit} \cdot A_{\mathrm{ADC}}\right) + \left(A_{\mathrm{FXP\text{-}add}}^{(n_{b3})} + A_{\mathrm{FXP\text{-}mul}}^{(n_{b3}, n_{bit})} + A_{\mathrm{FXP\text{-}div}}^{(n_{b3}+n_{bit})}\right)$$

$$A_{\mathrm{BPA2\text{-}MEM}} = n_{CBz} \cdot A_{grid}$$

$$A_{\mathrm{BPA3\text{-}ARL}} = \left(A_{\mathrm{FXP\text{-}add}}^{(n_{b4})} + n_{bit} \cdot A_{\mathrm{ADC}}\right) + \left(A_{\mathrm{FXP\text{-}add}}^{(n_{b4})} + A_{\mathrm{FXP\text{-}mul}}^{(n_{b4}, n_{bit})} + A_{\mathrm{FXP\text{-}div}}^{(n_{b4}+n_{bit})}\right)$$

$$A_{\mathrm{BPA3\text{-}MEM}} = n_{CBy} \cdot A_{grid}$$

$$n_{membits} = n_{bit} \cdot (4.5 n_{CBy} + n_c \cdot n_{CBy} + 1.5 n_{CBz}) + \ldots$$
$$\left((n_c + 1) \cdot n_c \cdot n_{bit} \cdot n_{CBy} + n_{b3} \cdot n_{CBz} + n_{b4} \cdot n_{CBy} + 2 n_{bit} \cdot n_{CBy}\right)$$

$$A_{\mathrm{ARL}} = A_{\mathrm{BPA1\text{-}ARL}} + A_{\mathrm{BPA2\text{-}ARL}} + A_{\mathrm{BPA3\text{-}ARL}} + A_{\mathrm{BPA4\text{-}ARL}} + A_{\mathrm{BPA6\text{-}ARL}}$$

$$A_{\mathrm{MEM}} = A_{\mathrm{CMOL\text{-}MEM}}^{(n_{membits})} + A_{\mathrm{BPA2\text{-}MEM}} + A_{\mathrm{BPA3\text{-}MEM}}$$

4.9.4.4 Power

$$P_{\mathrm{BPA2\text{-}ARL}} = \left(P_{\mathrm{FXP\text{-}add}}^{(n_{b3})} + n_{bit} \cdot P_{\mathrm{ADC}}\right) + \left(P_{\mathrm{FXP\text{-}add}}^{(n_{b3})} + P_{\mathrm{FXP\text{-}mul}}^{(n_{b3}, n_{bit})} + P_{\mathrm{FXP\text{-}div}}^{(n_{b3}+n_{bit})}\right)$$

$$P_{\mathrm{BPA2\text{-}MEM}} = n_{CBz} \cdot P_{grid}$$

$$P_{\mathrm{BPA3\text{-}ARL}} = \left(P_{\mathrm{FXP\text{-}add}}^{(n_{b4})} + n_{bit} \cdot P_{\mathrm{ADC}}\right) + \left(P_{\mathrm{FXP\text{-}add}}^{(n_{b4})} + P_{\mathrm{FXP\text{-}mul}}^{(n_{b4}, n_{bit})} + P_{\mathrm{FXP\text{-}div}}^{(n_{b4}+n_{bit})}\right)$$

$$P_{\mathrm{BPA3\text{-}MEM}} = n_{CBy} \cdot P_{grid}$$

$$P_{\mathrm{ARL}} = P_{\mathrm{BPA1\text{-}ARL}} + P_{\mathrm{BPA2\text{-}ARL}} + P_{\mathrm{BPA3\text{-}ARL}} + P_{\mathrm{BPA4\text{-}ARL}} + P_{\mathrm{BPA6\text{-}ARL}}$$

$$P_{\mathrm{MEM}} = P_{\mathrm{CMOL\text{-}MEM}}^{(n_{membits})} + P_{\mathrm{BPA2\text{-}MEM}} + P_{\mathrm{BPA3\text{-}MEM}}$$

4.9.5 *Example: Use of Architecture Assessment Methodology for Associative Memory Model*

We briefly discuss the use of our architecture assessment methodology for implementing an associative memory model (for further details on the model/implementation, refer to [3], [54]). Each step in Fig. 4.2 is briefly summarized below:

Step-1: Assume the (Palm and Willshaw) associative memory model [3].

Steps-2 and 3: Assume that both the input (*X*) and output (*Y*) vectors are binary, and a fixed number of active elements (or '1's), *l* and *k* respectively. Assume that the weight matrix is trained using the summation-rule (instead of the simple OR-rule), and hence consists of multi-bit weight values. During recall, input vector X is applied to the network; each intermediate output element is obtained as (inner-product) $\tilde{y}_i = \sum_j w_{ij} x_j$; then a global threshold (θ) is selected such that exactly k number of output elements are '1', i.e. $\forall_i$, set $y_i = 1$ if $\tilde{y}_i - \theta > 0$, else set $y_i = 0$.

Step-4: Major equations/operations during recall are: the inner-product for $\tilde{y}_i$, and the k-winner take all (k-WTA) to get y_i.

Step-5: The inner-product consists of multiplication and addition type computations, and the k-WTA consists of comparison type computations. For the inner-product, multiplication is simple, because input X is binary, hence a multi-input AND-gate can be substituted for a multiplier.

Steps-6 and 7: To accomplish the above computations (and storage) various circuit components (CMOS/Nano, digital/analog) can be used, hence leading to different design configurations. Some of these proposed configurations are: digital CMOS, digital CMOL, mixed-signal (MS) CMOS, and MS CMOL. (There are other possible configurations, but they not covered here. In addition, the MS designs proposed here are different from [3].)

Step-8: For a worst case analysis, assume that the weight matrix is full density (not sparse), and that the performance is estimated using sequential operations and/or worst case timing paths.

Step-9: In digital CMOS design, the weight matrix is stored as SRAM/eDRAM, the inner-product uses digital components, and k-WTA uses digital k-WTA circuit component [3]. Digital CMOL design is the same as digital CMOS, except that SRAM is replaced with CMOL-nanogrid based digital memory. In MS CMOS design, the weight matrix can be stored in an analog floating-gate (FG-cell) array [59]; the k-WTA is done using analog k-WTA circuit [3]. In MS CMOL design, the weight matrix can be stored on 1R nanodevices (or nano-memristors [60]) in a nanogrid; the k-WTA is done using analog k-WTA circuit [3]. In both MS CMOS/CMOL designs, for the inner-product, the multiply occurs in each FG-cell/nanodevice, and the addition occurs along the vertical nanowires as (analog) summation of charge/currents [3]. The size (and performance/price) of digital and analog components depend on the size of the input and output vectors, and the number of active '1's, etc., as selected by the user. The size of the CMOL nanogrids also depends indirectly on the size of the input and output vectors.

Step-10: A system-level block diagram needs to be generated for each design, using the constituent circuit components discussed earlier.

Step-11: The performance/price measures of all digital/analog CMOS circuit components can be derived from (example) published sources. For new components such as the CMOL nanogrid structures, the performance/price measures can be modeled using fundamental area/power equations along with Elmore-delay analysis [3]. These measures have to be scaled to the appropriate/desired CMOS/CMOL technology node using VLSI scaling rules [45].

Step-12: The final performance/price for the complete design (for each particular design configuration) needs to be evaluated considering the system-level block diagram, using the individual performance/price measures of the circuit components, and based on the computational requirements of the operations (i.e. according to the size of the input/output vectors, etc.).

Acknowledgment Useful discussions with many colleagues, including Prof. K.K. Likharev, Dr. Changjian Gao, and Prof. G.G. Lendaris are gratefully acknowledged.

References

1. M.S. Zaveri, D. Hammerstrom, CMOL/CMOS implementations of Bayesian polytree inference: digital & mixed-signal architectures and performance/price. IEEE Trans. *Nanotechnology* **9**(2), 194–211 (2010). DOI: 10.1109/TNANO.2009.2028342
2. D. Hammerstrom, M.S. Zaveri, Prospects for building cortex-scale CMOL/CMOS circuits: a design space exploration, in *Proceedings of IEEE Norchip Conference* (Trondheim, Norway, 2009)
3. C. Gao, D. Hammerstrom, Cortical models onto CMOL and CMOS – architectures and performance/price. IEEE Trans Circ. Syst-I **54**, 2502–2515 (2007)
4. S. Borkar, Electronics beyond nano-scale CMOS, in *Proceedings of 43rd Annual ACM/IEEE Design Automation Conf.* (San Francisco, CA, 2006), pp. 807–808
5. R.I. Bahar, D. Hammerstrom, J. Harlow, W.H.J. Jr., C. Lau, D. Marculescu, A. Orailoglu, M. Pedram, Architectures for silicon nanoelectronics and beyond, *IEEE Computer* **40**, 25–33 (2007)
6. D. Hammerstrom, A survey of bio-inspired and other alternative architectures, in *Nanotechnology: Information Technology-II*, ed. by R. Waser, vol. 4 (Wiley-VCH Verlag GmbH: Weinheim, Germany, 2008), pp. 251–285
7. Intel, 60 years of the transistor: 1947–2007, Intel Corp., Hillsboro, OR (2007), http://www.intel.com/technology/timeline.pdf
8. V. Beiu, Grand challenges of nanoelectronics and possible architectural solutions: what do Shannon, von Neumann, Kolmogorov, and Feynman have to do with Moore, in *Proceedings of 37th IEEE International Symposium on Multiple-Valued Logic*, Oslo, Norway, 2007
9. D.B. Strukov, K.K. Likharev, CMOL FPGA: a reconfigurable architecture for hybrid digital circuits with two-terminal nanodevices. Nanotechnology **16**, 888–900 (2005)
10. Ö. Türel, J.H. Lee, X. Ma, K. K. Likharev, Architectures for nanoelectronic implementation of artificial neural networks: new results, *Neurocomputing* **64**, 271–283 (2005)
11. K.K. Likharev, D.V. Strukov, CMOL: devices, circuits, and architectures, in *Introduction to molecular electronics*, ed. by G. Cuniberti, G. Fagas, K. Richter (Springer, Berlin, 2005), pp. 447–478
12. D.B. Strukov, K.K. Likharev, Reconfigurable hybrid CMOS/nanodevice circuits for image processing. IEEE Trans. Nanotechnol. **6**, 696–710 (2007)
13. G. Snider, R. Williams, Nano/CMOS architectures using a field-programmable nanowire interconnect. Nanotechnology **18**, 1–11 (2007)
14. NAE, Reverse-engineer the brain, Grand challenges for engineering (The U.S. National Academy of Engineering (NAE) of The National Academies, Washington, DC, [online], 2008), http://www.engineeringchallenges.org. Accessed 15 February 2008
15. R. Ananthanarayanan, D.S. Modha, Anatomy of a cortical simulator, in *ACM/IEEE Conference on High Performance Networking and Computing: Supercomputing*, Reno, NV, 2007

16. D. George, J. Hawkins, A hierarchical Bayesian model of invariant pattern recognition in the visual cortex, in *Proceedings of International Joint Conference on Neural Networks* (Montreal, Canada, 2005), pp. 1812–1817
17. T.S. Lee, D. Mumford, Hierarchical Bayesian inference in the visual cortex. J. Opt. Soc. Am. A. Opt. Image Sci. Vis. **20**, 1434–1448 (July 2003)
18. T. Dean, Learning invariant features using inertial priors, *Annals of Mathematics and Artificial Intelligence* **47**, 223–250 (2006)
19. G.G. Lendaris, On Systemness and the problem solver: tutorial comments. IEEE Trans. Syst. Man Cy. **16**, 604–610 (1986)
20. M.S. Zaveri, CMOL/CMOS hardware architectures and performance/price for Bayesian memory – The building block of intelligent systems, Ph.D. dissertation, Department of Electrical and Computer Engineering, Portland State University, Portland, OR, October 2009
21. K.L. Rice, T.M. Taha, C.N. Vutsinas, Scaling analysis of a neocortex inspired cognitive model on the Cray XD1, *J. Supercomput.* **47**, 21–43 (2009)
22. D. George, A mathematical canonical cortical circuit model that can help build future-proof parallel architecture, Workshop on Technology Maturity for Adaptive Massively Parallel Computing (Intel Inc., Portland, OR, March 2009), http://www.technologydashboard.com/adaptivecomputing/Presentations/MPAC%20Portland__Dileep.pdf
23. C. Gao, M.S. Zaveri, D. Hammerstrom, CMOS / CMOL architectures for spiking cortical column, in *Proceedings of IEEE World Congress on Computational Intelligence – International Joint Conference on Neural Networks*, Hong Kong, 2008, pp. 2442–2449
24. E. Rechtin, The art of systems architecting, *IEEE Spectrum* **29**, 66–69, 1992
25. D. Hammerstrom, Digital VLSI for neural networks, in *The Handbook of Brain Theory and Neural Networks*, ed. by M.A. Arbib (MIT Press, Cambridge, MA, 1998), pp. 304–309
26. J. Bailey, D. Hammerstrom, Why VLSI implementations of associative VLCNs require connectionmultiplexing, in *Proceedings of IEEE International Conference on Neural Networks* (San Diego, CA, 1988), pp. 173–180
27. J. Schemmel, J. Fieres, K. Meier, Wafer-scale integration of analog neural networks, in *Proc. IEEE World Congress on Computational Intelligence – International Joint Conference on Neural Networks* (Hong Kong, 2008), pp. 431–438
28. K.A. Boahen, Point-to-point connectivity between neuromorphic chips using address events. IEEE Trans. Circ. Syst. II: Anal. Dig. Sig. Process. **47**, 416–434 (2000)
29. D. George, B. Jaros, The HTM learning algorithms (Numenta Inc., Menlo Park, CA, Whitepaper, March 2007), http://www.numenta.com/for-developers/education/Numenta_HTM_Learning_Algos.pdf
30. K.L. Rice, T.M. Taha, and C.N. Vutsinas, Hardware acceleration of image recognition through a visual cortex model, *Optics Laser Tech.* **40**, 795–802 (2008)
31. C.N. Vutsinas, T.M. Taha, K.L. Rice, A neocortex model implementation on reconfigurable logic with streaming memory, in *IEEE International Symposium on Parallel and Distributed Processing* (Miami, FL, 2008), pp. 1–8
32. R.C. O'Reilly, Y. Munakata, J.L. McClelland, *Computational Explorations in Cognitive Neuroscience: Understanding the Mind by Simulating the Brain, 1st edn.* (MIT Press, Cambridge, MA, 2000)
33. J. Hawkins, D. George, Hierarchical temporal memory: Concepts, theory and terminology (Numenta Inc., Menlo Park, CA, Whitepaper, March 2007), http://www.numenta.com/Numenta_HTM_Concepts.pdf
34. J. Hawkins, S. Blakeslee, *On Intelligence* (New York: Times Books, Henry Holt, 2004)
35. D. Hammerstrom, M.S. Zaveri, Bayesian memory, a possible hardware building block for intelligent systems, AAAI Fall Symp. Series on Biologically Inspired Cognitive Architectures (Arlington, VA) (AAAI Press, Menlo Park, CA, TR FS-08-04, Nov. 2008), p. 81
36. J. Pearl, *Probabilistic Reasoning in Intelligent Systems: Networks of Plausible Inference* (Morgan Kaufmann, San Francisco, CA, 1988)
37. R. Genov, G. Cauwenberghs, Charge-mode parallel architecture for vector–matrix multiplication, *IEEE Trans. Circ. Syst.-II* **48**, 930–936 (2001)

38. B. Murmann, Digitally assisted analog circuits, *IEEE Micro* **26**, 38–47 (2006)
39. C. Johansson, A. Lansner, Towards cortex sized artificial neural systems, *Neural Networks* **20**, 48–61 (2007)
40. R. Granger, Brain circuit implementation: high-precision computation from low-precision components, in *Replacement Parts for the Brain*, ed. by T. Berger, D. Glanzman (MIT Press, Cambridge, MA, 2005), pp. 277–294
41. S. Minghua, A. Bermak, An efficient digital VLSI implementation of Gaussian mixture models-based classifier, *IEEE Trans.VLSI Syst.* **14**, 962–974 (2006)
42. D.B. Strukov, K.K. Likharev, Defect-tolerant architectures for nanoelectronic crossbar memories, *J. Nanosci. Nanotechnol.* **7**, 151–167 (2007)
43. K.K. Likharev, D.B. Strukov, Prospects for the development of digital CMOL circuits, in *Proceedings of International Symposium on Nanoscale Architectures* (San Jose, CA, 2007), pp. 109–116
44. J.M. Rabaey, *Digital Integrated Circuits: A Design Perspective* (Prentice Hall, Upper Saddle River, NJ, 1996)
45. N. Weste, D. Harris, *CMOS VLSI Design - A Circuits and Systems Perspective, 3rd edn.* (Addison Wesley/Pearson, Boston, MA, 2004)
46. M. Haselman, M. Beauchamp, A. Wood, S. Hauck, K. Underwood, K.S. Hemmert, A comparison of floating point and logarithmic number systems for FPGAs, in *13th Annual IEEE Symposium on Field-Programmable Custom Computing Machines* (Napa, CA, 2005), pp. 181–190
47. K.K. Parhi, *VLSI Digital Signal Processing Systems: Design and Implementation* (Wiley, New York, 1999)
48. K. Seungchul, L. Yongjoo, J. Wookyeong, L. Yongsurk, Low cost floating point arithmetic unit design, in *Proceedings of IEEE Asia-Pacific Conference on ASIC* (Taipei, Taiwan, 2002), pp. 217–220
49. D.M. Lewis, 114 MFLOPS logarithmic number system arithmetic unit for DSP applications, *IEEE J. Solid-St. Circ.* **30**, 1547–1553 (1995)
50. P.C. Yu, H.-S. Lee, A 2.5-V, 12-b, 5-MSample/s pipelined CMOS ADC, *IEEE J. Solid-St. Circ.* **31**, 1854–1861 (1996)
51. T.E. Williams, M.A. Horowitz, A Zero-overhead self-timed 160-ns 54-b CMOS divider. IEEE J. Solid-St. Circ. **26**, 1651–1662 (1991)
52. G. Gielen, R. Rutenbar, S. Borkar, R. Brodersen, J.-H. Chern, E. Naviasky, D. Saias, C. Sodini, Tomorrow's analog: just dead or just different? in *43rd ACM/IEEE Design Automation Conference* (San Francisco, CA, 2006), pp. 709–710
53. J.N. Coleman, E.I. Chester, A 32-Bit logarithmic arithmetic unit and its performance compared to floating-point, in *Proceedings of 14th IEEE Symposium on Computer Arithmetic* (Adelaide, Australia, 1994), pp. 142–151
54. C. Gao, Hardware architectures and implementations for associative memories – The building blocks of hierarchically distributed memories, Ph.D. dissertation, Department of Electrical and Computer Engineering, Portland State University, Portland, OR, Nov 2008
55. P. Narayanan, T. Wang, M. Leuchtenburg, C.A. Moritz, Comparison of analog and digital nanosystems: Issues for the nano-architect, in *Proc. 2nd IEEE International Nanoelectronics Conference* (Shanghai, China, 2008), pp. 1003–1008
56. D. George, J. Hawkins, Belief propagation and wiring length optimization as organizing principles for cortical microcircuits (Numenta Inc., Menlo Park, CA, 2007), http://www.stanford.edu/~dil/invariance/Download/CorticalCircuits.pdf
57. K.K. Likharev, Hybrid CMOS/nanoelectronic circuits: opportunities and challenges. *J. Nanoelectron. Optoelectron.* **3**, 203–230 (2008)
58. C. Gao, D. Hammerstrom, CMOL based cortical models, in *Emerging brain-inspired nano-architectures*, ed. by V. Beiu and U. Rückert (Singapore: World Scientific, 2008 in press)
59. M. Holler, S. Tam, H. Castro, R. Benson, An electrically trainable artificial neural network (ETANN) with 10240 "floating gate" synapses, in *International Joint Conference on Neural Networks* (San Diego, CA, 1989), pp. 191–196
60. S.H. Jo, K.-H. Kim, W. Lu, Programmable resistance switching in nanoscale two-terminal devices. *Nano Lett.* **9**, 496–500 (2009)

Chapter 5
A Hybrid CMOS-Nano FPGA Based on Majority Logic: From Devices to Architecture

Garrett S. Rose and Harika Manem

Abstract From the final years of the twentieth century, nanotechnology has emerged in various forms with the promise of enabling new applications among a variety of disciplines. The field of digital integrated circuit design is one such discipline within which researchers are continually seeking ways of leveraging novel nanoscale technologies to develop next generation circuits and architectures. A major motivating factor for this research is the expected end of device scaling for more conventional bulk silicon technologies, specifically CMOS. Many nanoscale devices have been proposed as replacements for the MOS transistor, from spintronic devices to molecular switches, each coming with their own pros and cons. In fact, it can be argued that, for many applications, nanoscale CMOS remains as viable as any other nanoelectronic device family, at least in the near term. Thus, this chapter explores the concept of hybrid CMOS-nano circuit design for leveraging the best of scaled CMOS alongside of the best of novel nanoelectronics. This is accomplished by describing some novel nanoelectronic devices and comparing them to the traditional MOSFET. After some discussion about circuit level considerations when integrating CMOS with nanoelectronics, a hybrid CMOS-nano field programmable gate array (FPGA) based on nanoscale hysteretic switches and negative differential resistance (NDR) is also described and explored in detail.

Keywords VLSI • Nanoelectronics • FPGA • Majority logic

5.1 Introduction

The infant field of molecular scale nanoelectronics is often defined as including any technology whose device feature sizes are on the scale of single molecules. With many devices in existence today, a great interest from the VLSI design community has

---

G.S. Rose (✉) and H. Manem
Department of Electrical and Computer Engineering, Polytechnic Institute of New York University, Five MetroTech Center, Brooklyn, NY 11201
e-mail: grose@poly.edu; hmanem01@poly.edu

K. Iniewski (ed.), *CMOS Processors and Memories*,
Analog Circuits and Signal Processing, DOI 10.1007/978-90-481-9216-8_5,

emerged where several researchers are considering how such nanoelectronic devices can be used as eventual replacements or at least extensions of existing integrated circuit technologies (i.e., CMOS). From the perspective of VLSI circuit design, experiments have shown how such devices could be fabricated with useful properties such as rectification, hysteresis and negative differential resistance (NDR) [1–6]. In addition to the many device level experiments performed, researchers have also made great strides toward realizing electronic circuits based on these technologies [4, 7–11].

For the most part, the various nanoscale architectures explored have been based on crossbar arrays of several nanowires with active switches at the crosspoints. Nanoscale crossbar arrays have been shown to be useful for both digital logic and memory [3, 4, 7, 10, 11]. This chapter will describe some of the opportunities and challenges associated with the nanoscale crossbar array. Furthermore, as many novel nanoelectronic technologies continue to emerge it is also apparent that CMOS technology is likely to play an important role in integrated circuits for some time to come. One of the several reasons CMOS isn't going away anytime soon is the large investment already made by the IC industry itself. Thus, it seems reasonable to assume that any new nanotechnology should be CMOS compatible before being considered commercially viable. This observation has led several in the field to consider what are known as hybrid CMOS-nano architectures, composed of both conventional CMOS circuitry as well as novel nanoelectronics [11–14]. Research has demonstrated the potential for CMOS-nano systems for memory, reconfigurable logic, and even neural networks [15, 16]. Early results suggest that CMOS-nano systems may not only provide a transition from CMOS to something new but that the hybrid systems themselves tend to be more efficient that either pure CMOS of even pure nanoelectronics. It is this best of both worlds approach to hybrid CMOS-nano circuit design that is explored here and used to develop the systems presented.

Later in this chapter we describe one possible design, the programmable majority logic array (PMLA), for implementing logic at the nano-scale based on nanoelectronic device characteristics – specifically hysteresis (conductance switching with memory) and negative differential resistance [12–14]. Furthermore, the particular design is extended into a hybrid CMOS-nano design for a field programmable gate array (FPGA) consisting of several nano PMLA units integrated with more conventional CMOS logic. Given the hybrid nature of this PMLA based FPGA, the circuit design itself must be partitioned such that any design can take advantage of the best of each technology, i.e., CMOS and nanoelectronics. Furthermore, the CMOS logic is also designed to be reconfigurable such that any design mapped to the FPGA must also be partitioned into components to be mapped in the nanoscale logic and those to be mapped in CMOS.

The programmable nanoelectronic architecture discussed here, the PMLA, utilizes the programmable nature of nanoscale switches, just as in most crossbar array designs. Specifically, this circuit consists of an array of nanoelectronic switches driving an NDR based circuit known as the Goto pair, which implements majority logic – a particular class of threshold logic [12, 17]. Since the switches are programmable, the resulting system can be considered a programmable majority logic array. The use of this Goto pair in conjunction with nanoscale crossbar arrays has also been considered elsewhere, but the proposed use of the Goto pair in other designs is only for signal restoration in a typical PLA (AND-OR, NOR-NOR, or NAND-NAND) [18].

The originality of the PMLA is in the realization that the Goto pair not only provides signal restoration but also leads to majority logic functionality.

This work takes the previously designed PMLA a step further by considering circuit-level issues that exist when such circuits are integrated with more conventional CMOS. We will show a system consisting of tiles of nano PMLA units tied together through a CMOS backplane, which consists of configurable circuits in the form of lookup tables. Thus, this overall hybrid CMOS-nano system implements an FPGA built from NDR based majority logic arrays [12, 13]. Thus, the majority logic PMLA based FPGA differs from most other systems which tend to focus on more traditional logical constructs such as NOR-NOR or NAND-NAND PLA units [7]. Another important property of the CMOS-nano PMLA based FPGA studied in this work is that each nanoscale PMLA consists of multiple logic levels between each interface to and from CMOS. By limiting the flow of data between the two technology layers, the additional load at the interfaces is reduced allowing for more robust performance. This is a distinctive feature when compared to some other systems which require an interface between CMOS and nanoscale devices at each stage of logic.

5.2 Nanoscale Technologies and Devices

In this section we describe some common devices and fundamental circuits employed in hybrid CMOS/nano architectures. More specifically, this section begins with a discussion of the basic operation of nanoscale hysteretic switches. Switching behavior has been observed for a range of technologies from molecular electronics to phase change devices. In the context of this chapter, hysteretic switching refers to the ability of a nanoelectronic device to switch from one conductivity state to another and remain in that state, even when powered down, until a switching condition is met. Nanoelectronic switches have proven useful in the design and implementation of nanoscale memory and logic. In addition to switching behavior, another useful property observed for some nanoscale devices is negative differential resistance (NDR). As described in these pages, hysteretic switching and NDR can be integrated together to develop robust nanoelectronic systems.

5.2.1 Nanoscale Switches and the Crossbar Array

The illustration in Fig. 5.1 shows a generic I-V curve observed for many two terminal nanoelectronic switching devices. From this I-V curve, it is apparent that the device can be set to conduct in one of two possible states with either high or low conductivity. For small voltages, the device remains in one of the two possible states until the voltage applied exceeds some toggle point. From the figure, the positive toggle voltage would be around the knee of the low conductivity curve in the upper-right quadrant while the negative toggle point is defined by the peak in the lower-left

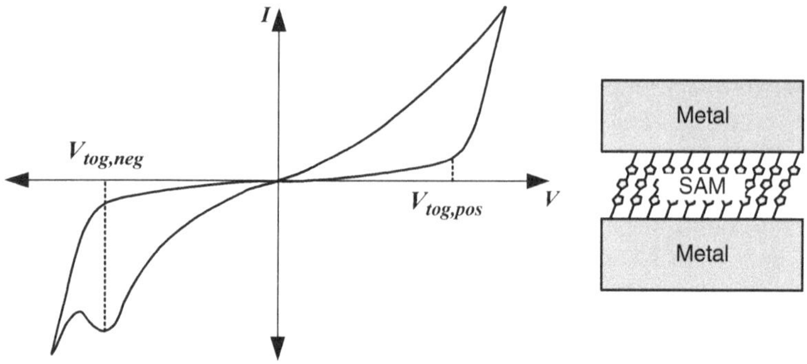

Fig. 5.1 Current versus voltage characteristic of a typical molecular hysteretic switch. These characteristics are common among SAM based devices (*shown to the right*)

quadrant [14, 19]. If the voltage applied across the device is greater than the positive toggle point, the device switches into the high conductivity state and will switch back into the low conductivity state only when the voltage becomes more negative than the negative toggle voltage. In the simulations detailed in this paper, these switching devices are modeled simply as tunable resistors that can be programmed into one of two possible resistivity states. Recent results have demonstrated how many of the same nanoscale structures discussed here can also be used to fabricate memristors which can be made to operate in any one of several possible conductivity states [20, 21].

5.2.1.1 Nanoscale Switches

Most nanoscale memory and logic circuitry devised so far are based on the property of hysteretic switching [7, 10, 12]. More specifically, non-volatile switches operating in two or more conductivity states have been developed and utilized to form these nanoelectronic circuits. Several technologies ranging from metal oxide switches to molecular switches have exhibited such switching behavior and are discussed in brief here [22].

Self Assembled Molecular Electronics

Self assembly and self alignment in molecular electronics are very interesting concepts that have been played upon in the implementation of molecular scale devices. One class of these devices are based on self-assembled monolayers (SAM) of organic molecules that are sandwiched between two conducting layers [2, 19]. More specifically, a SAM is a layer of organic molecules each with one end that has an affinity for attaching to the surface of a given substrate (e.g. Au, Cu, Si, etc.). As these molecules attach to the surface, they eventually push one another into a standing position until they all self-align like trees in a forest. This is a desired property as it can facilitate a robust bottom-up fabrication approach. As the top-down processes are reaching their limits such bottom-up techniques as this are very endearing to future fabrication processes.

Molecular devices based on hysteretic switching have been popular in several nanoscale implementations of both memory and programmable logic [3, 4]. They can be seen simply as switchable fuses in that they can be switched between two or more conductivity states. Figure 5.1 shows a generic I-V curve for a typical molecular switch alongside an illustration of the device structure. If the power is off, the switch remains in its present conductivity state, but when a voltage greater than its toggle voltage is applied the device is forced from one conductivity state to another. If a voltage greater than its positive toggle voltage is applied the device switches to the higher conductivity state and vice versa for a voltage less than the negative toggle voltage [12, 19].

Phase Change Devices

Phase-Change Memory (PCM) has been an area of considerable interest in the recent past as a potential next-generation non-volatile solid state memory technology [23–25]. It possesses many attributes such as high resistance contrast, a small number of new process integration steps, potential for multilevel storage, and better endurance and write speeds than flash memory that make it appealing for usage in high density memory system designs [26].

A phase-change bridge (PCB) devised by IBM and partners was implemented with a promising phase-change material, GeSb. The phase-change bridge (PCB) device consists of a narrow line of ultra-thin phase-change material bridging two underlying electrodes. The left side of Fig. 5.2 shows a potential integration scheme, in which one of the two device cell electrodes connects to the drain of the underlying access transistor, while the other is connected by a via up to an overlying bitline. Here the electrodes are formed very close together to obtain a reasonable threshold voltage, separated by a small oxide gap that defines the bridge length *L*. The thickness of the phase-change material deposited on this planarized surface defines the bridge height *H*, and the width *W* is defined with a subsequent patterning step. The I-V characteristics for the PCB device are as shown in the right of Figure 5.2. The red solid line indicates the “OFF”, i.e. lower conductivity state of the device and the blue dotted line shows the “ON”, i.e. higher conductivity state of the device. Although these devices provide a good ON-OFF ratio, their scalability to desired nano scales is still questionable.

Generic MRAM Device

Another class of devices that have gained interest in the field of nano electronics are spintronics based switches. The left side of Fig. 5.3 shows a spintronics based magnetic tunnel junction (MTJ) device for magnetoresistive random access memory

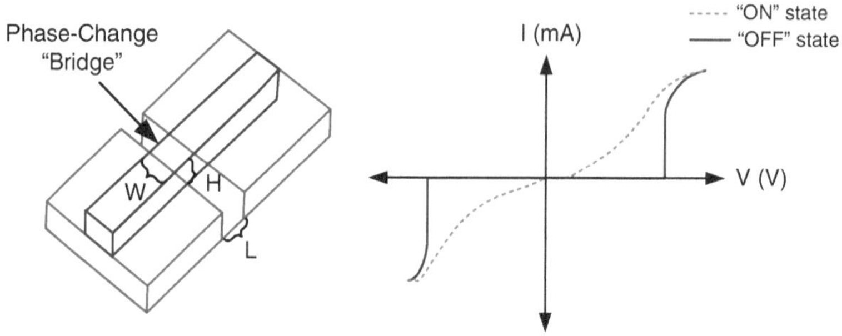

Fig. 5.2 Integration scheme for a phase-change bridge (PCB) device. The cross-sectional area ($W \times H$) depends linearly on lithographic patterning, offering both reduced sensitivity to variations in critical dimension and an alternative path for scaling to future technology nodes via ultra-thin films [26]. Current versus voltage characteristic of a typical PCB device

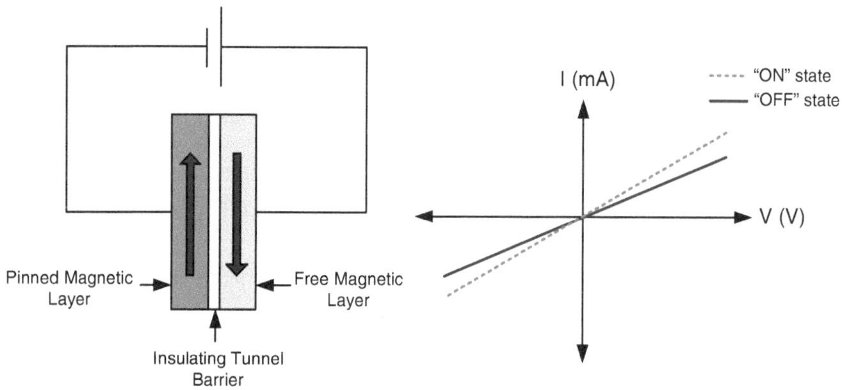

Fig. 5.3 A magnetic tunnel junction (MTJ) consisting of two layers such as cobalt-iron, separated by an ultrathin layer of insulator, typically aluminum oxide with a thickness of 1 nm. The insulating layer is so thin that the electrons can tunnel through the barrier if bias voltage is applied between the two electrodes. Current vs. voltage characteristic of a typical magnetic tunnel junction (MTJ) device for magnetoresistive random access memory (MRAM) systems

(MRAM) systems. As shown in the figure, the MTJ consists of two magnetic metal layers separated by an ultrathin insulating layer. The MTJ is composed of a pinned magnetic layer, a tunnel barrier, and a free magnetic layer. Electrons spin polarized by the magnetic layers traverse the tunnel barrier. The tunneling current depends on the relative orientation of magnetizations of the two ferromagnetic layers, which can be changed by the applied magnetic field. This phenomenon is called tunneling magneto resistance (TMR). A parallel alignment of the free layer with respect to the pinned layer results in a low resistance state i.e. "ON", while an antiparallel alignment results in a high resistance state i.e. "OFF".

Studies have shown that the MTJs in a 4-Mb MRAM device [27] are slightly non-linear over an applied voltage range, however these non-linearities are typically small, approximately 5 kΩ/V, and both states shift their impedances proportionally (i.e. if the on-impedance increases, the off-impedance correspondingly increases) [28]. The device impedances of 15 and 18 kΩ/V were measured for "ON" state and "OFF" state respectively [27]. The major constraint for these switching devices is their ON-OFF ratio. As can be observed from the I-V curve in Fig. 5.3, the slopes of the high and low conductivity curves are not very different. Therefore, the size of PMLA circuits would be greatly restricted if these devices were to be utilized.

Metal Oxide Device

Binary metal oxides have simple structures and are compatible with CMOS processing [29]. There has been significant progress in the development of Metal-Insulator-Metal (MIM) memory based on near-stoichiometric cuprous oxide (Cu_xO) and models have been devised to explain the switching mechanism and performance. Within a solid a location that restricts the movement of electrons or holes is termed as a trap. It consists of either a chemical impurity or an imperfection in the spacing of the atoms that constitute the solid. In a solid material with unfilled deep-level traps (i.e. traps that have an energy level near the centre of the band gap, making them neither donors nor acceptors), Space-Charge-Limited-Conduction (SCLC) current is significantly lowered from the trap-free case [29]. A dramatic current increase (and resistance reduction) occurs when the deep traps are filled at the traps-filled-limit voltage (VTFL) that is determined by the unfilled deep trap density. This process leads to the material switching from a high-resistance state ("OFF") into a low-resistance state ("ON"). Both "OFF" and "ON" states are described by SCLC model. Deep traps lower the "OFF" state current, while the "ON" state current approaches the trap-free limit conduction as the deep traps get filled. The "ON" state retention is determined by the "thermal release time" (detrapping through thermal processes). Long retention is expected on materials with the appropriate density of deep level traps. Figure 5.4 shows the I-V characteristics of a typical metal oxide switching device.

Drawing a comparison between these nanoscale switches reveals that metal oxide and phase change devices have higher driving currents and ON-OFF ratio when compared to their counterparts. This "ON-OFF" ratio is of major importance as the size of the circuitry that can be implemented is determined by it, as revealed in the next chapter. The only concern is compatibility with the rest of the circuitry and scaling. In the PMLA design molecular switching devices were chosen for their power and area economy; but the design can be modified in the future to utilize devices with better characteristics for improved circuit scaling and superior CMOS interface.

5.2.1.2 Resonant Tunneling Diodes

Another device required for the PMLA is based on the property of NDR that typically results from electron tunneling through an energy barrier [1, 2]. These barriers

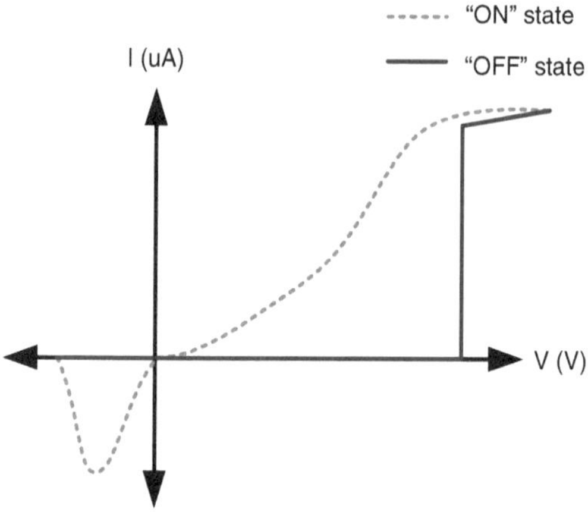

Fig. 5.4 Current vs. voltage characteristic of a typical metal oxide switching device model based on near-stoichiometric cuprous oxide (Cu_xO)

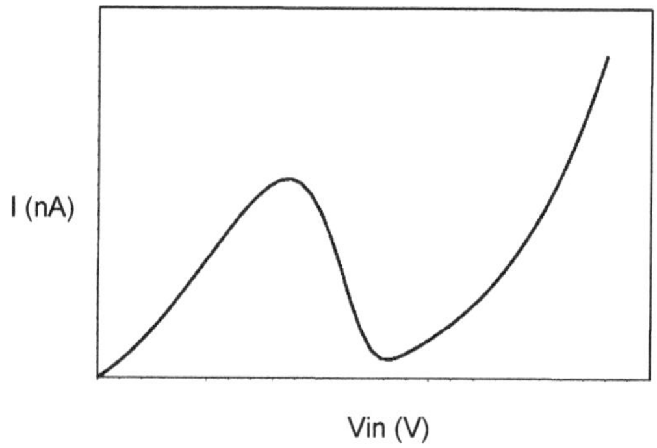

Fig. 5.5 Current versus voltage characteristic of a resonant tunneling diode

can be a consequence of the separation of a conducting island and the source and drain of the device by some insulating material. The electron has a higher tunneling probability when the energy of an electron at the source of the device is that of a vacant energy level inside the island. A bias potential can be applied across the device to modify the electron energy level to be within the range of energies for the source conduction band. This causes the device to alternate between high and low current densities with the variation in voltage. These devices are referred to as resonant tunneling diodes (RTD). An example I-V characteristic of such a device is shown in Fig. 5.5. From the perspective of circuit design, such devices are interesting due to the region of negative resistance (decreasing current with increasing voltage) that enables signal restoration at the nanoscale [17].

5.2.2 Fundamental Circuits

As described in the previous section, there are several contenders for nanoscale device technologies that can be utilized in nanoelectronic circuitry. Some fundamental nanoelectronic circuits that can be constructed from these devices are described in this section.

5.2.2.1 Crossbar Array

A nanoscale crossbar array consists of two sets of nanowires running perpendicular to one another [7]. At the intersection of each set of two perpendicular nanowires is a hysteretic switch, such as a molecular electronic switch. Many nanoscale circuits devised thus far have employed this array scheme for both logic and memory [3, 7, 10, 11]. Figure 5.6 shows a molecular crossbar array. Furthermore, the switches at each crosspoint of the crossbar can be considered tunable resistors that can be tuned to one of two possible conductivity states, as described in Section 5.2.1. This architecture is attractive for several reasons including its simplicity in design. One major drawback with such a scheme, in and of itself, is the lack of inverting functionality and signal restoration.

5.2.2.2 Programmable Logic Array (PLA)

A brief description of array logic is presented to provide a greater understanding of the PMLA. In the past PLAs were popularly used for implementing combinational logic. They are less frequently used now as modern logic synthesis often produces faster circuits. However, they have been resurrected for implementation in nanoscale structures for their high logic density and ease of usage.

Any logic function can be expressed in sum-of-products form, where each output is composed of the OR (sum) and the ANDs (products) of the inputs [30]. The PLA

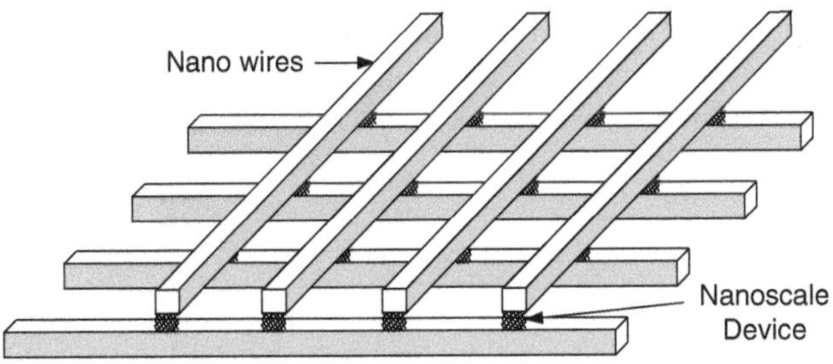

Fig. 5.6 The nanoscale crossbar array composed of nanowires crisscrossing perpendicularly, with SAM based molecular devices at each crosspoint

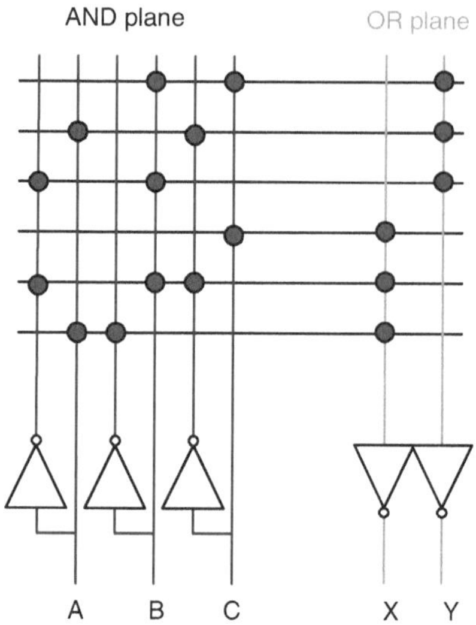

Fig. 5.7 A dot representation of the programmable logic array (PLA)

consists of an AND plane to compute the minterms and an OR plane to compute the outputs. Figure 5.7 shows a basic representation of the PLA. In the PMLA a nanowire crossbar with molecular switches at their crosspoints drive Goto pairs to implement array based majority logic. Thus the PMLA architecture is a form of PLA.

5.2.2.3 Goto Pair

As can be seen in Fig. 5.8, a Goto pair is a circuit with two RTDs (load and drive) stacked in series between a clock and ground. This circuit can be used to latch incoming data and provide signal restoration [12, 17]. Figure 5.8 also shows the load line diagram for the Goto pair circuit. From the figure it is apparent that the circuit has three operating points, two stable and one unstable. The two stable operating points, Vlo and Vhi in the figure, are used to represent logical '0' and logical '1'. The unstable operating point occurs at the intersection of the negative resistance regions for the two RTDs. When both the RTDs are biased in the NDR region the resistances will both be negative and the circuit is unstable. The output in this case will then fall to either Vlo or Vhi.

$$Id = Il + Iin \tag{5.1}$$

$$Vd = Vout = Vclk - Vl \tag{5.2}$$

Equations 5.1 and 5.2 represent the currents and voltages through the Goto pair, where d and l are subscripts denoting drive and load respectively. To ensure that the circuit operates in a bistable state the high voltage of the clock must be greater than $2V_p$.

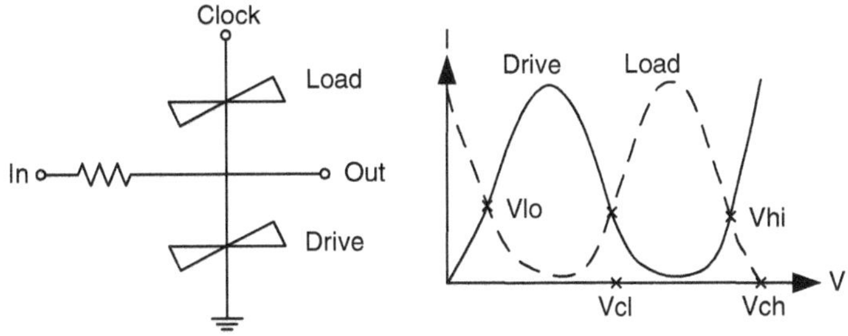

Fig. 5.8 Schematic of a Goto pair circuit (*left*) along with its load line diagram (*right*)

The other criterion for meeting this requirement is for the low voltage of the clock to be higher than V_p. V_{ch} and V_{cl} represent the high and low voltages of the clock respectively.

$$\frac{1}{2}Vch = Vcl > Vp, \text{where } Vch > 2Vp \tag{5.3}$$

Goto pairs are appealing as they can provide signal restoration and are very fast [17]. In fact, Goto pairs can bypass CMOS most of the time and can provide several levels of logic within the nano layer itself. The Goto pair can serve as both a majority gate and a latch. If at a given instant only one molecular switch drives the input, the Goto pair serves as a latch for any signal through that input. If several switches at the input are "ON" and drive the Goto pair then it functions as a majority logic gate to those inputs.

$$M(A,B,C) = AB + BC + CA \tag{5.4}$$

The above expression represents the output of a majority gate. From this expression or intuitively it can be observed that by forcing one of the inputs to either VLO or VHI, AND-OR logic can be implemented. If we were to set one of the inputs high we essentially have an OR gate and vice versa for an AND gate. This illustrates that logic can be realized akin to a PLA in nanoelectronics circuits.

In the three input majority gate shown in Fig. 5.9 it can be observed that by forcing one of the inputs to either V_{LO} or V_{HI}, AND-OR logic can be implemented. If we were to set one of the inputs high we essentially have an OR gate and vice versa for an AND gate.

5.2.3 *Device Modelling*

The devices considered for the nanoelectronic circuit considered here have been modeled using the universal device model (UDM) for circuit simulation [31].

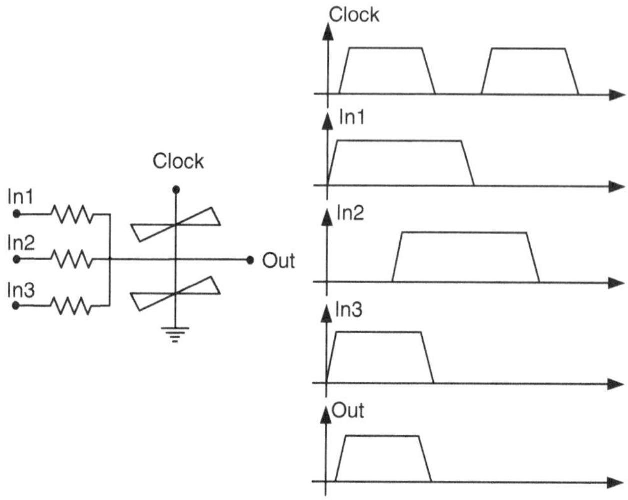

Fig. 5.9 Goto pair functioning as a three-input majority gate

This particular modeling approach allows for an accurate representation of the devices themselves as each model is developed empirically from experimental data. Thus, an important point concerning the approach presented here is that circuits are simulated using a fine-grained approach where the behavioral details of each individual device is explicit. In this way the low-level circuit simulations performed provide a sound representation of what is to be expected once such circuits are actually fabricated. From this perspective, this work is useful not only for circuit designers and architects but also device engineers working toward commercially viable products. It is in this way that we are able to determine a range of device behaviors that lead to acceptable and even optimal performance of the circuit.

In the analyses of the CMOS-nano FPGA, this work has taken a bottom-up approach where low-level circuit simulations using Cadence Spectre are performed on circuits built from a universal device model (UDM). The UDM is a empirical modeling approach based on fitting device data to allow for accurate modeling of devices observed experimentally [31]. In this way, the analyses are based on the effects of actual nanoscale devices. Using the UDM, the particular properties of the nanoelectronic switches are considered from both a pessimistic and optimistic perspective in terms of the magnitude of the effective impedance of the devices themselves. Furthermore, the effects of how logic is partitioned between the nano PMLA and CMOS are also considered. Given a particular function to map to the FPGA, such as a multiplier, logic can be implemented entirely in the nano PMLA, mostly in CMOS with the nano PMLA used as a "helper", or evenly between the nano PMLA and the CMOS circuitry included. Thus, we consider how different partitioning schemes affect the overall performance of the hybrid CMOS-nano PMLA based FPGA.

5.3 The Programmable Majority Logic Array

The programmable majority logic array (PMLA) is a combination of the functionality of both nanoelectronic switches and RTDs. It is the amalgamation of crossbar array and Goto pair circuits, built with the ability to implement logic as a network of latches and majority gates (Fig. 5.10). The voltage inputs are converted to currents as they flow through the switches which behave as input resistors to the Goto pairs at the outputs of the crossbar arrays. The original motivation for using Goto pair latches at the outputs of a crossbar array was to provide signal restoration at the nanoscale [8]. Though the use of Goto pairs does provide signal restoration the PLA is also altered such that it is based on majority logic as opposed to the usual sum-of-products form. Specifically, in the case where three switches in the row are in the 'on' state, the input currents are added up at the input of the respective Goto pair and the output value latched is the majority of the three inputs.

The majority function has been consistently used over time for the implementation of Boolean algebra. Also known as the median operator, the majority gate realizes a function from "n" inputs to one output. It returns a true or "1" if and only if at least (50% + 1) of its inputs are true. To ensure this condition is satisfied the inputs are usually odd in number. As mentioned in the previously in Section 5.2.2.3, Eq. 5.4 represents the output of a three-input majority gate. Majority gates may be scaled up to any number of inputs depending on the robustness of the devices utilized to implement it. For the case studies presented here, the input number is limited to three although the results of recent work has demonstrated the use of a similar system with Goto pairs driven by more than three inputs [32].

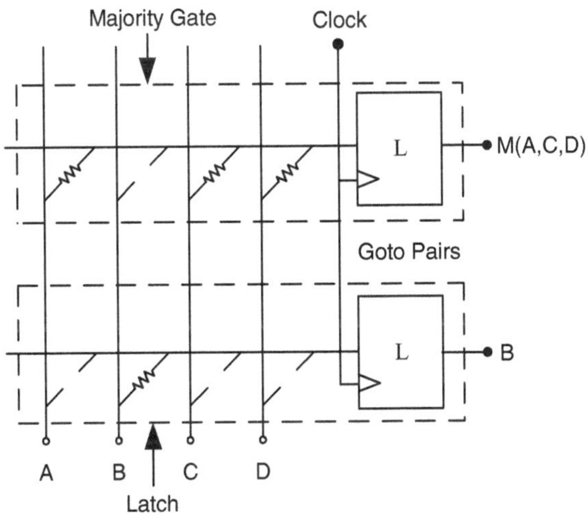

Fig. 5.10 Molecular switches can be used as input resistors to a Goto pair. *Top row* shows the Goto pair as a three-input majority gate, *bottom row* shows it functioning as a latch

Majority based logic is not usually explored with standard CMOS technologies, essentially due to the hardware inefficiencies in creating majority gates. However, due to recent innovations in nanoelectronics and molecular sciences as discussed earlier in this chapter, devices have been devised that are able to make the realization of majority logic conceivable. Due to their ease of design and implementation, majority gates have hence been greatly desired in the implementation of nanoscale circuitry. Beyond the PMLA, architectures such as quantum cellular automata and the NanoFabric are also based on majority logic circuit design. The PMLA is another breed of nanoscale circuitry that utilizes majority logic for the implementation of functions.

The layout for a PMLA with eight inputs and eight outputs can be seen in Fig. 5.11. It can be discerned from the figure that the output from one logic level can feed another level of logic by setting the molecular switches at the appropriate cross points into the 'on' state. This style of architecture is a version of whirlpool PLA logic where the outputs from one stage flow into the next stage in a circular fashion [12, 33].

Besides these inputs and outputs, the architecture also requires three clock signals, as shown in Fig. 5.12. It is very important that we implement the clocks in a suitable manner due to the potential for data indirection in the Goto pairs. This occurs because the Goto pair uses the same node for input as it does for output. Thus, backdriving could occur where the output switches the state of the circuit if we do not take care to prevent indirection. To avoid this, the clocks are

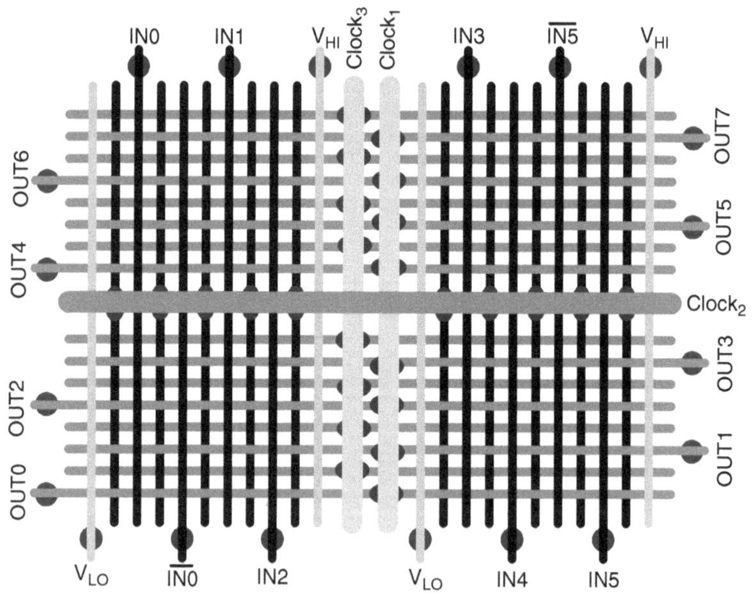

Fig. 5.11 Layout of a PMLA. The *ovals* between the clock lines and nanowires represent the stacked NDR devices making up the Goto pairs. At each junction between nanowires is a programmable molecular switch. The *circles* represent contacts from the nano circuitry to a CMOS layer

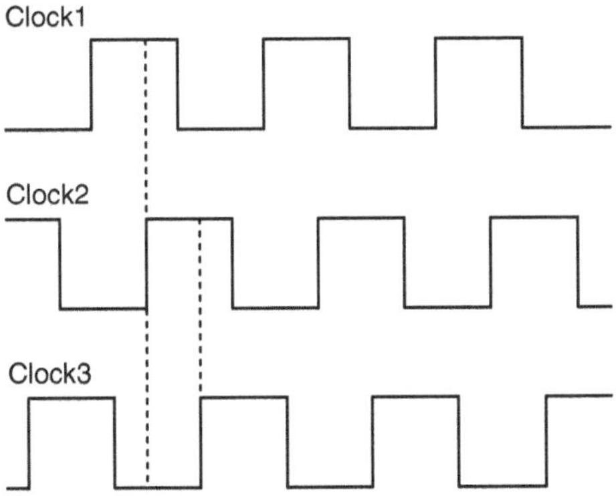

Fig. 5.12 Overlapping clocks employed for the PMLA

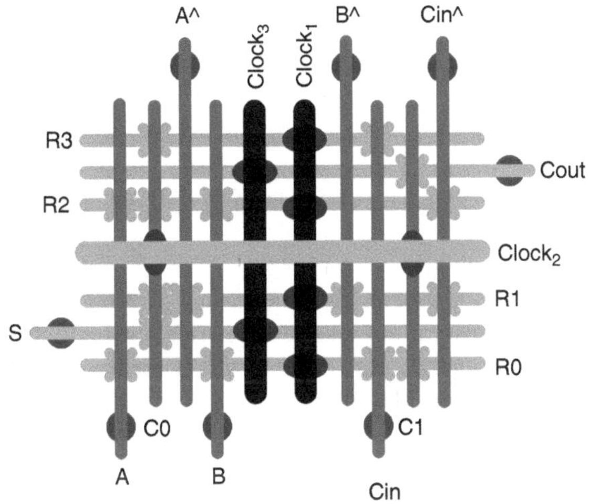

Fig. 5.13 Physical layout of a PMLA mapped to function as a full adder

chosen in such a way that they are out of phase but overlapping [12, 13]. Since a Goto pair from a particular stage is only able to drive a high output when its respective clock is high, the clock edge should arrive when data is valid only on the stage driving the input and not the output. Considering the clocking scheme in Fig. 5.12, a Goto pair controlled by Clock2 samples only on its rising edge which only occurs when data is valid on the Goto pair driven block Clock1 and not Clock3. Thus, data is forced to flow only from stage 1 to stage 2 and from stage 2 to stage 3.

Figure 5.13 shows an example of a PMLA programmed to be a full adder. There are three levels of logic implemented in the PMLA for the full adder. As the outputs of one logic level feed into the inputs of another, these cascaded arrays can be used to implement multiple levels of logic. As can be observed from the figure, the third

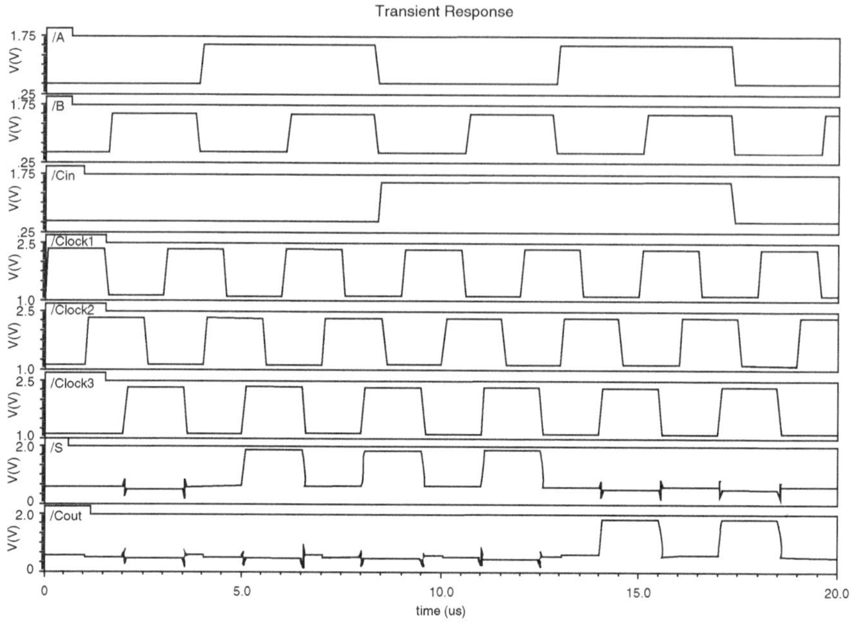

Fig. 5.14 Simulation results from Cadence Spectre of the PMLA programmed to function as a full adder

array is connected to the input of the first, thus forming a whirlpool PLA [33]. In the PMLA different quadrants can be reused to provide for multiple levels of logic where the signals "ping-pong" around the quadrant for the three levels of logic. Figure 5.13 shows the physical layout of the PMLA as a full adder with the results of the circuit shown in Fig. 5.13. As can be seen from the results shown in Fig. 5.14 the output is observed on Clock3.

5.4 A CMOS-Nano PMLA Based FPGA

In conventional nanoscale circuit and system designs, nanowires are generally used for interconnect [34]. In our design, we utilize the CMOS layer to serve as the interconnect between the nanoscale logic blocks. This is due to the fact that based on experimental results, nano arrays (e.g. PMLA) are somewhat rigid in that the time to program them may be longer than that of CMOS. Another parameter that influences the size and scalability of a PMLA is the "on-off" ratio of hysteretic switching devices which directly impacts the maximum size of the array. Too large an array provides for a larger number of sneak paths in parallel that could corrupt the signal output from the Goto pair [35].Therefore, the switch block for this FPGA architecture is in the CMOS layer interconnecting banks of appropriately sized PMLAs, as illustrated in Fig. 5.15. A high level illustration of the hybrid FPGA

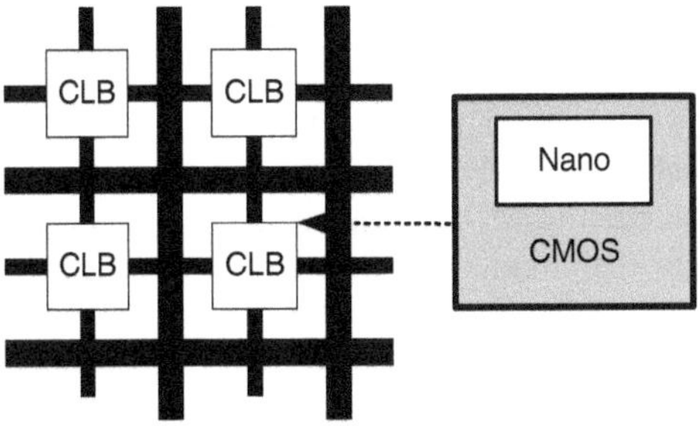

Fig. 5.15 High level illustration of the hybrid FPGA architecture with CMOS interconnection network connecting hybrid CMOS-Nano CLBs

architecture is shown in Fig. 5.15. It consists of hybrid CMOS-Nano configurable logic blocks (CLBs) interconnected by a CMOS network. The CLB has both CMOS and nano circuitry with distributed logic to optimize various parameters such as area, time and power, with respect to the application.

5.4.1 Partitioning Logic Between CMOS and Nano

A logic function can be realized in the proposed FPGA structure in one of two methods: (1) the entire logic mapped onto nano PMLAs with the CMOS layer functioning as just the interconnect, and (2) an alternative mapping that minimizes logic and parasitic delays by mapping logic onto both CMOS and nano layers (the CMOS layer also acts as the interconnect). For the purpose of studying these two options, an array multiplier was mapped onto the FPGA in both methods for the purpose of comparison.

5.4.1.1 All Nano PMLA Mapping

As a simple case study, a 3 × 3 array multiplier was mapped entirely onto nano PMLAs, with the CMOS used exclusively for interconnect and for inverting signals. The entire array multiplier can be mapped onto two PMLAs. This circuit is very conservative in area and power but not very fast when simulating with models based on pessimistic experimental results. The logic delay in obtaining an accurate output is 10 clock cycles and the single stage delay is 234.7 ps, where the single stage delay is the time delay from when the clock transitions to when the output is obtained for a particular Goto pair. The maximum power consumption for the entire circuit is 405.02 nW and the average power required for the computation is 276.18 nW. It should be noted that the final output is obtained after 10 clock cycles or an execution time of 2.347 ns for a clock period of 234.7 ps.

5.4.1.2 Equal Partitioning Between CMOS and Nano

An alternative to the all nano mapping is to partition the circuit onto both CMOS and nano layers equally. The CMOS-Nano mapping is analyzed with a 4 × 4 array multiplier mapped onto both CMOS and nano layers with the aid of a controller that controls the inputs to the circuits and determines when the final output is to be sampled at the last stage (ripple adder). After the signals are sent through the first two rows of nano and CMOS logic the two LSBs are sampled and stored by the controller. The remaining four outputs are then looped to the first nano layer for another stage of computation, i.e., the two LSBs of B are multiplied during the first stage and the last two bits during the second stage. The major advantage of this mapping technique is in the number of clock cycles it takes to obtain the output. The final output can be sampled after four cycles, which is less than half the time required to obtain the output when the entire design is mapped to the PMLA. One interesting observation is that there is an increase in the single stage delay, 4.276 ns (for a nominal Ron value of 40 MΩ), as the nano circuits are driving CMOS gates with higher capacitances than those of the PMLA nanowires. The major tradeoff in this scheme is the increase in power consumption as the CMOS layer is more power hungry. The maximum power consumed is 593.68 μW while the average power is 435.04 μW.

5.4.2 Results and Comparisons

Table 5.1 shows a detailed comparison between various metrics for both mappings for nominal device parameters. It should be noted that for simple applications the utilization of the CMOS layer for logic might result in worse system metrics. This is due to the large load at the interface between the CMOS and nano layers. The large load capacitance coupled with the large resistance of the nanoscale switch could result in considerable single stage delay. This was tested for varying device parameters considering both optimistic and pessimistic device metrics. Single stage delay in the PMLA was measured for increasing Ron values for two technology nodes of 22 nm (with a load of C = 250 aF) and 45 nm (with a load of C = 1 fF). The different Ron values can be attributed to several nanoscale switching devices that exhibit hysteretic switching ranging from metal oxide devices to molecular devices, in increasing order of resistance [22]. Figure 5.16 shows a graph comparing

Table 5.1 Comparison between various metrics for both the mappings

Metrics	Entirely nano mapping	Nano and CMOS mapping
No: of clock cycles	10	4
Single stage delay	0.2347 ns	4.276 ns
Maximum power	405.02 nW	593.68 μW
Average power	276.18 nW	435.04 μW

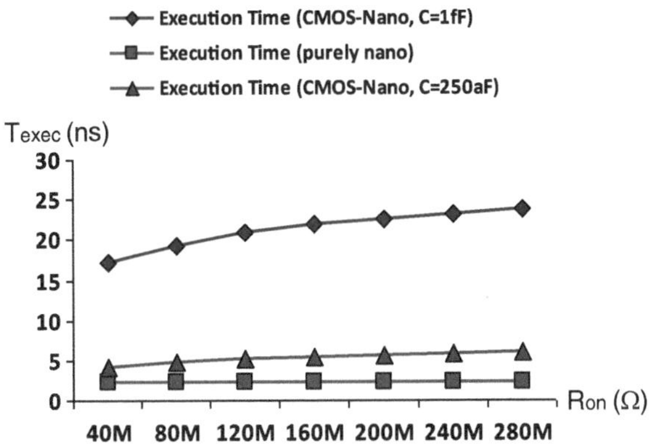

Fig. 5.16 Execution times (ns) versus Ron (Ω) for different mapping schemes

the execution times of the multiplier for the different mapping schemes and for the two technology nodes for the CMOS-nano mapping. The higher curves show the execution times for CMOS-nano mappings for loads 1 fF (45 nm node) and 250 aF (22 nm node) respectively, and the lowest delay curve shows the execution time for a purely nano mapping. Here the execution time is the product of the number of clock cycles and the minimum clock period required (twice the single stage delay, which increases linearly with the load and the Ron value). It can be observed that the execution time increases with increasing Ron values for the CMOS-nano mappings whereas the purely nano mapping has an almost steady execution time irrespective of the varying Ron value. Therefore, considering the application being implemented on a particular CLB it can be determined whether the CMOS layer should be utilized exclusively for interconnect and I/O and only limited logic mapping.

5.4.3 Impact: Considerations for Designing CMOS/Nano Circuits

The results above certainly suggest CMOS should be used sparingly for a hybrid CMOS/Nano mapping based on the PMLA. For the particular example considered, a simple 3-bit multiplier, this makes sense as the overall system is small. In fact, to implement the multiplier a PMLA with only 52 inputs and 52 outputs is required which means the nanowires in this circuit are short. These short nanowires lead to a small single stage load capacitance, and hence delay, in comparison to the capacitive load from CMOS logic. Of course, for more complex logic a larger PMLA will be required meaning the nanowires and the associated single stage delay will not be so small. Furthermore, more complex designs will require more clock cycles for the PMLA which will add to the total execution time of the system, especially for a pure nano mapping. Thus, it is reasonable to expect there to be a set of designs that benefit more in terms of delay when mapped to both CMOS and the nanoscale PMLA.

Based on these observations a hybrid CMOS/Nano design is certainly beneficial as it strikes a nice balance between the best of CMOS and the best of nanoelectronics. For the PMLA, a system design consisting of some CMOS logic can be beneficial in terms of latency and overall execution time. Furthermore, compared to driving off-chip circuits, deep sub-micron CMOS provides a smaller load for a nanoscale circuit such as the PMLA. Thus, CMOS logic should be included in any system consisting of nanoelectronic circuits to provide a soft I/O interface, if nothing else.

5.5 Future Prospects (Memristors)

The memristor, or "memory resistor," was first theorized by Leon Chua at the University of California-Berkeley in 1971 [36]. Only theoretical for more than 30 years, researchers at Hewlett-Packard recently announced the discovery of memristors fabricated in their lab [37, 38]. In terms of its behavior, a memristor is a device whose resistance changes under give toggle conditions (e.g., exceeding some voltage) and then holds that resistance until another toggle condition is met. In this way, memristors can be thought of as reconfigurable resistors with memory. However, given the nature at which Chua arrived at this particular switching property, relating charge (q) and flux linkage (ϕ), the memristor is a new fundamental electronic device in the same way resistors, capacitors and inductors are fundamental.

A natural extension to the aforementioned PMLA is the replacement of nanoscale two state hysteretic switches with multilevel memristors in the implementation of an NDR based threshold logic gate. A threshold logic gate is a gate that takes the weighted sum of several inputs and measures the result against a given threshold value. If the weighted sum of all inputs is greater than the threshold then the output is a logic 1 and is a logic 0 otherwise. Threshold logic actually provides the foundation for most artificial neural network designs. Since a memristor can operate with any one of a set of possible conductivity states, a memristive input to a circuit like the Goto pair acts as a weight for the respective input. As described in Section 5.2.2.3, the Goto pair by its very nature latches a logic 1 or a logic 0 based on the sum of the input currents at the node between the two RTDs. Thus, the combination of memristors and RTDs naturally leads to threshold logic gates and even artificial neural networks where the memristors act as synapses and the Goto pair acts as a neuron.

Considering the extension to threshold logic provided by the use of memristors, the PMLA becomes a more generic programmable threshold logic array. Work is already underway to better understand the benefits and limitations of such a threshold logic array including simple demonstrations for applications such as image processing [32]. Given that these circuits can implement threshold logic entirely at the nanoscale a natural next step is to further develop the overall CMOS/Nano system such that learning is made possible in order to realize PMLA based neural networks.

5.6 Summary

This chapter provides a detailed overview of several flavors of emerging technologies and how they can be leveraged in the realization of hybrid CMOS-nano logic architectures. In particular, this work describes a hybrid FPGA structure built from PMLAs that aims to incorporate the best of both emerging nanoscale technologies and nanoscale CMOS technology for an area efficient architecture. Results provided can aid in choosing an appropriate mapping scheme to either yield better performance or lower power consumption. It needs to be mentioned that the devices considered for these analyses are pessimistic in their approach. They provide the minimum conditions required for circuits considered. They can be further scaled to provide higher currents and thereby provide for better time utilization. Also, most simulators are not very accommodating of NDR based devices and hence do not provide flexibility in design. Future work should also include developing and using simulating tools better suited to NDR to analyze the complete potential of such an FPGA architecture. That being said, it is widely acknowledged that NDR based devices are very fast and thereby circuits incorporating these devices are very conservative with respect to time. Hence, this hybrid FPGA structure shows great promise to be a very efficient design, with respect to area, time and power.

Also mentioned in this chapter are future trends in logic and memory design made possible by recent discoveries in emerging nanoscale devices. The memristor is one such device that has made possible the design of multilevel memory and logic and especially neural networks in the nanoscale. This has opened up a plethora of options in the area of high logic and memory density system design. As the realization of fully nano systems is still currently not feasible, hybrid CMOS-nano systems are an appealing alternative for the implementation of such systems in the near future.

References

1. M.A. Reed, Molecular-scale electronics. Proc. IEEE **87**(4), 652–658 (April 1999).
2. J. Chen, W. Wang, M.A. Reed, A.M. Rawlett, D.W. Price, J.M. Tour, Room-temperature negative differential resistance in nanoscale molecular junctions. Appl. Phys. Lett. **77**(8), 1224–1226 (August 2000).
3. P.J. Kuekes, J.R. Heath, R.S.Williams, Molecular wire crossbar memory, U.S. Patent # 6 128 214, 3 Oct 2000.
4. Y. Chen, G. Jung, D.A.A. Ohlberg, X. Li, D.R. Stewart, J.O. Jeppesen, K.A. Nielsen, J.F. Stoddart, R.S. Williams, Nanoscale molecular-switch crossbar circuits. Nanotechnology **14**, 462–468 (2000).
5. A.R. Pease, J.O. Jeppesen, J.F. Stoddart, Y. Luo, C.P. Collier, J.R. Heath, Switching devices based on interlocked molecules. Accounts Chem. Res. **34**(6), 433–444 (2001).
6. C.P. Collier, J.O. Jeppesen, Y. Luo, J. Perkins, E.W. Wong, J.R. Heath, J.F. Stoddart, Molecular-based electronically switchable tunnel junction devices. J. Am. Chem. Soc. **123**(50), 12632–12641 (2001).
7. A. DeHon, Array-based architecture for FET-based, nanoscale electronics. IEEE Trans. Nanotechnol. **2**(1), 23–32 (2003).

8. S.C. Goldstein, M. Budiu, Nanofabrics: Spatial computing using molecular nanoelectronics, in *Proceedings of 28th Annual International Symposium on Computer Architecture*, Göteberg, Sweden, June 2001, pp. 178–189.
9. Y. Luo, C.P. Collier, J.O. Jeppesen, K.A. Nielson, E. Delonno, G. Ho, J. Perkins, H. Tseng, T. Yamamoto, J.F. Stoddart, J.R. Heath, Two-Dimensional molecular electronics circuits, *Chem. Phys. Chem.* **3**(6), 519–525 (2002).
10. M.R. Stan, P.D. Franzon, S.C. Goldstein, J.C. Lach, M.M. Ziegler, Molecular electronics: From devices and interconnect to circuits and architecture. Proc. IEEE **91**(11), 1940–1957 (2003).
11. D.B. Strukov, K.K. Likharev, CMOL FPGA: a reconfigurable architecture for hybrid digital circuits with two-terminal nanodevices. Nanotechnology **16**(6), 888–900 (2005).
12. G.S. Rose, M.R. Stan, Jr., A programmable majority logic array using molecular scale electronics, *IEEE Trans. Circuits Syst-I* **54**(11) (2007).
13. H. Manem, P.C. Paliwoda, G.S. Rose, A hybrid CMOS/nano FPGA architecture built from programmable majority logic arrays, in *Proceedings of the ACM Great Labes Symposium on VLSI*, Orlando, FL, USA, May 2008.
14. M.R. Stan, G.S. Rose, M.M. Ziegler, "Hybrid CMOS/molecular electronic circuits," in *Proceedings of Joint 19th International Conference VLSI Design 5th International Conference on Embedded System Design*, Hyderabad, India, pp. 703–708, January 2006.
15. O. Turel, K. Likharev, Crossnets: possible neuromorphic networks based on nanoscale components. Int. J. Circ. Theor Appl. **31**, 37–53 (2003).
16. K.K. Likharev, A. Mayr, I. Muckra, O. Turel, CrossNets: high-performance neuromorphic architectures for CMOL circuits. Ann. NY Acad. Sci. **1006**(1), 146–163 (2003).
17. E. Goto, K. Murata, K. Nakazawa, K. Nakagawa, T. Moto-Oka, Y. Ishibashi, T. Soma, E. Wada, Esaki diode high-speed logical circuits. IRE Trans. Elect. Comp. **EC-9**, 25–29 (1960).
18. S.C. Goldstein, D. Rosewater, Digital logic using molecular electronics, in *Proceedings of International Solid-State Circuits Conference*, San Francisco, CA, USA, February 2002, vol. 1, pp. 204–459.
19. N. Gergel, N. Majumdar, K. Keyvanfar, N. Swami, L.R. Harriott, J.C. Bean, G. Pattanaik, G. Zangari, Y. Yao, J.M. Tour, Study of room temperature molecular memory observed from a nanowell device, *J. Vac. Sci. Technol.* **23**(4), 880–885 (2005).
20. L.O. Chua, Memristor-the missing circuit element. IEEE Tran. Circ. Th. **5**, 507–519 (1971).
21. D.B. Strukov, G.S. Snider, D.R. Stewart, S.R. Williams, The missing memristor found. Nature **453**, 80–83 (2008).
22. A.C. Cabe, G.S. Rose, M.R. Stan, Data encoding Parasitic current paths in molecular memory, *in Proceedings of the IEEE Conference on Nanotechnology*, Hong Kong, China, August 2007, pp. 70–75.
23. S. Lai, T. Lowrey, Characteristics os Si-Sb-Te films for phase change memory, *IEDM Tech. Digest*, 803–806 (2001).
24. M. Lankhorst, B. Ketelaars, R. Wolters, Phase-change materials towards a universal memory, *Nature Mater.* **4**(4), 347 (2005).
25. T.D. Happ, M. Breitwisch, A. Schrott et. al., Novel one-mask self-heating pillar phase change memory, *Symp. VLSI Tech.*, 120–121 (2006).
26. Y.C. Chen, C.T. Rettner, S. Raoux et al., Ultra-thin phase-change bridge memory device using GeSb, in *Proceedings of IEEE Electron Devices Meeting*, San Francisco, CA, USA, December 2006, pp. 1–4.
27. B.N. Engel, J. Akerman, B. Butcher et al., A 4-mb toggle MRAM based on a novel bit and switching method. IEEE Trans. Magnetics **41**(1), 132–136 (January 2005).
28. C.H. Cho, J.H. Ko, D. Kim, A CMOS macro-model for mtj resistor of MRAM cell, *Physica Status Solidi (a)* **8**, 1653–1657 (2004).
29. A. Chen, S. Haddad, Y.C. Wu, T.N. Fang et al., Non-volatile resistive switching for advanced memory applications, in *Proceedings of IEEE Electron Devices Mtg.*, Washington, DC, USA, December 2005, pp. 746–749.
30. N. Weste, D. Harris, CMOS VLSI Design, Addison Wesley, 3rd edn. (2005), pp. 750–756.

31. M.M. Ziegler, G.S. Rose, M.R. Stan, A universal device model for nanoelectronic circuit simulation, in *Proceedings of the 2nd IEEE Conference on Nanotechnology*, Washington, DC, August 2002, pp. 83–88.
32. J. Rajendran, H. Manem, G.S. Rose, NDR based Threshold Logic Fabric with Memristive Synapses, in *Proceeding of the IEEE Conference on Nanotechnology*, Genoa, Italy, July 2009.
33. F. Mo, R.K. Brayton, Whirlpool PLAs: A regular logic structure and their synthesis, in *Proceedings of 2002 IEEE/ACMInternational Conference on Computer Aided Design (ICCAD)*, San Jose, CA, November 2002, pp. 543–550.
34. G.S. Snider, R.S. Williams, Nano/CMOS architectures using a field-programmable nanowire interconnect, *Nanotechnology* **18**(3) (2007).
35. H. Manem, G.S. Rose, The Effect of Device Parameter Variation on Programmable Majority Logic Arrays, in *Proceedings of the IEEE Conference on Nanotechnology*, Arlington, Texas, August 2008.
36. L.O. Chua, Memristor-the missing circuit element. IEEE Tran. Circ. Th. **5**, 507–519 (1971)
37. D.B. Strukov, G.S. Snider, D.R. Stewart, R.S. Williams, The Missing Memristor Found. Nature **453**(7191), 80–83 (May 2008).
38. R.S. Williams, How we found the missing memristor, *IEEE Spectrum* **45**(12), 28–35 (2008).

Part II
Memories

Chapter 6
Memory Systems for Nano-computer

Yong Hoon Kang

Abstract In this chapter, memory systems will rigorously be discussed with the bottom-up approach, starting from device and circuit level up to system and software's memory usage model. The goal of this chapter is to understand the effects of innovative idea in device level on the application level performance. We will cover basic elements of semiconductor technology like the memory cells, transistors, interconnect, and so on. And then, analog and logic circuits will be discussed, where state machines, sense amplifiers, DC generators, charge pump, current driver, etc. would be considered with the logic synthesis and analog design methodologies. Thus new memory systems and new memory hierarchies would be comprehensively discussed with all aspects of device, circuit, system, and software.

Keywords Memory systems • Future memory systems • Memory devices • Memory chip • Memory hierarchy • Memory technologies • DRAM • SRAM • NAND • Flash PRAM • RRAM • STT-MRAM

6.1 Introduction

6.1.1 The Value of a Computer

Why do you need computer? If you can answer this question, you will be able to understand the value of investigating future computer, which is better than the computer today. Looking at history, the elementary automatic machine gave humanity a huge rise in productivity, making plenty of food, clothes, and necessities of life. Now with the aid of computer, it became possible to investigate a new drug, to design more comfortable shelter and transportation, and to forecast weather, and to

Y.H. Kang (✉)
Samsung Electronics Co. Ltd., Hwasung city, Kyungki-Do South Korea
e-mail: yonghoon.kang@samsung.com

K. Iniewski (ed.), *CMOS Processors and Memories*,
Analog Circuits and Signal Processing, DOI 10.1007/978-90-481-9216-8_6,

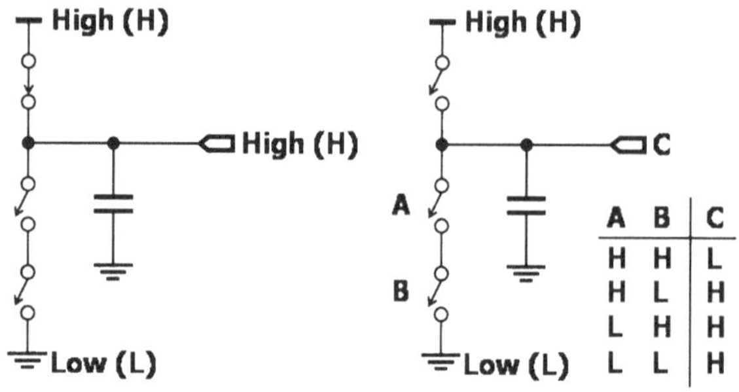

Fig. 6.1 Example on how to implement the NAND gate using generic switches. *Left* figure represents the preparation phase and *right* figure shows the evaluation phase for determining the value of the logic output

make more convenient communication and agricultural activities, and to make more usable life necessities. Altogether it is now possible to live in more safe and rich environment beyond what is needed to survive [1].

6.1.2 The Origin of a Computer Body

Not too different materials compose both of the human body and computer machine. Even it has been veiled of what is the origin of the elementary particles, it is well known that the quarks were gathered to form the nucleus and the atoms could be completed with the electron cloud that is surrounding the nucleus. After the atoms of the high atomic number (Silicon, Carbon) were created from the nuclear fusion in a star, eventually, those atoms are being gathered (after the explosion of the stars) at the low temperature on the surface of a cold planet (the earth), which allows the foundation materials of the human body and computer machine [2]. Within small temperature range from –40°C to 130°C, the chunk of the Silicon atoms reveals the characteristics of a semiconductor according to the quantum mechanical phenomenon [3]. By utilizing the semiconductor, we could step up to the automation of intelligence in the replacement of the logical thinking capability. Interestingly, in a cosmic point of view, the semiconductor and organic machineries can exist only within a very small range of temperature. Here, the semiconductor machinery is simply based on the three making processes of the forming conductor with doping technology and metal interconnect, forming nonconductor with oxidation technology, and forming switches with both previous technologies [4, 5].

An arbitrary logic circuit can be constructed only using switches. Figure 6.1 illustrates how two input NAND gate can be realized with several switches. Similarly, we can obtain the NOR gate using the switches. Since the sum-of-product form enables any Boolean logic function with the composition of the inverter, NAND gate, and NOR gate, in principle, the switches are sufficient to obtain an arbitrary logic circuit [6, 7]. If we use the semiconductor technology to realize the logic circuit in Fig. 6.1, the switches and wires will have very small feature size.

And so, we can take a great gain that the logic circuit will be able to run extremely fast. This is because of the small resistance on the short path and the small input capacitance from the small switch size. (Circuit speed is totally dependent on the electrical node's RC product regardless of the logic circuit or analog circuit.) Surely, we cannot devaluate the gain of the small size itself that leads to the market penetration of the modern small electronics.

6.1.3 The Birth of a Memory Device

Inside computer, logical operations and mathematical calculations are being made constantly. In principle, it is possible that all of the computer operations are implemented with the hardwired circuit. However, that is impossible in reality. It might not be unattainable that the movies or other fixed mediums in a DVD are realized with the hardwired circuit. But, the variable contents of the computer game or most of the user specific applications are too diverse and randomized to be dealt with the hardwired circuits. It is better that such variable contents are stored in the memory device as an information form, and we process the information by using a CPU. Thus, computer can be so flexible to many applications because of the adoption of the memory device. Information is not a physical substance, but an abstract substance. To take the significance of the information, therefore, the physical object to store the information is indispensable. In the case of the binary information in a computer, the entity of the binaries is in the memory device. If we can say that the every moment change of the memory is the core for a man to be thought as human, this is because that the abstract information like experience is chemically stored in the synapses as a physical form [8, 9].

Memory is a device that utilizes "memory effect" originated from the semiconductor physics, which is controlled by the electronic circuits. Analog and logic circuits are all used to accomplish the basic memory cell operation like read and write (and erase) in addition to the advanced operations like the error detection and correction, data buffering, and several security features. Since various memory effects require the high voltage or shaped current for some memory cells, the specific circuits like charge pump, current shaper, and other dedicated circuits are required. The operations of memory cells are being done in pure analog domain, but the function like reading, writing, and erasing must be completed on the predetermined time, respectively. For this purpose, the delay circuit and the state machine are used for a fast memory and a slow memory, respectively, and the output signals from those circuits controls the timings for a memory cell operation. Thus, all of the operations of the memory devices can be separated into two different functions of the core operation (memory cell operation) and the I/O operation (data input/output transactions between memory controller and memory device). Since the minimum timing of the core operations is highly dependent on the device physics of a memory cell, the timings cannot be much improved without the fundamental technology breakthroughs. Instead, the I/O operation speed could be enhanced with the application of the high performance transistors (transistor performance is easily improved by shrinking their feature size). Consequently, overall memory performance could have been increased, because the memory response time is the sum of the core and I/O operation times. In addition, the modular technique with many memory devices or the multi-bank method in a memory device can hide the core operation time by the use of the interleaving technique between devices or banks [10, 11].

6.1.4 The Brief on a Computer System

To managing computing environment by users themselves (to manipulate computer with machine language) is very complicated and time consuming, it is usual that user works on operating system (OS). The user can use simple intuitive commands (abstraction of the machine-level instructions) provided by OS, and such orders are translated to more complex bundle of machine language, which is stored in the OS, and it controls the CPU and all the hardware components connected to it [12, 13]. The user basically wants to run his application programs by their own computer. The application program is the software, made by long codes, and generally stored at hard disk drive (HDD) of PC and server or at NAND Flash memory of mobile electronic devices. By using file system serviced by OS, the user can search all the software inside HDD (or NAND Flash), and by using the relevant command sets of the OS (system call), the user can intuitively manage arbitrary software like the file read and the file write. File read command loads the codes in a file to the memory device and then the meaning of the codes are interpreted by the CPU. As a result, CPU generates the data from its execution or the control signals to reach to the connected hardware (actually overall computer system is tightly under the CPU's control) [14, 15]. In this process, CPU's control signal should be translated to the words of the each hardware, and so the intermediate control devices between CPU and hardware devices should be added. In the case of the memory device, the binaries from CPU are going through several circuit components where the binaries are translated to fit to the input of the memory device. Most important circuit block in this signal path is the memory controller. Memory controller is a logic circuit that is interfacing between CPU and memory device, which is also optimize the performance of a memory device using the algorithms and policies like interleaving, address remapping, command reordering, page modes, reflesh, and so on [16].

6.1.5 The Brief on a Memory Hierarchy

Each optimized memory device described above satisfies the user's requirements about the system performance by utilizing the memory hierarchy. The users want to have as many application programs as possible in their computer, so they needs a memory with large capacity and also needs a memory of fast access to run their application programs quickly. But because of some economical and technical reasons, we do not use one memory device to satisfy both conditions in above, instead, we makeup the memory hierarchy inside computer [17]. Figure 6.2 shows the common memory hierarchy. The capacity is increased (so, the cost per GB is decreased) in the order of the cache (SRAM) in CPU, main memory (DRAM), and storage (HDD). But, the

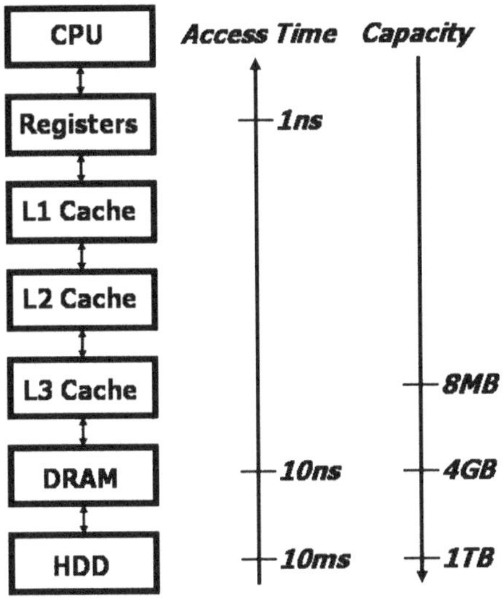

Fig. 6.2 A typical memory hierarchy and the element specific features whose difference is caused by the capacity (bits per unit cost) and access time (performance per unit cost)

access time for a data is also highly grown in that order. The storage of large capacity is used to storing many user application programs, and then a part of the programs are cached into the main memory and caches in CPU to achieve the statistical performance enhancement (relevantly many caching algorithms have been introduced). When user uses computer, the statistical characteristic of the accessed memory data reveals both of the spatial and temporal localities. It is noticeable that the large gaps exist in the capacity and access time between HDD and DRAM as shown in Fig. 6.2. As will be discussed later, to cope with the gaps, new memory hierarchies have been actively explored with the insertion of NAND Flash or PRAM as a storage cache [18, 19].

6.2 Memory Devices and Circuits

6.2.1 The Foundation of a Memory Core

Memory is made up by two-dimensional array of numerous memory cells (Recently there are active investigations about three-dimensional array) [20]. Memory cells are classified into the volatile and nonvolatile types. DRAM and SRAM are examples of a volatile memory and NAND/NOR Flash, STT-MRAM, RRAM, and PRAM are examples of a nonvolatile memory. To access a cell from the two-dimensional array, two lines of x-axis and y-axis should be able to represent an arbitrary cell. In all memories, cell data is always transferred through the bit line (y-axis), and the word line (x-axis) is used to select the cell. The read process of all memory devices is common, so, after settling the predefined voltage to the bit line,

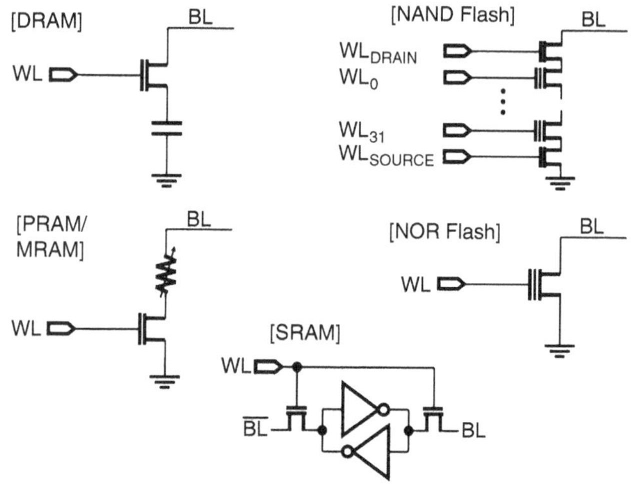

Fig. 6.3 Different types of memory cells and their equivalent circuit models

the bit line is electrically connected with the memory cell. And then the stored bit information of the cell will modulate the bit line voltage (or current). The modulated analog value (voltage or current) of a bit line is converted to the digital value (namely, GND or V_{DD}) by the sense amplifier (SA) circuit. For instance, since the DRAM cell is a capacitor, the charge sharing process can modulate the bit line voltage. And the NAND/NOR Flash cell acts like the switch between the bit line and GND where the bit information of the cell will turn on or off the switch. Therefore if the switch is on, the bit line voltage will be lowered (or the bit line current will be increased). The memory cells of the STT-MRAM, RRAM, and PRAM are utilizing the resistance changes, so, the high and low resistance states will correspond to the different cell data. The read mechanism of those memories is similar with that of the Flash memories. However, when the resistance ratio was compared to that of the transistor on/off, the resistance change is much smaller than that of the Flash memories. More sensitive SA, therefore, should be designed to resolve the bit line's small analog values. In Fig. 6.3, several memory cells are described. Even though the lumped circuit models of the cells look like very simple, state-of-the-art technologies are involved to advance the device shrink and relevant process development. For example, a DRAM cell has evolved to the new structures and dielectric materials to maximize the cell capacitance while shrinking the capacitor size. In addition, special transistors have been deployed to reduce the charge leakage from the capacitor [20]. In the cases of NAND/NOR Flash, new memory cells, dielectric materials, and transistors for the higher voltage endurance, smaller size, and lower interference with adjacent cells have been investigated [20]. SRAM has continuously been progressing with the help of device level improvements, for example, the transistor characteristics like threshold voltage, saturation current, and NMOS-PMOS ratio are controlled to optimize the leakage current, cell (latch) size, p-well to n-well distance, and stable core operations [21]. In the case of the new memories, new materials for variable resistance have been intensively explored. For a PRAM, many researchers have explored to find an optimum ratio of the GST ($Ge_2Sb_2Te_5$) compound material [22].

To complete a memory chip, not only memory cell devices but also MOSFET devices are indispensable for designing the analog and digital circuits. For the MOSFET device, the most important thing is to obtain the highest saturation current at the lowest gate–source voltage while maintaining very low level of leakage current. For this purpose, the strained silicon technology for higher channel mobility [21] and the silicon on insulator (SOI) technology for the minimization of the junction parasitic capacitance have been applied [7, 23]. Both of the metal gate and high-k technologies were introduced to avoid the gate depletion and the gate leakage, respectively, and are already employed to the high performance devices like the microprocessors. In the short channel MOSFET, the pocket doping techniques have been widely used to alleviate the peak electric field at the boundary of drain and channel (it reduces the trapped oxide on the channel, resulting in higher reliability), and the doping has also been useful to avoid the punch-through condition [4, 5, 21].

Once the memory cell is prepared, we have to consider the second important metric that is strongly influencing on the memory devices' performance: a bit line capacitance. Since many memory cells are sharing a bit line and the short distance between adjacent bit lines exposes the parasitic capacitance, a bit line capacitance is appeared from the accumulation of those capacitances. Notably, in many cases, memory cell array can occupy more than half of the memory chip size and the long bit line along with the y-axis of the cell array can yield large parasitic capacitance. Memory devices can have several different bit line architectures depending on the required performance of a memory like speed, cost, number of parallel accessible cells, and power consumption. But, NAND Flash normally reveals the huge parasitic capacitance because the bit line is across whole y-axis of the cell array with minimum interval. Thus, the large bit line capacitance will lag the driving speed and consume more power while developing a bit line voltage. But, more importantly, this memory architecture has been the best fit to the demands for high capacity memories. On the other hands, DRAM and PRAM

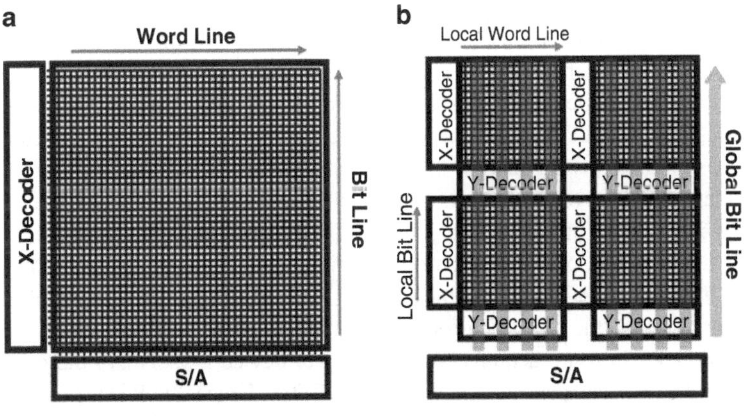

Fig. 6.4 Comparison of two different bit line architectures: Plain bit line and hierarchical bit line structure were chosen for a simple example

have applied two separated bit lines in their architectures, one is a global bit line and another is a local bit line. As a result, we could reduce the parasitic capacitance on a bit line and so the access time (for the read and write) to the memory cells. An example is presented in Fig. 6.4. Global bit line is only connected to the Y-decoders of each memory blocks, so, we can reduce the number of memory cells per a bit line. In addition, the increased pitch between global bit lines and the adoption of the higher metal layer can remove most of the parasitic capacitance. Y-decoder selects a local bit line from an input address, by which a global bit line is connected to a memory cell. For reference, DRAM uses bit line SA located on the Y-decoder to elevate the core performance (in this case, SA in Fig. 6.4 should be renamed to I/O SA).

At previous paragraph, we explained the bit line structures of memory devices. In Fig. 6.4a, the number of SA is either the same as or half of the number of bit lines. The large numbers of memory cells along with x-axis correspond to the large numbers of S/A (8 KB cells exist for a NAND Flash). On the other hands, in Fig. 6.4b, the number of SA is the same as the number of global bit lines, so much smaller number of SA is required to that (in the case of PRAM, the number of global bit lines are typically about 8–16B). On the same silicon process, more sensitive SA needs more silicon area, in opposite, less sensitive SA saves the silicon area. In the case of NAND Flash, such a large number of SA will result in the unsustainably large chip size. To avoid such a non-economical situation, we must use the smaller size SA. As a result, we should achieve more sensing margin (the voltage or current difference between "1" and "0") by increasing the read time of a bit line. In that way, the weak memory cell signal is amplified and becomes sensible with the small size and insensitive SA. That is considered as one of the important metrics for trading-off between read speed and chip size. X-decoder in Fig. 6.4 will be discussed at the next paragraph.

6.2.2 The Foundation of a Memory Design

6.2.2.1 Analog Circuits for a Memory Device

As was stated, to complete a memory chip from the memory cells, the analog and digital circuits are necessary. So the memory cells should be connected electrically to the circuits that control the cells. After drawing the analog and digital circuits, many circuit simulations should be performed to the verification of the designed circuits, and the Spice simulation is the basis of those circuit simulations. It is a computer program to calculate the value of current and voltage at each electrical node inside the circuits, which is based on the numerical analysis of the Kirchhoff's equations with the current–voltage relationships of many electronic devices like bipolar transistor, MOSFET, resistor, and capacitor. To increase the accuracy of the Spice simulation, therefore, it is important to obtain the precise current–voltage relationships of the electronic devices, and the relationships are acquired with the novel device models and the extraction process of the models' parameters. For the MOSFET device, the BSIM3 (or BSIM4) model is most well known and most commonly used nowadays [24].

Now let's takes a look at the circuits needed for the memory devices. Figure 6.5 shows two examples of SA. Here, (a) is a small sized latch type voltage SA, and

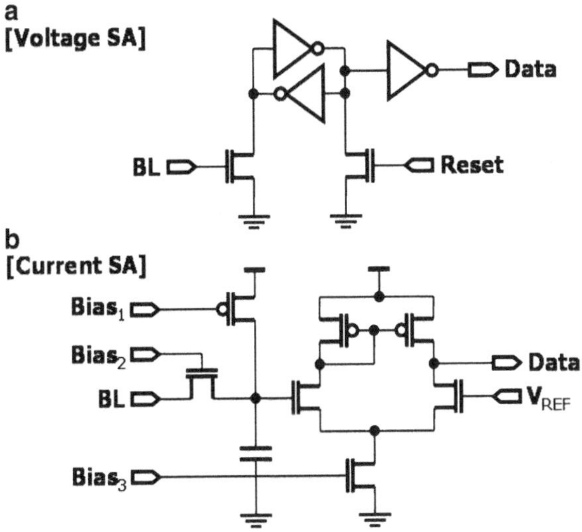

Fig. 6.5 Example of the two major sense amplifier (SA) circuits: The selection criterion is determined by the circuit area and operating speed

(b) is a differential amplifier type current SA. In (b), $Bias_1$ and $Bias_3$ are for the current sources, and $Bias_2$ is a control voltage to regulate the bit line voltage. Latch in a SA will be working as the logic circuit not the analog circuit, so its operation is very robust to the process variations. As a consequence, we could have made a latch type SA with a minimal feature size. However, since many numbers of a SA will bring the large leakage current, we need to control some device parameters related to the threshold voltage like doping density, channel length and width, etc. Differential amplifier is most commonly used as a basic block for a majority of analog circuits like input buffer, comparator, voltage generators, oscillators, amplifiers, and so on. It has a symmetrical shape between two inputs, and so, several layout techniques have been developed to minimize the manufacturing mismatch between two inputs. Where, we optimize the transistors' lengths and widths to pursue the maximum gain at higher frequency. Fundamentally, the operation can intuitively be explained with the simple relationship of $Q = C \cdot V$. When a small + signal is applied to one N-MOSFET gate, the N-MOSFET flows + charge from drain to source (in reality, electron is flowing from source to drain). And then the charges are being piled up at the source node, because the bottom N-MOSFET is a current source that flows the constant current while changing the drain–source voltage (MOSFET device of longer than 1 μm channel length allows that dI_{DS}/dV_{DS} is approximately zero at the saturation mode). Therefore, the piled + charge on the source node will decrease the gate–source voltage of another N-MOSFET as the relationship of $V = Q/C$ (where C is just a parasitic junction capacitance, and smaller C will result in the faster and larger voltage change). At the saturation mode, the channel current of an N-MOSFET is highly sensitive to the gate–source voltage, but the P-MOSFET current mirror will flow the constant current (see Fig. 6.5b). As a result, the channel current of the N-MOSFET of the smaller gate–source voltage is largely decreased and so the drain node voltage of the N-MOSFET will be highly increased. In here, V = Q/C relationship of the drain node is also a dominant factor to determine the speed and amplitude of the output voltage change. Notably, it was well known that the junction capacitance and the gate capacitance are similar in the values on the ordinary CMOS process. In summary, the initial small + voltage on the input node will be amplified with finite gain to the large output voltage at the output node. As was explained, final voltage gain is cumulatively affected by several nodes' capacitances on the signal path. In reality, since MOSFET switches reveal the resistance, the R · C products on the signal path sequentially modulate the amplitude and phase of the input voltage. Primary pole is occurred from the biggest R · C value, which diminishes the voltage gain mainly by −20 dB/decade as a function of frequency. This simple analysis is also adequate for more complicated amplifiers have the multiple gain stages. To find more advanced theory based on the equations, see the many reference books or papers [25, 26].

Other important circuits for basic memory operations are the voltage and current generators to drive the memory cell operation and the timing generators to issue the procedures for a memory operation. Lots of those circuits have been designed with differential amplifiers. First, analog circuits including DC generation circuits will be introduced, and the overall logic circuits containing timing generators will be discussed from the next paragraph. A memory chip generally requires two different external supply voltages. One is the normal V_{DD} voltage for the CMOS circuits another is the V_{DDQ} voltage which dedicated for the I/O input buffer and output driver. Since isolating the large I/O noise occurred by rapid data transactions from the main CMOS circuits and the less power consumption by the use of particularly low I/O voltage are possible, the separation of the supply voltage have been widely employed as a reasonable policy. Various other voltages are needed for the operations of a memory cell. In the case of the voltages lower than V_{DD}, arbitrary voltages can be generated using the internal voltage generator which is operating at V_{DD} supply voltage. But, to generate the higher voltages than V_{DD}, special circuits like a charge pump should be used (for an example, NAND Flash uses 20 V generated from 1.8 V supply voltage). Charge pump is an analog circuit of which the input clock swinging between GND and V_{DD} drives many capacitors regularly to force the unidirectional current flow. The charge pump circuit is illustrated in Fig. 6.6. The unidirectional current characteristic of a charge pump is implemented by using the diode or its equivalent circuits. Here, large inverters are driving the one side of capacitors, and then the voltage on the opposite side of capacitors will be coupled up or down at the in-phase with the input clock voltages of V_{DD} or GND, respectively. As shown in Fig. 6.6, the capacitive coupled voltages result in two different bias conditions to a diode: forward bias and reverse bias. Thus, the diode flows a current only in a forward bias condition of one clock phase, but it does not flow a current at another clock phase. In Fig. 6.6, the charge pump is driving a current from left to right and many positive charges are being stored at C_L, of which it will generate the high voltage with the relationship of $V = Q/C_L$. We can assume that the forwarding voltage across the diode is $\alpha \cdot V_{DD}$ (here, α is a coupling coefficient, and the reason of not the value of $2\alpha \cdot V_{DD}$ is that only half of the entire diode will contribute to current flow at one point). By subtracting the diode threshold voltage,

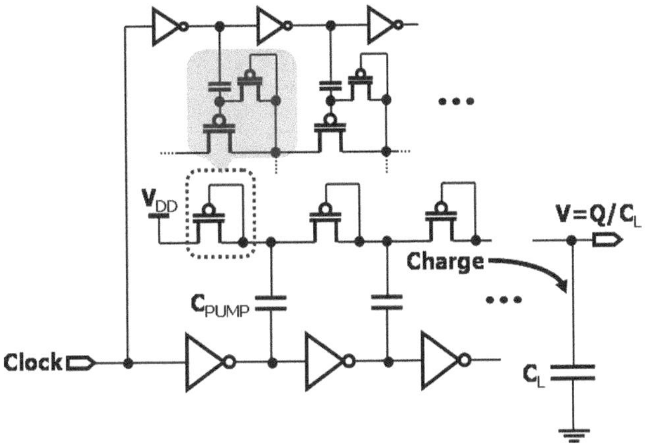

Fig. 6.6 Example of the charge pumping circuit. The main operation is to move the positive charges to the right direction

we achieve a term of $\alpha \cdot V_{DD} - V_{TH}$ as an effective voltage to induce a charge sharing process. Thus, the forwarding current will be continued by the charge sharing process until $\alpha \cdot V_{DD} - V_{TH}$ will be zero. Where, we presumed that the clock half period is sufficiently long to complete the charge sharing process. Therefore, the charges in an amount of $C_{PUMP} \cdot (\alpha \cdot V_{DD} - V_{TH})$ will be delivered to the load capacitor of C_{PUMP}. The charge pump of N diodes connected in series can generate the output voltage up to $N \cdot (\alpha \cdot V_{DD} - V_{TH})$ in maximum. After reaching the maximum output voltage, the charge pump cannot drive the output current anymore. As such, the current driving capability of a charge pump (is related to the pumping efficiency) is inversely proportional to the output voltage. As was explained, the charge pump using a diode will diminish the maximum output voltage because of the diode's threshold voltage, and so, the pump efficiency will also be suffered. To improve such inefficiency, P-MOSFET is used in place of the diode and its gate voltage is actively controlled by clock, whose circuit is depicted on the upper side of the diodes in Fig. 6.6. The substitution circuit instead of a diode will turn on and off the P-MOSFET switch dynamically at a proper timing with the application of the capacitive coupling similarly in the earlier mechanism. That leads to the unidirectional current flowing though the P-MOSFET like the case of a diode, but, without the voltage drop due to the threshold voltage. More detailed explanations and relevant circuit theories will be found in several references of [27]. For the generation of a specific high voltage, the charge pump needs the regulation circuit that is basically a voltage comparator based on the differential amplifier (the schematic is illustrated in Fig. 6.7). The comparator, here, has a role of gating the input clock to the charge pump dynamically to obtain $V_{REF} \cdot (R_1 + R_2)/R_2$ as an output voltage. If the input clock is gated to be blocked by the comparator, then the charge pump cannot drive the output current anymore.

DC voltage generator of lower than V_{DD} has been designed by combining the comparator circuit, i.e., differential amplifier and operational amplifier, with the

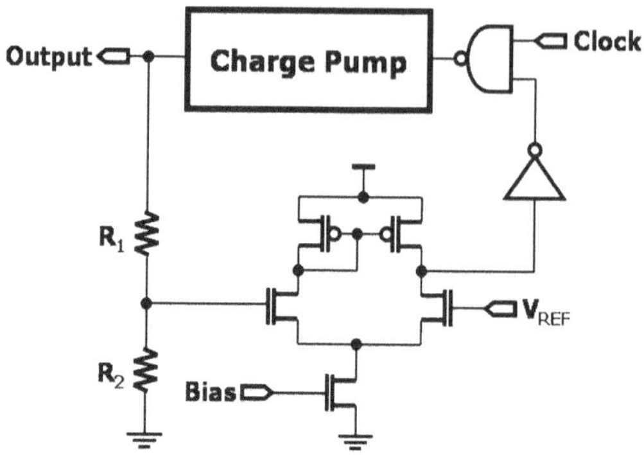

Fig. 6.7 Example of the high voltage regulator circuit. It is with a charge pump circuit to generate a specific high voltage

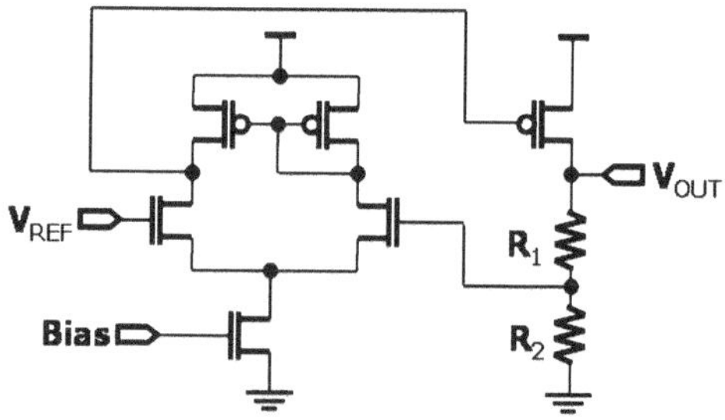

Fig. 6.8 Example of the conventional DC voltage generator for less than V_{DD}. Using P-MOSFET driver is particularly beneficial to the supply of positive charges to the output node. In contrast, N-MOSFET driver is appropriate in the supply of negative charges to the output node. Except for its size, the internal voltage converter (IVC) circuit is very similar to this schematic

feedback loop from a resistor voltage divider [26]. Typical example of a DC generator is represented in Fig. 6.8, in which the output voltage will be able to be calculated simply from the condition that both inputs of a differential amplifier are same voltage as V_{REF}. From the $V = I \cdot R$ relationship, the output voltage of a DC generator is straightforwardly achieved with the equation of $V_{OUT} = V_{REF} \cdot (1 + R_1/R_2)$. During a memory device operation, various DC voltages are alternatively used, and which requires the fast DC voltage changes to the large capacitance node between two sequential operation modes. The DC generator in Fig. 6.8 is designed with the P-MOSFET driver, which is very adequate for the fast pull up of the output node voltage (V_{OUT}). On the contrary, if the following DC voltage has to be lowered than the presently applied voltage, N-MOSFET driver is necessary to achieve the fast pull down speed at the output voltage node. To obtain both of the fast pull up and the fast pull down the output node voltages, we sometimes use P-MOSFET and N-MOSFET drivers at the same time.

In the case of PRAM, high cell current is indispensable for the write operation. High current density raises the cell temperature and it melts the cell to an amorphous state.

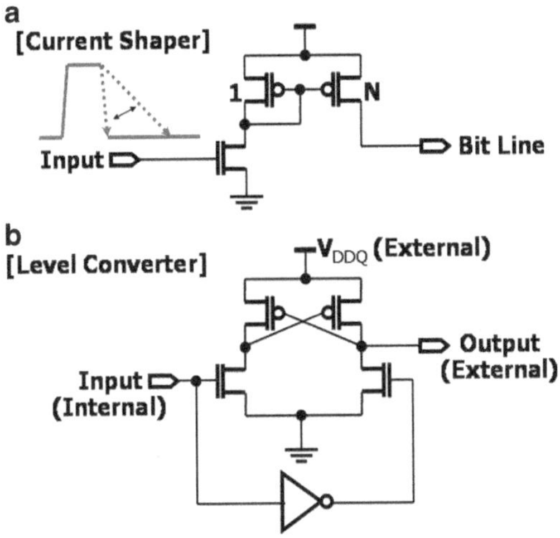

Fig. 6.9 Example of the current shaping circuit and the level converter circuit. Level converter is only suitable for signal transmission to the high voltage domain in the low voltage zone

Then the steep current off cools the cell to keep an amorphous state (large resistance state) but the slow current tail crystallizes the PRAM cell and it makes the cell to the low resistance state, wherein special current driver circuit to control the current shape is required [28]. In Fig. 6.9a shows the current shaping circuit that is based on the standard current mirror circuit [25, 26]. By controlling the gate voltage of the N-MOSFET and by amplifying its drain current by N, we can achieve the high current source with the shaping capability that is called the write current driver in the PRAM.

Internal voltage converter (IVC) is another important circuit. The circuit is for generating the uniform internal power even when the external power is disturbed by several noise sources. Nowadays, NAND Flash is designed to generate the constant 2.1 V internal voltage from the external voltage that is in the range of 2.7–3.6 V. DRAM uses the internally generated 1.3 V power from the varying external voltage of 1.8 ± 0.1 V. The IVC circuit is very similar with the DC voltage generators in the previous paragraph. But, the transistor size of the IVC circuit has to much larger than that of the DC generators to supply the full chip power, and the response time after detecting the internal peak current consumption should be fast enough to stabilize the internal supply voltage (namely, IVC has to rapidly transfer the charges from the external power node to the internal power node, which prevents the internal power drop occurred by the sudden current consumption by operating the internal circuits). Stabilizing the internal power is becoming challenge while getting smaller the difference between internal and external voltage. Using large size MOSFET and its implementation for faster response time result in the higher current consumption to the IVC circuits themselves, which takes the limitation on minimum response time using any size of the MOSFET. The problem can easily be dealt with the adoption of a power capacitor that is just a large capacitor connected between the internal power node and GND. Once many charges are stored in the capacitor, the power drop from the incident current consumption will be diminished, and so possibly it can make the sufficient time to respond to the drop using smaller IVC circuit. Moreover, in prior to the input of a command to initiate the internal circuit operations (for example, a chip enable signal), we disable the main IVC circuit to reduce the power consumption of its own but a small IVC circuit is enabled only to cover the leakage current at the minimum supply voltage. From the separation between internal and external power, the data on the internal power domain should be converted to the data on V_{DDQ} domain to be delivered over the PCB board. Therefore, the level converter circuit is necessary for an interface between V_{DD} and V_{DDQ} domain as shown in Fig. 6.9. So the output nodes of internal circuits on V_{DD} domain can be connected to the outer world circuits only through the level converter circuit operating in V_{DDQ}. In the case of an input buffer, the level converter is not an indispensable part. But, the difference between the external trip point (trip point means the threshold input voltage can flip the logical state.) and the internal trip point could occur the timing violation when an input voltage slope from an outer driver circuit is changed by several reasons like supply voltage and temperature variation.

In addition to the foregoing circuits, memory cell operation basically requires several other analog circuits, i.e., reference voltage generator, oscillator, power level (and noise) detector, temperature sensor, digital-to-analog converter (DAC), and delayed locked loop (DLL) or phase locked loop (PLL). Those circuits are so sophisticated and difficult to describe all through this chapter. If one needs more information, refer [25, 26].

6.2.2.2 Logic Circuits for a Memory Device

All the voltage and current generations for the memory operation are responsible for the analog circuits as discussed earlier. But, as was stated in the section "the birth of memory devices," to operate the memory devices, various DC voltages have to be applied to a memory cell at the proper timing. In general, the ns timing uses the analog delay circuit, because the timing is much smaller than the clock cycle time. In the case of the μs timing, by contrast, is usually implemented with the state machine that is a logic circuit synchronized to the clock signal [6, 7, 29]. All memory devices utilize those two timing circuit from very fast core operations to slow transitions between operating modes. For instance, the fast core timings of the SRAM have been effectively controlled with the delayed pulses generated by the analog delay circuit, the DRAM has used the state machine to implement the mode transitions among active, standby, read, write, and refresh mode, and the NAND Flash has used the state machines for the core timings (namely, program, read, and erase) and relevant bias conditions as well as for the mode transitions.

Any logic circuits can be created through a composition of NAND, NOR, INV (inverter), MUX (multiplexer), and registers (or flip-flops). The logic implementation processes are highly automated with a number of relevant computer aided design (CAD) tool. Thus, we can easily describe arbitrary logic functions using the hardware description language (HDL), i.e., VHDL or Verilog, resembling a

C programming language in some way. We can express logic functions with the language from the very detailed gate level to the complicated subsystem level. In many cases, the language has to describe the functional behaviors based on the register level operations, so, the register transfer level (RTL) code are commonly used to design the optimum logic circuits while maintaining high flexibility. The resulting RTL codes will finally be converted to the specific logic circuits (hardware) by using the logic synthesis tools like the design compiler [6, 7, 30]. Since there are many ways to express the same logic function, the synthesis tools are providing many options to find an optimized hardware on the trade-off relationships among area, speed, and power [30]. To all this possible, the building block of NAND, NOR, INV, MUX, and registers should be identified as a fundamental element, in which each component's logical functionality, circuit area (layout information), and timing information including propagation delay are characterized at various conditions in supply voltage, temperature, and process, respectively. (Those information could be obtained by using direct measurements or indirect Splice simulations.) The CAD tools recognize such characteristics with the predetermined data format in a library. In reality, recent trend in VLSI design is biased primarily to the CAD related issues. Even though the CAD based logic design seems to be always allowing highly reliable logic circuits, in fact such reliable circuits are limited only to the static type logic circuits. Therefore the specialty logic circuits like dynamic and domino logic need highly customized design similarly in the analog circuit design in order to achieve the state-of-the-art performance on speed and low power over that of the automated tools. And, the hardware theory of the arithmetic circuits, i.e., adder, multiplier, and divider has been well established for obtaining the smallest size, lowest power, and fastest speed, respectively. In this case, the synthesis tools allow that the qualified arithmetic circuits are directly embedded into the predefined area, which is performed by means of several synthesis options [30, 31]. Figure 6.10 shows an example of the RTL code corresponding to the flow chart representing the operation of a state machine, where we can see that the RTL coding is not greatly hard work and is intuitive just like a C programming: Next states (next value of States [1:0]) are determined by the triggering signal (Next) and the current statuses (current value of States [1:0]). Also we can see that all state transitions are occurred at positive clock edge by following the statement "posedge Clock." The ease of coding encourages that someone can design the generic logic circuits, i.e., discrete logic components easily with just a little effort. On the other hand, noticeably, designing the logic circuits for high performance, low power, and size optimization require special coding skills and associated experiences. More and more memory devices are gradually requiring the error correcting code (ECC) circuits in the forms of on-chip integration or off-chip support to cope with the increasing bit error rate in the deeply scaled memory cells. Such ECC circuits are usually implemented by the following common-sense process: determining the specification of the ECC, RTL coding to meet the specification, verifying the behavior of the function level operations, generating the hardware via logic synthesis, and checking the timings and performance violations.

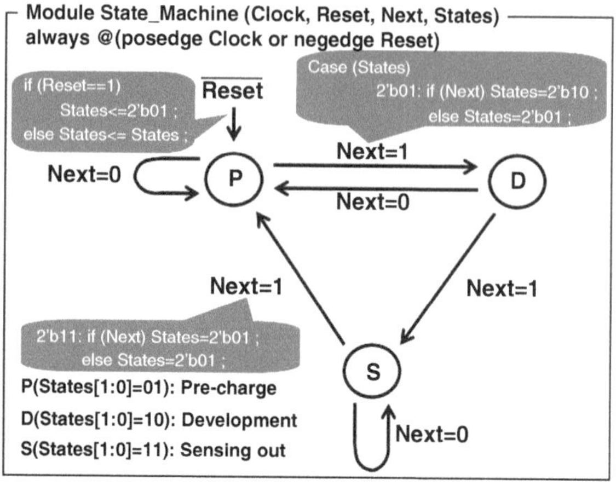

Fig. 6.10 Example of the simple state machine and its flowchart. The register transfer level (RTL) codes in response to the flowchart are shown in together

Another important feature included in a memory device is the ability to interpret the input command and to decode the input address. Using "if–else" statement of the HDL, the meaning of the input commands will easily be interpreted. Four times the input address with eight pins can provide a 32-bits address space corresponding to 2^{32} bytes of memory space. In this case, as an example, by the presence of an address decoding circuit, the entered address will select one word line and eight column lines to access one byte (8 bits) data (It should be emphasized that this example differs from the actual situation like DRAM and NAND Flash. Both of them can latch one page, 2–8 KB, data by selecting one word line). After all, the address decoding circuit is basically the role of a translator transforming any binary number into the one-hot encoded form [6]. As mentioned previously, the logic circuit design often requires the customized design process, in particular for implementing the critical data paths. In memory devices, the decoder circuitry is often considered as the main critical path. Essentially, the propagation of the digital signals within the logic circuits is just to drive successive capacitors present in each logic circuit. For instance, we can imagine that some signal propagates through the logic gates in the order that two-input NOR, three-input NAND, two-input NAND, and INV to drive 10 pF load capacitor (a variety reasons can cause such a large capacitor: a long word line connected with many memory cells or a input node in the large output driver), where the signal propagation speed is dependent on how fast the circuits drive the intermediate load capacitors mostly from junctions and gates. The theory to calculate the sizes of transistors for the fastest logic circuits was introduced in the name of the "logical effort" [7, 32]. For reference, the channel width is mainly used for adjusting the size of the transistors in the logic circuits because the channel length is already chosen as the smallest size can be provided by the fabrication process. Many books are devoted to the contents of the "logical effort" theory in their digital circuit chapter.

6.2.3 The Brief on the Interconnect Issues

So far we explored overall devices and circuits for memory chips. Now the interconnect-related issues will be briefly introduced. Integrating the electronic devices into a semiconductor chip brings great benefits in the design of interconnection wires, which allows very short length of the interconnection wires. The importance of this point is obvious accordingly as data transfer speed is almost beyond the level of radio frequency (RF). At high frequency, if its wavelength is less than the length of an interconnection wire, then the wire acts like an antenna and so the electrical signal will be disappeared by radiation in free space. To avoid this issue, the bulky, expensive, and complex cables (for example, coaxial, twisted pair, and micro-strip line cable) should be used for an interconnection wire [33]. In fact, DRAM module minimizes such a signal loss by making the wave-guide in the form of the micro-strip line onto the printed circuit board (PCB). Aggressive scaling of semiconductor devices has naturally solved the radiated loss problem while maintaining the high speed data operation. Instead, the signal interference problem is increasingly deteriorated by narrowing the distance between interconnection wires. Many design tools have been developed to model and simulate the cross talk phenomenon, which is still an ongoing subject should be resolved by whoever designs a chip.

6.3 Memory Hierarchy and Hardware Compositions

6.3.1 The Types of Memories

Generally the term of "memory" in the computer architecture means the semiconductor memory of a fast access time. For instance, all of mobile, laptop, desktop, and server computing system use the EEPROM (or Flash memory) to store the basic input/output system (BIOS) codes and use the DRAM to execute system/user codes. In the cases of laptop, desktop, and server, DRAM has been used as a module form is composed of 4–16 DRAM chips (for reference, the module generally divided into four different types: SODIMM, UDIMM, RDIMM, and FB-DIMM), and every program can be an executable form with the CPU only after the codes were kept on the DRAM. C (or C++) codes, for example, layout the memory space into the stack, heap, code, and data area, in other words, all the programs written by C (or C++) codes are running through the use of the DRAM address space. Specifically, C (or C++) pointer is a way to refer to the specific address within the DRAM. As a consequence, the RAM, available for a byte access, is essential for the execution of the computer program. Nowadays, since the fastest and cost effective RAM is just a DRAM, all of the computers have used the DRAM for the main memory in a hundred percent. The DRAM, however, is a volatile memory, and so, it lost the stored data when the power is off. Therefore, the nonvolatile storage medium is essential for the durable use of their data. Hard disk drive (HDD) is the most popular storage device because of the superiority in a term of the capacity per unit cost. Nonetheless, the HDD can read and write data only for a fixed 512 bytes (this is a sector size), and so it cannot provide a random accessible nature such as a DRAM. Eventually, we cannot execute the program directly on HDD, but, as was stated, we have to copy the program into the DRAM to make the codes an executable one. For some laptops and mobile electronics use the NAND Flash memory as a storage device. Like HDD, NAND Flash is allowed to read and write the data only for a fixed 2–8 KB (this is a page data size). Because of the property, NAND Flash cannot replace the DRAM to be able to make the program run.

6.3.2 The Memory Architectures

By shrinking the semiconductor devices, CPU speed was rapidly increasing. This is acquired by increase in clock frequency, adoption of a coprocessor, enhancement on micro-architecture, and parallel processing using multi-core, etc. On the other hands, DRAM could not improve the performance as much as CPU, because the DRAM capacity has been dramatically increased at the same time. Thus, the number of DRAM cells per a word line and a bit line has been continuously increased with the development of a smaller DRAM cell. As a result, the parasitic capacitance at the core nodes has not been scaled down as much as the geometrical device shrinking, and so, the core performance could not be much improved. But, by increasing the interface speed (historically DRAM interfaces have been evolved in the order that SDR, DDR, DDR2, DDR3, and DDR4), higher throughput could be partially achieved. Fundamentally all of the memory devices including a DRAM were limited in the performance by the core architecture, namely, the memories cannot allow the independent core operations using the different word lines if they are sharing the bit line. If some word line is in the core operation, the operation will occupy the shared bit lines with the bit information. And other word lines should not be activated until the previous core operation is completed. To prevent this performance degradation, different memory cores (it called "plane" for a NAND Flash and "Bank" for other memories including a DRAM) are separately embodied into a memory chip. Each core in here provides the capability of the independent core operations. Moreover, memory chips are often used with a module form. For example, DRAM module is used to the main memory for the PC and server, and NAND module is employed for the solid-state drive (SSD). Those modules take the advantages of a parallelism by extending the number of ranks (ways) and channels as shown in Fig. 6.11. The memory modules should be eventually connected to the memory controllers. For example, DRAM controller manipulates all of the channels, ranks and banks in a DRAM module, and NAND controller is responsible for utilizing the channels, ways, and planes inside a NAND module. In Fig. 6.11, each channels have the dedicated buses and so their operations are totally independent each other. But, in the cases of ranks (ways), they are sharing a bus, and it cannot support several data transactions simultaneously. However, the core operations of every rank (way) are allowing the

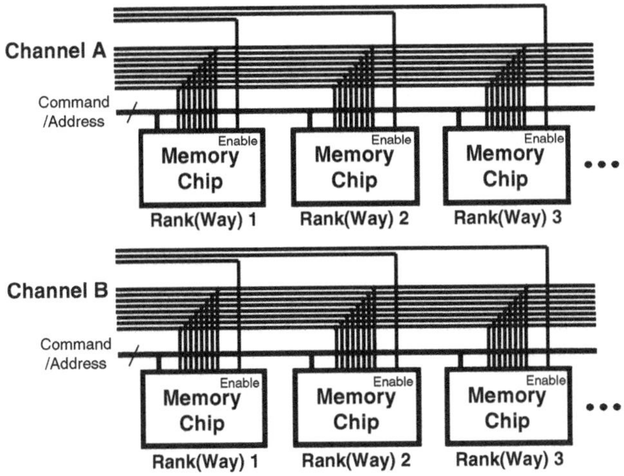

Fig. 6.11 Example on how to implement DRAM and NAND Flash module. A common address bus and a separate data bus were supposed to the schematic

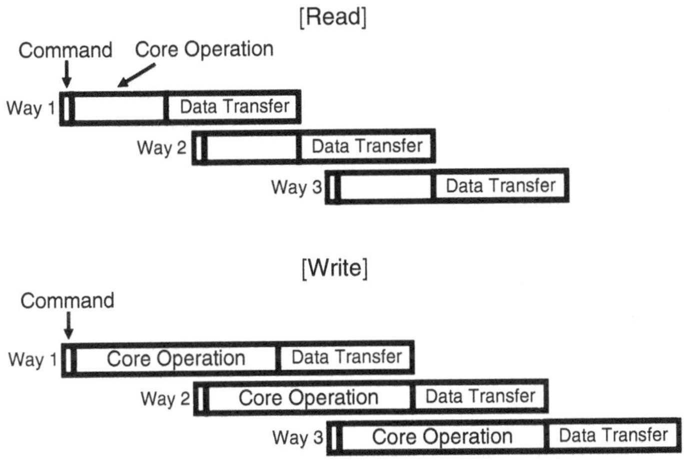

Fig. 6.12 Interleaving timings of reading and writing operation were indicated, respectively

timing overlaps (notably, both address and command buses are common to the ranks or ways). Thus some rank (way) can use the data bus while other ranks (ways) are executing the core operations. Total latency of some memory chip is obtained by the sum of the command transaction, data transaction, and core operation times. By making a pipelining between the data transaction time and the core operation time, we can effectively increase the throughput of a memory module (this technique called "interleaving"). In an optimum case, core timings can completely be hided by the data transaction timings as illustrated in Fig. 6.12, where the module's throughput is determined only by the data transaction speed (In Fig. 6.12, we used the example of a NAND module, here, both of the read and write operations were considered).

6.3.3 The Brief on a Memory Controller

A role of a memory controller was shortly introduced in the previous paragraph. In reality, it participates in many other control jobs [16]. DRAM controller, as an example, is also responsible for the paging mode policies (open and close page modes), refresh operation, and command reordering to maximize the module's performance. When the CPU commands write and read operation in the mixed order, the memory controller will reorder the commands and classify that into the command groups binding the same commands. And so, the command reordering can minimize the timing mismatches between the read and write operations by lowering the command change frequency. As a result, the memory module will allow the optimum performance with the efficient interleaving. Basically the operation of a NAND controller is very similar with that of a DRAM controller. However, NAND controller embodied in a NAND SSD extends its role to the small computer system, where it executes the firmware with the dedicated CPU and DRAM. NAND has several peculiar properties compared to the DRAM. For instance, it reveals the limitation on the number of the program and erasing operations, and the error correction code (ECC) is indispensable for a NAND controller due to the relatively higher bit error rate of the cell. In addition, the electrical characteristic of a NAND cell is highly disturbed by the electrical states of adjacent cells, and so, it requires the special algorithms to ensure the written value in a victim memory cell.

6.3.4 The Memory Hierarchy

Human brain is divided into the long-term and short-term memories. Hippocampus is likely storing the short-term memories, and the strengthen memories by the continued stimulus are preserved by the gray substance as long-term memories [34]. The memory hierarchy in a modern computer is similar with the division model of a human memory. As was illustrated in Fig. 6.2, fast and volatile memories are near the CPU and it manipulates the temporal data processing. The storage of the high memory capacity far from the CPU is storing the long-term data [17]. Since the CPU can process the data only with the registers, the data should be previously copied to the registers from somewhere other memories. Therefore some codes and data to be processed by the CPU will be copied from the HDD to all of the memory devices on the memory hierarchy like the L1 cache, L2 cache, L3 cache, and main memory (DRAM). In the modern computer architecture, all caches in the memory hierarchy and the memory controller are integrated in a CPU chip, and the main memory is connected to the CPU through the PCB bus. Moreover, the I/O controller hub (ICH) chip on PCB is used to bridge between HDD and CPU. Thus, CPU contains the controllability that is spreading over all computer hardware including the main memory and storage system. And so, the CPU, including ICH chip, coupled with a computer in the entire structure, the instructions for reading and writing data on the main memory and storage system could be implemented within the CPU (in other words, the hardware-level design is done for such instructions). For example, by using the load and store assembly instructions, we can readily execute the data transactions between CPU registers and main memory. For the data transaction between HDD and main memory, additional instructions are required to specify the source and target address and the data size to be transferred. Thus, the sector address, memory address, and the number of sectors define the data transaction between HDD and main memory and the information should be previously set to the specific registers, and then the I/O interrupt assembly instruction will start the data transaction according to the transaction description. It would be noticeable that each assembly instructions correspond to each hardware configurations. Figure 6.13 shows the capacity differences among cache, DRAM, and HDD as different square sizes. Where, the cache has a piece of programs copied from the main memory, and the main memory holds a part of files (light gray) as a program form (dark gray: an executable form of a file controlled by the operating system) copied from the HDD. Figure 6.13 shows also an example how the code and data of a file are mapping into the memory layout as an executable form (namely a process), where heap and stack are mainly used for dynamical data and function calls, respectively. Memory hierarchy was developed to gain the higher performance at the lower cost. The contents addressable memory (CAM) and SRAM are the fastest memories, but they are much expensive memories compared to the other memories and storages like DRAM, Flash, and HDD. Therefore, they are only used to the CPU cache while minimizing its size. Even though the physical size of the CPU cache is much smaller than that of the main memory, we can effectively obtain that the performance of the cache from the main memory can be implemented. The CPU's memory access pattern shows that a small part of the program is used to run for a long time, after running, next chunk of the program is quickly loaded to the

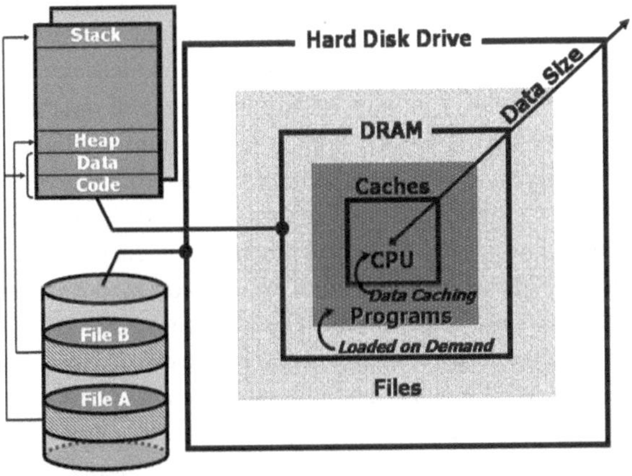

Fig. 6.13 Data flows in the conventional memory hierarchy. Files stored on the disk are copied into the DRAM in the form of the executable program. Parts of the program are copied into the caches and registers and then it can be executed by the CPU

cache by utilizing the full bandwidth of a DRAM. Such an intensive access pattern to the small data chunk (is called the spatial locality) is partly originated from the software technique optimized to the small cache size [17], and recently accessed data is likely to be accessed again (it is called the temporal locality). Consequently, most of the running time of the CPU is to the access the CPU cache, and so, the time to utilize the main memory (DRAM) is small enough when compared to the total execution time. The SRAM is generally used for the CPU cache, and the CAM is widely used for the timing critical cache application like the translation look-aside buffer (TLB) [17]. To enhance the access time of the SRAM based cache, i.e., for a L1 cache, the SRAM is divided into several banks to be simultaneously accessed with the index field of an input address and the output data from the banks are tagged with the tag field to select correct bank data. By that cache architecture, we can lookup the correct data only with the small index field without the full address search of a CAM.

6.3.5 *The Future of the Memory Hierarchy*

When we execute a new application program, it requires that some of the previous programs in DRAM should be replaced with a new program because of the limited capacity of the DRAM. And so, here, many HDD sectors should be read out to execute a new program. In this case, the arm in the HDD may be moved in zigzags, whose process eventually delays the startup time of a new program (the startup time is generally called the responsiveness), thus the HDD shows a slow responsiveness. The interest for the execution of a program is changed to that of an environment running multiple programs, in which each program's responsiveness is becoming increasingly important. This is not only because of the environment familiar with using multiple applications at the same time, but also because of the slow improvement in the latency of the HDD [35]. So, over time, the latency gap between the HDD and the CPU and DRAM has been worse. To resolve this problem, several approaches by changing both of computer architecture and operation system have been aggressively explored. The first way is to change the HDD to another fast storage device. For example, now the change from the HDD to the NAND SSD is about five times more expensive, but that can improve the random read performance several times. The second way is to insert a new cache layer between HDD and main memory. As shown in Fig. 6.2, if we employ the storage cache that has the 10–100 μs access time and 10–100 GB capacity between HDD and main memory, overall system performance will be improved as was discussed in the previous paragraph. NAND Flash is a nonvolatile memory satisfying the conditions to be a storage cache in the access time and capacity, so a new memory hierarchy employing the NAND Flash is one of the hot research items [18, 36]. Intel, for example, has tried with a codename "Braidwood" for the commercialization of a NAND based storage cache. However, the software development to overcome the limited endurance of a NAND Flash (namely, the maximum number of programs and erases is limited) is highly required to the success of the "Braidwood." And, now the project is postponed indefinitely. Nevertheless, the low power and high capacity characteristics of a NAND Flash, so many researchers are looking hopefully to the NAND Flash's role in the future computer architecture [18, 36]. The third way is to make an innovation on the memory device technology for improving the responsiveness. New memory like the phase change RAM (PRAM) and the resistive RAM (RRAM) is a nonvolatile memory that will be obtainable the sub-microsecond access time and large capacity with the three-dimensional memory cell technology. Since they are showing a RAM characteristic, we can apply the new memories to the main memory as well as the storage cache. IBM storage class memory (SCM) is taking the same approach where the nonvolatile RAM of large capacity is composing the tiered main memory architecture with the DRAM [37]. Here, the primary programs are stored in the SCM and one of them is loaded to the DRAM as required. Since the access time of the SCM is close to that of the DRAM, the responsiveness of some program will be much improved wherever the program is in either of DRAM or SCM. But, the SCM requires many changes to the computer hardware and software, because the existing storages (for example HDD and SSD) are highly optimized to the block (512 bytes) based operation in the transaction data size, but the SCM is the RAM based storage should be supported by the new hardware components like SCM controller and bus. In addition, the OS should be largely reconstructed to utilize the byte-addressable capability [12, 13, 19]. This is a good example how the device level innovation can lead to whole system changes. The web server computer of lower power and faster response time is already being introduced using the DRAM-NAND hybrid main memory [18]. IBM and Microsoft present that the DRAM-PRAM hybrid main memory can improve both of durability and responsiveness [19, 38]. PRAM of a high capacity enables that the new file system is implemented to the hybrid computing system. The file system can bypass the sophisticated software overhead by utilizing the byte addressable capability of a PRAM. For example, the multiple transfers of a data from the HDD to the visible user memory space, address calculations to find the data blocks in a HDD, and inefficient search algorithms and data structures adequate only in the block-based storage can be much reformed with the RAM-based file system and its direct addressing capability as if we can access arbitrary data using pointers in a C (or C++) program [19, 38]. Currently, new computer architectures suited for new memory devices are being reported in many papers to pursue the performance improvement on the system and application levels.

6.3.6 The Device-Level Innovations

What is more, many efforts to innovate the devices in the existing computer architecture are being made. General CMOS logic, for example, achieves the low power and high performance gains from the multi threshold CMOS (MTCMOS) technology (low threshold CMOS is applied only to the critical speed path to minimize the power consumption), and the silicon on insulator (SOI) technology has been widely explored to optimize both of the performance and consuming power [7, 29]. Multi-gate transistors including the FinFET are actively employed to the general logic and current driver circuits [28, 29]. In the case of the memory technology, the embedded DRAM is being developed to increase the capacity of a CPU cache [39], and NAND Flash increases the capacity using four-bits multi-level cell (MLC) technology and three-dimensional memory cell technologies [40]. Moreover, NOR Flash is now being replaced by the PRAM to increasing the performance and reliability at the lower cost. And tens of nm DRAM cell enables that one chip alone provides sufficient DRAM capacity for mobile applications, which led to the new DRAM chip architecture that two independent ports are formed on a chip will be connected to two different CPUs, and so, the chip could allow the lower power and compact memory solution. Especially STT-MRAM is potentially one of the promising universal memories due to the nonvolatile, faster, and lower power characteristics. In the future, various computer architectures will be appeared in response to the different applications from the super computer, data center, and web server to the laptop, smart phone, electrical appliance, and smart card.

6.4 Software Interfaces of Memory Devices

6.4.1 The Memory as a System Resource

It is generally thought that the memory performance is only equivalent to the timing parameters written in the memory specification. It is not very wrong story if we are focusing on the memory response time shortly after the data request of a memory controller. In reality, however, additional memory accesses are common every time to access the specific memory data, because the OS codes should be performed to manage the memory resources to coincide with the execution of an application program [12, 13]. Today the environment is commonplace that one computer can run multiple programs at the same time (namely, multitasking environment), thus, the OS should allocate independent memory spaces to each programs and arrange the map table what files are opened by some program and schedule the priorities between programs. And, the OS has to manage the memory spaces to be evicted using some dedicated algorithms and respond to the several interrupts and faults, and also be in charge of the security issues, networking, and managing some buffers for a storage. Those things require the redundant computing power and DRAM space. One of the biggest management overheads occurs in the virtual memory system that allocates memory spaces to each program and translates the logical addresses of each program to the physical addresses in a real DRAM. OS divides the memory space into 1–4 KB (or up to several MB depending on the options) pages, and the page tables are introduced to translate from logical address to physical address and to describe the page's properties like access right, dirty/accessed bits, memory attributes, and so on. When a program is starting to run, the OS lookups the page table contents to find the usable pages and then it allocates the required memory space to the program. Since the lookup process is very time consuming, bitmap is usually used to reduce the searching time. The bitmap is a simple map stored in a DRAM composed of many bits where one bit corresponds to one page, and bitmap's small data size enables fast lookup time [12, 13]. Then the logical to physical address translation is performed with the page table. Both of the hardware (CPU) and software (OS) are involved in the translation process (hardware part is typically called memory management unit (MMU)). The table includes the physical addresses and we find a physical address by indexing the table with the logical address. An example is illustrated in Fig. 6.14. This shows how two programs' logical addresses (each programs use six pages, respectively) can be mapping to the single physical address space (12 pages). When a process is scheduled by the OS kernel to start a program execution, many high-level routines should be performed in prior to allocate a memory resource to the process as was depicted in Fig. 6.14. And so, they yield the considerable software overhead that will delay the program's response time [12, 13]. Using the page table, we can extend the logical address space more than the physical DRAM space. The key is that only a small portion of large logical address space can be mapped to the DRAM space as required, and the rest of logical address is mapped to the storage (HDD). For this purpose, the page swapping operation should be implemented in a virtual memory system, which swaps the pages between the page in a DRAM and the page in a storage device (HDD) [14, 15]. Notably, Intel uses the segment to divide the memory space roughly into the several areas like code, data, stack, and heap and it is once again split into the pages that was discussed.

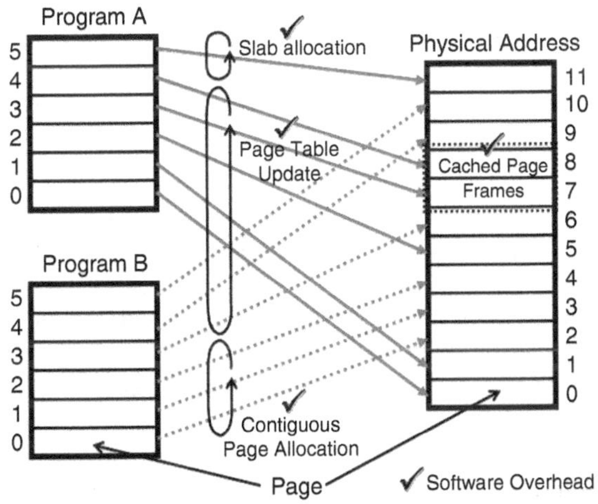

Fig. 6.14 Example on how the address translations and various management routines (the kernel routines) are serviced by the operating system (OS). Whenever the memory resources are accessed, large software overhead will be incurred and it significantly degrades overall system performance

6.4.2 The Software Overhead in a Memory-related Performance

As was stated, both of page table and bitmap should be stored somewhere in the DRAM and they are referred by OS whenever a memory access. To reduce this overhead, TLB is used as the dedicated CPU cache for a page table. But, the table should be totally replaced in prior to start the new program in a new address space. The software overheads, in conclusion, are very diverged and redundant in the modern computing systems. In the typical case, the OS creates the process, allocates memory space, swaps the pages, writes back the victim pages to the HDD, and updates the page table and the TLB contents. In here, each relevant algorithm and high-level routines should be performed in an optimal way to manage a memory resource, and finally the OS loads the actual running programs into the DRAM [12–15]. Therefore, so ironically enough, the high performance CPU can enhance the actual memory performance, because it reduces the software overheads by the shorter running time (but, interestingly the CPU performance is highly dependent on the cache memories' performance embedded in the CPU). This point is very important as much as the memory chip performance. Such software overheads are becoming increasingly important for the many-core computing system and the virtualized memory system.

6.4.3 The Future Computing System

6.4.3.1 Evolution of a Memory System

The software improvement is not only solution to reduce the software overheads in the previous section. Since the overheads are caused by the frequent use of a memory resource, there may be some ways to resolve it through the memory side innovation. Actually, but, the overheads will be more worsened over time, results from increasing the performance gap between CPU and memory that is called "memory wall." To cope with this wall, we have increased the data rate on the memory bus pursuing to the higher throughput. This method, however, will be limited in the near future for two reasons. First, the higher data rate requires the lower load capacitance on the memory bus to assure the signal integrity (SI), so we have to reduce the number of DRAM modules at the shared PCB bus (channel). As a result, we cannot scale up the DRAM capacity freely for some computing system, which will degrade the availability of the system. Second, the higher data rate consumes more powers. Nowadays, the I/O power of the total memory power ratio is approximately 40% in typical. But this ratio can be much increased if we do not scale down the I/O supply voltage. Many silicon devices, however, cannot operate at any low voltage, and so, the power constraint will eventually limit the data rate. All of the device and system level approaches may be possible to resolve those problems. One example of the device level innovation will be in the development of the wide, low power, and fast I/O bus. The best way to achieve this is by integrating the DRAM into the CPU die, thus the sufficient bus width can be freely achievable with the minimum power. But, the approach is really impractical due to the several obstacles: the different process between DRAM and CPU and huge chip size will significantly lower the yield and will make the chip to an ultimately expensive one. On the other hands, the backend process of the through silicon via (TSV) will be the possible alternative. TSV is a technology that the interconnection wires are passing through the chip and so the stacked chips can be electrically connected each other [41]. In here, the manufacturing process for wafer thinning, deep via, vertical metal wire, and die attachment is highly dependent on the device and packaging technologies. Or in other way, an embedded logic circuit in a memory chip may reduce the simple repetitive software workloads, i.e., data searching, map table management, simple loop operation, and simple calculation for the addressing and graphic acceleration. Thus, we can hide the data transaction latency between CPU and DRAM and can reduce the power consumption by eliminating a number of transactions. Although increasing the design and test complexity may become a challenge, the matured logic technology will easily be able to implement such simple logic functions without the significant overheads on them. For an example, a commercial memory product was already introduced for a mobile application, where NAND, SRAM, and logic circuits were integrated on a chip to provide some embedded functions (it called "fusion memory"). Since the NAND Flash shows the high bit error rate and is only sequentially accessible, we cannot use it for a code execution thus a working memory. In order to be able to run a code with the high density NAND Flash, an idea was developed that the error correction code (ECC) circuit and the embedded SRAM are used to resolve the high bit error rate and the only sequentially accessible characteristic of a NAND Flash, respectively. And so, the associated interfacing logic circuits to transfer the data between NAND Flash and SRAM should be involved to complete the idea. Since a NAND Flash is a nonvolatile memory, we can store the firmware to be used to manage the fusion memory into the NAND Flash. For a code execution, the codes stored in the NAND Flash have to be copied rapidly to the SRAM. Depending on the application, such a firmware should be prepared inside the RAM before the power on process is finished to boost the start-up time of some mobile device. In this case, the fusion memory has to guarantee the operation at the lower supply power than the normal power, so, the device is highly dependent on several low power technologies.

6.4.3.2 Evolution of a Storage System

In the storage devices, the software performance has become increasingly important. Emerging storages like the NAND SSD cannot use the legacy commands and algorithms optimized to the HDD. However, the major OS providers are not so aggressive to support such new storage devices, and so HDD is only a storage device supported by the OS. Therefore, the interfacing software that can translate the HDD commands to the NAND SSD commands should be implemented inside the SSD. This is a firmware called the "flash translation layer" (FTL) that optimizes the SSD performance with the command translation, channel and way control, error correction, wear leveling (an algorithm for flattening the number of program or erase per a memory cell), logging system (a software safety against the sudden system failure), address mapping, and other algorithms like the merge operation of the fragmented memory spaces and garbage collection. The firmware is stored in the nonvolatile memory and is starting to execute when the power on is detected. As was shown in the previous examples, many controlling operations and algorithms of the NAND SSD are highly requiring the general purpose CPU and working memory (namely, DRAM). Eventually, the SSD is very similar with the small computing systems. So we can improve the SSD performance by taking advantage of new device, circuit, and software to optimize the general computing system (it has been the main subject through the previous sections) as well as the firmware level improvements. According to the progress in the host interface performance (for example, the maximum I/O speed of the serial advanced technology attachment (SATA) and serial attached SCSI (SAS) interfaces is 6 Gbps today but it will jump to 12 Gbps in the near future according to the roadmap), the computing performance will be more important because it can become a real performance bottleneck on the NAND SSD.

6.5 The Top-down Approach: Wrap-up

Finally, let's try the top-down approach to describe the whole computing system. The purpose of this approach is to show that we can really understand the computing system at a glance, and thus nanometer-scale device engineers can have a perception about the whole. Application programs are using the software services provided by the OS (but, some performance critical services are supported by the lower level routines), i.e., data operations such as write, read, execute, and delete some files. Most of the OS and other service routines are written by C (or C++) program language [42]. The language is highly readable because it has designed to provide an intuitive programming. And so someone can express any desired computing job by using the C (or C++) language. For an execution, the written C (or C++) program will be converted to the assembly lines (some interpreter software, compiler, is used to help this process), and each assembler command line corresponds to specific CPU instructions, respectively. Actually CPU instructions represent the associated hardware logic circuits, so, we can describe a set of instructions by using the hardware description language (HDL). HDL is much similar with the C (or C++) language in its excellent readability, and so we can easily implement arbitrary hardware instructions what we want using the HDL (HDL codes will eventually be converted to the hardware logic circuits with the help of an interpreter software). For increasing the flexibility and availability of a computing machine, the memory device has been incorporated into the machine, and so, the huge logic circuit blocks will be replaced by the flexible codes in a memory device, and it should be transferred to the special chunk of logic circuits (CPU) every time in prior to execute the computing job using the hardwired processing units. After all, the logic and analog circuit building blocks should be prepared, which will be used by the HDL interpreter software. Here, the circuit building blocks should be proven with many Spice simulations and silicon measurements. The Spice simulation requires the device models of MOSFET, bipolar transistor, diode, capacitor, resistor, and interconnection wires in addition to several noise models at each electrical node. The device model is a set of mathematical equations and describes all electrical characteristics of an electronic device at various operating conditions and process variations. Essentially the modeling process is to develop the mathematical equations and to determine the model parameters in the equations (is named by the extraction procedure of model parameters). However, it is noticeable that the device model is just a mimic the behavior of an actual device as precise as possible. Since the device technology is the foundation of all other technologies, the device innovation occupies a large proportion to achieve the best attributes of electronic devices. The innovation is based on an understanding of the device physics and relevant manufacturing process. For the design and implementation of the actual device, we need the help of a large study area such as chemistry, material science, physics, optics, mechanics, and electrical engineering. In addition, we like to include a system level insight in the list, which is the purpose of this chapter.

6.6 Conclusion

In this chapter we presented a broad but brief explanation about the memory system, including device technology, circuit technology, system technology, and software's memory usage manner. What we wanted to do in this chapter was to give a brief observatory view about the whole to the device engineers, who spend much of their time in seeking details, so give them an opportunity to think about the meaning of their work for a while. As we attempted a broad but brief explanation, we added as many references as possible, so this chapter can serve as an introduction or "a table of contents" to many new fields.

According to the continued scaling, the minimum feature sizes of logic transistors and memory cells are already reached tens of nanometers level. This shrinking trend has been coincided with enabling higher speed, additional functionality, and larger memory capacity on a single die, but the chip-to-chip performance through the PCB scale has been not much of that. Therefore, the independent multiple chips, i.e., memory management unit, graphic processing unit, north-bridge (or MCH), and maybe south-bridge (or ICH) in the future, have continued to be integrated into a single chip. The memory chips, however, will not easily be combined to the CPU chip as was discussed, but will find some opportunities to boost up the performance in the stacking technologies of different chips. As a result, the future computing system will show a simplified form that CPU, memory (DRAM), storage (namely, NAND Flash or PRAM or other new memories), and communication chip are stacked using short and wide interconnect inputs/outputs, and putted into a small package. This technology trend needs much of the device engineers' contribution as was discussed, which is a prerequisite to implement such a future computing system.

Finally, there is a memory system of higher level, we were not considered yet, beyond what is presented in this chapter. It is a distributed system, which is combining many computers under a network, so making the whole system to work as a single computer can provide the huge DRAM capacity adequate for the processing of large working data set. And as virtualization technology was introduced, it becomes possible to run the application programs of an arbitrary size without regard to the kind of the computer and OS. Since such system technologies are relying primarily on a software technology, whole system performance for a target will be able to be attained from the software optimization process not by the hardware improvements. On the other hands, the device and circuit innovations will improve all of the computing systems fundamentally without restriction to particular systems or specific applications. The memory system performance in here is relying more on the device-level performance.

References

1. J.J. Ritsko, D.I. Seidman, Applications of Information Technology. IBM J. Res. Dev. Special Report: Celebrating 50 years of the IBM Journals, http://www.research.ibm.com/journal/50th/ (2006)
2. M. Rees, *Just Six Numbers: The Deep Forces That Shape the Universe* (Basic Books, New York, 2000)
3. C. Kittel, *Introduction to Solid State Physics* (Wiley, New York, 1996)
4. Y. Taur, T.H. Ning, *Fundamentals of Modern VLSI Devices* (Cambridge University Press, UK, 1998)
5. S.M. Sze, *Physics of Semiconductor Devices*, 2nd edn. (Wiley, New York, 1986)
6. S. Brown, Z. Vranesic, *Fundamentals of Digital Logic with Verilog Design* (McGraw-Hill, New York, 2003)
7. N.H.E. Weste, D. Harris, *CMOS VLSI Design: A Circuits and Systems Perspective*, 3rd edn. (Addison-Wesley, Boston, 2005)
8. B. Alberts, A. Johnson, J. Lewis, M. Raff, K. Roberts, P. Walter, *Molecular Biology of the Cell*, 4th edn. (Garland Science, New York, 2002)
9. R. Ananthanarayanan, S.K. Esser, H.D. Simon, D.S. Modha, The cat is out of the bag: Crtical simulations with 10^9 neurons, 10^{13} synapses, in IEEE Supercomputing Conference Dig. Tech. Papers, Portland, Oregon, Nov. 2009
10. C. Dirik, B. Jacob, The performance of PC Solid-State Disks (SSDs) as a function of bandwidth, concurrency, device architecture, and system organization, in *Proceedings of the 36th International Symposium on Computer Architecture (ISCA),* Austion, TX, June 2009

11. V. Cuppu, B. Jacob, Concurrency, latency, or system overhead: Which has the largest impact on uniprocessor DRAM-system performance? in *Proceedings of the 28th International Symposium on Computer Architecture (ISCA)*, Goteborg, Sweden, June 2001
12. A. Silberschatz, P.B. Galvin, G. Gagne, *Operating System Concepts*, 8th edn. (Wiley, New York, 2008)
13. A.S. Tanenbaum, A.S. Woodhull, *Operating Systems: Design and Implementation*, 2nd edn. (Prentice-Hall, New Jersey, 1997)
14. Intel(R) 64 and IA-32 Architectures Software Developer's Manuals. Intel Corp., http://www.intel.com/products/processor/manuals/
15. ARM(R) Architecture Reference Manual. ARM, Ltd.
16. B. Jacob, S.W. Ng, D.T. Wang, *Memory Systems: Cache, DRAM, Disk* (Morgan Kaufmann, Brulington, 2008)
17. J.L. Hennessy, D.A. Patterson, *Computer Architecture: A Quantitative Approach*, 4th edn. (Morgan Kaufmann, Brulington, 2006)
18. D. Roberts, T. Kgil, T. Mudge, Integrating NAND Flash Devices onto Servers. Communications of ACM **52**(4), 98–106 (Apr 2009)
19. J. Condit, E.B. Nightingale, C. Frost, E. Ipek, B. Lee, D. Burger, D. Coetzee, Better I/O through byte-addressable, persistent memory, in *Proceedings of the ACM SIGOPS 22nd Symposium on Operating Systems Principles*, Montana, Oct. 2009
20. K. Kim, G. Jeong, Memory Technologies for sub-40nm Node, in IEEE International Electron Devices Meeting (IEDM) Dig. Tech. Papers, Washington, DC, Dec 2007
21. Introduction to Intel's 32nm Process Technology, White Paper. Intel Corp., http://download.intel.com/pressroom/kits/32nm/westmere/Intel_32nm_Overview.pdf, 2009
22. D. Ha, K. Kim, Recent Advances in High Density Phase Change Memory (PRAM), in IEEE International Symposium on VLSI Technology Dig. Tech. Papers, Kyoto, June 2007
23. J.L. Pelloie, Y. Laplanche, T.F. Chen, Y.T. Huang, P.W. Liu, W.T. Chiang, M.Y.T. Huang, C.H. Tsai, Y.C. Cheng, C.T. Tsai, G.H. Ma, 65nm CMOS BULK to SOI comparison, in Proceedings of the IEEE International SOI Conference, Foster City, CA, Oct 2009
24. BSIM3v3.2.2 MOSFET User's Manual, Regents of University of California, 1999
25. D.A. Johns, D. Martin, *Analog Integrated Circuit Design* (Wiley, New York, 1997)
26. B. Razavi, *Design of Analog CMOS Integrated Circuits* (McGraw-Hill, New York, 2001)
27. Y.H. Kang, J.-K. Kim, S.W. Hwang, J.Y. Kwak, J.-Y. Park, D. Kim, C.H. Kim, J.Y. Park, Y.-T. Jeong, J.N. Baek, S.C. Jeon, P. Jang, S.H. Lee, Y.-S. Lee, M.-S. Kim, J.-Y. Lee, Y.H. Choi, High-Voltage Analog System for a Mobile NAND Flash. IEEE J. Solid State Circuits **43**(2), 507–517 (Feb 2008)
28. K.-J. Lee, B.-H. Cho, W.-Y. Cho, S. Kang, B.-G. Choi, H.-R. Oh, C.-S. Lee, H.-J. Kim, J.-M. Park, Q. Wang, M.-H. Park, Y.-H. Ro, J.-Y. Choi, K.-S. Kim, Y.-R. Kim, I.-C. Shin, K.-W. Lim, H.-K. Cho, C.-H. Choi, W.-R. Chung, D.-E. Kim, K.-S. Yu, G.-T. Jeong, H.-S. Jeong, C.-K. Kwak, C.-H. Kim, K. Kim, A 90nm 1.8V 512Mb Diode-Switch PRAM with 266MB/s Read Throughput, in IEEE ISSCC 2007 Dig. Tech. Papers, San Francisco, CA, Feb 2007, pp. 472–473
29. N.H.E. Weste, K. Eshraghian, *Principles of CMOS VLSI Design: A System Perspective*, 2nd edn. (Addison-Wesley, Boston, 1994)
30. Design Compiler™ User Guide, Synopsys, Inc.
31. B. Parhami, *Computer Arithmetic: Algorithms and Hardware Design* (Oxford University Press, New York, 1999)
32. I. Sutherland, B. Sproull, D. Harris, *Logical Effort: Designing Fast CMOS Circuits* (Academic, San Diego, 1999)
33. D.M. Pozar, *Microwave Engineering* (Wiley, New York, 1998)
34. S. Greenfield, *Brain Story: Why Do We Think and Feel as We Do?* (BBC Books, UK, 2000)
35. D.A. Patterson, Latency lags bandwidth. Commun. ACM **47**(10), 71–75 (Oct 2004)
36. G. Graefe, The Five-Minute Rule 20 Years Later (and How Flash Memory Changes the Rules). Commun. ACM **52**(7), 48–59 (July 2009)

37. R.F. Freitas, W.W. Wilcke, Storage-class memory: The next storage system technology. IBM J. Res. Dev. **52**(4) (July 2008)
38. M.K. Qureshi, V. Srinivasan, J.A. Rivers, Scalable High Performance Main Memory System Using Phase-Change Memory Technology, in *Proceedings of the 36th International Symposium on Computer Architecture (ISCA)*, Austin, TX, June 2009
39. IBM Power7 CPU, IBM, Corp. http://en.wikipedia.org/wiki/POWER7
40. H. Tanaka, M. Kido, K. Yahashi, M. Oomura, R. Katsumata, M. Kito, Y. Fukuzumi, M. Sato, Y. Nagata, Y. Matsuoka, Y. Iwata, H. Aochi, A. Nitayama, Bit Cost Scalable Technology with Punch and Plug Process for Ultra High Density Flash Memory, in IEEE International Symposium on VLSI Technology Dig. Tech. Papers, Kyoto, June 2007
41. S. Naffziger, Microprocessors of the future: Commodity or engine of growth? IEEE Solid State Circuits Mag. winter, 76–82 (2009)
42. The Linux Kernel Archives, http://www.kernel.org/

Chapter 7
Flash Memory

Taku Ogura

Abstract Non-volatile memory, especially Flash memory is a key device in various digital equipments. NOR Flash memory for code/data usage and NAND Flash memory for data storage have been industry-standard products by making the most of each feature, respectively. This chapter will mainly focus on these two types of Flash memory and compare from the view point of memory array architecture and operation schemes (Program, Erase and Read) with discussing reliability issues and cell scaling issues for next generation Flash technology.

Keywords Non-volatile memory • NOR flash memory • NAND flash memory

7.1 Introduction to Flash Memory

7.1.1 Introduction

Our life is filled by various modern electronic products. Semiconductor memories are essential parts of these products and have been growing in performance and density in accordance with Moore's law like all silicon technology. In this past decade, the most remarkable phenomenon in the field of semiconductor memories has been the explosive growth of the Flash memory market, driven by cellular phone and other types of electronic portable equipment (USB memory, digital audio player, digital still camera and so on). In recent years, SSD (Solid-State Drive) as mass storage for personal computers and enterprise servers is expected as a next killer application of Flash memory.

Flash memory is classified into NVM (Non-Volatile Memory) by its unique feature "non-volatility" that the data of memory is retained when power supply is

T. Ogura (✉)
GENUSION Inc., 7-1-3, Douicho, Amagasaki, Hyogo, 660-0083 Japan
e-mail: ogura.taku@genusion.co.jp

K. Iniewski (ed.), *CMOS Processors and Memories*,
Analog Circuits and Signal Processing, DOI 10.1007/978-90-481-9216-8_7,

turned off. Today Flash memory is outstanding presence in many types of NVM in terms of production volume and magnitude of sales.

Until now, many types of Flash-cell and architecture have been proposed, but today two of them can be considered as industry standard; NOR Flash memory, which is suited for code/data usage, and NAND Flash memory, which is optimized for data storage.

This chapter will mainly focus on these two types of Flash memory: NOR Flash memory and NAND Flash memory, and compare each Flash memory from the viewpoint of memory array architecture and operation schemes (Program, Erase and Read). In addition to basic SLC (Single Level Cell: 1bit/cell) technology, MLC (Multi Level Cell: 2 bits/cell) technology for reducing the cost per bit will be introduced. Furthermore the main reliability issues, such as endurance and data retention, will be discussed, together with the understanding of the basic physical mechanisms, which is valid for each Flash memory because these are based on the same concept of floating-gate MOS (Metal Oxide Semiconductor) transistor. Finally, cell scaling issues of each Flash memory will be covered, pointing out the main challenges.

7.1.2 Semiconductor Memory

Semiconductor memories are divided into the following two branches as shown in Fig. 7.1.

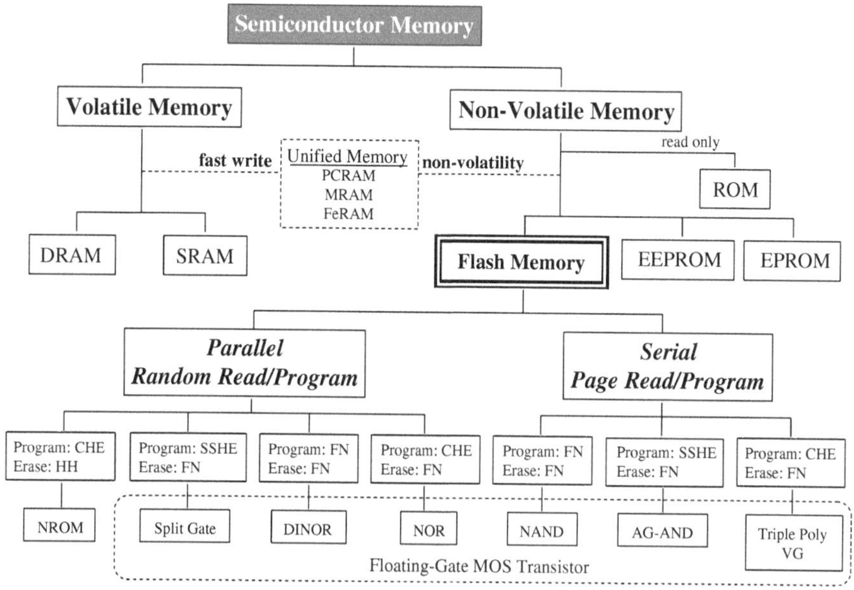

Fig. 7.1 Family tree of semiconductor memories

- *Volatile Memory*

DRAM (Dynamic Random Access Memory) and SRAM (Static Random Access Memory) are included in this category. These memories have been utilized as a work memory and cache memory for various electronic products by making the most of its fast read/write performance. However the data of these memories is not retained when power supply is turned off.

- *Non-Volatile Memory*

EPROM (Erasable and Programmable Read Only Memory), EEPROM (Electrically Erasable and Programmable Read Only Memory) and Flash memory are included in this category. Although these memories are less competitive than DRAM and SRAM in terms of read/write performance, the data of memory is retained without power supply.

Each non-volatile memory is classified by erase scheme. The EPROM cell is erased by UV (Ultra Violet) irradiation. On the other hand, both EEPROM and Flash memory offer the possibility to be erased on system without UV irradiation. The term "Flash" implies block erasure rather than byte-erasure capability of EEPROM.

Besides, so-called "unified memory" such as PCRAM (Phase Change RAM), MRAM (Magnetic RAM) and FeRAM (Ferroelectric RAM) which have both non-volatility and fast write performance have been proposed and developed. Particularly, PCRAM has been expected as a substitute for NOR Flash memory by making the most of its cell scalability. On the other hand, both MRAM and FeRAM have larger-sized cell than that of Flash memory, therefore these memories mainly have been used in embedded memory system like MCU and ASIC which needs lower density (<1 MB).

7.1.3 Flash Memory

Many types of Flash memory have been developed since the first Flash EEPROM has been proposed in 1984 [1]. As shown in Fig. 7.1, Flash memories can be divided in terms of access type, parallel or serial, and in terms of program/erase scheme, CHE injection, SSHE (Source-Side Hot Electron) injection, FN (Fowler–Nordheim) tunneling and HH (Hot Hole) injection. Each scheme has both advantage and disadvantage from the viewpoint of performance, power consumption, reliability and so on.

A parallel-access type of Flash memory is represented by NOR Flash memory. Its origin is "ETOX (EPROM with Tunnel Oxide)" cell proposed in 1988 [2] following "single transistor EEPROM" cell proposed in 1985 [3]. NOR Flash memory is characterized by its fast random access (<100 ns) and reliable storage. Code execution requires fast random access because code jumps from one memory location to another when branching. Furthermore code stored in Flash memory for direct execution must be retained with high reliability, since any errors will cause a system fault. NOR Flash memory generally guarantees 100% good bits without any need for ECC (Error Correcting Code) or controllers. For these reasons, NOR Flash memory is suited for code/data usage in cellular phone and embedded applications.

On the other hand, a serial-access type of Flash memory is represented by NAND Flash memory. Its origin is "Flash EEPROM with NAND structure" proposed in 1987 [4]. The advantages of NAND Flash memory are fast page programming and high density arrays, which lower die costs at the expense of random-access capability. The data in NAND Flash memory is accessed sequentially because the first access in a page is slow (<50 us). Therefore it is suited for data storage like audio and video player or data logging. NAND Flash memory typically does not guarantee 100% good bits, so NAND Flash memory requires ECC or controllers to fix the error data, it leads up to complexity of the memory system.

The main characteristics of semiconductor memories are summarized in Table 7.1. As has been mentioned above, each memory has both strong and weak points in terms of non-volatility, cell size, program/erase speed, read time, endurance and so on. Above all, cell size is the most important factor to realize high density memory at small cost. For this reason, in the field of non-volatile memory, middle-density NOR Flash memory for code/data usage and high-density NAND Flash memory for data storage have been developed by making the most of each feature, respectively. This fact is supported by Fig. 7.2 which shows the memory-density trend of each Flash memory; calculated results by using a die size demonstrated at ISSCC (International Solid-State Circuits Conference) in each year [5–24]. The memory density of each Flash memory has been growing every year in accordance with Moore's law. In general, the memory density of NAND Flash memory is about three times higher than that of NOR Flash memory in a same technology node based on the difference of each cell size; $4F^2$ or $10F^2$. The representation of the cell size in terms of number of "F^2" is a usual way to compare different technology node with the same metric, where "F" refers to the minimum feature on the technology node. Each Flash memory has available MLC technology for reducing the cost per bit. This MLC technology is a distinctive feature of each Flash memory which has contributed to driven the memory density aggressively since 2000s.

Table 7.1 Main characteristics of semiconductor memories

	Volatile memory	Unified memory		Non-Volatile memory	
	DRAM	PCRAM	MRAM	NOR Flash	NAND Flash
material	SiO_2 or Ta_2O_5	GeSbTe	CoFeB	SiO_2/Poly-Si	
cell size (F^2)	6~8	6~20	20~30	10	4
memory density	middle	low ~ middle	low	middle	high
MLC technology	unavailable	under exam.	unavailable	available	
program speed / data size	~100ns / 1W	~100ns / 1W	~10ns / 1W	~10us / 1W	200us ~ 800us/ 1KW ~ 8KW
erase speed / block size		not need		100ms~ls / 32KW~ 128KW	1ms~10ms 64KW~ 512KW
read time	~ 100ns	~100ns	~10ns	~100ns	~50us
endurance	> 1E+16	1E+6 ~ 1E+12	> 1E+16	1E+5	1E+5

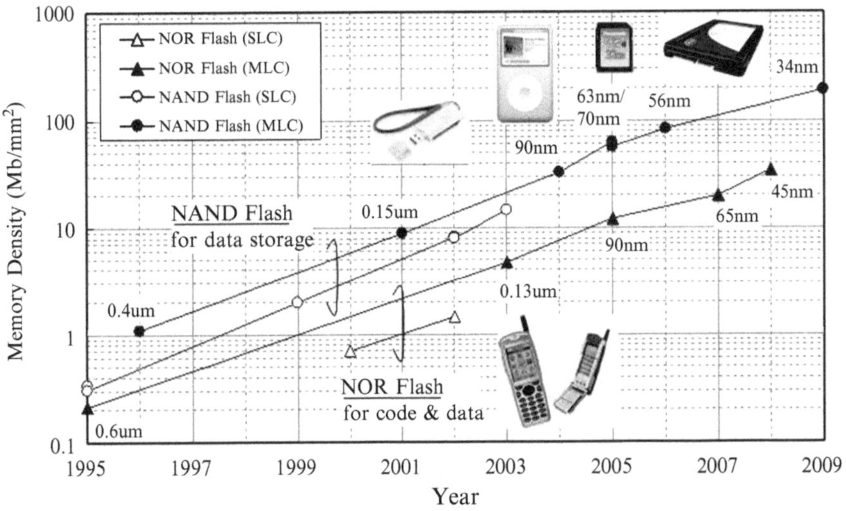

Fig. 7.2 Memory-density trend of Flash memory(calculated results by using a die size demonstrated at ISSCC in each year)

7.2 Flash Memory Architecture

7.2.1 *Chip Architecture*

The basic chip architecture is mostly common to each Flash memory. For reference, die photographs of each Flash memory are shown on a same scale in Fig. 7.3a, b, respectively. As shown in these photographs, each Flash memory array is divided into several partitions called "Bank" or "Plane" which includes individually erasable blocks. In general, NOR Flash memory has smaller-sized partitions than that of NAND Flash memory. In case of NOR Flash memory, a fast random-access capability (<100 ns) is required, therefore both word-line and bit-line length in memory array must be shortened to reduce a parasitic capacitance which has a significant influence on read time.

Figure 7.4 shows the block diagram of Flash memory. Each partition is controlled by individually allocated peripheral circuits. This divided-partition architecture has several merits as follows.

- *RWW (Read While Write)*

This means that Flash memory allows the host processor to read from one Bank while the other one is being programmed or erased. This configuration significantly increases the memory data throughput since it reduces the time the host processor is idle waiting for program/erase operations to be completed [7, 8, 10]. This kind of feature is often referred to as BGO (Back Ground Operation) [25] and generally

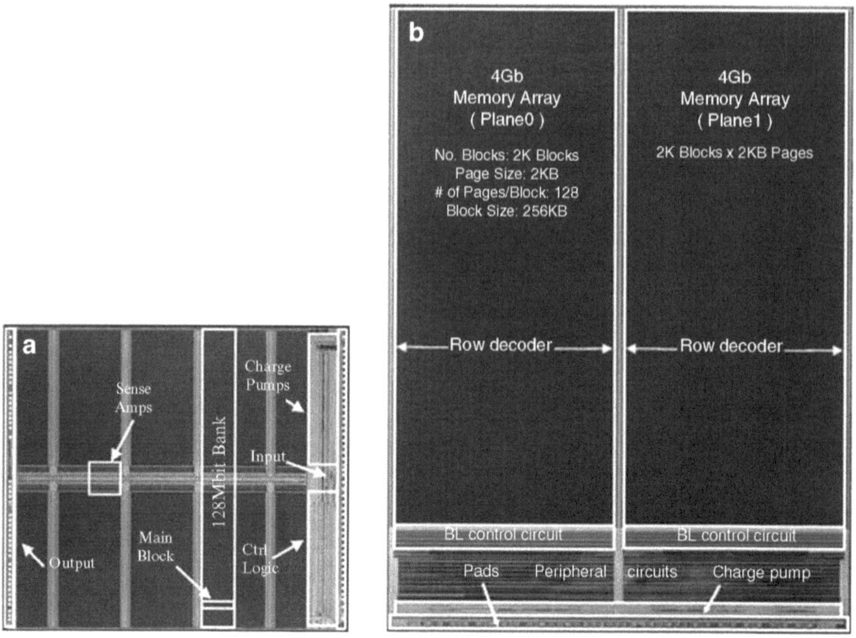

NOR Flash memory
65nm/1Gb/MLC/52.1mm^2
@2007 ISSCC, ST

NAND Flash memory
70nm/8Gb/MLC/145.5mm^2
@2005 ISSCC, Toshiba

Fig. 7.3 Die photographs of Flash memory

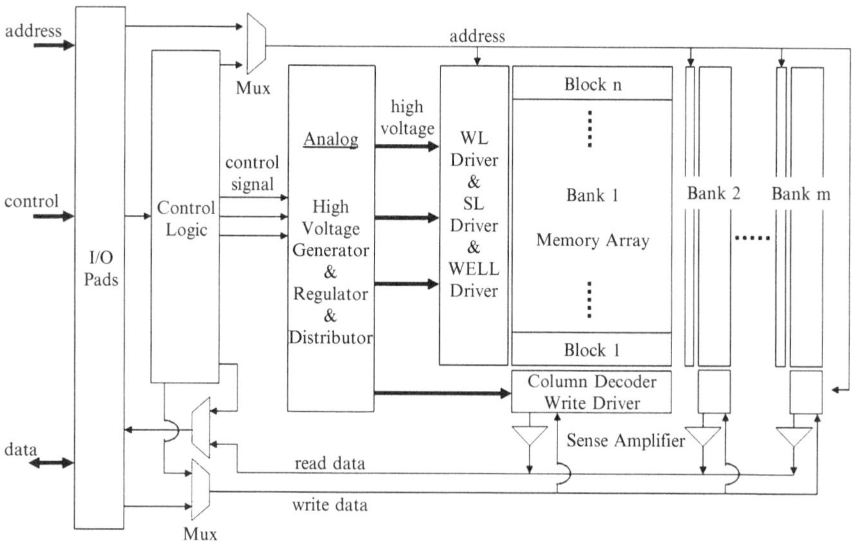

Fig. 7.4 Block diagram of Flash memory

adopted in NOR Flash memory as required by application platforms such as wireless communication systems.

- *Multi-Bank Program*

This means that several Banks can be programmed simultaneously, so it significantly increases the program throughput. This feature is usually adopted in NAND Flash memory, because FN-based programming current is very small and it allows increasing the parallel programming cells [15, 16, 23]. The programming scheme of each Flash memory will be described in detail later (Section 7.2.4).

As described in Section 7.1.3, various program/erase schemes such as CHE-injection and FN-tunneling are utilized in each Flash memory. These mechanisms require high voltage over power supply, therefore internal high-voltage generator (charge-pump circuit) and distributor to convey the generated high voltage to the memory array are required in a chip. For this reason, high-voltage transistor with graded junction and thick gate oxide (~15 nm@NOR Flash, ~40 nm@NAND Flash) is indispensible for these analog circuits and driver circuits connected to the memory array. This is one of the most distinctive features of Flash memory in comparison with other semiconductor memories. In addition to analog circuits, logic circuits which can control the Flash memory array for programming and erasing are indispensable.

The role of various circuit blocks in Flash memory can be summarized as follows.

- *Memory Array*

This is an array of Flash-cells, and the cell array is often divided into several blocks that can be individually erased.

- *WL Driver, SL Driver and WELL Driver*

These are the driver circuits connected to the memory array. Each driver applies the accurate voltage according to the operation mode to the gate, source and back-gate of Flash-cell, respectively.

- *Column Decoder, Write Driver and Sense Amplifier*

In read mode, this block connects the selected columns (bit-lines) to the sense amplifiers for judging the cell data. In program mode, write drivers are connected to the selected columns (bit-lines) to apply the programming voltage depending on the write data to the drain of Flash-cell.

- *Analog*

This block includes all the analog circuits to execute program, erase and read; especially high-voltage generator (charge-pump circuit), regulator to control the generated high voltage to the target level, distributor to convey the generated high voltage to the memory array and so on.

- *Control Logic*

This is the internal controller which decodes the commands issued by the host processor and executes the programming and erasing algorithm.

7.2.2 *Basic Operating Principles of Flash-Cell*

The basic Flash-cell structure is common to each Flash memory; it is a floating-gate MOS transistor formed in P-well surrounded by N-well in triple-well structure so that the various voltages for any operation can be applied to the back-gate of Flash-cell. As shown in Fig. 7.5a, FG (Floating Gate) is a poly-silicon completely surrounded by high-quality dielectrics, and electrically governed by a capacitively coupled CG (Control Gate). Usually the dielectric between the FG and the transistor channel is a silicon dioxide in the range of 9–10 nm and called "tunnel oxide" since FN-tunneling occurs through it. The dielectric that separates the CG from the FG is typically formed by a triple layer of ONO (silicon dioxide/silicon nitride/silicon dioxide) in the range of 15–20 nm of equivalent oxide thickness and called "interpoly dielectric".

A schematic symbol and a capacitor model of a floating-gate MOS transistor are shown in Fig. 7.5b, c, respectively. The voltage applied to the CG, S (source), D (drain) and B (back-gate) will be capacitively coupled to the FG. Therefore the voltage on the FG can be calculated as follows.

$$Vfg = Kg\ Vg + Ks\ Vs + Kd\ Vd + Kb\ Vb + Qfg / Ct \tag{7.1}$$

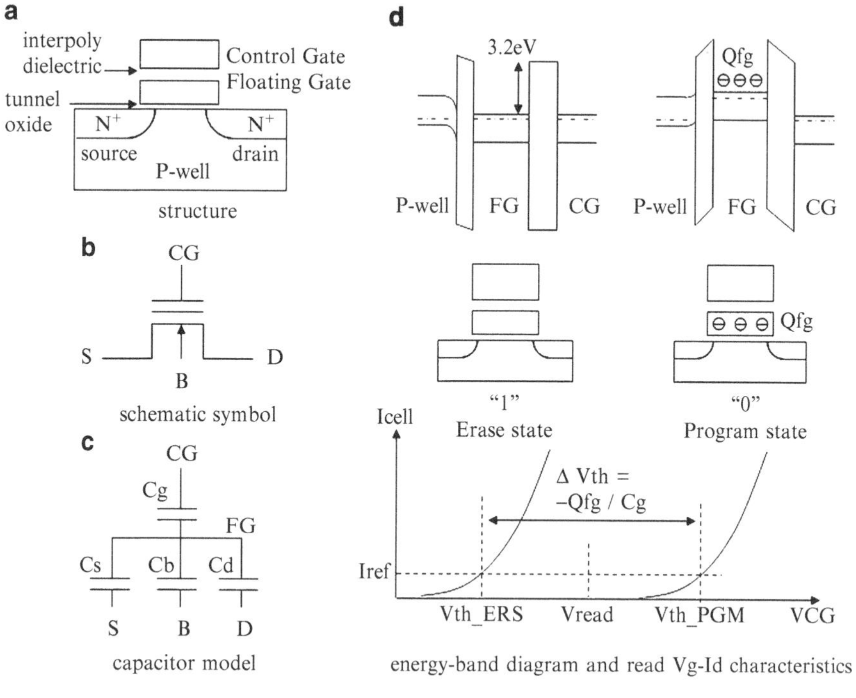

Fig. 7.5 Basic operating principles of Flash-cell

where

$$Kg = Cg / Ct, Ks = Cs / Ct, Kd = Cd / Ct, Kb = Cb / Ct$$
$$Ct = Cg + Cs + Cd + Cb$$
$$Qfg: \text{ stored charge in the FG}$$

Equation 7.1 means that Vfg changes in proportion to Qfg on condition that all of Vg, Vs, Vd and Vb are constant, in other words, read-cell current of Flash-cell varies according to the stored charge in the FG.

Figure 7.5d shows the energy-band diagram and read Vg-Id characteristics of Flash-cell in each state. As shown in this energy-band diagram, the FG is isolated by the high-energy barrier (3.2 eV) between the conduction band in the poly-silicon and the conduction bands in each dielectric; tunnel oxide and ONO interpoly dielectric. These barriers, much larger than the thermal energy, provide nonvolatile retention of the stored charge in the FG, consequently non-volatility of Flash-cell.

The neutral (or positively charged) state is associated with the logical state "1" (erase state) and negatively charged state, corresponding to electrons stored in the FG, is associated with the logical state "0" (program state). In general, each state can be determined by measuring the threshold voltage of Flash-cell which is defined by Vg applied to the CG on condition that the read-cell current "Icell" is equal to the reference current "Iref" as follows.

$$Vth = Vg\left(@Icell = Iref = constant\right) \tag{7.2}$$

Each state exhibits the same transconductance Vg-Id curve but shifted by a quantity that is proportional to the stored charge in the FG as follows.

$$\Delta Vth = Vth\_PGM - Vth\_ERS = -Qfg / Cg \tag{7.3}$$

In practical read operation using sense amplifiers, each state is determined by comparing the read-cell current with the reference current at a fixed control-gate voltage (= Vread). As shown in Fig. 7.5d, Icell of logical state "1" is larger than Iref, and Icell of logical state "0" is smaller than Iref on condition that Vread is applied to the CG.

7.2.3 Memory Array Architecture

As described in Section 7.2.2, each Flash memory is similar in terms of cell structure using a floating-gate MOS transistor, but memory array architectures differ widely from each other. Figure 7.6a, b shows the memory array architecture of each Flash memory, respectively, comparing the equivalent circuit diagram, top view and cross section views along bit-line and word-line.

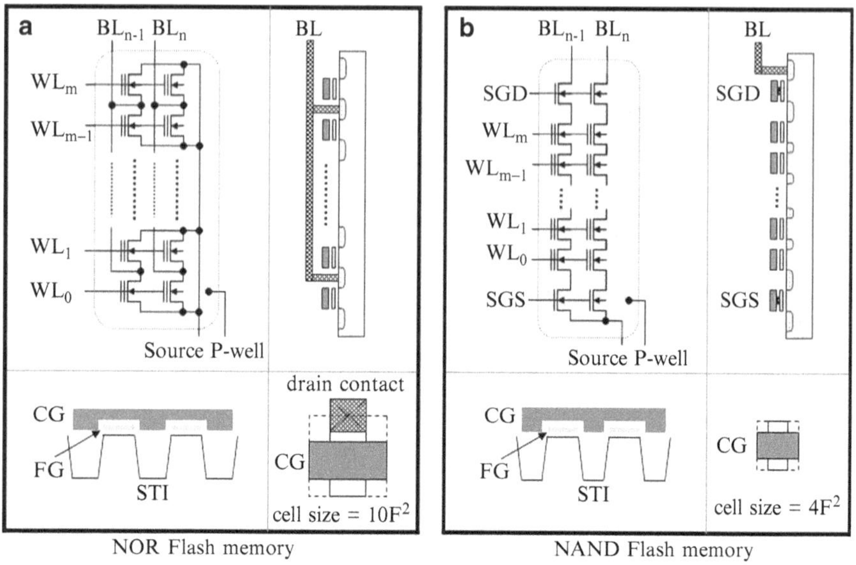

Fig. 7.6 Memory array architecture of Flash memory

- *NOR Flash Memory*

The most distinctive feature of NOR Flash array is that all the Flash-cells are directly connected to each metal bit-line with low resistance for realizing fast random access.

In the direction of bit-line, typically 128–512 cells share the same metal bit-line and each bit-line is repeated in parallel with a regular space. The drain contact is shared by two cells and all the drain contacts for a given column are connected to a common bit-line. This drain contact area is a main cause of larger-sized cell ($10F^2$) than that of NAND Flash memory ($4F^2$).

In the direction of word-line, typically 1,024–4,096 cells are connected to a common CG. Each word-line is repeated in parallel too, but the space between the word-lines is irregular since a drain contact is required for each pair of cells, whereas the source is not contacted as frequently (since it is common to all the Flash-cells in the array). It is not shown in this diagram, but the source regions are connected together for all the Flash-cells by diffusion, so-called "SAS (Self Aligned Source)" structure. In general, diffusion-source resistance is high (~200 Ω/cell), so it is periodically strapped with metal to prevent a voltage fluctuation due to large cell current during program/read operation.

Including above-mentioned drain-contact formation and high source-resistance problem, the scaling issues of NOR Flash memory will be described in detail later (Section 7.5).

- *NAND Flash Memory*

The most distinctive feature of NAND Flash array is that drain-contactless architecture with smaller-sized cells ($4F^2$) is adopted for realizing higher density memory at smaller cost than NOR Flash memory.

In the direction of bit-line, typically 16–64 cells are serially connected between two select transistors, which is called "NAND-string". NAND-string is connected to a metal bit-line thorough the SGD select transistor and a common source-line thorough the SGS select transistor. There is only one contact hole per two NAND-strings in NAND Flash array, which is suitable for miniaturizing the array and cell size. Each NAND-string is repeated in parallel with a regular space "2F" defined by STI (Shallow Trench Isolation) lithography.

In the direction of word-line, typically 32–64 K cells are connected to a common CG, which determines the page size (2–4 KB) for programming and reading. Each word-line is repeated in parallel with a regular space "2F" defined by CG wiring lithography.

As a result, NAND Flash memory can realize a smaller cell size "$4F^2$" than that of NOR Flash memory ($10F^2$).

In each Flash array, the CG is continuous for all the Flash-cells along the word-line, but each cell is isolated by STI and has its own unique FG to store the information for that cell location. The CG is shown to wrap around the FG, and it helps to improve the coupling coefficient "Kg" and lower the gate voltage "Vg" for any operation (ref. Eq. 7.1).

7.2.4 Program Operation

This section refers to program operation in SLC technology comprehensibly, MLC technology will be described later (Section 7.3). In each Flash memory, writing data "0" equals to raising the Vth of Flash-cell to higher program state, but writing data "1" equals to keeping it to lower erase state. Therefore the program operation means that electrons are injected to the FG of Flash-cell according to the write data.

Figure 7.7a, b shows a program-sequence flow and operating conditions in each Flash memory, respectively. As shown in this diagram, the basic sequence flow is common to each Flash memory. The program operation starts by input of "program" command and write data from external I/O pins. Then the program pulses are applied to the selected Flash-cell according to the write data until the Vth of selected Flash-cell reaches the target level with PV (Program Verify).

In general, programming characteristics of Flash-cells are different from each other, therefore the Vth of selected Flash-cell must be controlled precisely by combining a program-and-verify scheme with a staircase VCG ramp, so-called "ISPP (Incremental Step Pulse Programming) [10, 12, 17, 21, 22, 25]. This scheme should theoretically lead to Vth-distribution width of program state equal to ΔVCG. In fact, it is larger than ΔVCG due to several reasons given by the following equation.

$$\text{Vth-distribution width} \approx \Delta\text{VCG} + \Delta\text{Vread} + \Delta\text{Vsense} \tag{7.4}$$

where

ΔVCG: control-gate voltage step

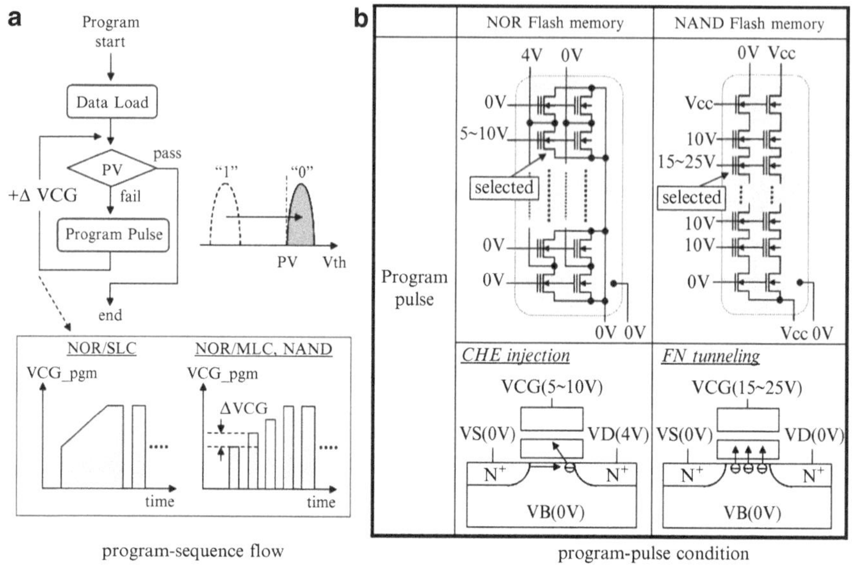

Fig. 7.7 Program operation of Flash memory

ΔVread: read (or verify) voltage fluctuation
ΔVsense: sense-amplifier inaccuracy

As shown in Eq. 7.4, the Vth-distribution width of program state can be tightened by decreasing ΔVCG. Obviously, this is paid in terms of a larger number of programming pulses together with verify phases, therefore the tradeoff relationship between Vth-distribution width and programming time must be chosen for each case considering the types of Flash memory (SLC or MLC) and application specification.

- *NOR Flash Memory*

The most distinctive feature of program operation in NOR Flash memory is that the CHE-injection mechanism is utilized for data programming.

In case of SLC technology, a smaller number of program-and-verify (equal to large ΔVCG) is preferable for fast programming without unnecessary tightening of Vth distribution. For this reason, in the "Program pulse" sequence, the selected word-line is biased at a gradual VCG ramp (5–10 V) for the first pulse, while unselected word-lines are set to 0 V. This gradual VCG ramp plays an important role in decreasing program-and-verify steps, also leveling of programming current by adjusting the VCG to the Vth of programming Flash-cell. The selected bit-line for writing data "0" is biased at 4 V by write driver, while the selected bit-line for writing data "1" and unselected bit-lines are set to 0 V. The source and P-well of memory array are set to 0 V.

On these bias conditions, the electrons gain energy from the lateral field at the drain side of the channel, and ones which have enough energy to surmount the Si-SiO_2 energy barrier are injected to the FG of the selected Flash-cell. Several models such as "lucky electron model" [26] and "effective electron temperature model" [27] have been proposed to describe this CHE-injection mechanism, but no closed-form analytical expression exists due to the complex two-dimensional nature of this phenomenon and many unknown physical parameters.

In general, CHE-injection programming is very fast (<10 us) but the programming current is large (~100 uA/cell) with low electron-injection efficiency, so the Flash-cells for parallel programming are restricted to at most 16–64 cells in terms of operation current and it determines a lower program throughput (~1 MB/s) in NOR Flash memory than that in NAND Flash memory (10–100 MB/s).

To increase the CHE-injection efficiency, CHISEL (Channel Initiated Secondary Electron) injection scheme in which P-well is biased at a negative voltage and allows the secondary impact ionization effect by the vertical field in the silicon has been proposed [28, 29]. CHISEL operation needs lower drain and gate bias for similar programming speed of conventional CHE programming.

- *NAND Flash Memory*

The most distinctive feature of program operation in NAND Flash memory is that the FN-tunneling mechanism is utilized for data programming with low power.

In contrast to NOR Flash memory, NAND Flash memory needs narrower Vth distribution of program state from the viewpoint of tradeoff relationship between a series-channel resistance of NAND-string and a read-disturb margin, therefore program operation is usually executed by ISPP scheme with appropriate ΔVCG (~0.5 V).

In the "Program Pulse" sequence, the selected word-line is biased at a staircase VCG ramp (15–25 V), while unselected word-lines are biased at 10 V. The selected bit-line for writing data "0" is set to 0 V by write driver, while the selected bit-line for writing data "1" and unselected bit-lines are set to VCC. The source and P-well of memory array are set to VCC and 0 V, respectively.

On these bias conditions, a strong electric field (in the range of ~10 MV/cm) across a tunnel oxide of the selected Flash-cell induces the FN-tunneling current, so the electrons are injected from the whole cell channel area to the FG of the selected Flash-cell.

In general, FN-based programming is slow (~1 ms) but the programming current is very small, so the Flash-cells for parallel programming can be increased to 16–128 K cells and more. This feature enables an extremely high program throughput (~100 MB/s@SLC, ~10 MB/s@MLC) for data storage with low power [23, 30].

During program operation in each Flash memory, two types of program-disturb mode must be taken into account.

1. *Drain Disturb*

In case of NOR Flash memory, all the Flash-cells connected to the selected bit-line biased at 4 V and the unselected word-line biased at 0 V are stressed by this drain-disturb mode. Some of the programmed cells might lose charge due to FN-tunneling or HHI (Hot-Hole Injection) at drain edge [31] and some of the erased cells might gain charge due to CHE-injection. In fact, the memory array is divided into several blocks and bit-lines in each block are isolated from each other by using a select transistor, so that total effective stress time is quite short and its influence is negligible.

In case of NAND Flash memory, all the Flash-cells connected to the selected bit-line biased at 0 V and the unselected word-line biased at 10 V are stressed by this drain-disturb mode. Some of the programmed cells might lose charge due to leakage current through the ONO dielectric layer from the FG to the CG and some of the erased cells might gain charge due to FN-tunneling through the tunnel oxide from the whole cell channel area to the FG.

2. *Gate Disturb*

In case of NOR Flash memory, all the Flash-cells connected to the selected word-line biased at a positive high voltage (5–10 V) and the unselected bit-line biased at 0V are stressed by this gate-disturb mode. This gate-disturb mode is similar to the drain-disturb mode in NAND Flash memory from the viewpoint of bias condition.

In case of NAND Flash memory, all the Flash-cells connected to the selected word-line biased at a positive high voltage (15–25 V) and the unselected bit-line biased at VCC are stressed by this gate-disturb mode. On these bias conditions, unselected NAND-string is electrically isolated from the bit-line and source-line, because both SGD and SGS select transistors are turned off. Therefore the cell-channel potential is boosted up by coupling effect with all the word-lines in NAND-string, so-called "channel-boost scheme". This channel-boost scheme can reduce the electric field across a tunnel oxide and prevent this gate-disturb mode [12].

As mentioned above, in case of NAND Flash memory, unselected word-line level determines the both modes of program-disturb immunity with tradeoff relationship. The higher level makes drain-disturb immunity worse, and the lower level makes gate-disturb immunity worse, respectively. Therefore it must be optimized to satisfy both program-disturb margins. In addition to above-mentioned "channel-boost scheme", "local self-boost scheme" [14] and "source-line programming scheme" [32] have been proposed to improve the program-disturb immunity.

7.2.5 Erase Operation

This section refers to erase operation in SLC technology comprehensibly, MLC technology will be described later (Section 7.3). In each Flash memory, erase operation is executed by lowering the Vth of all the Flash-cells in the selected block at once. It means that electrons are extracted from the FG of Flash-cell, typically by FN-tunneling mechanism.

Figure 7.8a, b shows an erase-sequence flow and operating conditions in each Flash memory, respectively. As shown in this diagram, the basic sequence flow is common to each Flash memory. The erase operation starts by input of "erase" command from external I/O pins. Then the erase pulses are applied to the selected block until the Vth of selected Flash-cell reaches the target level with EV (Erase Verify).

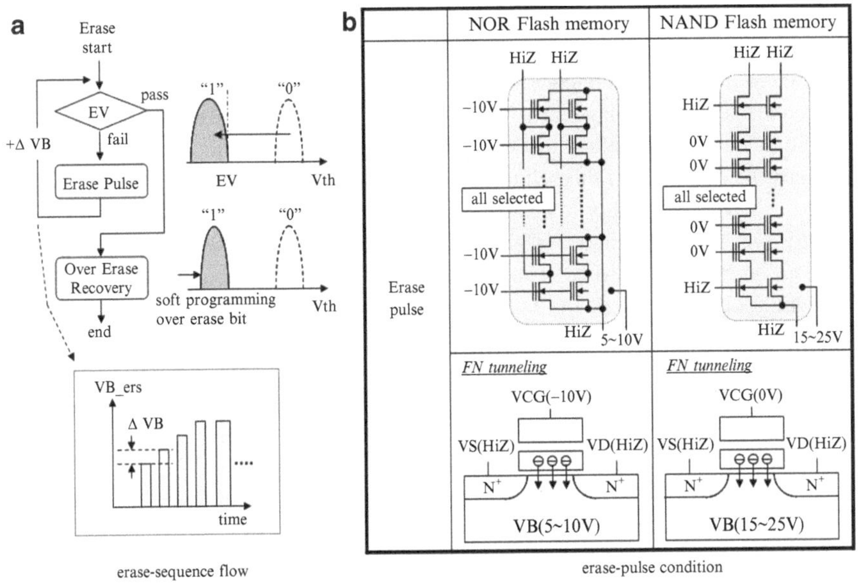

Fig. 7.8 Erase operation of Flash memory

The Vth of selected Flash-cell can be controlled by combining an erase-and-verify scheme with a staircase VB ramp like ISPP scheme in program operation. This staircase VB ramp scheme has a role of leveling the electric field across a tunnel oxide of the selected Flash-cell during FN-tunneling erase operation rather than tightening the Vth-distribution width [33]. As shown in the following equations, the electric field across a tunnel oxide can be kept constant if the VB ramp "ΔVb" is equal to the Vth shift of Flash-cell "–ΔVth".

$$\text{Eox} = (\text{Vb} - \text{Vfg})/\text{Tox} \tag{7.5}$$

$$\begin{aligned}\Delta\text{Eox} &= (\Delta\text{Vb} - \Delta\text{Vfg})/\text{Tox} \\ &= \{\Delta\text{Vb} - (1-\text{Kg})\,\Delta\text{Vb} - \Delta\text{Qfg}/\text{Ct}\}/\text{Tox} \quad (\text{from eq.}(1)) \\ &= (\Delta\text{Vb} + \Delta\text{Vth})\text{Kg}/\text{Tox} \quad (\text{from eq.}(3))\end{aligned} \tag{7.6}$$

where

Eox: electric field across a tunnel oxide
Tox: thickness of tunnel oxide

It is important to keep a lower constant electric field across a tunnel oxide for realizing the better reliability, because the generated traps in a tunnel oxide due to FN-tunneling stress is an increasing function of the maximum electric field. Including this cycling-induced degradation of tunnel oxide, the reliability issues of Flash memory will be described in detail later (Section 7.4).

In each Flash memory, the "Over Erase Recovery" sequence follows the "Erase Pulse" sequence during erase operation. As has been described in Section 7.2.3, NOR Flash-cells are directly connected to each bit-line, so the leakage current of unselected cells connected to same bit-line as selected cell influences the read current for judging the cell data, which is called "over-erase problem". Therefore, in case of NOR Flash memory, the main role of this sequence is to avoid this "over-erase problem", so over-erased cell with extremely lower Vth state is softly programmed by CHE-injection [7, 25]. On the other hand, NAND Flash memory does not have this "over-erase problem" through its NAND-string structure, therefore this sequence is adopted for restricting the influence of FG-FG interference by tightening the Vth distribution of erase state as narrower as possible by applying FN-based programming pulse to all the Flash-cells [20]. Including this FG-FG interference problem, the scaling issues of Flash memory will be described in detail later (Section 7.5).

- *NOR Flash Memory*

Until now, several FN-based erase schemes have been proposed and adopted in each generation NOR Flash memory as follows.

1. *Source Erase (generation 1)*

Flash-cells are erased by applying a positive high voltage to the source-diffusion area then extracting electrons from the FG to the source. This scheme has many problems in terms of operation current due to operating condition which is very close to junction breakdown, and source-structure scaling which needs high breakdown voltage.

2. *Negative-Gate-Source Erase (generation 2)*

Erasing voltage for source-erase scheme is divided between the CG and source. Flash-cells are erased by applying a negative voltage to the CG and a positive voltage to the source-diffusion area, then extracting electrons from the FG to the source. This scheme has reliability issues of data retention due to trapped holes in tunnel oxide generated by BTBT (Band-To-Band Tunneling) current during erase operation [34].

3. *Negative-Gate-Channel Erase (generation 3: now)*

This scheme is available by adopting the triple-well structure for memory array, and has several strong points in terms of operation current, reliability and cell scaling in comparison with other old schemes, therefore it is standard erase scheme for now NOR Flash memory [6, 7, 25].

In the "Erase Pulse" sequence using this "negative-gate-channel erase scheme", all the word-lines in the selected block are biased at −10 V and P-well of memory array is biased at a staircase VB ramp (5–10 V), while all the bit-lines and source of memory array are kept floating, in fact, charged through parasitic PN diode consisted of P-well and n+ diffusion.

On these bias conditions, a strong electric field (in the range of ~10 MV/cm) across a tunnel oxide of the selected Flash-cell induces the FN-tunneling current, so the electrons are extracted from the FG to the whole cell channel area of the selected Flash-cell.

- *NAND Flash Memory*

The erase scheme in NAND Flash memory is similar to that in NOR Flash memory from the viewpoint of utilizing FN-tunneling mechanism through whole cell channel area, but the operation voltages differ from each other. In case of NAND Flash memory, only positive voltage is utilized for any operation to simplify the peripheral circuits, particularly WL drivers (consisted of only one NMOS transistor) connected to the memory array. For this reason, erasing voltage is not divided between the CG and P-well. Moreover the P-well is common to all the blocks in NAND Flash memory to decrease the area penalty due to block isolation.

In the "Erase Pulse" sequence, all the word-lines in the selected block are set to 0V and P-well of memory array is biased at a staircase VB ramp (15–25 V), while all the word-lines in the unselected block are kept floating, in fact, charged to nearly VB by coupling effect with P-well.

On these bias conditions, all the Flash-cells in the selected block are erased by FN-tunneling mechanism like that in NOR Flash memory. On the other hand, in the unselected block, the electric field across a tunnel oxide is very low by coupling effect between the CG and P-well. Therefore there is no erase-disturb problem, although all the blocks are formed in a same P-well.

7.2.6 Read Operation

This section refers to read operation in SLC technology comprehensibly, MLC technology will be described later (Section 7.3). Each Flash memory has a common basic read concept which determines the cell data according to read-cell current, but read operation schemes in memory array differ from each other. Figure 7.9a, b shows the Vth-setting and operating conditions in each Flash memory, respectively.

- *NOR Flash Memory*

The most distinctive feature of read operation in NOR Flash memory is that fast random access (<100ns) can be achieved.

In case of NOR Flash memory, the Vth of data "1" must be set over 1.5 V to lower the leakage current and avoid the "over-erase problem". The Vth-distribution width of data "1" by erasing is about 2 V, so the highest Vth of data "1" is about 3.5 V. Typically the Vth of data "0" by programming is set over 6.0 V, therefore the data of selected cell can be judged by applying the middle voltage (typically ~5 V) between 3.5 and 6.0 V to the CG during read operation. This selected word-line voltage for read operation must be optimized to satisfy a read margin and prevent a read disturb.

A read operation is started by ATD (Address-Transition Detection) pulse generated by transition of internal address signal. The ATD pulse activates

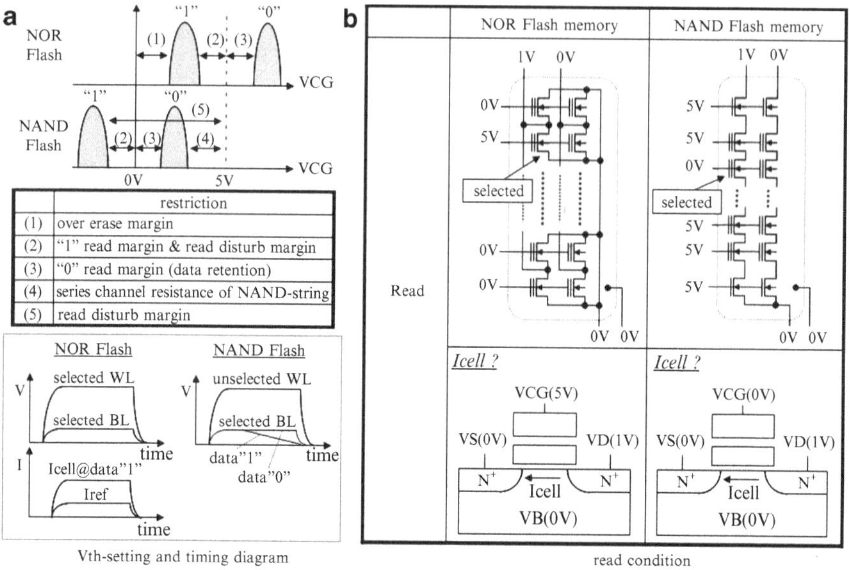

Fig. 7.9 Read operation of Flash memory

word-line drivers and sense amplifiers. The selected word-line is biased at 5 V, while unselected word-lines are set to 0 V. The selected bit-line is biased at 1 V by drain-bias control circuit in sense amplifier to prevent a read disturb, while unselected bit-lines are set to 0 V. The source and P-well of memory array are set to 0 V.

On these bias conditions, the selected read-cell current is compared with the reference current (typically ~5 uA) and judged by sense amplifier. The read-cell current of data "1" is typically over 10 uA and read time is within 100 ns.

In general, a read time in NOR Flash memory depends on the following factors.

1. *Word-line setup time*

It is mainly determined by CR delay of word-line (CG), so it is desirable to realize the shorter word-line length and lower-resistance material for CG.

2. *Bit-line setup time*

It is mainly determined by parasitic capacitance of bit-line, so it is desirable to realize the shorter bit-line length.

3. *Sense time*

It is mainly determined by the read-cell current, so it is desirable to realize the larger transconductance of Flash-cell. For this reason, it is important that the coupling coefficient "Kg" should be improved as shown in the following equation.

$$\text{Sense time} \approx \text{Cload}\,\Delta\text{Vsense}/\left(\text{Icell} - \text{Iref}\right)$$

$$\approx \text{Cload}\ \Delta\text{Vsense} / \text{Gm}\left(\text{Vread} - \text{Vth\_ERS}\right) \tag{7.7}$$

$$\text{Gm} = \Delta\text{Icell}/\Delta\text{Vg} = \text{Kg}\ \Delta\text{Icell} / \Delta\text{Vfg} \tag{7.8}$$

where

Cload: parasitic capacitance in sense amplifier
ΔVsense: voltage sensitivity in sense amplifier
Gm: transconductance of Flash-cell

- *NAND Flash Memory*

The most distinctive feature of read operation in NAND Flash memory is that large-sized data (2–16 KB) can be judged by sense amplifiers at once (~50 us), and data is outputted sequentially to I/O pins by one byte within 50 ns.

The Vth-setting for NAND Flash memory is different from that for NOR Flash memory. The Vth of data "1" can be set under –1 V, because NAND Flash memory does not have the "over-erase problem" through its NAND-string structure. The Vth of data "0" is set in the range of 1.0–2.5 V by ISPP scheme, and the unselected word-line level during read operation is set about 5.0 V so that it can satisfy the tradeoff relationship between a series-channel resistance of NAND-string and a read-disturb margin. The selected cell data can be judged by applying the middle voltage (typically ~0 V) between –1 and 1 V to the CG during read operation.

A read operation is started by input of "read" command from external I/O pins. By decoding its command, internal signal activates word-line drivers and sense amplifiers. The selected word-line is set to 0 V, while unselected word-lines are biased at 5 V. The selected bit-line is pre-charged at 1 V by drain-bias control circuit in sense amplifier to prevent a read disturb, while unselected bit-lines are set to 0 V. The source and P-well of memory array are set to 0 V.

On these bias conditions, the selected read-cell current is judged by sense amplifier. In contrast to NOR Flash memory, the read-cell current of data "1" is very small (<1 uA) due to its NAND-string structure with high series-channel resistance, therefore cell data is judged by the voltage swing of bit-line within constant sense time as following equation.

$$\Delta\text{Vswing} \approx \text{Tsense Icell} / C_{BL} \tag{7.9}$$

where

Tsense: sense time (constant)
C_{BL}: parasitic capacitance of bit-line
(*) 0.5 V = 10 us × 100 nA/2pF (for example)

In case of data "1", pre-charged level of selected bit-line is discharged thorough the on-state NAND-string, so ΔVswing is large. On the other hand, in case of data "0", pre-charged level is kept constant, so ΔVswing is nearly zero.

As mentioned above, a read-disturb problem exists in each Flash memory. In case of NOR Flash memory, the selected Flash-cell is stressed with gate voltage of 5 V and drain voltage of 1 V during read operation. If the Vth of selected Flash-cell is erase state, the channel is strongly on and a small amount of electrons may be injected to the FG by CHE-injection even with drain voltage of only 1 V. In case of NAND Flash memory, the unselected cells which are included in same NAND-string as selected cell may be stressed for same reason. Therefore it is important to optimize not only Vth-setting for each state but also read-bias condition for less stress. In general, a read-disturb immunity gets worse after program/erase cycling, because the generated traps in the tunnel oxide due to cycling stress induce the electron-trapping [35] and SILC-related charge gain [36] on read-bias condition, which result in Vth shift to the higher program state. Including this SILC (Stress Induced Leakage Current) problem, the reliability issues of Flash memory will be described in detail later (Section 7.4).

7.3 MLC Technology

7.3.1 Concept of MLC Technology

MLC technology is essential for realizing higher density memory with smaller cost in each Flash memory. Up to now, MLC technology usually has meant 2 bits/cell, but recently both 3 and 4 bits/cell have been proposed and developed in NAND Flash memory for mass data storage, respectively. In this section, the basic concept of 2 bits/cell will be introduced to represent the MLC technology.

As has been described in Section 7.2.2, the read-cell current varies according to the stored charge in the FG. This means that if the stored charge in the FG can be accurately placed to one of four charge states (one erase state and three program states), the Flash-cell can be said to store 2 bits. Each of the four charge states is associated with a 2-bit data pattern, and generally each bit is called LSB (Least Significant Bit) and MSB (Most Significant Bit), respectively.

In general, two types of MLC technology concept exist from the viewpoint of how to define the cell state as shown in Fig. 7.10a, b, respectively. One of them is "Vth-control scheme" adopted in each Flash memory, in which the cell state is defined by the threshold voltage of Flash-cell on the basis of constant reference current. On the other hand, "Icell-control scheme", it has the feature that the cell state is defined by the read-cell current on the basis of constant control-gate voltage. This scheme is adopted in NOR Flash memory only. Each scheme has both strong and weak points in terms of random-access capability and sensing accuracy as described later.

7.3.2 Precise Charge Placement in MLC Technology

In MLC technology for 2 bits/cell, one erase state and three program states must be placed precisely. Therefore it is important for ISPP scheme to be optimized in

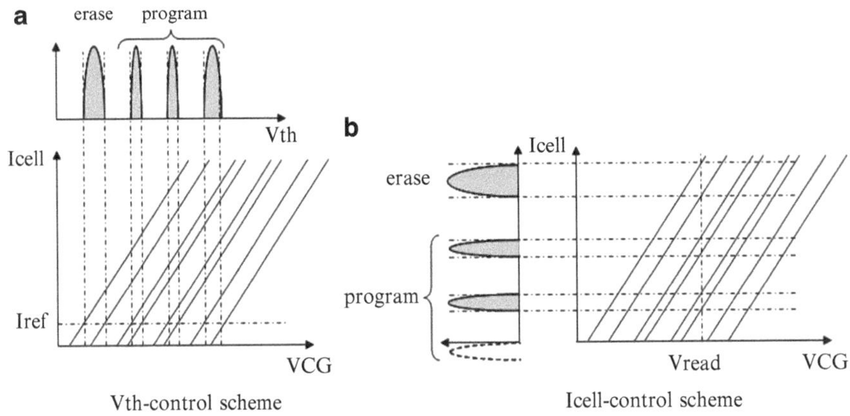

Fig. 7.10 Concept of MLC technology

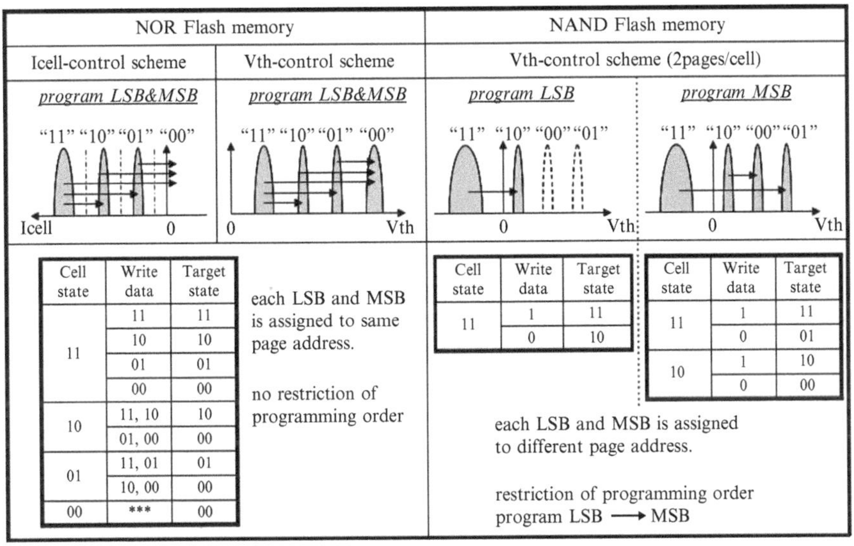

Fig. 7.11 Program operation in MLC technology

consideration of the tradeoff relationship between Vth-distribution width and programming time as described in Section 7.2.4. Program operation in MLC technology is summarized in Fig. 7.11.

- *NOR Flash Memory*

In case of NOR Flash memory, the lowest erase state and the highest program state are determined from the viewpoint of over-erase problem and data retention characteristics, respectively. Therefore these four states must be placed within the restricted range, typically in order of Vth (or Icell) associated with data "11", "10",

"01" and "00". Above all, data "00" is associated with the highest Vth state (equal to the smallest Icell state), because overwriting is allowed without a restriction of programming order in NOR Flash memory. Each LSB and MSB is assigned to different I/O data in same page address.

Data "11" is associated with erase state and narrower distribution width (typically <1 V or <20 uA) than that in SLC technology (typically ~2 V) must be achieved. Therefore the "Over Erase Recovery" sequence using CHE-injection mechanism with a staircase VCG ramp is executed for practically all the Flash-cells during erase operation. For this reason, erase-operation time in MLC technology is much longer (typically ~1 s) than that in SLC technology (typically ~200 ms).

Data "10", "01" and "00" are associated with program states, each state is set all at once by using ISPP scheme according to both write data and present cell state. Both states of "10" and "01" have adjacent states in both sides, therefore the distribution width must be tightened (typically 0.3–0.4 V or 6–8 uA) to improve a read margin. On the other hand, data "00" is associated with the highest Vth state (equal to the smallest Icell state), therefore wider distribution width is allowable. For these reasons, ISPP scheme is executed by two steps sequentially with a different ΔVCG, first step for both states of "10" and "01" with a smaller ΔVCG (typically 0.1–0.2 V) and second step for state "00" with a larger ΔVCG (typically 0.2–0.4 V), respectively [25].

- *NAND Flash Memory*

In case of NAND Flash memory, the lowest erase state and the highest program state are mainly determined from the viewpoint of read-disturb margin. Therefore these four states must be placed within the restricted range like NOR Flash memory, typically in order of Vth associated with data "11", "10", "00" and "01". In contrast to NOR Flash memory, each LSB and MSB is assigned to different page address with a restriction of programming order, in which MSB must be programmed after programming LSB [20, 22].

Data "11" is associated with erase state and wider distribution width than that in NOR Flash memory is allowable, because NAND Flash memory does not have the "over-erase problem" through its NAND-string structure. Therefore erase operation for MLC technology in NAND Flash memory is almost similar to that for SLC technology.

Data "10", "00" and "01" are associated with program states, but each state is not set all at once like NOR Flash memory. At first, state "10" is set by programming LSB from state "11", and next, both states of "00" and "01" are set by programming MSB from state "10" and state "11", respectively. In case of NAND Flash memory, the distribution width of each program state must be tightened (0.3–0.4 V) to lower the unselected word-line level during read operation to improve a read-disturb margin, keeping lower series-channel resistance. Therefore each program state is set by using ISPP scheme with a small ΔVCG (typically 0.1–0.2 V).

These program scheme and state arrangement in NAND Flash memory have several merits as follows.

1. *Fast program*

By assigning 2 bits of MLC into different page address, program-operation time for each page can be shortened, because it is unnecessary for three program states to be programmed all at once like NOR Flash memory. One state in case of programming LSB and two states in case of programming MSB only have to be programmed, respectively.

2. *Fast read*

By arranging the order of states "11", "10", "00" and "01", it is unnecessary for selected word-line level to be changed three times during read operation like "Vth-control scheme" in NOR Flash memory. Read operation can be executed by changing word-line level two times in case of reading LSB and one time in case of reading MSB, respectively. This reading scheme for MLC technology will be described later (Section 7.3.3).

7.3.3 Precise Charge Sensing in MLC Technology

Up to now, several sensing schemes for MLC technology have been proposed in each Flash memory. Read operation in MLC technology is summarized in Fig. 7.12.

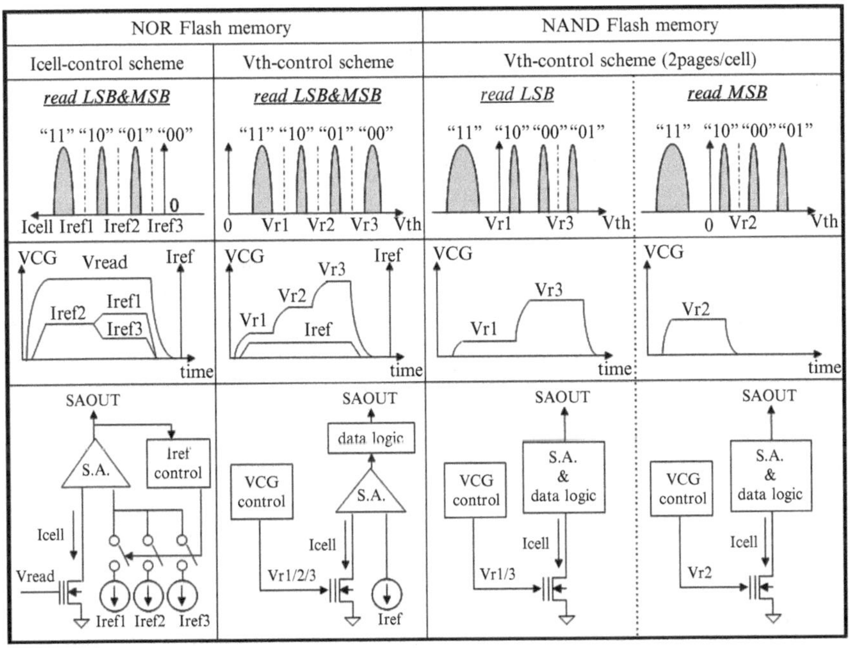

Fig. 7.12 Read operation in MLC technology

- *NOR Flash Memory*

In case of NOR Flash memory, typically 2 bits of MLC are assigned to different I/O data in same page address, therefore MLC data is outputted at once in each sensing scheme.

1. *Icell-control scheme*

In this scheme, the cell state is defined by the read-cell current on the basis of constant control-gate voltage. Therefore the cell states are determined by switching the reference current and comparing the read-cell current with each reference current two times sequentially such as binary search (1st-sense@Iref2 and 2nd-sense@ Iref1 or Iref3 selected by the 1st-sense result) [5, 8, 25]. Then final data is logically operated from these two output data as follows.

Data "11": "1"@Iref = Iref2 & "1"@Iref = Iref1
Data "10": "1"@Iref = Iref2 & "0"@Iref = Iref1
Data "01": "0"@Iref = Iref2 & "1"@Iref = Iref3
Data "00": "0"@Iref = Iref2 & "0"@Iref = Iref3

This sensing scheme has both merits and demerits as follows.

<Merits>

The selected word-line level is kept constant during read operation, so it is easy to realize a fast random access.

<Demerits>

At least three reference currents for read operation and four reference currents for PV & EV are necessary for each sense amplifier, so many reference currents must be set precisely at manufacturing test. Therefore this scheme needs to take account of read margin loss due to the distribution of many reference currents, it would strongly affect sensing accuracy when read-cell current gets smaller and smaller due to cell scaling.

2. *Vth-control scheme*

In this scheme, the cell state is defined by the threshold voltage of Flash-cell on the basis of constant reference current. Therefore the cell states are determined by applying 3-step read voltage (Vr1/Vr2/Vr3) to the selected word-line and comparing the read-cell current with constant reference current three times sequentially [9]. Then final data is logically operated from these three output data as follows.

Data "11": "1"@VCG = Vr1
Data "10": "0"@VCG = Vr1 & "1"@VCG = Vr2
Data "01": "0"@VCG = Vr1 & "0"@VCG = Vr2 & "1"@VCG = Vr3
Data "00": "0"@VCG = Vr1 & "0"@VCG = Vr2 & "0"@VCG = Vr3

This sensing scheme has both merits and demerits as follows.

<Merits>

Only one reference current is necessary for all the sense amplifiers, so this scheme does not need to take account of read-margin loss due to the distribution of many reference currents.

<Demerits>

The selected word-line level must be changed three times sequentially, so word-line setup time which is determined by mainly CR delay of word-line (CG) affects a random access.

As mentioned above, each control scheme has both merits and demerits. In general, Icell-control scheme has been adopted until 130 nm technology node and not after 90 nm technology node, because it becomes difficult to keep enough read-cell current to sense precisely due to cell scaling. Therefore Vth-control scheme which does not need a large cell current for judging the cell data has been adopted after 90nm technology node. It is not described, but another control scheme which is suitable for fast read by utilizing ramp read voltage instead of 3-step read voltage has been proposed [37].

- *NAND Flash Memory*

In case of NAND Flash memory, typically Vth-control scheme is adopted and 2 bits of MLC are assigned to different page address, therefore MLC data is outputted separately according to the accessed page address.

1. *Read LSB*

LSB data is determined by applying 2-step read voltage (Vr1/Vr3) to the selected word-line and judging the read-cell current two times sequentially. Then final data is logically operated from these two output data as follows.

Data "1": "1"@VCG = Vr1 or "0"@VCG = Vr1 & "0"@VCG = Vr3
Data "0": "0"@VCG = Vr1 & "1"@VCG = Vr3

2. *Read MSB*

MSB data is determined by applying read voltage (Vr2) to the selected word-line and judging the read-cell current as follows.

Data "1": "1"@VCG = Vr2
Data "0": "0"@VCG = Vr2

As described in Section 7.3.2, each read operation can be executed by changing the word-line level at most two times, therefore it can improve the read performance in NAND Flash memory.

During read operation in each Flash memory, read-cell current varies due to fluctuation of applied voltage to Flash-cell as given by the following equation.

$$\Delta \text{Icell} \approx \text{Gm}(\Delta \text{Vg} - \Delta \text{Vs}) + G_0(\Delta \text{Vd} - \Delta \text{Vs}) \tag{7.10}$$

where G_0: output conductance of Flash-cell

Furthermore read-cell current varies according to ambient temperature. Even small fluctuation of read-cell current can cause a fatal read error in MLC technology, because the current difference between each state is very small. For these reasons, it is important for variation of read-cell current on any conditions to be compensated so that the cell state can be judged correctly.

In case of NOR Flash memory which utilizes reference current for judging the cell data, the best solution is that the reference Flash-cell which has same structure and dimension as normal Flash-cell is utilized as a source of reference current. Reference Flash-cells are placed in array like normal memory array, and they are programmed to the target level at manufacturing test. In this way, both read-cell current and reference current varies almost equally even though read condition would change, and cell state can be judged correctly.

In case of NAND Flash memory which does not utilize reference current for judging the cell data, the fluctuation of read-cell current should be compensated by adjusting "ΔVswing" or "t_sense" (as shown in Eq. 7.9) according to the read condition.

In addition to these compensation schemes, it is essential for voltage fluctuation in memory array to be decreased. In each Flash memory, many Flash-cells are accessed simultaneously during read operation (16–256 cells @NOR Flash, 16–32 K cells @ NAND Flash), and total read-cell current flowing through common source depends on read-data pattern. This means that the source-voltage of Flash-cell fluctuates according to read-data pattern. Therefore common source must be periodically strapped with metal to lower the total resistance and decrease the voltage fluctuation [25, 38].

7.4 Flash Memory Reliability

7.4.1 Endurance

For the state-of-the-art Flash memory, generally 100 K program/erase cycling can be required. However Flash memory reliability is dominated by oxide-degradation effect, notably trap buildup in the tunnel oxide, which occurs as a result of program/erase cycling. This cycling-induced degradation appears in terms of program/erase speed. Figure 7.13a shows the cycling-induced Vth variation of each state while applying constant program/erase pulse, so Vth-window closing means the degradation of program/erase speed. As shown in this diagram, particularly the degradation of erase speed stands out in comparison with that of program speed.

In each Flash memory, erase is executed by FN-tunneling thorough the tunnel oxide. This FN-tunneling mechanism requires high electric field in the tunnel oxide, and this high electric field stress creates traps in the tunnel oxide and generates the interface states. The initial lowering of the Vth of erase state is due to pile-up of positive charge which enhances FN-tunneling efficiency, while the long-term increase of Vth is mainly due to negative charged traps. As shown in Fig. 7.13b, energy band is bended by these negative charged traps in the tunnel oxide during erase pulse, therefore the tunneling distance increases and FN-tunneling current decreases. Moreover these negative charged traps in the tunnel oxide raise the native Vth of Flash-cell, so it weakens the electric field in the tunnel oxide during erase pulse. In addition to above-mentioned degradation mechanism, the

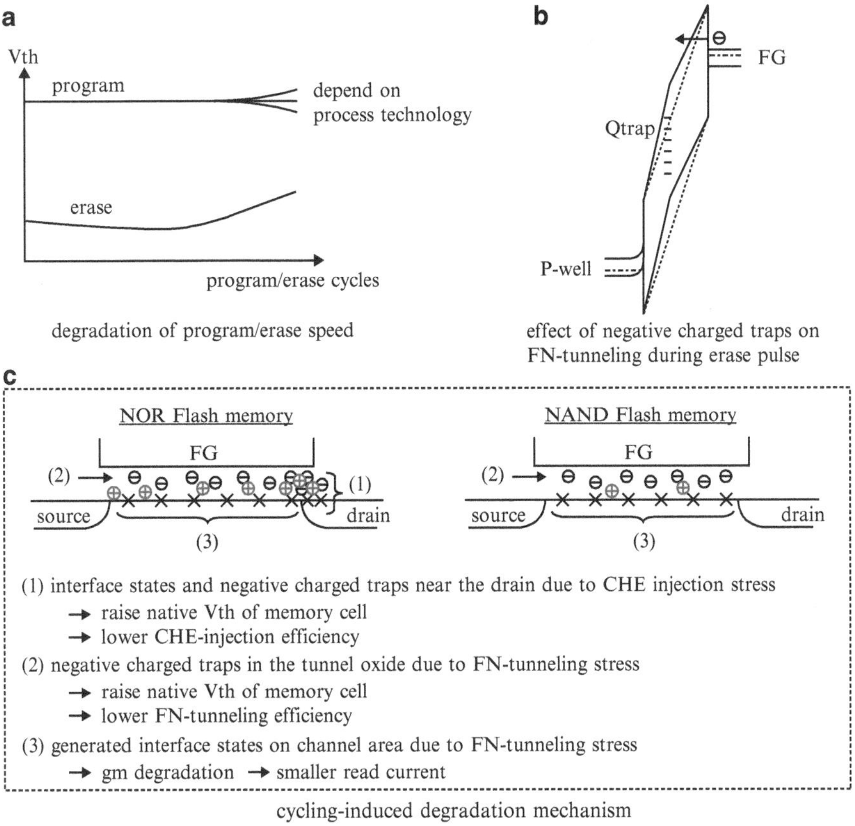

Fig. 7.13 Reliability issues of Flash memory – endurance

increase of interface states leads to the degradation of transconductance of Flash-cell due to mobility reduction. This means that Flash-cell must be erased further to lower the Vth to obtain enough read-cell current. For these reasons, erase speed slows down with program/erase cycling.

On the other hand, the degradation of program speed is usually less impact than that of erase speed. In each Flash memory, program is executed by either CHE-injection or FN-tunneling. In case of NOR Flash memory, the main degradation mechanisms by CHE-injection are generation of interface states and negative charged traps in the tunnel oxide near the drain junction. These negative charged traps not only raise the native Vth of Flash-cell but also lower the CHE-injection efficiency, however these effects compensate each other in terms of program speed. In case of NAND Flash memory, as in case of NOR Flash memory, the higher native Vth of Flash-cell and the lower FN-tunneling efficiency due to negative charged traps in the tunnel oxide compensate each other too. Like this, the degradation of program speed is less than that of erase speed, and it depends on each process technology. These cycling-induced degradation mechanisms of each Flash memory are summarized in Fig. 7.13c.

7.4.2 Data Retention

Non-volatile memory is defined by its unique feature "non-volatility" that the data of memory is retained when power supply is turned off. Data retention is the ability for the memory to retain the correct data over a prolonged period of time under storage conditions. Practically the stored charges in the FG can leak away thorough the tunnel oxide or interpoly dielectric, and trapped charges in the tunnel oxide can be detrapped. Both phenomena cause Vth shift of Flash-cell and can result in fatal read error. Figure 7.14a shows an example of data retention characteristics after program/erase cycling in case of MLC NOR Flash memory, mainly dominated by the charge-detrapping mechanism and SILC mechanism as described in detail later. Up to now, various types of charge-loss mechanism have been observed and evaluated in each non-volatile memory as shown in Fig. 7.14b. Each charge-loss mechanism is summarized as follows.

- *Intrinsic Charge Loss*

When charge is stored in the FG, electric field exists across the barrier oxides, which will tend to pull the charge out of the FG. This is the intrinsic charge loss.

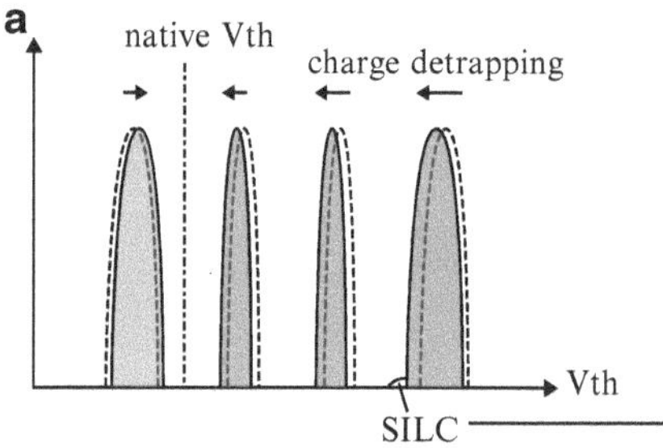

data-retention characteristics of MLC NOR Flash memory

b

mechanism	activation energy (eV)
intrinsic charge loss	1.4
oxide defects	0.6
ionic contamination	1.2
ONO	0.35
charge detrapping	1.1
SILC	0.25

charge-loss mechanisms and associated activation energies

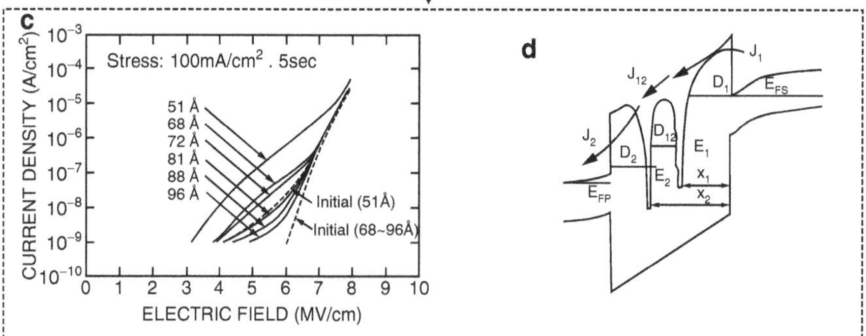

SILC (Stress Induced Leakage Current) K. Naruke et al. IEDM 1988

SILC-conduction model P. Cappelletti et al. IEDM 2004

Fig. 7.14 Reliability issues of Flash memory – data retention

In general, this mechanism is not dominant since the thermal emission barrier height of 1.4 eV is considerably higher than the activation energies for the other charge-loss mechanisms [39].

- *Charge Loss due to Oxide Defect*

This mechanism is related to defects in the tunnel oxide. It is typically a single bit and can be screened out at manufacturing test. The activation energy for this mode has been found to be about 0.6 eV [39].

- *Charge Loss due to Contamination*

This mechanism is related to contamination by positive ions. If the FG is not electrically isolated by surrounding dielectrics, these positive ions may compensate a part of negative charge in the FG. The activation energy for this mode has been found to be about 1.2 eV [39].

- *Charge Loss through ONO*

It is considered that there are three dominating charge-loss mechanisms through the ONO film as follows [40]. At the first phase, initial fast Vth shift which increases with nitride thickness is observed. It is considered that the physical origin could be carrier movement in the nitride or a nitride-polarization effect. The second phase is caused by movement of trapped electrons in the nitride film which are introduced during programming. It is considered that trap-to-trap hopping could be the transport mechanism of this phase. The activation energy for this phase has been found to be 0.35 eV. At the third phase, a long-term charge loss due to leaking thorough the top oxide is observed.

- *Charge Loss due to Charge Detrapping*

As described in Section 7.4.1, the charged traps in the tunnel oxide due to program/erase cycling shift the native Vth of Flash-cell. In other words, if these trapped charges can be detrapped under storage conditions, the Vth of Flash-cell changes and results in data retention problems [41]. The activation energy for this mode has been found to be about 1.1 eV.

- *Charge Loss due to Cycling-Induced Oxide Damage*

It is well known that high-field stressing of the gate oxides increases the low-field leakage current, so-called "SILC" as shown in Fig. 7.14c. SILC can cause serious limitations on data retention and read disturb in each Flash memory, which become more severe with tunnel-oxide thickness scaling down to 7–8 nm. SILC has been modeled as a trap-assisted tunneling process caused by the generated traps due to high-field stressing as shown in Fig. 7.14d [42]. This model is based on random placement of traps in the tunnel oxide, and only unlucky cell which has traps lined up to easy to tunnel through results in data loss even at room temperature. Furthermore it has been reported that SILC-related leakage-path can be deactivated at temperatures above 100°C, so the activation energy is very small (0.25 eV).

In general, when charge is stored in the FG, electric field exists across a tunnel oxide as given by the following equation.

$$\mathrm{Eox} = \mathrm{Vfg} / \mathrm{Tox} = \mathrm{Qfg} / \mathrm{Ct\ Tox} \quad (\text{from eq.}(1)) \tag{7.11}$$

As shown in Eq. 7.11, the electric field is directly proportional to the stored charge in the FG. This means that the more the charge in the FG, the larger the leak current due to built-in electric field by itself. Therefore this charge-loss mechanism is mainly related to the Flash-cells with the highest Vth state as shown in Fig. 7.14a.

7.5 Flash Memory Scaling

7.5.1 Cell Scaling Issues

Up to now, the memory density of each Flash memory has been growing every year in accordance with Moore's law. It has been based on the cell scaling with a size of $10F^2$ (NOR Flash) or $4F^2$ (NAND Flash) in each technology node. However, now beyond 45 nm technology node, Flash-cell scaling faces many challenging issues as shown in Fig. 7.15.

	scaling item	restriction	NOR Flash	NAND Flash
(1)	gate length	lithography	---	●
		voltage for CHE programming	●	---
(2)	contact size	lithography	●	---
(3)	contact/ gate space	voltage for FN erasing	●	---
(4)	source resistance	program/read current	●	---
(5)	trench isolation	lithography and voltage for programming	●	●
(6)	FG space	lithography and FG-FG interference	●	●

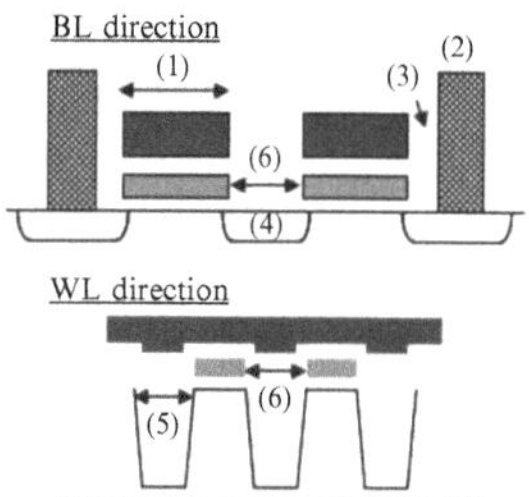

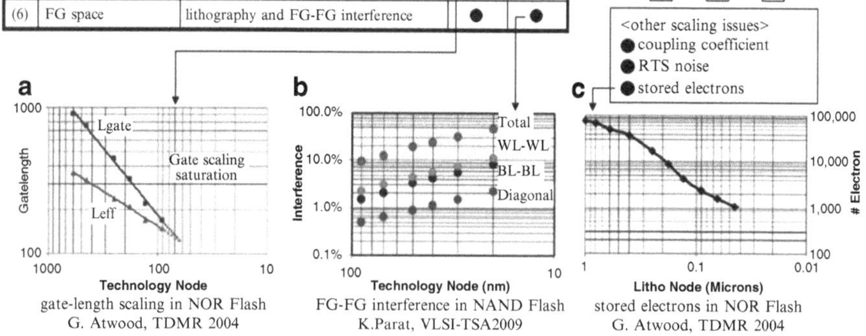

Fig. 7.15 Scaling issues of Flash memory

- *NOR Flash Memory*

1. *Gate Length*

In case of NOR Flash memory, program is executed by CHE-injection, therefore about 4 V is required between drain and source to generate electrons with high energy to surmount the 3.2 eV Si-SiO_2 barrier height. For this reason, the gate length of Flash-cell will be limited around 100 nm to withstand the required programming voltage as shown in Fig. 7.15a [43].

2. *Contact Size*

The NOR Flash array has a feature that the drain contact is shared by two cells, and this contact-formation process will be limited by lithography and etching process technology.

3. *Contact-Gate Space*

During erase pulse, about 20 V is applied between the CG and drain contact, therefore this space will be limited by the electrical isolation immunity of side-spacer dielectric.

4. *Source Resistance*

In case of NOR Flash memory, a large cell current (~100 uA@program, ~10 uA@ read) flows from drain to common source during each operation. The higher source resistance causes in the larger voltage fluctuation, and could result in the less CHE-injection efficiency and less read margin, respectively. For these reasons, LIC (Local Inter Connect) structure which connects each source of Flash-cell by metal with low resistance has been proposed instead of high-resistance SAS structure [25, 44, 45]. Furthermore this LIC structure can prevent the FG-FG interference between the Flash-cells placed along bit-line, therefore it could be strong candidate for NOR Flash array in the next generation.

5. *Trench Isolation*

Each Flash-cell is isolated by STI and about 4V is applied to STI during program operation. Therefore this space will be limited by not only the lithography but also the electrical isolation immunity of STI dielectric.

- *NAND Flash Memory*

1. *Gate Length*

In case of NAND Flash memory, the voltage difference between drain and source is very small (<1 V) during any operation. Therefore the gate length of Flash-cell will be limited by lithography only.

2. *Trench Isolation*

Each Flash-cell is isolated by STI and about 10 V is applied to STI during program operation. Therefore its restriction is same as that of NOR Flash memory.

In addition to above-mentioned several scaling issues in each Flash memory, there are common issues as follows.

1. *Floating Gate Space*

This space will be limited by FG-FG interference due to the increased proximity of neighboring cells. The Vth of a given Flash-cell becomes dependent on the Vth of the surrounding cells due to the capacitive coupling between each cell. This influence is given by the following equation on condition that all the neighboring cells shift from the erase state to the program state.

$$\Delta \mathrm{Vth} = (\mathrm{Vth\_PGM} - \mathrm{Vth\_ERS})\mathrm{Cfg} / (\mathrm{Ct} + \mathrm{Cfg}) \tag{7.12}$$

$$\mathrm{Kfg} = \Delta \mathrm{Vth} / (\mathrm{Vth\_PGM} - \mathrm{Vth\_ERS}) = \mathrm{Cfg} / (\mathrm{Ct} + \mathrm{Cfg}) \tag{7.13}$$

where

Cfg: total capacitance between a given FG and neighboring FGs
Kfg: coefficient of FG-FG interference

Equation 7.12 shows that ΔVth increases according to increase of Cfg due to cell scaling. For example, in case of NAND Flash memory, Kfg becomes as much as 40% by the 20 nm technology node as shown in Fig. 7.15b [46], which is quite unacceptable. Therefore several countermeasures against this FG-FG interference have been taken in terms of cell structure such as charge trapping memory [47] and program-sequence flow with smaller Vth swing [48].

2. *Coupling Coefficient*

As has been described in Section 7.2.3, generally the CG wraps around the FG, and helps to improve the coupling coefficient "Kg" and lower the gate voltage "Vg" for any operation. However it becomes more difficult for the CG to wrap the FG according to reduction of FG space in word-line direction due to cell scaling. Therefore the development of high-k interpoly dielectric which can take place of ONO film has been wanted in order to maintain a good coupling coefficient [49].

3. *RTS Noise*

RTS (Random Telegraph Signal) is well known as a fluctuation of channel current between discrete levels. In general, two-level fluctuation of channel current in deep-submicron MOS transistor is attributed to the capture and emission of single electron by oxide traps or interface states. In case of floating-gate MOS transistor like Flash memory, this influence is given by the following equation.

$$\Delta \mathrm{Vth} = \mathrm{q\,Tox} / \mathrm{Kg Weff\,Leff}\,\varepsilon\mathrm{ox} \tag{7.14}$$

Equation 7.14 shows that ΔVth increases according to reduction of channel width "Weff" and channel length "Leff". This means that RTS noise could be serious problem in terms of Flash-cell scaling, because the thickness of tunnel oxide "Tox" cannot be scaled down due to SILC-related problems, and the coupling coefficient "Kg" magnifies this effect. Furthermore program/erase cycling increases the traps in the tunnel oxide and interface states, so this RTS noise increases after cycling in Flash memory [50].

4. *Stored Electrons*

As the Flash-cell scales down, the tolerance to charge loss deteriorates, because the stored electrons in the FG decrease according to reduction of cell capacitance due to cell scaling as shown in Fig. 7.15c [43]. For example, in case of NOR Flash memory, the number of stored electrons in the FG is approximately 1,000 in a 45 nm technology node, therefore it necessitates an electron loss of a few electrons per year for 10-year data retention requirement.

7.5.2 Alternative Method for High Density

As has been described in Section 7.5.1, there are many issues of Flash-cell scaling due to two-dimensional area reduction. In this section, alternative method for realizing a high-density Flash memory will be described.

- *Three-Dimensional Approaches*

Up to now, two types of three-dimensional approach have been proposed in NAND Flash memory. One of them is memory array architecture which has two planes, one plane is stacked on another plane and each plane can be controlled independently [51]. This architecture can double the memory density without area penalty. Another approach is memory array architecture which has vertically connected NAND-string named "P-BiCS (Pipe-shaped Bit Cost Scalable)" [52]. In this architecture, SONOS-type Flash-cell is formed with a cylindrical shape in which cell channel is surrounded by the poly-gate. Theoretically, the higher density can be achieved by stacking the more cells in NAND-string.

- *MLC Technology (3bits/cell, 4bits/cell)*

MLC technology for 2bits/cell has been introduced in Section 7.3. Recently, both 3 and 4 bits/cell have been proposed and developed energetically for realizing high-density memory [53–55]. In principle, the concept of precise charge placement and sensing is similar to that of 2 bits/cell. Whether or not, eight states for 3 bits/cell and 16 states for 4 bits/cell must be placed and sensed more precisely. This means that longer program/read time is indispensable considering the tradeoff relationship between performance and sensing accuracy. It can cause more strict reliability issues due to small Vth window, therefore it would require more ECC bits, intelligent wear-leveling scheme, restriction of program/erase cycling and so on.

7.6 Conclusions

Flash memory has its unique feature "non-volatility" and has driven the various memory markets by growing in performance and density in accordance with Moore's law since 1980s. Especially, NOR Flash memory for code/data usage and NAND Flash memory for data storage have been industry-standard products, respectively. Each Flash memory has different memory architectures, operation schemes and electrical characteristics, but has been adopted for each application by making the most of each feature. In future, there are many issues for cell scaling beyond 45 nm technology node, but various approaches to solve these problems have to continue the Flash technology.

References

1. F. Masuoka et al., A new flash E^2PROM cell using triple polysilicon technology. *IEEE Dig. Tech. Papers, IEDM*, **30**, 464–467 (1984)
2. V. Kynett et al., An in-system reprogrammable 32K × 8 CMOS flash memory. IEEE J. Solid-St. Circ. **23**, 1157–1163 (October 1988)
3. S. Mukherjee et al., A single transistor EEPROM cell and its implementation in a 512K CMOS EEPROM. *IEEE Dig. Tech. Papers, IEDM*, **31**, 616–619 (1985)
4. F. Masuoka et al., New ultra high density EPROM and flash EEPROM with NAND structure cell. *IEEE Dig. Tech. Papers, IEDM*, **33**, 552–555 (1987)
5. M. Bauer et al., A multilevel-cell 32Mb flash memory. *IEEE Dig. Tech. Papers, ISSCC*, 132–133 (1995)
6. S. Atsumi et al., A channel-erasing 1.8V-only 32Mb NOR flash EEPROM with a bit-line direct-sensing scheme. *IEEE Dig. Tech. Papers, ISSCC*, 276–277 (2000)
7. T. Tanzawa et al., A 44mm^2 4-bank 8-word page read 64Mb flash memory with flexible block redundancy and fast accurate word-line voltage controller. *IEEE Dig. Tech. Papers, ISSCC*, **2**, 78–79 (2002)
8. D. Elmhurst et al., A 1.8V 128Mb 125MHz multi-level cell flash memory with flexible read while write. *IEEE Dig. Tech. Papers, ISSCC*, **1**, 286–287 (2003)
9. M. Taub et al., "A 90nm 512Mb 166MHz Multilevel Cell Flash Memory with 1.5MByte/s Programming", *IEEE Dig. Tech. Papers, ISSCC*, **1**, 54–55 (2005)
10. C. Villa et al., A 65nm 1Gb 2b/Cell NOR Flash with 2.25MB/s program throughput and 400MB/s DDR interface. *IEEE Dig. Tech. Papers, ISSCC*, 476–477 (2007)
11. J. Javanifard et al., A 45nm self-aligned-contact process 1Gb NOR flash with 5MB/s program speed. *IEEE Dig. Tech. Papers, ISSCC*, 424–425 (2008)
12. K. Suh et al., "A 3.3V 32Mb NAND flash memory with incremental step pulse programming scheme. *IEEE Dig. Tech. Papers, ISSCC*, 128–129 (1995)
13. K. Imamiya et al., A 35ns-cycle-time 3.3V-only 32Mb NAND flash EEPROM. *IEEE Dig. Tech. Papers, ISSCC*, 130–131 (1995)
14. T. Jung et al., A 3.3V 128Mb multi-level NAND flash memory for mass storage applications. *IEEE Dig. Tech. Papers, ISSCC*, 32–33 (1996)
15. K. Imamiya et al., A 130mm^2 256Mb NAND flash with shallow trench isolation technology. *IEEE Dig. Tech. Papers, ISSCC*, 112–113 (1999)
16. T. Cho et al., A 3.3V 1Gb multi-level NAND flash memory with non-uniform threshold voltage distribution. *IEEE Dig. Tech. Papers, ISSCC*, 28–29 (2001)
17. J. Lee et al., A 1.8V 1Gb NAND flash memory with 0.12um STI process technology. *IEEE Dig. Tech. Papers, ISSCC*, **1**, 104–105 (2002)

18. H. Nakamura et al., A 125mm^2 1Gb NAND flash memory with 10MB/s program throughput. *IEEE Dig. Tech. Papers, ISSCC*, **1**, 106–107 (2002)
19. J. Lee et al., A 1.8V 2Gb NAND flash memory for mass storage applications. *IEEE Dig. Tech. Papers, ISSCC*, **1**, 290–291 (2003)
20. S. Lee et al., A 3.3V 4Gb four-level NAND flash memory with 90nm CMOS technology. *IEEE Dig. Tech. Papers, ISSCC*, **1**, 52–53 (2004)
21. T. Hara et al., A 146mm^2 8Gb NAND flash memory with 70nm CMOS technology. *IEEE Dig. Tech. Papers, ISSCC*, **1**, 44–45 (2005)
22. D. Byeon et al., An 8Gb multi-level NAND flash memory with 63nm STI CMOS process technology. *IEEE Dig. Tech. Papers, ISSCC*, **1**, 46–47 (2005)
23. K. Takeuchi et al., A 56nm CMOS 99mm^2 8Gb multi-level NAND flash memory with 10MB/s program throughput. *IEEE Dig. Tech. Papers, ISSCC*, 507–508 (2006)
24. R. Zeng et al., A 172mm^2 32Gb MLC NAND flash memory in 34nm CMOS. *IEEE Dig. Tech. Papers, ISSCC*, 236–237 (2009)
25. T. Ogura et al., A 1.8-V 256-Mb multilevel cell NOR flash memory with BGO function. IEEE J. Solid-St. Circ. **41**, 2589–2600 (2006)
26. C. Hu, Lucky-electron model of channel hot electron emission. *IEEE Dig. Tech. Papers, IEDM*, **25**, 22 (1979)
27. E. Takeda et al., Submicrometer MOSFET structure for minimizing hot-carrier generation. IEEE J. Solid-St. Circ. **17**, 241–248 (April 1982)
28. S. Mahapatra et al., CHISEL flash EEPROM PART I: performance and scaling. IEEE Trans. Electron Dev. **49**, 1296–1301 (July 2002)
29. S. Mahapatra et al., CHISEL flash EEPROM PART II: reliability. IEEE Trans. Electron Dev. **49**, 1302–1307 (July 2002)
30. D. Nobunaga et al., A 50nm 8Gb NAND flash memory with 100MB/s program throughput and 200MB/s DDR interface. *IEEE Dig. Tech. Papers, ISSCC*, 426–427 (2008)
31. D. Ielmini et al., A study of hot-hole injection during programming drain disturb in flash memories. IEEE Trans. Electron Dev. **53**, 668–676 (April 2006)
32. K. Takeuchi et al., A source-line programming scheme for low-voltage operation NAND flash memories. IEEE J. Solid-St. Circ. **35**, 672–681 (May 2000)
33. R. Bez et al., A new erasing method for a single-voltage long-endurance flash memory. IEEE Electr. Dev. Lett. **19**, 37–39 (February 1998)
34. M. Suhail et al., Effects of fowler Nordheim tunneling stress vs. channel hot electron stress on data retention characteristics of floating gate non-volatile memories. Proc. IRPS, 439–440 (2002)
35. M. Kato et al., Read-disturb degradation mechanism due to electron trapping in the tunnel oxide for low-voltage flash memories. *IEEE Dig. Tech. Papers, IEDM*, 45–48 (1994)
36. S. Satoh et al., Stress-induced leakage current of tunnel oxide derived from flash memory read-disturb characteristics. IEEE Trans. Electr. Dev. **45**, 482–486 (February 1998)
37. C. Villa et al., A 125MHz burst-mode flexible read-while-write 256Mbit 2b/c 1.8V NOR flash memory. *IEEE Dig. Tech. Papers, ISSCC*, **1**, 52–53 (2005)
38. K. Takeuchi et al., A double-level-Vth select gate array architecture for multilevel NAND flash memories. IEEE J. Solid-St. Circ. **31**, 602–609 (April 1996)
39. R. Shiner et al., Data retention in EPROMs. Proc. IRPS, 238–243 (1980)
40. K. Wu et al., A model for EPROM intrinsic charge loss through oxide-nitride-oxide (ONO) interpoly dielectric. Proc. IRPS, 145–149 (1990)
41. R. Yamada et al., Analysis of detrap current due to oxide traps to improve flash memory retention. Proc. IRPS, 200–204 (2000)
42. P. Cappelletti et al., What we have learned on flash memory reliability in the last ten years. *IEEE Dig. Tech. Papers, IEDM*, 489–492 (2004)
43. G. Atwood, Future directions and challenges for ETox flash memory scaling. IEEE Trans. Dev. Mater. Reliab. **4**, 301–305 (September 2004)
44. M. Wei et al., A scalable self-aligned contact NOR flash technology. *IEEE Dig. Tech. Papers, VLSI Technol.*, 226–227 (2007)

45. R. Fastow et al., A 45nm NOR flash technology with self-aligned contacts and 0.024um^2 cell size for multi-level applications. *IEEE Dig. Tech. Papers, VLSI-TSA*, 81–82 (2008)
46. K. Parat, Recent development in NAND flash scaling. *IEEE Dig. Tech. Papers, VLSI-TSA*, 101–102 (2008)
47. C. Lee et al., Multi-level NAND flash memory with 63nm-node TANOS (Si-Oxide-SiN-Al_2O_3-TaN) cell structure. *IEEE Dig. Tech. Papers, VLSI Technol.*, 21–22 (2006)
48. K. Park et al., A zeroing cell-to-cell interference page architecture with temporary LSB storing and parallel MSB program scheme for MLC NAND flash memories. IEEE J. Solid-St. Circ. **43**, 919–928 (April 2008)
49. C. Chung et al., Electrical field dependence of data retention in high-k interpoly dielectrics. Proc. IRPS, 280–283 (2009)
50. H. Kurata et al., Random telegraph signal fin flash memory: its impact on scaling of multilevel flash memory beyond the 90-nm node. IEEE J. Solid-St. Circ. **42**, 1362–1369 (2007)
51. K. Park et al., A 45nm 4Gb 3-dimensional double-stacked multi-level NAND flash memory with shared bitline structure. *IEEE Dig. Tech. Papers, ISSCC*, 510–511 (2008)
52. R. Katsumata et al., Pipe-shaped BiCS flash memory with 16 stacked layers and multi-level-cell operation for ultra high density storage devices. *IEEE Dig. Tech. Papers, VLSI Technol.*, 16–18 (2009)
53. S. Chang et al., A 48nm 32Gb 8-level NAND flash memory with 5.5MB/s program through-put. *IEEE Dig. Tech. Papers, ISSCC*, 240–241 (2009)
54. T. Futatsuyama et al., A 113mm^2 32Gb 3b/cell NAND flash memory. *IEEE Dig. Tech. Papers, ISSCC*, 242–243 (2009)
55. C. Trinh et al., A 5.6MB/s 64Gb 4b/cell NAND flash memory in 43nm CMOS. *IEEE Dig. Tech. Papers, ISSCC*, 246–247 (2009)

Chapter 8
CMOS-based Spin-Transfer Torque Magnetic Random Access Memory (ST–MRAM)

B.C. Choi, Y.K. Hong, A. Lyle, and G.W. Donohoe

Abstract Recently it has been demonstrated that the vortex core magnetization can be switched by applying a short (<100 ps) magnetic field pulse. This is an important step towards utilizing the two stable states of the core magnetization in memory device elements. For practical applications, however, it is desirable to find a more practical method, i.e., current-driven switching of the magnetization state of the vortex. In this chapter we discuss detailed studies of the spin-torque transfer effect on the vortex magnetization in a multilayered $Ni_{80}Fe_{20}$/Cu/Co nanoelement, in which the chirality in the NiFe can be controllably switched by injecting a current pulse. The controlled switching of chirality can be achieved by flipping the polarity of current pulses and probed by detecting the changes in the electrical resistance due to the giant magnetoresistance (GMR). From an application point of view, the current pulse induced switching suggests a new way to implement the vortex magnetization in memory device applications without magnetic fields. Our approach will lead to a new integration of magnetic components into CMOS technology, giving rise to an all-electrically controlled magneto-electronic device.

Keywords Magnetic random access memory • Spin transfer torque • Nanomagnets • Magnetization dynamics

B.C. Choi (✉)
Department of Physics and Astronomy, University of Victoria,
Victoria, BC V8W3P6, Canada
e-mail: bchoi@uvic.ca

Y.K. Hong
Department of Electrical and Computer Engineering, University of Alabama,
Tuscaloosa, AL 35487-0118, USA

A. Lyle
Department of Electrical and Computer Engineering, University of Alabama,
Tuscaloosa, AL 35487-0118, USA

G.W. Donohoe
Department of Electrical and Computer Engineering, University of Idaho,
Moscow, ID 83844, USA

K. Iniewski (ed.), *CMOS Processors and Memories*,
Analog Circuits and Signal Processing, DOI 10.1007/978-90-481-9216-8_8,

8.1 Introduction

The magnetism in ultrathin magnetic elements, where the thickness is on the nanometer scale, has become one of the most vigorous research areas in condensed matter physics and materials science [1–3]. The subject of thin film magnetism is stimulated both by discoveries of a steadily increasing range of magnetic phenomena and by the increasing interest in advanced magnetic information storage and data process technology. Over the last decades, novel magnetic phenomena have been found in ultrathin magnetic materials. These include the two-dimensional (2D) magnetic phase transition, magnetic interactions, surface magnetic anisotropies and 2D magnetic ordering phenomena, which have all yielded important new fundamental insights. Particularly important findings from the application point of view include the discovery of giant magnetoresistance (GMR) effect [4], interlayer exchange coupling and spin dependent tunneling. Tremendous possibilities of novel magnetic phenomena for technological applications become immediately apparent by considering that most of the information we deal with is processed and stored magnetically, from audio and video products to information storage on computer hard disks. One example of such practical applications is GMR effect, in which its extreme sensitivity to the magnetic field is used in computer hard disks to enhance information storage density.

Recently attention has focused on the dynamic behavior of the magnetization in small patterned magnetic elements, fabricated by using sophisticated lithography technique [5–8]. This is mainly because magnetization dynamics in thin continuous films on short time scales is different in many aspects from the static case [3, 9]. Moreover, the magnetization dynamics in small patterned elements significantly differs from that in continuous films due to the magnetostatics of element edges, thus modifying the equilibrium states of the element in terms of the magnetic moment distribution [10, 11]. From a practical point of view, understanding magnetization dynamics on nano- and pico-second time scales in small elements with dimensions in the micrometer size regime and below has become crucial, owing to the increasing demands on conventional storage technologies and for newer approaches such as magnetic random access memories (MRAM) [12, 13]. For example, the data rate for current hard disk magnetic recording device is approaching 1 Gbit/s. Motivated by all of these accumulated interests, dynamic behavior in micro- and nanosized magnets is being actively studied by a number of groups [14–18], and there has been substantial progress in recent years in the understanding of the magnetic domain configuration and its correlation to magnetization dynamics in micro- and nano-sized patterned elements [19–23]. In particular the high-speed dynamics in the vortex magnetization state has drawn increasing attention due to its possible applications in high-density magnetic storage devices [24–27]. In Ref. [25, 26], it was found that the vortex core magnetization can be switched by applying a very short (<100 ps) magnetic field pulse with an appropriate strength and duration. The switching occurs by means of the formation of additional vortex–antivortex pair and the subsequent annihilation of the initial vortex core with the antivortex. Very recently, the switching of the polarity of the vortex core was experimentally observed in Permalloy ($Ni_{80}Fe_{20}$) elements by excitation with short bursts of an alternating magnetic field [24]. For practical applications, however, it is desirable to find a more convenient method, i.e., current-driven switching of the magnetization state of the vortex. The spin-transfer effect has been theoretically predicted by Slonczewski [28] and Berger [29], and has been verified by numerous experiments with the rapid development of nanofabrication technology [30–32]. Recently, Caputo et al. [33] reported the effect of the spin-polarized current on the vortex core magnetization, in which the polarity switching of a vortex core was predicted. Another topological index characterizing the state of the vortex structure is the vortex chirality, which refers to the clockwise (CW) or the counterclockwise (CCW) rotational direction of the in-plane curling magnetization. It is known that the vortex chirality in a magnetic disk is hard to control by applying a magnetic field [34]. According to the result in Ref. [33], the switching process of the core polarization does not affect the vortex chirality. Very recently, however, Choi et al. demonstrated that the chirality of the vortex state in multilayered magnetic nanopillar structures could be controllably switched by applying spin-polarized current pulse with appropriate amplitude, polarity, and duration [35].

In this chapter, a new paradigm in the magnetoelectronic application, i.e., spin-transfer torque induced magnetization switching in multilayered nanopillar magnetic memory elements, is discussed. In order to detail the dynamics of magnetization switching in multilayered nanopillar elements, numerical and experimental studies are carried out with varying size and chemical composition of the magnetic elements. In particular, we discuss a new type of random access memory (RAM) based on the vortex structure in magnetic multilayers. As discussed in Ref. [35], if all of the layers initially have their curling magnetization parallel, the device will be in the low-resistance MR state and the high resistance MR state will be achieved when the soft magnetic layer switches to the antiparallel magnetization. This type of magnetic elements will provide an excellent opportunity to implement the toggle switching mechanism of the vortex chirality in memory device applications, and will lead to a newly designed magnetoelectronic memory device operating exclusively by spin-polarized current pulses without applying magnetic fields. This approach will lead to a new integration of magnetic components into CMOS technology, giving rise to an all-electrically controlled magneto-electronic device.

8.2 CMOS-based ST–MRAM Elements

8.2.1 Background

There is an increasing demand for nonvolatile, high density, high endurance, and high speed memory for applications that range from cell phones to satellites. A promising candidate to meet these criteria is magnetic random access memory (MRAM), which utilizes the magnetic resistance (MR) between a free and fixed magnetic layer to store information in the form of zeros and ones.

Table 8.1 compares the technologies currently in use for memory application [36]. Dynamic random access memory (DRAM) is widely used due to large density, high endurance, and low manufacturing cost. However, DRAM is a volatile memory which means that when power is lost the information is lost. Static RAM (SRAM) is a semiconductor memory that uses bi-stable latching circuitry to store information. It offers faster read/write times and lower power consumption than DRAM. A few of the problems with SRAM are that it has low density and it is a volatile memory. The most popular nonvolatile memory is Flash which has been used for a variety of applications such as jump drives and digital cameras. Some of the problems with Flash are that it has low endurance, long read/write times, and high active power. Spin-transfer torque MRAM (ST–MRAM) offers low power consumption, fast read/write times, and high endurance which makes it a promising candidate for memory applications.

MRAM technology is based on the giant magnetoresistance (GMR) effect, i.e., the change of a material's electrical resistance caused by an applied magnetic field, which was first discovered by Fert and Grünberg in 1988 [4, 37]. A simple structure that utilizes GMR for MRAM is shown in Fig. 8.1a. For simplicity, all electron spins are assumed to be either parallel or anti-parallel to the magnetization. When the magnetization (shown with the arrows in Fig. 8.1a) in two magnetic layers are in parallel, half of the electrons will be reflected from the first layer and the other half will pass through both layers resulting in a relatively low resistance. In the case of the anti-parallel state neither spin orientation is allowed to pass through the junction resulting in a relatively high resistance. The high and low resistances are used in MRAM to represent '1' and '0', respectively. This effect is also used in hard disk drives in the read heads which sense the small fields generated by the disk's magnetization domains.

For practical applications, however, it is desirable to find a more convenient method to switch the magnetization state of the MRAM elements. In 1996, Slonczewski and Berger [28, 29] predicted that a spin-polarized electron current could exert a torque on the local magnetization of a ferromagnetic material due to spin momentum transfer. Since then, it has been experimentally proven that a spin-polarized electron current exerts a torque on the local magnetization of a ferromagnetic material due to spin momentum transfer (SMT) [38–41]. This torque can be used to manipulate nano-scale level ferromagnetic elements by reversing the direction of the magnetization or by inducing microwave oscillations. When applied to a magnetic tunneling junction (MTJ) or giant magneto-resistance (GMR) stack, SMT can be used to control magnetic memory or oscillator devices. The use of current-driven switching has allowed simpler more reliable structures and lower manufacturing cost for MRAM, as shown in Fig. 8.1b.

The most common method to polarize current is to pass it through ferromagnetic materials as shown in Fig. 8.2a, b. Initially, the electrons that comprise the current will have their spin oriented with an angle. However, when the current is passed through a ferromagnetic material, the electron's spin will oriented itself with the magnetic moment of the material. This causes the current to become spin-polarized

Table 8.1 List of technologies currently in use for memory devices

	DRAM	SRAM	FLASH	ST–MRAM
Cell area	6F2	70F2	4F2/10F2	6–8F2
Density (50 nm)	4 Gb	32 Mb	16 Gb/2 Gb	>1 Gb
Shrink factor	Cap area	Cell area	Litho (Tr)	Litho (Tr)
R/W access time	<70 ns	<20 ns	100 ns–10 μs	<25 ns
Clock	500 MHz (Sync)	500 MHz/66 MHz	100 MHz (read only)	143 MHz
Active power	<5 V	< 5 V	>10 V	<5 V
Stand by current	~mA	0.1 μA–mA	~10 μA	~10 μA
Endurance	1E15	1E15	1E16	1E15
Data retention	×	×	>10 year at 85°C	10 year at 85°C
Application	Main memory	Cash memory	Mobile storage	Mobile main memory

S. C. Oh, paper No. BA-02, International Symposium on Advanced Magnetic Materials and Applications, May 28–June 1, 2007, Jeju, South Korea

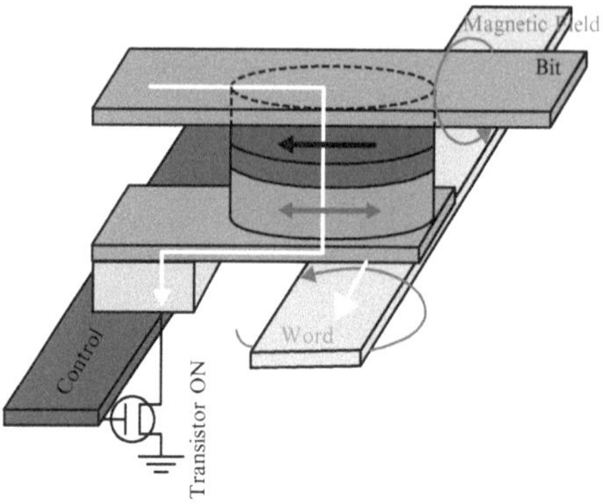

Fig. 8.1 Magnetic memory element for magnetic field switching (**a**) and current switching (**b**)

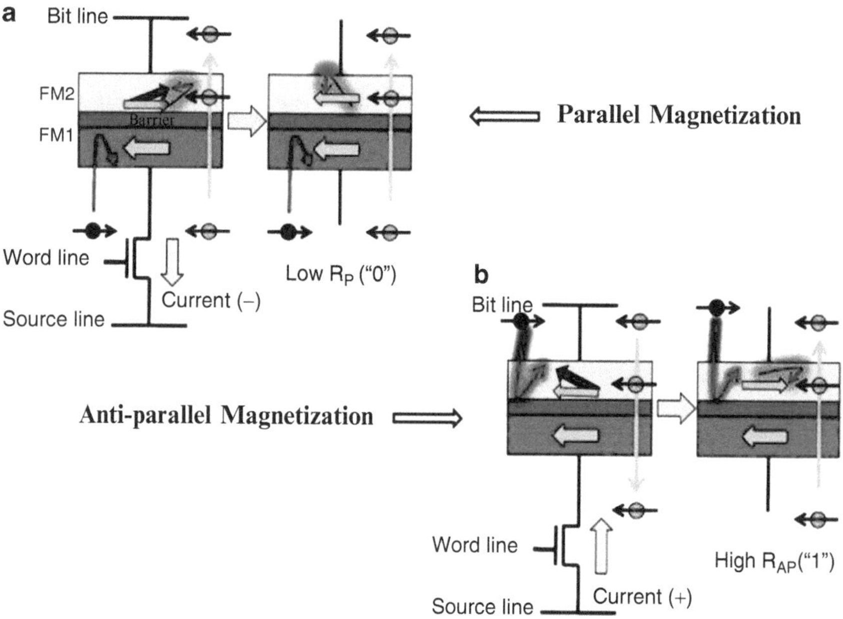

Fig. 8.2 Principle of spin-polarized current induced magnetization switching. Switching to (**a**) parallel magnetization configuration and (**b**) anti-parallel magnetization configuration. Source: T. Kawahara, R. Takemura, K. Miura, J. Hayakawa, S. Ikeda, Y.M. Lee, R. Sasaki, Y. Goto, K. Ito, T. Meguro, F. Matsukura, H. Takahashi, H. Matsuoka, H. Ohno, IEEE J. Solid State Circuits, **43**(1), 109 (2008)

and the electrons to transfer torque to the magnetic moment of the material. This torque is what is used to manipulate magnetization in small magnetic elements.

In order to flip the free layer's magnetic moment, current will run up or down through the stack depending on the orientation of the free layer as shown in Fig. 8.2a, b. If the free layer is anti-parallel to the fixed layer the current is passed through the free layer first. The electron's spin will be aligned with the magnetic moment of the fixed layer as shown in Fig. 8.2a. This does exert torque on the magnetic moment of the fixed layer. However, since the fixed layer is harder to flip than the free layer the amount of torque required to flip the fixed layer will not be reached before the free layer flips. When the electrons pass through the free layer they will orient their spins with the magnetic moment of the free layer. This exerts significant amount of torque on the magnetic moment of the free layer, which will flip the free layer so that it is aligned with the fixed layer. For the second case, when the free layer is parallel to the fixed layer, the electron is passed through the fixed layer first as shown in Fig. 8.2b. As a result the electron's spin will align with the magnetic moment of the free layer. When the electrons hit the fixed layer, the electrons that align parallel to the magnetic moment of the layer will be transmitted through, but the portion that is anti-aligned will be reflected back toward the free layer. These reflected electrons will exert a torque on the free layer causing it to flip anti-parallel to the fixed layer [38].

8.2.2 Current Issues

Current research on spin-transfer torque MRAM (ST–MRAM) has focused on an increasing aerial density and reducing the critical current density [38–44]. To increase the aerial density, closed and semi-closed magnetization configurations have been studied in place of traditional linear magnetization processes to allow closer spacing between elements by limiting undesired interactions [44–55]. To reduce the critical current density, various multilayer configurations, geometries, and materials have been researched [38–41]. Currently, to flip in the 1–10 ns range requires currents on the order of 10 mA [38]. However, in order to utilize this spin-torque technology with minimum-area metal-oxide-semiconductor field-effect transistor (MOSFET) requires the switching current to be on the order of 0.1–0.2 mA [38].

Another critical question in the development of ST–MRAM is: what will the dynamic pathway be for the magnetization in the laterally confined elements to develop a nonequillibrium configuration in response to a sub-nanosecond electrical current pulse? The underlying physical mechanism is that the spin-polarized currents apply a torque on the magnetization when the spin direction of the conduction electrons has a relative angle to the local magnetization. However, the precise mechanism of the spin-torque transfer to magnetization has been a subject of considerable debate. In our previous research the complexity of the nonequilibrium magnetization configuration, triggered by a magnetic field pulse, has been demonstrated [5, 23]; an increasing complexity in the spatial structure of the magnetization evolution is found to accompany the increasing switching speed, when a ferromagnetic element is driven by progressively faster switching magnetic field pulses applied antiparallel to the initial magnetization direction. Similarly, one would expect a transition from quasi-static to dynamic behavior of the vortex magnetization in response to a current pulse by varying the rise time and density of the electric current pulse.

8.3 Magnetization Dynamics in ST–MRAM Elements

To gain a deeper understanding of spin-torque induced magnetization dynamics in ST–MRAM elements embedded in CMOS, micromagnetic modeling and magneto-transport measurements in prototype ST–MRAM elements are carried out.

A schematic of such ST–MRAM elements is shown in Fig. 8.3a. The diameter of the multilayered nanopillar is 100 nm, and the hard magnetic layer (20 nm Co) is separated from the soft magnetic layer (6 nm $Ni_{80}Fe_{20}$) by a 4 nm thick Cu spacer layer. The thickness of the Cu spacer layer is less than the spin diffusion length, and therefore the depolarization of spin-polarized current due to spin-flip scattering can be ignored. Previously it has been reported that the vortex configuration is energetically favorable for 100 nm diameters and above for film thicknesses of 5 nm or thicker [12, 13]. This is confirmed in our modeling, shown in Fig. 8.3b, c, in which the vortex magnetization configurations in both layers are uniform, and each layer contains a single vortex core located at the center of the disk. The initial vortex

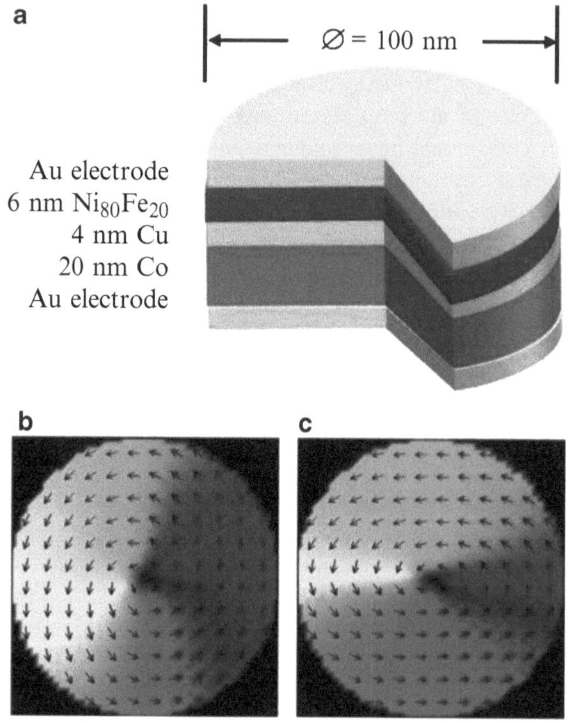

Fig. 8.3 (**a**) Schematics of Co/Cu/$Ni_{80}Fe_{20}$ multilayered nanopillar ST-MRAM element. The pillar has a diameter of 100 nm. (**b**) and (**c**) Initial magnetization state of $Ni_{80}Fe_{20}$ and Co layers. The vortex structures in both layers are uniform, and each layer contains a single vortex core located at the center of the disk. The initial vortex states of the $Ni_{80}Fe_{20}$ (**b**) and Co (**c**) layers are identified as CCW-positive based on their chirality and positive polarities of the core magnetizations

states of the $Ni_{80}Fe_{20}$ (b) and Co (c) layers are identified as CCW-positive based on their chirality and positive polarities of the core magnetizations. The details of the method of micromagnetic modeling are described below.

8.3.1 Method of Micromagnetic Modeling

The micromagnetic simulations have been carried out using a numerical code based on the Landau–Lifshitz (LL) equation, $d\mathbf{M}(t)/dt = -|\gamma|[\mathbf{M}(t) \times \mathbf{H}_{eff}(t)] - \lambda\{\mathbf{M}(t) \times [\mathbf{M}(t) \times \mathbf{H}_{eff}(t)]\}$, which is a phenomenological description of magnetization dynamics. Here the gyroscopic constant γ represents the precessional frequency, and the phenomenological damping factor λ drives the system towards an energy minimum after stopping energy input to a system, i.e. energy relaxation. In general, λ is small compared to γ/M_s (where M_s is the saturation magnetization), which implies that the energy relaxation of the system takes much longer than a few full precessional cycles of magnetization **M**. The $\mathbf{H}_{eff}$ is the internal effective magnetic field of a system. Generally, the total energy density ε_{tot} in a system mainly includes Zeeman energy ε_z when a external magnetic field is applied to the system, exchange interaction energy ε_{ex}, magnetocrystalline anisotropic energy ε_{ani}, and demagnetization energy ε_d, i.e. $\varepsilon_{tot} = \varepsilon_z + \varepsilon_{ex} + \varepsilon_{ani} + \varepsilon_d$. All these energies are functions of the magnetization distribution **M**, i.e. $\varepsilon_i = \varepsilon_i(\mathbf{M})$. The change of each energy with respect to magnetization **M** represents the corresponding magnetic field $\mathbf{H}_i$ ($H_i = -d\varepsilon_i/dM$). Therefore, the $\mathbf{H}_{eff}$, is the vector sum of the applied external field $\mathbf{H}_{app}$, the exchange interaction $\mathbf{H}_{ex}$, the magnetocrystalline anisotropy field $\mathbf{H}_{ani}$, and the demagnetizing field $\mathbf{H}_d$, i.e. $\mathbf{H}_{eff} = \mathbf{H}_{app} + \mathbf{H}_{ex} + \mathbf{H}_{ani} + \mathbf{H}_d$. Since the coherent rotation reversal time is very short comparing with the energy relaxation time after the reversal, the precessional motion of the magnetization vector is primarily governed by the first term of the LL equation. The spin-transfer torque effect is taken into account by including the Slonczewski term, $d\mathbf{M}_{1,2}/dt = \beta[\mathbf{M}_{1,2}\times(\mathbf{M}_1\times\mathbf{M}_2)]$, to the LL equation [28]. Here the magnetizations, $\mathbf{M}_1$ and $\mathbf{M}_2$, are sequential along positive z, and β represents the driving term of the spin-momentum transfer and includes the degree of spin polarization of the current (η). In the simulations the degree of spin polarization of the current $\eta = 0.26$ (0.4) for $Ni_{80}Fe_{20}$ (Co) are used, and the element is subdivided into unit cells of the dimension of $2 \times 2 \times 2$ nm^3. The other parameters used are the saturation magnetization $M_s = 800$ (1414) emu/cm^3 for $Ni_{80}Fe_{20}$ (Co), exchange constant $A = 1.05 \times 10^{-6}$(3.05×10^{-6}) erg/cm for $Ni_{80}Fe_{20}$ (Co), and zero magnetic anisotropy constant.

8.3.2 Spin-Polarized Current Pulse Switching of $Ni_{80}Fe_{20}$/Cu/Co Nanopillar Elements

In this section we discuss details of the spin-transfer effect on the vortex magnetization in multilayered nanopillar elements. We demonstrate that the vortex chirality can be controllably switched by injecting a spin-polarized current pulse with appropriate amplitude and duration.

Figure 8.3b, c shows the initial magnetization configurations in the $Ni_{80}Fe_{20}$ and Co layers, where each layer contains a single vortex core located at the center of the disk. The core magnetizations are out-of-plane in the positive z-direction whereas the magnetizations are in-plane away from the core. The arrows in the images are to help visualize the circulating magnetization around the disk center; each arrow indicates the $\mathbf{M}_{NiFe}$ and $\mathbf{M}_{Co}$ averaged from four unit cells. The initial vortex states are identified as CCW–positive ($V_{\uparrow}^{CCW}$) based on their chirality, and positive polarities of the core magnetizations. The dipolar interaction between Co and $Ni_{80}Fe_{20}$ layers at the pillar edges is assumed negligible due to the closed flux structures in both magnetic layers.

To study the dynamic response of the vortex magnetization to the spin-transfer torque, the equilibrium configurations of $\mathbf{M}_{NiFe}$ and $\mathbf{M}_{Co}$ are excited by injecting a short current pulse. The current direction is denoted positive when it flows from $Ni_{80}Fe_{20}$ to Co layer. Figure 8.4 shows a series of the nonequilibrium states of $\mathbf{M}_{NiFe}$ captured at selected times after a negative current pulse of 30 mA in amplitude with the duration of 200 ps is applied. The rise and fall time of the current pulses is assumed to be 2 ps. Following the temporal evolution of $\mathbf{M}_{NiFe}$ in Fig. 8.4, one finds that individual $\mathbf{M}_{NiFe}$ undergoes the CW rotation, while the whole vortex configuration appears to circulate CCW around the disk center. The image (b), captured at 84 ps after the current pulse is applied, reveals that $\mathbf{M}_{NiFe}$ rotates by ~90°, and with increasing time $\mathbf{M}_{NiFe}$ further rotates by ~180° (c), i.e., the chirality of $\mathbf{M}_{NiFe}$ switches from CCW (a) to CW (c) after about a half cycle of rotation. $\mathbf{M}_{NiFe}$ of the switched vortex state in (c), however, is still in a nonequilibrium state, and the dynamics of the energy dissipation after the initial chirality switching involves a series of complex magnetization processes. The magnetization configuration in (d), for instance, reveals a high degree of complexity both due to the nonunifrom distribution of $\mathbf{M}_{NiFe}$ and the creation of an additional vortex ($V^{\uparrow}$)–antivortex (V^{A}) pair. The pair vanishes with increasing time without undergoing an annihilation process. We note that the vortex core dynamics in Fig. 8.4 is different from the previously reported results [26], in which the polarity of vortex core was reversed via the explosion-like annihilation of vortex–antivortex pair after the vortex magnetization was excited by an in-plane magnetic field pulse. The details of the vortex core dynamics in Fig. 8.4 are difficult to analyze due to the high nonuniformity of $\mathbf{M}_{NiFe}$ distribution during the chirality switching process, but the polarity of the core magnetization remains unswitched under the given current pulse condition. A full relaxation of $\mathbf{M}_{NiFe}$ is reached after about 2 ns after the current pulse excitation (e). By contrast, the magnetization distribution of $\mathbf{M}_{Co}$ is only slightly perturbed from the initial state through the entire process of the chirality switching of $\mathbf{M}_{NiFe}$, and no switching of vortex chirality of $\mathbf{M}_{Co}$ occurs (f). Therefore, the curling in-plane magnetizations of $\mathbf{M}_{NiFe}$ and $\mathbf{M}_{Co}$ is aligned nearly antiparallel to each other after the reversal of the vortex chirality of $\mathbf{M}_{NiFe}$ is completed.

The switching of the vortex chirality of $\mathbf{M}_{NiFe}$ discussed in Fig. 8.4 is understood in terms of the spin-transfer torque exerted on $\mathbf{M}_{NiFe}$. With the negative electric current flow, the conduction electrons are spin-polarized when they are reflected from the Co layer. The spin-torque exerted on the $Ni_{80}Fe_{20}$ layer destabilizes the equilibrium

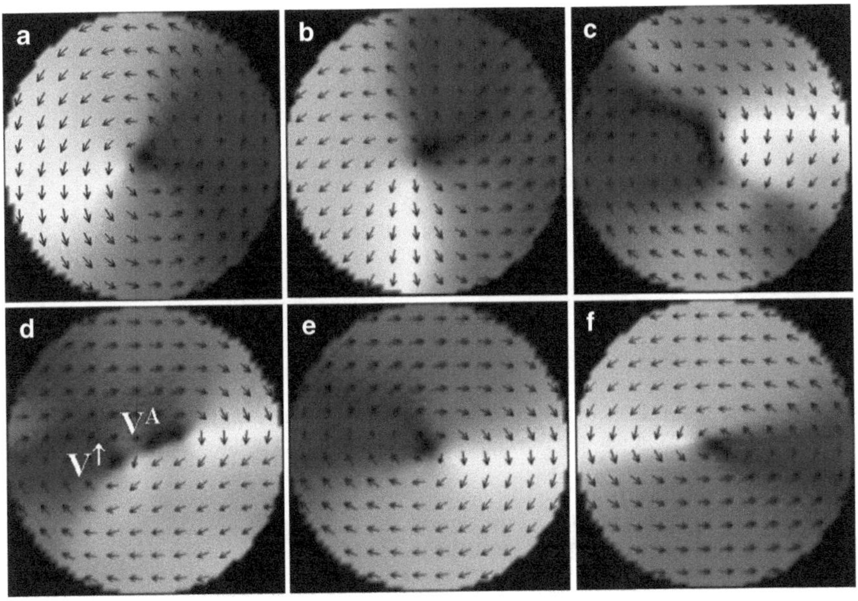

Fig. 8.4 Nonequilibrium magnetization configurations of $\mathbf{M}_{NiFe}$ at selected time points after the onset of a current pulse of −30 mA with 200 ps duration: (**a**) 0 ps, (**b**) 84 ps, (**c**) 114 ps, (**d**) 178 ps, and (**e**) 2 ns after the current pulse. (**f**) $\mathbf{M}_{Co}$ distribution captured at 2 ns after the current pulse. Each arrow indicates the averaged local magnetization from four unit cells (Copyright © AIP 2007)

configuration of $\mathbf{M}_{NiFe}$, and initiates the precessional motion of $\mathbf{M}_{NiFe}$ that eventually leads to the chirality switching. We note that the modeling does not lead to a reversal of the chirality if the spin-transfer torque term is excluded in the calculation. When the Amperian current-induced magnetic field is solely taken into account, the evolution of nonequilibrium magnetization configuration shows quite different dynamics, in which the vortex–antivortex pair is created and subsequently annihilated to a sudden spin-wave excitation without switching the chirality [25, 26]. This confirms that the spin-transfer torque plays a key role in the dynamic process described in Fig. 8.4.

In order to obtain a more detailed understanding of the magnetization dynamics induced by spin-torque transfer to the vortex magnetization, the actual direction of $\mathbf{M}_{NiFe}$ and $\mathbf{M}_{Co}$ during the chirality switching is reconstructed. Figure 8.5 shows the view of three-dimensional magnetization trajectory and corresponding projections of the M_x, M_y, and M_z components of $\mathbf{M}_{NiFe}$ and $\mathbf{M}_{Co}$. The magnetization components are averaged over the area with the diameter of 10 nm, as indicated by the shaded region in the inset. For $\mathbf{M}_{NiFe}$, the trajectory reveals an initial rapid reversal of M_x–component immediately after the current pulse. The spin-transfer torque also tilts $\mathbf{M}_{NiFe}$ out of the *x*–*y* plane. After the main reversal process, i.e., $(M_x/M_s)^{NiFe}$ from −1 to +1, the trajectories of M_y^{NiFe} and M_z^{NiFe} follow rotational paths. Therefore, it is concluded that the magnetic response to the spin-torque leads to an

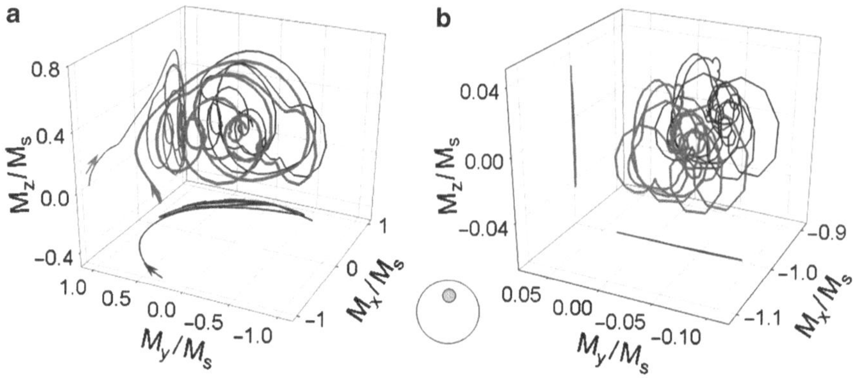

Fig. 8.5 Three-dimensional trajectory of the magnetization averaged over the area of 10 nm in diameter, shaded area in the inset. (**a**) The trajectory reveals an initial rapid rotation of $\mathbf{M}_{NiFe}$ from $-Mx$ to $+Mx$, and the spin torque induced by the current pulse tilts the magnetization vector out of the x–y plane. (**b**) Dynamic response of $\mathbf{M}_{Co}$ components. Note that the scale of the axis is an order of magnitude smaller than that in (**a**). This indicates that $\mathbf{M}_{Co}$ resists the spin-transfer torque by the spin-polarized electrons partly reflected from the Cu/$Ni_{80}Fe_{20}$ interface (Copyright © AIP 2007)

irreversible switching of M_x^{NiFe}, accompanied by the precession of $\mathbf{M}_{NiFe}$ around the x-direction. The trajectory of $\mathbf{M}_{NiFe}$ following a rotational path is reminiscent of the magnetization precession, observed in the magnetization reversal driven by short magnetic field pulses [18]. However, the dynamics of the precession in response to the spin-transfer torque is not of a uniform excitation mode, and develops a more complex pathway. The Fourier transform of the magnetization components of Fig. 8.5a yields a relatively broad peak centered around 16 GHz along with a large low frequency response at 4 GHz. The Fourier transform shows a quite different frequency spectrum, if the spin-torque effect is excluded in the modeling. In this case, no significant peaks are found, implying a high degree of incoherence in the precession due to spin wave excitation triggered by vortex–antivortex pair annihilation and subsequent complex behavior of magnetization dynamics [24–26].

Finally, the switched vortex chirality is reversed back to the initial vortex configurations by applying a positive current pulse, 25 mA in amplitude and 200 ps in duration. The critical current density J_c required for the back-switching is reduced to about 8×10^7 A/cm^2, which is about 20% lower than the J_c value needed for the negative current pulse discussed in Fig. 8.4. After flowing through the Co layer, the electrons have become spin-polarized along the direction of $\mathbf{M}_{Co}$, and exert the spin-torque on $\mathbf{M}_{NiFe}$ in the CCW direction. This is seen in Fig. 8.6b. After $\mathbf{M}_{NiFe}$ undergoes a half period of precession, the circulating in-plane components of $\mathbf{M}_{NiFe}$ are reversed (d), and reaches an alignment that is nearly parallel to $\mathbf{M}_{Co}$ at ~2 ns (e). The temporal evolution of $\mathbf{M}_{NiFe}$ distribution is similar to the previous switching process discussed in Fig. 8.4, except that the sense of rotation of $\mathbf{M}_{NiFe}$ around the vortex core

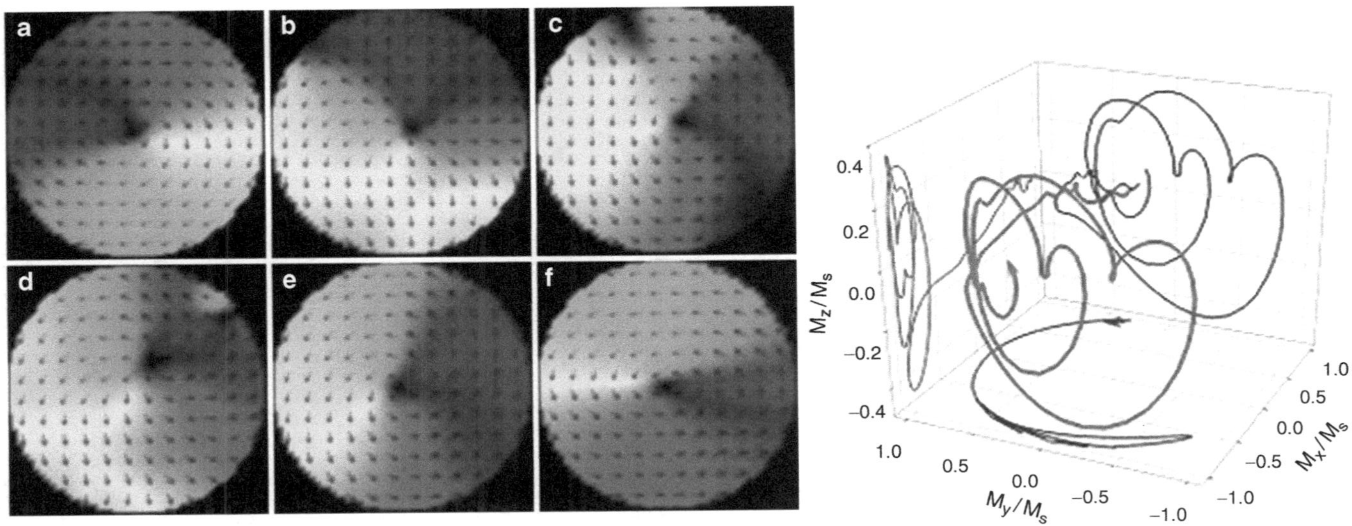

Fig. 8.6 Nonequilibrium magnetization distributions of $\mathbf{M}_{\mathrm{NiFe}}$ at selected time points after the onset of a vcurrent pulse of +25 mA with 200 ps duration: (**a**) 0 ps, (**b**) 200 ps, (**c**) 245 ps, (**d**) 343 ps, **Fig 8.6** (continued) and (**e**) 2 ns after the current pulse. Image (**f**) is captured for $\mathbf{M}_{\mathrm{Co}}$ at 2 ns. Each arrow indicates the averaged local magnetization from four unit cells. (Copyright © AIP 2007) (*Right panel*) Three-dimensional trajectory of the magnetization averaged over the area of 10 nm in diameter

is reversed. The dynamics of $\mathbf{M}_{NiFe}$ averaged over the local area with the diameter of 10 nm is illustrated in the right panel. The result reveals that the vortex chirality switching is mediated by the precession of $\mathbf{M}_{NiFe}$ initiated by the spin-transfer torque.

8.3.3 Fabrication and Characterization of Prototype ST–MRAM

The above descried new type of ST–MRAM element suggests a new way to implement the magnetization switching mechanism in memory device applications without using magnetic fields. In this section we describe the proof-of-the-concept experiments in which prototype 8 × 8 array of multilayered nanopillar ST–MRAM elements with diameters varying from 100 to 400 nm are fabricated and characterized.

8.3.3.1 Fabrication of 8 × 8 Array of ST–MRAM Nanopillar Elements

The fabricated 8 × 8 spin valve array consists of thirty-eight 100 nm pillars, eight 200 nm pillars, eight 300 nm pillars, eight 400 nm pillars, and two 5 μm pillars as shown in Fig. 8.7a. Electron-beam (E-beam) evaporation was used to deposit the thin films. Photolithographic patterning and ion milling were used to shape the entire stack into the electrode pads. Once the electrode pads were formed, e-beam lithography and ion milling were performed to pattern an 8 × 8 array of pillars into the ST–MRAM structure on top of the bottom electrode. Then, the structure was planarized using SiO_2 deposited with an E-beam evaporator and a liftoff process. Finally, photolithography, E-beam evaporation, and a lift-off process were used to shape the top Cu electrode. After fabrication, the array was characterized using a four point probe MR tester for magnetic field switching.

The inset in Fig. 8.7a shows 92 × 97 nm patterned resist which is used to form approximately 100 nm nanopillars. Figure 8.7b shows a schematic of the fabricated 100 nm ST–MRAM nanopillar. The Co/Cu/$Ni_{80}Fe_{20}$ portion of the fabricated nanopillar is based on the micromagnetic simulations for spin-polarized current switching of vortex chirality discussed in the Section 8.3.2. A FeMn layer was added to increase the thermal stability and to help pin the Co fixed layer. The bottom $Ni_{80}Fe_{20}$ layer was added as a texture layer to promote coupling between AFM layer and Co fixed layer. Finally, a gold capping layer was used to protect stack from oxidization during the SiO_2 liftoff procedure.

8.3.3.2 Characterization of ST–MRAM Nanopillar Elements

After fabrication of the array, the array was characterized using a four-point probe magnetoresistance (MR) measurement and Physical Properties Measurement System (PPMS) system. Figure 8.8a shows the four-point probe measurements system used to characterize the magnetic field switching of the spin valve elements. The system was developed by Rigel Systems.

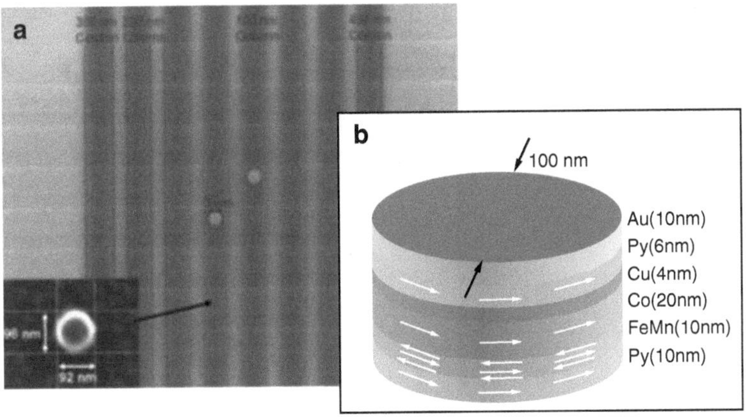

Fig. 8.7 (**a**) Optical micrograph of the layout of 8 × 8 array. Inset shows a typical SEM image of a 92 nm by 96 nm patterned resist. (**b**) Schematic of fabricated 100 nm ST–MRAM nanopillar element. White arrows represent the magnetization direction in each magnetic layers

Figure 8.8b shows the two probes used to source the 100 μA current and the two probes used to measure the change in voltage. A maximum field of 1 kOe can be applied in-plane in the longitudinal or transverse direction. A field sweep of 50 Oe with a 2 Oe step size was used to characterize the 8 × 8 spin valve arrays. A Physical Properties Measurement System (PPMS) system is also used to characterize the spin-polarized current stimulation of 100 nm ST–MRAM nanopillar elements. The temperature of the sample can be varied from 4.2 to 300 K using liquid helium. A magnetic field greater than 10 kOe can be applied in the vertical direction during measurement. The current could be sourced up to 30–40 mA with a step size down to 0.1 mA per step. For measurements a current is sourced while the change in voltage and current are measured.

Figure 8.9 shows the magnetoresistance (MR) signal for magnetic field switching of a ST-MRAM nanopillar element with ϕ = 100 nm. A 50 Oe in-plane magnetic field was applied with a step size of 2 Oe to switch the magnetization of the nanopillar element. Two distinct peaks are observed with a switching field of approximately 10 Oe. For the 50 Oe to −50 Oe (forward) sweep, the MR starts to increase at around 10 Oe. A peak in MR is observed at approximately −5 Oe. Finally, the magnetization reversal is completed at approximately −15 Oe. The −50 Oe to 50 Oe (reverse) sweep shows similar behavior. Similar MR data was observed for all diameters of ST–MRAM nanopillars in the 8 × 8 array. Figure 8.10a, b shows another example of magnetic field switching MR signal observed from 100 nm nanopillar elements. In Fig. 8.10a, one finds that the magnetoresistance decreases by applying in-plane external magnetic field, while a maximum MR value is found at zero field. This result implies that the initial vortex configurations in $Ni_{80}Fe_{20}$ and Co layers have opposite chiralities, in which the in-plane curling magnetizations in both layers are antiparallel at their initial state, as schematically shown in Fig. 8.10c. Under the applied external magnetic field,

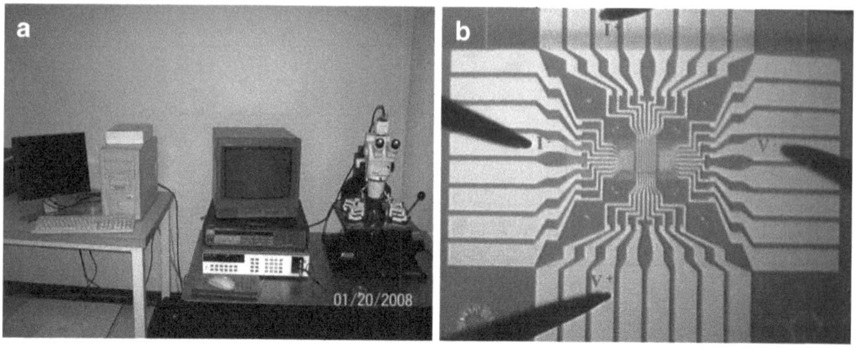

Fig. 8.8 (**a**) 4-Point probe MR tester and (**b**) 4-point probe placement on 8 × 8 array

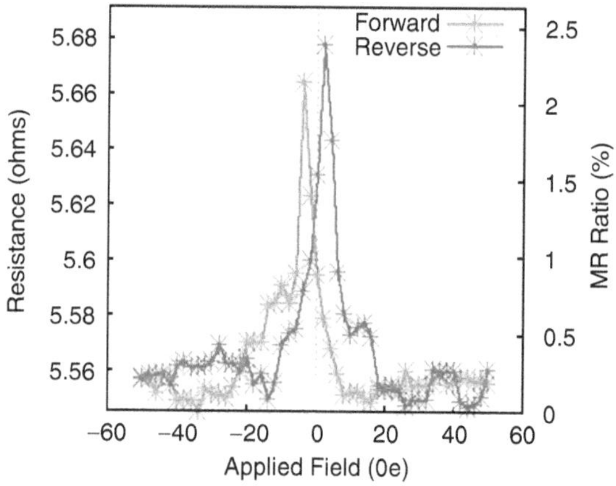

Fig. 8.9 Magnetoresistance (MR) measurement of ST-MRAM element (= 100 nm) at room temperature. An in-plane magnetic field of 50 Oe was applied to switch the magnetization of the nanopillar

the in-plane magnetization configuration in the magnetic layers deviates from its initial anti-parallel state, and this leads to decrease of the magnetoresistance. We note that for about half of the pillar structures MR measurements reveal the curves with the minimum value of the magnetoresistance at zero applied field, and the application of external magnetic field leads to the increase of MR (Fig. 8.10b). This is attributed to the same chiralities of the in-plane magnetization of the initial vortex configuration in $Ni_{80}Fe_{20}$ and Co layers, as schematically shown in Fig. 8.10d. For these multi-layered elements the initial in-plane magnetizations of $Ni_{80}Fe_{20}$ and Co layers are parallel, and it leads to a minimum magnetoresistance at remanence. In this case the application of external magnetic field leads to small changes of the magnetoresistance. The drastic change of the MR values by the relative vortex chiralities in $Ni_{80}Fe_{20}$ and Co layers strongly infers the potential of the suggested nanopillar elements as magnetic memory elements.

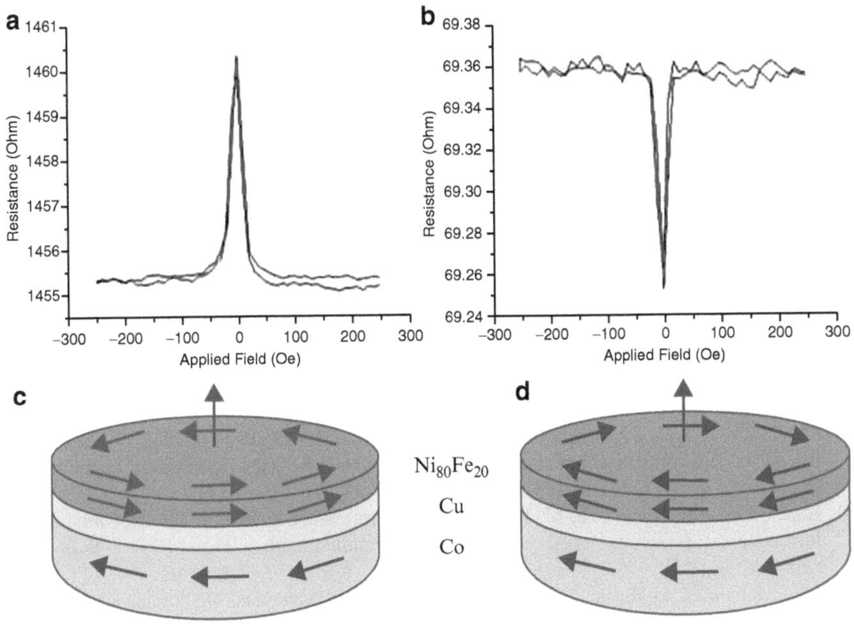

Fig. 8.10 (**a**) and (**b**) MR curves for the nanopillar elements with different initial vortex magnetization configurations. (**c**) and (**d**) schematics of the corresponding vortex magnetization states in $Ni_{80}Fe_{20}$ and Co layers for the MR measurements (**a**) and (**b**)

8.4 Conclusions

In conclusion, magnetization switching dynamics in ST–MRAM, a newly designed magnetoelectronic memory device, has been discussed. We demonstrate that the spin-transfer torque provides a new means to controllably switch the vortex magnetization within a few nanoseconds in multi-layered magnetic elements. From an application point of view, the results suggest opportunities to implement the magnetic vortex chirality switching mechanism to memory device applications, in which vortex states with controllable chirality switching in the free layer of a magnetic nanopillar may be used as memory bits. The great advantage of the ST–MRAM discussed in this chapter is the simplicity in the circuit design since no magnetic field generating lines are necessary. Furthermore highly selective magnetization switching is possible without worrying about the effect of stray magnetic fields on neighboring elements. Our approach will lead to a new integration of magnetic components into CMOS technology, giving rise to an all-electrically controlled magneto-electronic device. Referring to our results, the current density needed to switch the magnetization elements is in the range of 10^7–10^8 A/cm^2 with a duration of a few 100 ps.

Acknowledgement This work was supported by the Natural Sciences and Engineering Research Council (NSERC) of Canada and US Air Force Research Laboratory (AFRL) under Grant No. F29601-04-1-206.

References

1. S.A. Wolf, D.D. Awschalom, R.A. Buhrman, J.M. Daughton, S. von Molnár, M.L. Roukes, A.Y. Chtchelkanova, D.M. Treger, Spintronics: A spin-based electronics vision for the future. Science **294**, 1488 (Nov 2001)
2. P. Grünberg, Layered magnetic structures: History, highlights, applications. Phys. Today **54**(5), 31 (May 2001)
3. B. Heinrich, J.A.C. Bland (eds.), *Ultrathin Magnetic Structures* (Springer Verlag, Berlin, Heidelberg, 1994)
4. M.N. Baibich, J.M. Broto, A. Fert, F. Nguyen Van Dau, F. Petroff, P. Etienne, G. Creuzet, A. Friederich, J. Chazelas, Giant magnetoresistance of (001)Fe/(001)Cr Magnetic superlattices. Phys. Rev. Lett. **61**, 2472 (Nov 1988)
5. B.C. Choi, G.E. Ballentine, M. Belov, W.K. Hiebert, M.R. Freeman, Ultrafast magnetization reversal dynamics investigated by time domain imaging. Phys. Rev. Lett **86**, 728 (Jan 2001)
6. Y. Acremann, C.H. Back, M. Buess, O. Portmann, A. Vaterlaus, D. Pescia, H. Melchior, Imaging precessional motion of the magnetization vector. Science **290**, 492 (Oct 2000)
7. R.H. Koch, J.G. Deak, D.W. Abraham, P.L. Trouilloud, R.A. Altman, Lu Yu, W.J. Gallagher, R.E. Scheuerlein, K.P. Poche, S.S.P. Parkin, Magnetization reversal in micron-sized magnetic thin films. Phys. Rev. Lett. **81**, 4512 (Nov 1998)
8. W.K. Hiebert, A. Stankiewicz, M.R. Freeman, Direct observation of magnetic relaxation in a small Permalloy disk by time-resolved scanning Kerr microscopy. Phys. Rev. Lett. **79** (Aug 1997)
9. S.W. Yuan, H.N. Bertram, Domain-wall dynamics in thick Permalloy films. J. Appl. Phys. **73**, 5992 (May 1993)
10. A.F. Popkov, L.L. Savchenko, N.V. Vorotnikova, S. Tehrani, J. Shi, Edge pinning effect in single- and three-layer patterns. Appl. Phys. Lett. **77**, 277 (July 2000)
11. K.J. Kirk, J.N. Chapman, C.D.W. Wilkinson, Switching fields and magnetostatic interactions of thin film magnetic nanoelements. Appl. Phys. Lett. **71**, 539 (July 1997)
12. J.M. Daughton, Magnetoresistive memory technology. Thin Solid Films **216**, 162 (Aug 1992)
13. W.J. Gallagher, S.S.P. Parkin, Lu Yu, X.P. Bian, A. Marley, K.P. Roche, Microstructured magnetic tunnel junctions. J. Appl. Phys. **81**, 3741 (Apr 1997)
14. C.H. Back, D. Weller, J. Heidmann, D. Mauri, D. Guarisco, E.L. Garwin, H.C. Siegmann, Magnetization reversal in ultrashort magnetic field pulses. Phys. Rev. Lett. **81**, 3251 (Oct 1998)
15. C. Stamm, F. Marty, A. Vaterlaus, V. Weich, S. Egger, U. Maier, U. Ramsperger, H. Fuhrmann, D. Pescia, Two-dimensional magnetic particles. Science **282**, 449 (Oct 1998)
16. M. Hehn, K. Ounadjela, J.-P. Bucher, F. Rousseaux, D. Decanini, B. Bartenlian, C. Chappert, Nanoscale magnetic domains in mesoscopic magnets. Science **272**, 1782 (June 1996)
17. R.P. Cowburn, D.K. Koltsov, A.O. Adeyeye, M.E. Welland, D.M. Tricker, Single-domain circular nanomagnets. Phys. Rev. Lett. **83**, 1042 (Aug 1999)
18. B.C. Choi, J. Rudge, M.R. Freeman, Y.K. Hong, Q.F. Xiao, Nonequilibrium magnetic domain structures as a function of speed of switching process in $Ni_{80}Fe_{20}$ thin-film element. IEEE Trans. Magn. **43**(1), 2 (Jan 2007)
19. K. Yamada, S. Kasai, Y. Nakatani, K. Kobayashi, H. Kohno, A. Thiaville, T. Ono, Electrical switching of the vortex core in a magnetic disk. Nat. Mater. **6**(4), 270 (Apr 2007)
20. KYu Guslienko, X.F. Han, D.J. Keavney, R. Divan, S.D. Bader, Magnetic vortex core dynamics in cylindrical ferromagnetic dots. Phys. Rev. Lett. **96**, 067205 (Feb 2006)

21. F. Romanens, J. Vogel, W. Kuch, K. Fukumoto, J. Camarero, S. Pizzini, M. Bonfim, F. Petroff, Influence of topography and Co domain walls on the magnetization reversal of the FeNi layer in FeNi/Al_2O_3/Co magnetic tunnel junctions. Phys. Rev. B **74**, 184419 (Nov 2006)
22. V.V. Kruglyak, P.S. Keatley, R.J. Hicken, J.R. Childress, J.A. Katine, Dynamic configurational anisotropy in nanomagnets. Phys. Rev. B **75**, 24407 (Jan 2007)
23. B.C. Choi, J. Ho, G. Arnup, M.R. Freeman, Nonequilibrium domain pattern formation in mesoscopic magnetic thin film elements assisted by thermally excited spin fluctuations. Phys. Rev. Lett. **95**, 237211 (Dec 2005)
24. B. Van Waeyenberge, A. Puzic, H. Stoll, K.W. Chou, T. Tyliszczak, R. Hertel, M. Fähnle, H. Brückl, K. Rott, G. Reiss, I. Neudecker, D. Weiss, C.H. Back, G. Schütz, Magnetic vortex core reversal by excitation with short bursts of an alternating field. Nature **444**, 461 (Nov 2006)
25. Q.F. Xiao, J. Rudge, B.C. Choi, Y.K. Hong, G. Donohoe, Dynamics of vortex core switching in ferromagnetic nanodisks. Appl. Phys. Lett. **89**, 262507 (Dec. 2006)
26. R. Hertel, C.M. Schneider, Exchange explosions: Magnetization dynamics during vortex–antivortex annihilation. Phys. Rev. Lett. **97**, 177202 (Oct. 2006)
27. M. Buess, T. Haug, M.R. Scheinfein, C.H. Back, Micromagnetic dissipation, dispersion, and mode conversion in thin permalloy platelets. Phys. Rev. Lett. **94**, 127205 (Apr. 2005)
28. J.C. Slonczewski, Current-driven excitation of magnetic multilayers. J. Magn. Magn. Mater. **159**, L1 (June 1996)
29. L. Berger, Emission of spin waves by a magnetic multilayer traversed by a current. Phys. Rev. B **54**, 9353 (Oct 1996)
30. J.A. Katine, F.J. Albert, R.A. Buhrman, E.B. Myers, D.C. Ralph, Current-driven magnetization reversal and spin-wave excitations in Co/Cu/Co pillars. Phys. Rev. Lett. **84**, 3149 (Apr 2000)
31. Y. Jiang, S. Abe, T. Ochiai, T. Nozaki, A. Hirohata, N. Tezuka, K. Inomata, Effective reduction of critical current for current-induced magnetization switching by a Ru layer insertion in an exchange-biased spin valve. Phys. Rev. Lett. **92**, 167204 (Apr 2004)
32. N.C. Emley, I.N. Krivorotov, O. Ozatay, A.G.F. Garcia, J.C. Sankey, D.C. Ralph, R.A. Buhrman, Time-resolved spin-torque switching and enhanced damping in Permalloy/Cu/Permalloy spin-valve nanopillars. Phys. Rev. Lett. **96**, 247204 (June 2006)
33. J.-G. Caputo, Y. Gaididei, F.G. Mertens, D.D. Sheka, Vortex polarity switching by a spin-polarized current. Phys. Rev. Lett. **98**, 056604 (Feb 2007)
34. T. Taniuchi, M. Oshima, H. Akinaga, K. Ono, Vortex-chirality control in mesoscopic disk magnets observed by photoelectron emission microscopy. J. Appl. Phys. **97**, 10J904 (May 2005)
35. B.C. Choi, J. Rudge, E. Girgis, J. Kolthammer, Y.K. Hong, A. Lyle, Spin-current pulse induced switching of vortex chirality in permalloy/Cu/Co nanopillars. Appl. Phys. Lett. **91**, 022501 (July 2007)
36. A. Lyle, M.Sc. Thesis, Department of Electrical and Computer Engineering, University of Alabama, 2009
37. P. Grunberg, R. Schreiber, Y. Pang, M.B. Brodsky, H. Sowers, Layered magnetic structures: Evidence for antiferromagnetic coupling of Fe layers across Cr interlayers. Phys. Rev. Lett. **57**, 2442 (Nov 1986)
38. E.B. Myers, D.C. Ralph, J.A. Katine, F.J. Albert, R.A. Buhrman, Point-contact studies of current-controlled domain switching in magnetic multilayers. J. Appl. Phys. **87**, 5502 (May 2000)
39. I.N. Krivorotov, N.C. Emley, R.A. Buhrman, D.C. Ralph, Time-domain studies of very-large-angle magnetization dynamics excited by spin transfer torques. Phys. Rev. B **77**, 054440 (Feb. 2008)
40. P.M. Braganca, I.N. Krivorotov, O. Ozatay, A.G.F. Garcia, N.C. Emley, J.C. Sankey, D.C. Ralph, R.A. Buhrman, Reducing the critical current for short-pulse spin-transfer switching of nanomagnets. Appl. Phys. Lett. **87**, 112507 (Sept 2005)
41. O. Ozatay, N.C. Emley, P.M. Braganca, A.G.F. Garcia, G.D. Fuchs, I.N. Krivorotov, R.A. Buhrman, D.C. Ralph, Spin transfer by nonuniform current injection into a nanomagnet. App. Phys. Lett. **88**, 202502 (May 2006)
42. S. Tehrani, J.M. Slaughter, M. Deherrera, B.N. Engel, N.D. Rizzor, J. Salter, M. Durlam, R.W. Dave, J. Janesky, B. Buthcer, K. Smith, G. Grynkewich, Magnetoresistive random access memory using magnetic tunnel junctions. Proc. IEEE **91**(5), 703 (May 2003)

43. J. Janesky, N.D. Rizzo, L. Savtchenko, B. Engel, J.M. Slaughter, S. Tehrani, Magnetostatic interactions between sub-micrometer patterned magnetic elements. IEEE Trans. Magn. **37**(4), 2052 (July 2001)
44. J.G. Deak, Effect of vortex handedness on spin momentum torque dynamics in dual-vortex ferromagnetic nanopillar structures. J. Appl. Phys. **103**, 07A505 (Apr 2008)
45. B.C. Choi, Q.F. Xiao, Y.K. Hong, J. Rudge, G. Donohoe, Vortex core switching dynamics in submicron Permalloy disk. IEEE Trans. Magn. **43**(6), 2926 (June 2007)
46. Q.F. Xiao, J. Rudge, B.C. Choi, Y.K. Hong, G. Donohoe, Dynamics of vortex core switching in ferromagnetic nanodisks. Appl. Phys. Lett. **89**, 262507 (Dec 2006)
47. M. Schneider, H. Hoffman, J. Zweck, Magnetic switching of single vortex permalloy elements. Appl. Phys. Lett. **79**, 3113 (Nov 2001)
48. Y. Gaididei, D.D. Sheka, F.G. Mertens, Controllable switching of vortex chirality in magnetic nanodisks by a field pulse. Appl. Phys. Lett. **92**, 012503 (Jan 2008)
49. M. Schneider, H. Hoffman, S. Otto, T. Haug, J. Zweck, Stability of magnetic vortices in flat submicron permalloy cylinders. J. Appl. Phys. **92**, 1466 (Aug 2002)
50. P. Vavassori, M. Grimsditch, V. Metlushko, N. Zaluzec, B. Ilic, Magnetoresistance of single magnetic vortices. Appl. Phys. Lett. **86**, 072507 (Feb 2005)
51. M.H. Park, Y.K. Hong, S.H. Gee, D.W. Erickson, B.C. Choi, Magnetization configuration and switching behavior of submicron NiFe elements: Pac-man shape. Appl. Phys. Lett. **83**, 329 (July 2003)
52. M.H. Park, Y.K. Hong, S.H. Gee, D.W. Erickson, T. Tanaka, B.C. Choi, Effect of shape anisotropy on switching behaviors of Pac-man NiFe submicron elements. J. Appl. Phys. **95**, 7019 (June 2004)
53. B.R. Pujada, B.C. Choi, M.H. Park, Y.K. Hong, S.H. Gee, D.W. Erickson, Magnetic switching depending on as-patterned magnetization state in Pac-man shaped $Ni_{80}Fe_{20}$ submicron elements. J. Appl. Phys. **96**, 4362 (Oct 2004)
54. B.R. Pujada, B.C. Choi, M.H. Park, Y.K. Hong, S.H. Gee, H. Han, Micromagnetic configurations and switching mechanism in Pac-man-shaped submicron $Ni_{80}Fe_{20}$ magnets. J. Appl. Phys. **97**, 73904 (March 2005)
55. H. Han, Y.K. Hong, M.H. Park, B.C. Choi, S.H. Gee, J.F. Jabal, G. Abo, A. Lyle, B. Wong, G.W. Donohoe, Interaction effect on switching behaviors of paired Pac-Man array. IEEE Trans. Magn. **41**(11), 4341 (Nov. 2005)

Chapter 9
Magnetization Switching in Spin Torque Random Access Memory: Challenges and Opportunities

Xiaobin Wang, Yiran Chen, and Tong Zhang

Abstract Dynamic thermal magnetization switching and magnetization switching variability determine nano-scale magnetic device performance. As a magnetic device scales down, achieving fast nanosecond time scale magnetization switching and maintaining thermal stability at second to years time scale become increasingly challenging. At the same time, the increased variability due to device dimension shrinking results device performance degradation. In the chapter, these challenges will be illustrated through spin torque random access memory device (SPRAM), which integrates magnetic tunneling junction (MTJ) and CMOS to achieve non-volatility, fast writing/reading speed, almost unlimited programming endurance and zero standby power. In order to fully characterize SPRAM performance, experiment and theoretical studies are pursued at widely separated time and spatial scales. Magnetization switching behaviors from nano-second to second region are measured and modeled. Microscopic quantum electronic spin transport model and macroscopic stochastic magnetization dynamics model are combined to study spin torque induced magnetization switching. Coupled micro-magnetic model and dynamic circuit model are used to simulate SPRAM device performance. Based upon these studies and future SPRAM scaling down requirements, research and development opportunities are identified to reduce switching current magnitude and control device variability.

Keywords Spin transfer torques • Magnetic random access memory • Magnetization Switching • Switching current reduction • Variability control

---

X. Wang (✉)
Seagate Technology, Bloomington, MN, USA
e-mail: xiaobin.wang@seagate.com

Y. Chen
Seagate Technology, Bloomington, MN, USA

T. Zhang
Rensselaer Polytechnic Institute, Troy, NY, USA

K. Iniewski (ed.), *CMOS Processors and Memories*,
Analog Circuits and Signal Processing, DOI 10.1007/978-90-481-9216-8_9,

9.1 Introduction to Spin Torque Random Access Memory

Macroscopic magnetization dynamics and microscopic spin transport are of great interest for both fundamental physics and practical application. Research of understanding and manipulating the spin degrees of freedom of electrons in solid-state systems over the past decades has given birth to a field called "spintronics". The 2007 Nobel physics prize was awarded to Dr. Albert Fert and Dr. Peter Grunberg for their discovery of giant magnetoresistance, which leads to the first widely used spintronic device in industry. Today spintronics is one of the fastest-growing areas in nano-scale physics and provides promising solutions for post-silicon technology, ranging from on-market read head sensor device to emerging spin torque random access memory technology and to future spin quantum computer.

Modern nano-scale magnetoelectronics device has reached a stage where device performance is explicitly determined by coupling between macroscopic magnetization dynamics and microscopic spin and electronic transport. One representative example is the emerging spin torque random access memory (SPRAM) technology that integrates magnetic tunneling junction (MTJ) and CMOS to achieve non-volatility, fast writing/reading speed, almost unlimited programming endurance and zero standby power.

SPRAM basic device structure and working principle are shown in Fig. 9.1. The key component of SPRAM cell is Magnetic Tunneling Junction (MTJ), which includes two ferromagnetic layers and one oxide barrier layer, e.g., MgO. The resistance of MTJ depends on the relative magnetization directions of the two ferromagnetic layers: when the magnetization is parallel (anti-parallel), MTJ is in low (high) resistance state. In SPRAM design, the magnetization direction of one ferromagnetic layer (reference layer) is fixed by coupling to a pinned magnetization layer above, while the magnetization direction of the other ferromagnetic layer (free layer) can be changed. The 0/1 data information is stored as magnetization direction of the free layer in SPRAM. For reading process, when a current passing through MTJ, the magnetization information is transformed to electric information through MTJ resistance difference for different free layer magnetization orientation directions (magnetic tunneling resistance).

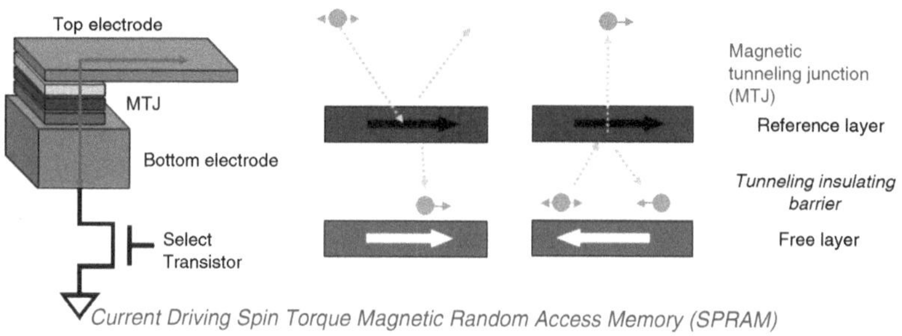

Fig. 9.1 Schematic pictures of structure and working principle of spin torque random access memory

The writing operation of SPRAM is based upon spin polarization current induced magnetization switching [1, 2]. To switch the free layer magnetization to the same direction of the reference layer magnetization, an electric current passes from the reference layer to the free layer. The injected current electrons have spins pointing to the same and the opposite directions of the reference layer magnetization. After passing through the reference layer, the electrons have a preferred spin orientation direction pointing to the same direction of the reference layer magnetization. This is because most of the electrons with spin pointing to the opposite direction of the reference layer magnetization are reflected back due interaction between itinerant electron spin and reference layer local magnetization. The polarized current electrons, with a net spin moment in the same direction of the reference layer magnetization, will switch the free layer magnetization to the same direction of the reference layer magnetization.

In order to switch the free layer magnetization to the opposite direction of the reference layer magnetization, electron current passes from the free layer to the reference layer. Based upon the same physics argument as before, the reflected electrons from the reference layer have preferred spin direction opposite to the direction of the reference layer magnetization. These will switch the free layer magnetization to the opposite direction of reference layer magnetization.

SPRAM is based upon macroscopic magnetization switching induced by microscopic spin moment transferring through polarized current. As device scales down, achieving fast macroscopic magnetization switching through microscopic spin transfer at nanosecond region and at the same time maintain stable macroscopic magnetization at room temperature at second to year time scale become increasingly challenging. At the same time, the increased variability due to device dimension shrinking results device performance degradation. In this chapter, we will study these challenges. SPRAM device characterization and system scaling down requirements will be pursued at widely separated time and spatial scales. Based upon device physics understandings and future SPRAM scaling down requirements, research and development opportunities are identified to reduce switching current magnitude and control device variability.

9.2 Magnetization Switching Challenges as SPRAM Scales Down

Dynamic thermal magnetization switching and magnetization switching variability are two fundamental challenges for magnetic device to scale down. These challenges are quite general for nano-scale non-volatile magnetic storage devices. In this section we will discuss these challenges for both spin torque random access memory (SPRAM) and hard disk drive (HDD). HDD and MRAM are the main technological implementations of non-volatile storage through magnetization orientations. The magnetization switching in HDD is through magnetic field and the magnetization switching in SPRAM is through spin torque. Examination of magnetization switching in both HDD and SPRAM could help us to understand more on fundamental challenges of the magnetization switching.

For current commercial hard disk drive, the data information is stored as magnetization orientations on continuous media. The media is composed of hard magnetic grains with grain size and magnetic properties distributions. The magnetization directions of these grains are switched by magnetic field from a writing head. Magnetization transitions are formed between opposite magnetization directions.

In order to increase magnetic recording density, magnetization transition widths between data bits must scale down. For continuous magnetic recording media with granular structure, given sufficient head field gradient, magnetization transition width ultimately saturates to a limit determined by the finite grain size [3, 4]. In order to further reduce the magnetization transition width, the grain size must shrink. However, media grain thermal stability at long times is proportional to the grain volume multiplying the media coercivity (Hc). Coercivity is the magnetic field required to switch media grain magnetization. For the same thermal stability, grain size reduction means media coercivity increasing. Media with increased coercivity requires increased magnetic field to write. Thus, in order to switch high coercivity media with smaller grain size, writer head field must be increased.

Above scaling arguments indicate that for the same thermal stability, the required magnetic field to switch media grains scales roughly as inverse of the grain volume. The exponential annual growth of area density and continuous scaling down of head dimensions already push conventional recording to the writability edge, where head fields may not be able to switch magnetization of media grains.

The magnetic field required to switch media magnetic grains (coercivity) is a dynamic concept. It depends upon the time scale [6]. Figure 9.2 shows media

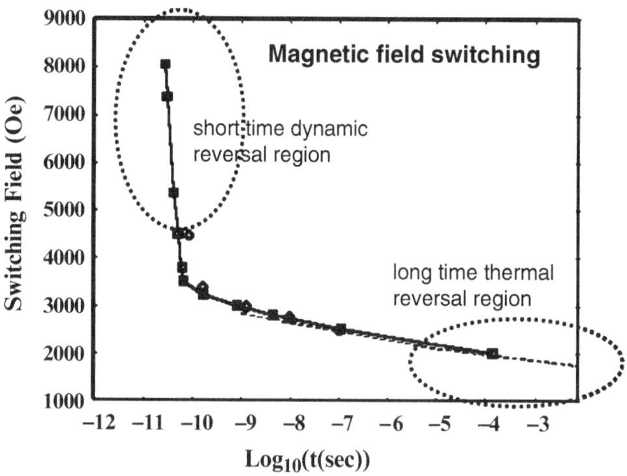

Fig. 9.2 Dynamic thermal switching of magnetic grains under external magnetic field [6]. Dots are experiment data and curves are theoretical calculations

dynamic coercivity versus switching time for the whole time range, from short time dynamic switching to long time thermal switching. As time scales down from second to nanosecond region, there is a rapid increasing in the required magnetic field to switch magnetic grains. This curve shows the challenge of achieving nano-second dynamic magnetization switching and at the same time maintaining thermal stability at second-to-year scales.

As HDD scales down, besides magnetization switching challenge, the increased variability brings device performance degradation. For example, the ultimate transition width for continuous media is not only determined by averaged grain size, but also determined by media grain size variation [5]. Increased grain size variation due to media grain size shrinking will lead to magnetization transition width broadening, resulting higher media noise.

For spin torque induced magnetization switching in SPRAM, the "writability versus thermal stability" challenge is shown as switching current density versus switching time in Fig. 9.3. Again we see this rapid increasing in the required switching current density as time scales down from second region to nanosecond region. With the same scaling arguments before, we can see that spin torque switching current density scales roughly as inverse of element volume for the same thermal stability constrain.

There is a scalability advantage of spin torque induced magnetization switching over magnetic field induced magnetization switching. As magnetic element shrinks, spin torque switching current for a given thermal stability requirement scales as inverse of the element thickness and is independent of the element surface area, because the switching current density scales as the inverse of the element volume. On the other hand for magnetic field induced element switching, the magnetic field scales with inverse of the grain volume. If the magnetic field is generated by a current wire, the magnetic field magnitude is proportional to the wire current magnitude.

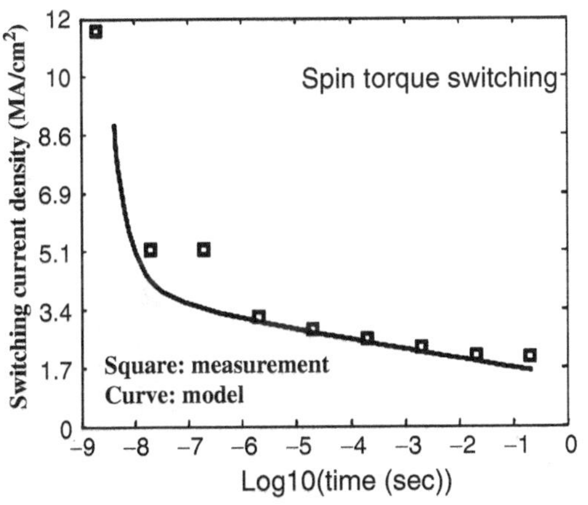

Fig. 9.3 Dynamic thermal switching of a magnetic element with spin torque current [7]. Dots are experiment data and curves are theoretical calculations

Thus the current magnitude scales with inverse of the element volume or inverse of the surface area for wire current generated magnetic field switching. This is the physics basis for better scalability of spin torque MRAM compared to magnetic field MRAM.

However, even with this scalability advantage of spin torque magnetization switching, there are till significant writability challenges for SPRAM to scale down. As will be illustrated in the next section, this is due to the decreased circuit current driving strength as CMOS size shrinks for SPRAM to scale down.

There are two sources of variability for SPRAM. The first variability is due to elements variation in processing. This gives a distribution in writing switching current and reading high to low resistance difference. As a result, SPRAM writing and reading margins are decreased. The second variability comes from random thermal fluctuations at finite temperature. Figure 9.4 shows the measured resistance versus current for a SPRAM magnetic element. The curves on the figure correspond to switching of the same element at different times. The switching variation is due to random thermal fluctuation at room temperature. We will give detailed experiment and model studies on this in the following SPRAM device characterization. Successful device variability control is critical for SPRAM to scale down.

Dynamic thermal magnetization switching and magnetization switching variability determine nano-scale magnetic device performance. Nanosecond switching, long time thermal stability and device variability will be examined in the following sessions for spin torque random access memory. Based upon SPRAM device understandings and future system scaling down requirements, we will discuss some research and development opportunities to achieve switching current magnitude reduction and device variability control.

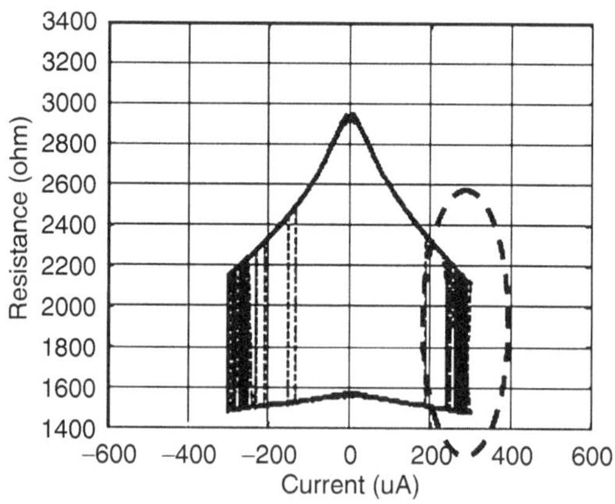

Fig. 9.4 Spin torque switching of a same magnetic element at different times. The different curves are switching variation due to thermal fluctuations at finite temperature

9.3 SPRAM Device Characterization and System Scaling Down Requirement

In order to fully characterize SPRAM performance, experiment and theoretical studies are pursued at widely separated time scales. Magnetization switching behaviors from nano-second to second region are measured and modeled. Microscopic quantum electronic spin transport model and macroscopic stochastic magnetization dynamics model are combined to study spin torque induced magnetization switching. Coupled micro-magnetic model and dynamic circuit model are pursued to simulate SPRAM device performance.

9.3.1 Characterization of Spin Torque Induced Magnetization Switching

When characterizing spin torque induced magnetization switching, both the averaged switching behavior and the switching variation are important. Here we study switching mean and variation for a single magnetic device through experimental measurement and theoretical modeling. The study covers a wide time range, from the short time scale of dynamic switching (approximately nanoseconds) to the long time scale of thermal switching (approximately seconds).

A special probing assembly was used, which operates from DC up to the gigahertz range. It included a Tektronix-Sony 710 arbitrary waveform generator (AWG710), a Picoprobe microwave probe, and a Keithley 2400 source meter. The AWG710 allowed pulse duration to vary from longer than 1 s to as short as 250 ps. The bandwidth of the Picoprobe was from DC to 40 GHz. After each pulse was applied, the device resistance was measured by using the Keithley 2400 source meter. The measurement for determining the switching current due to current pulses was carried out as follows: (1) a DC current was applied through the device, this current was sufficient to consistently set the device into the anti-parallel (or parallel) state; (2) small opposite-polarity current pulses at a certain pulse duration were applied through the device, and the device resistance was measured; (3) the applied pulse amplitude was increased until the device state changed. This process was repeated 100 times to get the switching current variation at that particular pulse duration. The switching current variation was determined in this way for a systematically varied group of pulse durations ranging from the long time scale thermal switching (0.1 s) to the short time scale dynamic switching (10 ns).

The magnetization dynamics in the free layer of a magnetic tunneling junction (MTJ) stack is described by the stochastic Landau–Lifshitz–Gilbert equation at a finite temperature with spin torque terms:

$$\frac{d\vec{m}}{dt} = -\vec{m} \times \left(\vec{h}_{eff} + \vec{h}_{th}\right) - \alpha \vec{m} \times \vec{m} \times (\vec{h}_{eff} + \vec{h}_{th}) + \frac{\vec{T}}{M_s} \tag{9.1}$$

where $\vec{m}$ is the normalized magnetization vector, time t is normalized by γM_s with γ being the gyromagnetic ratio and M_s being the magnetization saturation. $\vec{h}_{eff} = \vec{H}_{eff} / M_s = \frac{\partial \varepsilon}{\partial \vec{m}}$ is the normalized effective magnetic field with normalized energy density ε, and α is the damping parameter. $\vec{h}_{fluc}$ is the thermal fluctuation field , whose magnitude is determined by the fluctuation-dissipation condition at room temperature and whose formalism follows Ref. [8]. $\vec{T}_{norm} = \frac{T}{M_s V}$ is the normalized spin torque term with units of magnetic field. The net spin torque $\vec{T}$ can be obtained through microscopic quantum electronic spin transport model [9–15]. At macroscopic magnetization dynamics level, spin torque can be approximated through adiabatic term proportional to $\vec{m} \times \vec{m} \times \vec{p}$ and non-adiabatic term proportional to $\vec{m} \times \vec{p}$ where $\vec{p}$ is a unit vector pointing to the spin polarization direction.

An example of measured and modeled switching current density mean and standard deviation versus mean switching time is shown in Figs. 9.3 and 9.5. In the model calculation, the normalized spin torque is $\beta(\vec{m} \times \vec{m} \times \vec{p})$ $\beta = \eta \hbar I / 2eM_s^2 V$ is the normalized spin torque polarization magnitude with η being the polarization efficiency, V being the element volume, and I being the current magnitude. In model data calibration, the element's geometry and magnetic properties can be obtained through a combination of patterning procedures and magnetic measurements on sheet film. The main fitting parameters in the model data correlation are the damping parameter and polarization efficiency. The polarization efficiency is chosen to be consistent with measured tunneling magnetoresistance (TMR) value and the fitted damping parameter can be compared to Ferromagnetic Resonance (FMR) line width measurement. As shown in Ref. [7], model agrees well with the experimental measurement over a wide time range, from the 0.1 s thermal reversal region to the 10 ns dynamic reversal region.

Figure 9.6 shows an example of measured and modeled switching time standard deviation versus the mean switching time. There is the good agreement between theoretical model and experiment data. Figures 9.5 and 9.6 show switching current variation and switching time variation scale quite differently as time scales down from long time (~0.1 s) to short time (~10 ns). As time scales down, switching times variation decreases and switching current variation increases. The details analysis for this scaling difference are presented in Refs. [7, 16]. It is the result of the rapid increasing of switching current density as switching time scales down from second to nanosecond region. Due to this rapid increasing of switching current density versus switching time, a small switching time variation due to thermal fluctuations at nanosecond region implies a big switching current density variation.

Stochastic Landau–Lifshitz–Gilbert equation with phenomenological spin torque term can describe spin torque switching mean and variation behaviors when calibrated to experimental data. However, it is not sufficient to describe switching anomalies due to intrinsic spin transport. One such example is the intrinsic spin torque induced magnetization switching asymmetry. Figure 9.7 shows the measured and simulated spin torque magnetization switching for a magnetic tunneling

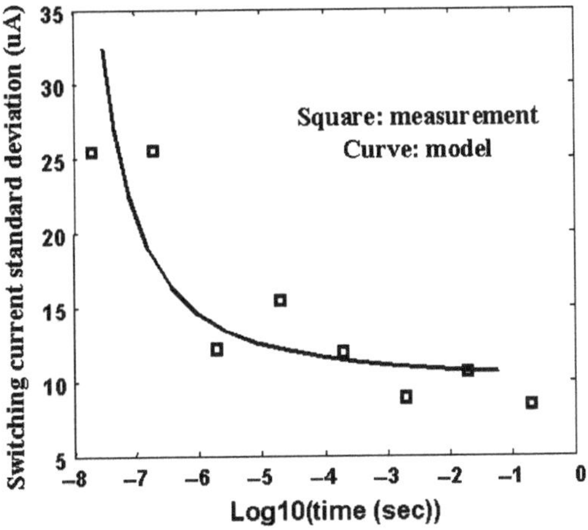

Fig. 9.5 Switching current density standard deviation versus mean switching time [7]. Mean switching current density is shown on Fig. 9.3

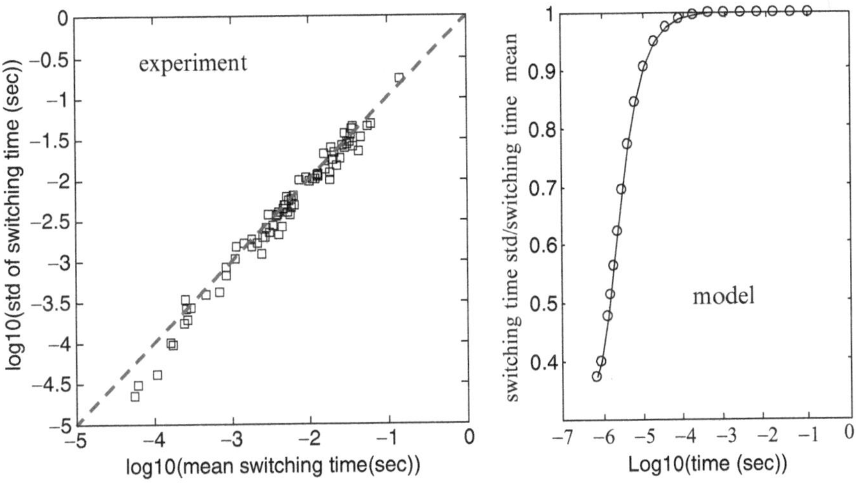

Fig. 9.6 Modeled and measured switching time standard deviation as a function of the mean switching time [7]

junction device. There is this asymmetry between switching from parallel to anti-parallel state and switching from anti-parallel state to parallel state. This switching asymmetry is not due to unbalanced magnetic field of orange-peel coupling and the demagnetization field of the MTJ stack. Figure 9.8 shows the measured magnetic field-induced free layer magnetization switching. The measurement shows a centered resistance versus external magnetic field loop, indicating that magnetic field is balanced in the stack.

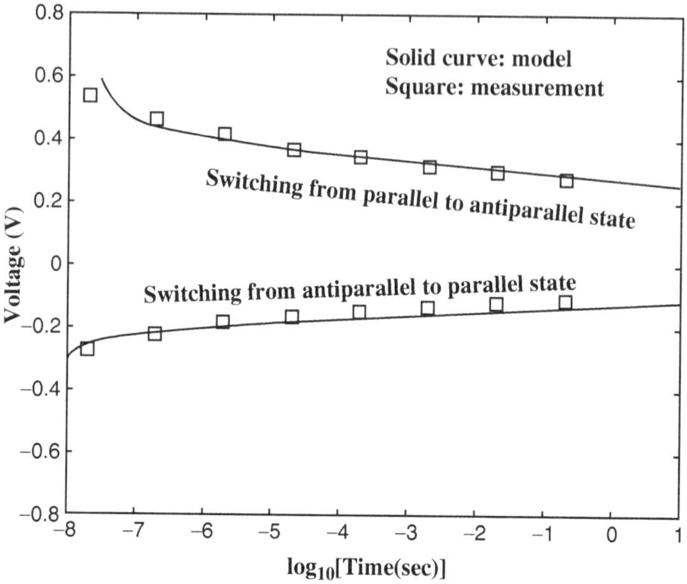

Fig. 9.7 Experimental measurement and theoretical calculation of switching voltage versus pulse duration time for switching from parallel magnetization state to anti-parallel magnetization state and switching from anti-parallel magnetization state to parallel magnetization state [17]

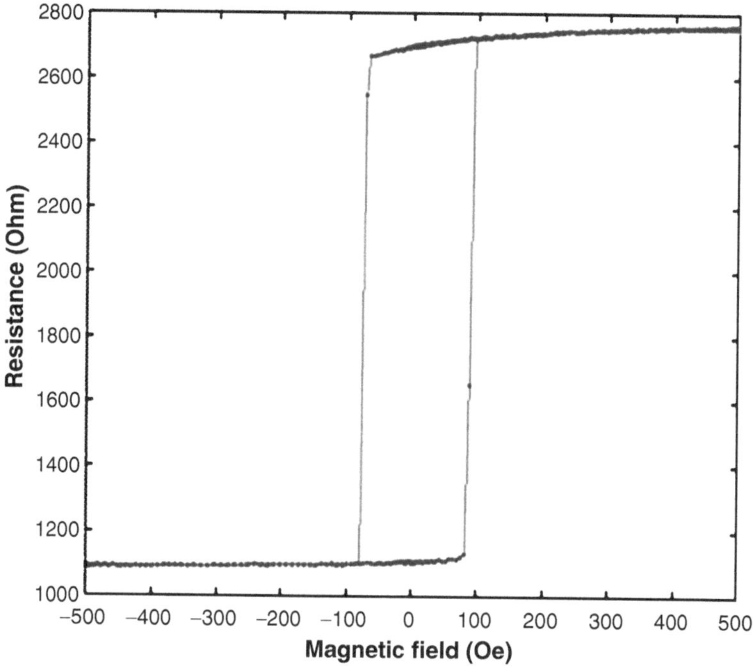

Fig. 9.8 Resistance vs magnetic field for a magnetic field balanced stack

Although there could be many sources of switching asymmetry besides unbalanced magnetic field (such as the non-uniform current distribution in the high TMR magnetic tunneling junctions), we believe this significant switching asymmetry is due to intrinsic spin torque transport asymmetry from parallel to anti-parallel state and from anti-parallel state to parallel state. In practical design, we need to understand the observed asymmetries of magnetic switching between the two resistance states. We developed a model combining quantum spin transport with stochastic magnetization dynamics [17] to study the intrinsic magnetization switching asymmetry due to spin transport across a magnetic tunneling junction.

The quantum spin transport is calculated through non-equilibrium Green's function approach with a tight binding Hamiltonian. The Hamiltonian for electrons in ferromagnetic thin films and insulating layer are:

$$H = \sum_{\sigma,i}(E_b^\sigma + 2t_h^\sigma)c_i^+ c_i - \sum_{\sigma,i} t_h^\sigma c_{i+1}^+ c_i + h.c. \tag{9.2}$$

where E_b^σ and t_h^σ are the band offset and hopping terms for spin channel. The charge density and spin current density transport between atomic sites are calculated through the nonequilibrium Green's function:

$$\begin{aligned} I_{i,i\pm1} &= \frac{et}{2\pi\hbar}\int Tr_\sigma[G_{i,i\pm1}^< - G_{i\pm1,1}^<]dE. \\ \vec{I}_{i,i\pm1}^s &= \frac{t}{4\pi}\int Tr_\sigma[\left(G_{i,i\pm1}^< - G_{i\pm1,1}^<\right)\vec{\sigma}]dE. \end{aligned} \tag{9.3}$$

where $\vec{\sigma}$ is the Pauli matrix, $G^<$ is the nonequlibrium Keldysh Green's function and subscript i refers to the individual atomic site. In this formalism, the system is partitioned into a channel and contacts. Contacts of ferromagnetic thin film are included through self-energy matrices describing the interactions between channel and contacts. The nonequilibrium Green's functions can be determined through:

$$G^< = iG_R \Sigma^< G_R^+ \tag{9.4}$$

where $G_R = \left[E - H - \Sigma_L - \Sigma_R\right]^{-1}$ is system retarded Green's function and $\Sigma^< = f_L(\Sigma_L^+ - \Sigma_L) + f_R(\Sigma_R^+ - \Sigma_R)$ is the nonequilibrium self-energy with $f_{L(R)}$ the L(R) contacts Fermi–Dirac distribution and Σ_L, Σ_R the self-energy for L(R) contacts. The self energy Σ_L, Σ_R can be obtained from the retarded Green's function of the isolated semi-infinite L(R) contacts and channel-contact coupling strength. Conservation of the total angular momentum implies that the spin torque current lost at an atomic site i is transferred to its local magnetic moment, thereby exerting a local spin transfer torque on site:

$$\vec{T}_i = \vec{I}_{i-1,i}^s - \vec{I}_{i,i+1}^s$$

In the model simulation of Fig. 9.7, the microscopic parameters of the simulation are chosen to give a voltage dependent TMR and RA (resistance multiplying surface) values close to experimental measurement. The theoretical details of the model data

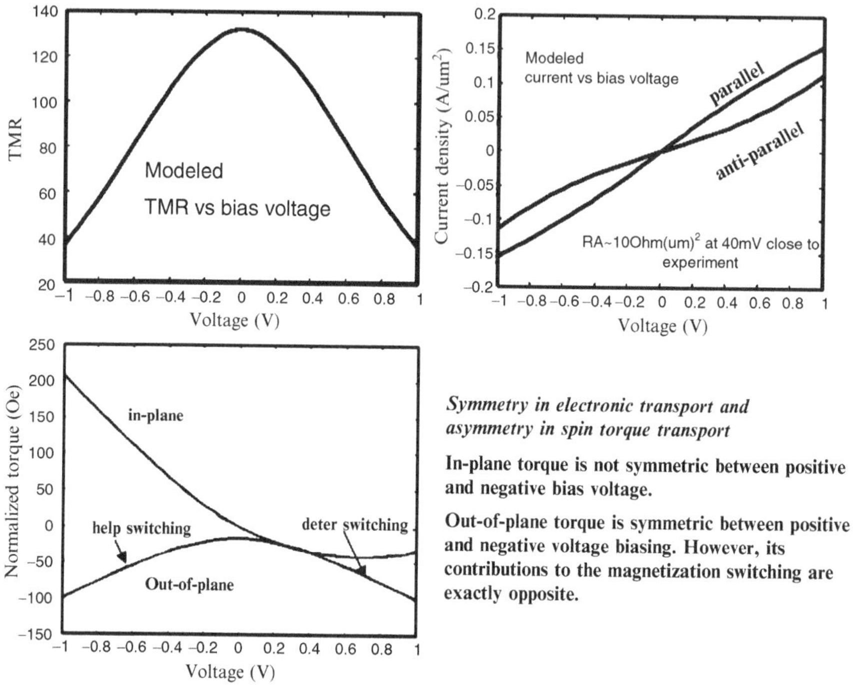

Fig. 9.9 Modeled electronic and spin transport for magnetic tunneling junction

comparison can be referred to [17]. Figure 9.9 shows the calculated electronic and spin transport. The simulation shows symmetry in electronic transport and asymmetry in spin torque transport. The voltage dependence TMR and current density is symmetric regarding positive and negative voltage. However, spin torque magnitude is not symmetric regarding positive and negative voltage.

There are two components of spin torque. In-plane torque corresponds to adiabatic term proportional to $\vec{m} \times \vec{m} \times \vec{p}$ and out-plane torque corresponds to non-adiabatic term proportional to $\vec{m} \times \vec{p}$. In the simulation, the torque for positive voltage corresponds to switching from parallel to anti-parallel states (or from low resistance to high resistance), and the torque for negative voltage corresponds to switching in the reverse order. It is clear that the in-plane torque is not symmetric between positive and negative bias voltage. Although the modeled out-of-plane torque is symmetric between positive and negative voltage biasing, its contributions to the magnetization switching from parallel state to anti-parallel state and those from anti-parallel state to parallel state are exactly opposite. The negative magnitude of the out-of-plane torque in this simulation helps magnetization switching from the anti-parallel to the parallel state and inhibits magnetization switching from the parallel to the anti-parallel state. Thus, both the in-plane and out-of-plane components of the torque contribute to the asymmetric feature of higher voltage and steeper slope for switching from the parallel state to the anti-parallel state.

The calculated spin torque term from quantum transport model is used to drive magnetization switching at finite temperature through Eq. (9.1). The modeled dynamic thermal switching behavior of the magnetic element compared well to experimental measurement and reveals the intrinsic spin torque asymmetry due to quantum spin transport (Fig. 9.7).

9.3.2 SPRAM System Dynamic Modeling

SPRAM operating is an intrinsic dynamic process. Device characterization in previous session shows that magnetic tunneling junction switching current magnitude and variation depend strongly upon the time scale or the switching speed. For SPRAM writing process, MTJ and CMOS are dynamically coupled: on the one hand, MTJ switching is highly dependent on the dynamics and histories of write current applied; on the other hand, MTJ shows as a time varying resistance in electric circuit and greatly affects the driving current provided by CMOS. In order to model the coupled dynamic behavior of MTJ-CMOS switching, we developed a SPRAM switching model by combining micro-magnetic simulation of MTJ and SPICE simulation of CMOS circuit [18, 19].

Stochastic Landau–Lifshitz–Gilbert (LLG) equation with spin torque term is used to describe magnetization dynamics of MTJ stack. The magnetization dynamics of MTJ stack including free layer, reference layer and pinning layer are simulated micro-magnetically. For modeling combining effects of dynamic magnetization switching behavior and dynamic electrical circuit behavior, driving current for MTJ is a time varying quantity and its magnitude at particular time is determined by effective resistance and capacitance of SPRAM cell. The AC response of CMOS transistor to the change of MTJ resistance is determined by the equivalent MOS capacitance. Here MTJ is represented by a variable resistance. CMOS transistor is modeled as voltage controls current source, which is characterized by SPICE simulation. For one-transistor-one-MTJ design in Fig. 9.10, the dynamic electrical behavior of SPRAM cell can be express by:

$$
\begin{aligned}
&I_{\mathrm{M}} + I_{\mathrm{GD}} = I_{\mathrm{DB}} + I_{\mathrm{T}} \\
&\qquad I_{\mathrm{GD}} = \mathrm{C}_{\mathrm{GD}} \cdot \left(-\frac{\mathrm{dV}}{\mathrm{dt}} \right) \\
&\qquad I_{\mathrm{DB}} = \mathrm{C}_{\mathrm{DB}} \cdot \left(\frac{\mathrm{dV}}{\mathrm{dt}} \right) \\
&I_{\mathrm{M}} \cdot R(t) = V_{\mathrm{DD}} - V \quad (\mathrm{BL} \rightarrow \mathrm{SL})\,\mathrm{or} \\
&I_{\mathrm{M}} \cdot R(t) = -V \quad (\mathrm{SL} \rightarrow \mathrm{BL})
\end{aligned}
\tag{9.5}
$$

Here CGD, CGS, CDB, CSB and CGB denote the capacitance between gate (G) and drain (D), gate (G) and source (S), drain (D) and bulk (B), source (S) and bulk (B), gate (G) and bulk (B), respectively. Usually CGB can be ignored.

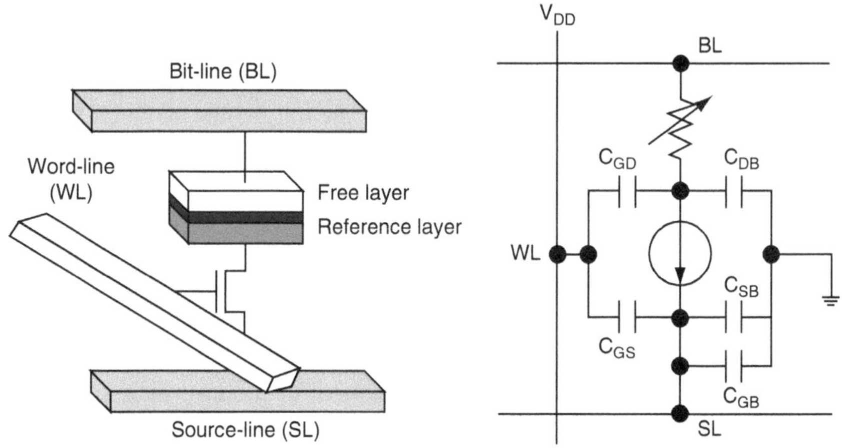

Fig. 9.10 Schematic picture and equivalent circuit of one MTJ and one CMOS ST-MRAM

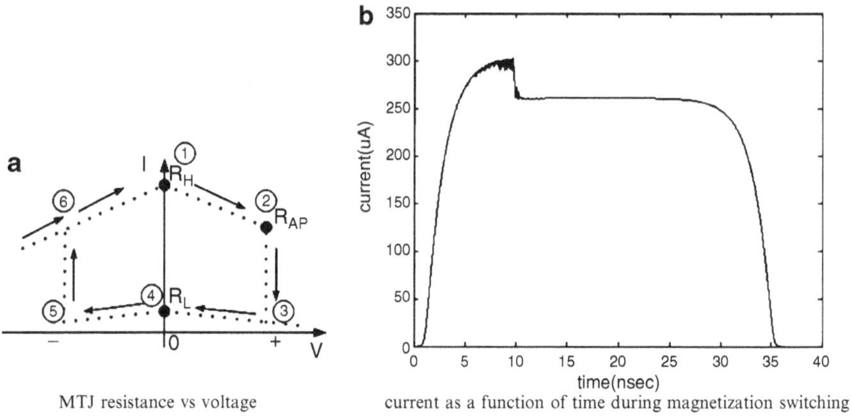

Fig. 9.11 MTJ Switching in SPRAM design

The effectiveness of dynamic coupling between MTJ and CMOS in SPRAM design has been explored in Ref. [18]. To illustrate the importance of coupling SPRAM dynamics to circuit dynamics, let's consider a SPRAM cell example at TSMC 90 nm technology. Figure 9.11a shows the resistance versus voltage obtained at micron-time scale. The resistance jump in the figure corresponds to magnetization switch from one equilibrium state to the other equilibrium state. In reality this resistance jump involves complex magnetization and circuit dynamics. Figure 9.11b shows the simulated dynamic response of the current through MTJ over time, when the MTJ resistance switches from low resistance state R_L to high resistance state R_H. The oscillation of the current is incurred by the fluctuation of MTJ resistance during the rotation of the magnetization direction of free layer. The current drops abruptly when the angle between the magnetization directions of the free layer and the reference layer crosses 90° at 10 ns. This current drop represents the jump of MTJ resistance at Fig. 9.11a and slows down the magnetization rotation, compared to driving MTJ with a constant current.

If dynamical coupling between MTJ and CMOS at nanosecond region are neglected, during magnetization switching from R_L to R_H, only a constant current can be considered for circuit design. The feedback between MTJ and CMOS in Fig. 9.11b is neglected. Figure 9.12 compares MTJ switching between dynamic case and static cases.

Scenario-I is the dynamic MTJ switching case, where CMOS transistor provides a driving current of 300 and 264 μA when MTJ resistance equals R_L or R_{AP}, respectively. Scenario-II corresponds to MTJ driven by a constant current of 300 μA. This scenario overestimates the average current applied to MTJ and results in an optimistic estimation of the switching time of MTJ. Scenario-III corresponds to MTJ driven by a constant current of 264 μA. This scenario underestimates the average current applied to MTJ and results in a pessimistic estimation of the switching time of MTJ. The corresponding switching times of MTJ in Scenario-I, -II and -III are 10, 9.17, and 12.05 ns, respectively. These results show that Scenario-III (pessimistically) overestimated the switching time of MTJ by twenty percent, and Scenario-II (optimistically) underestimated the switching time of MTJ by eight percent, compared to Scenario-I.

The details of SPRAM design margin exploration using coupled dynamic MTJ and CMOS circuit model can be referred in Ref. [18].

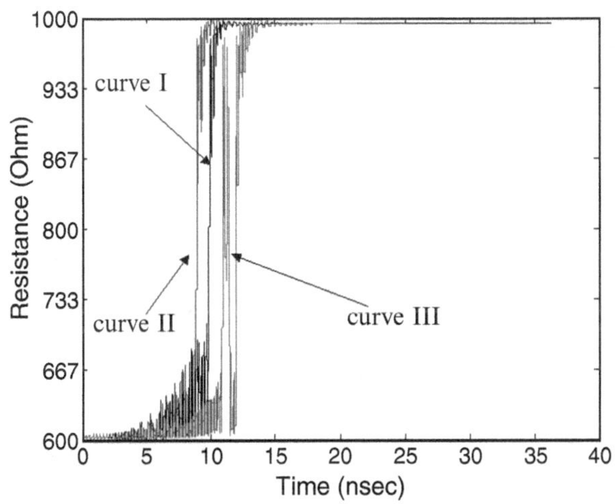

Fig. 9.12 Dynamic response of MTJ Switching

9.3.3 SPRAM Scale Down Requirments

Scaling down SPRAM requires shrinking both CMOS and MTJ dimensions. In general CMOS driving current scales with channel length and channel width as:

$$I_{DS} = K \cdot \frac{W}{L} \cdot f\left(V_{GS}, V_{DS}\right) \tag{9.6}$$

where K is the process trans conductance parameter. W is the channel width and L is the effective channel length. V_{GS} and V_{DS} are gate to source and drain to source voltage. For a particular technology node with fixed processing parameters, effective channel length and process conductance parameter are more or less constant. In order to achieve higher memory/logic density, CMOS transistor channel width needs to be scaled down with degraded CMOS driving strength. Figure 9.13 shows the current provided by CMOS as a function of MTJ resistance for 22 nm channel length with different channel width. The driving current not only depends upon channel width, but also depends upon MTJ junction resistance.

As CMOS scales down, MTJ writing current needs to be reduced correspondingly. Changing MTJ magnetic material properties and scaling down magnetic element can reduce MTJ writing current threshold. However, the competition between the short time dynamic writability and the long time thermal stability is a well-known physics factor limiting magnetic device scaling down. We have discussed this as a fundamental challenge in the previous sections.

Variations at nano-scale further degradate SPRAM performance. Processing variations and thermal fluctuations are two main sources of variability. MTJ processing variation in lateral dimension and tunneling barrier thickness result high and low state resistance variations. Figure 9.14 shows an example of SPRAM reading

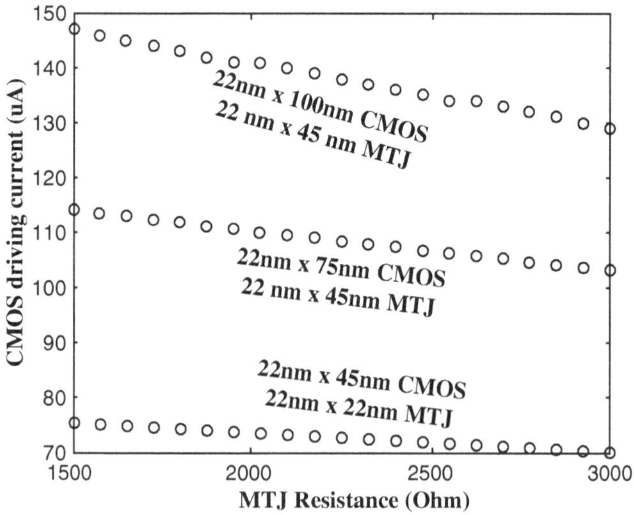

Fig. 9.13 CMOS driving strength versus MTJ resistance for 22 nm channel width with different channel length

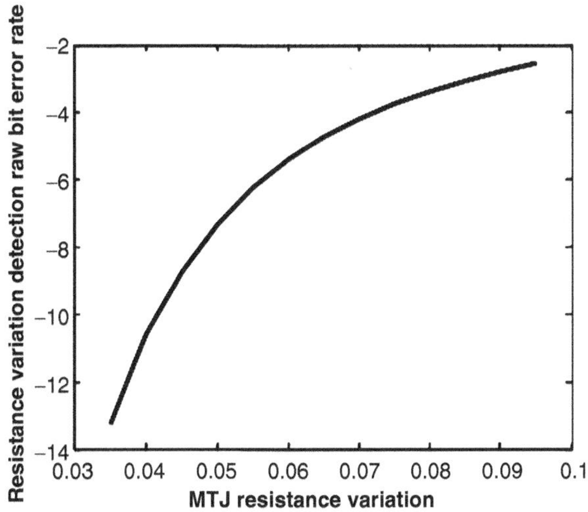

Fig. 9.14 Dependence of reading detection raw bit error upon resistance variation

detection raw bit error rate versus resistance standard deviation over mean. The detection scheme in Fig. 9.14 is based upon resistance difference between high and low state. Resistance variation decreases reading sensing margin.

Thermal fluctuations cause random flip of writing bit during the reading process. The probability of this random flip depends upon reading current magnitude. Higher reading current gives more disturbance and thus causes more random bit flips. Figure 9.15a shows the flip bit error rate caused by thermal fluctuation as a function of reading voltage. Higher reading voltage results higher random flip bit error rate. Figure 9.15b shows reading detection raw bit error rate as a function of reading voltage. This error rate is due to processing resistance variations discussed in previous paragraph. For resistance variation detection error rate, higher voltage provides higher signal window, thus results less detection raw bit error rate. The example here shows different dependence upon sensing voltage (or current) magnitude between reading errors caused by thermal fluctuations and reading errors caused by resistance variations.

Thermal fluctuations and processing variations also require more writing current strength due to the decreased writing margin. For example, in our design example of at TSMC 90 nm technology node (Fig. 9.12), the calculated standard deviation of switching time is shown in Fig. 9.16. For a worst-scenario 6-sigma design, if switching time is set to be 15 ns, a mean switching time of 10 ns is required because the standard deviation of switching time is 8% of the averaged switching time.

Scaling path down to 22 nm TSMC technology node has been proposed [19] and the main device requirements are shown in Table 9.1 [19]. Aggressive switching current reduction is required for MTJ device to scale down to 22 nm technology node. Another significant challenge is to the reduction of RA to keep MTJ resistance low as MTJ surface area scales down. SPRAM variation requirements are not explicitly shown in Table 9.1, because variation requirements depend upon trade-offs in system writing and reading scheme. However it can be seen from Figs. 9.14 through 9.16 that device variations need to be tightly controlled for SPRAM to scale down.

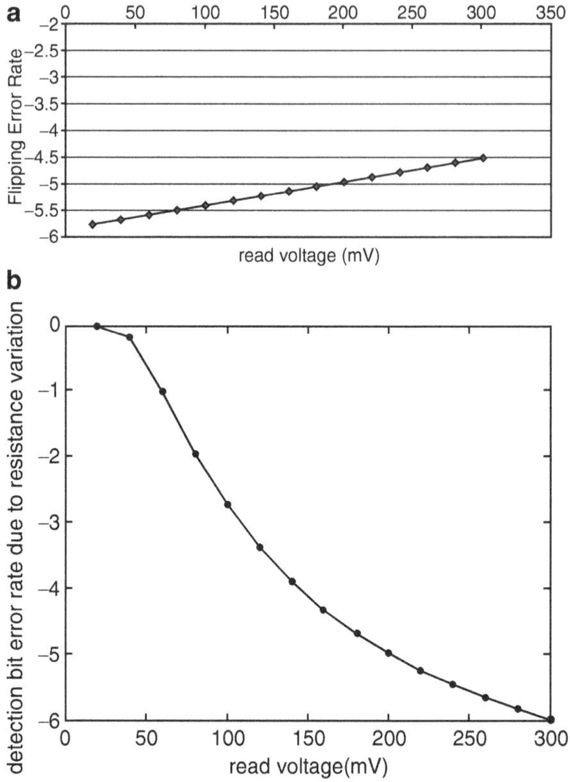

Fig. 9.15 Dependence of reading error upon sensing voltage (**a**) reading errors caused by thermal random bit flip (**b**) detecting error due to resistance variations

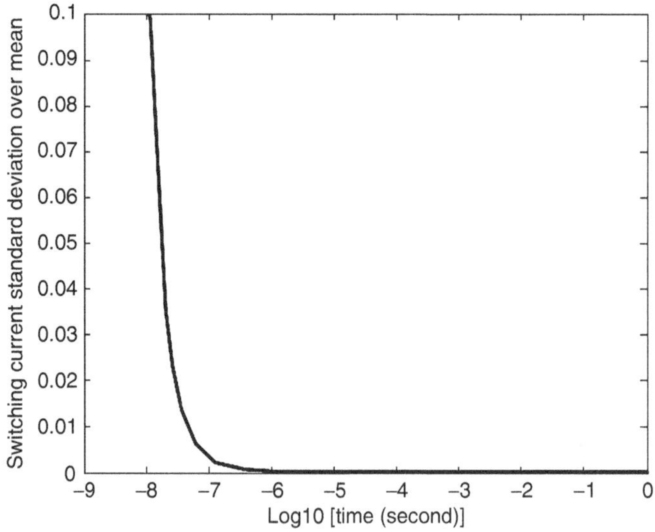

Fig. 9.16 Switching variation due to thermal fluctuations for SPRAM at TSMC 90 nm Technology node

Table. 9.1 Proposed SPRAM scaling down to 22 nm TSMC technology node

Tech node	TX width (nm)	VDD (V)	Area factor F2	MTJ length (nm)	MTJ width (nm)	MTJ thickness (nm)	Ms (emu/cc)	Rmax (ohm)	Rmin (ohm)	RA_min (ohm μm^2)	Ic (μA)	KV/kT
90	885.0	1.2	43.3	110	60	2.2	1,000	1,200	600	4.0	300	52
45	272.8	1.0	28.2	90	40	2.3	1,000	2,000	1,000	3.6	170	52
22	105.8	1.0	23.2	45	22	3.2	1,200	3,000	1,500	1.5	100	50
22	75.0	1.0	25.0	45	22	3.2	1,200	3,000	1,500	1.5	75	50
22	45.0	1.0	12.2	22	22	2.0	1,200	3,000	1,500	1.5	54	50
22	69.6	1.0	16.7	45	22	3.2	1,200	8,000	4,000	4.0	50–100	50

9.4 Reasearch and Development Opportunities for Switching Current Reduction and Variability Control

Switching current reduction and variability control are the two key issues for SPRAM to scale down. In the following, some research and development opportunities for current reduction and variability control will be identified and discussed.

9.4.1 *Current Reduction Through Changing Magnetic Properties*

Reduction of pin torque switching current must be pursued without sacrificing device thermal stability. Thermal stability is characterized by energy barrier over thermal fluctuation energy: $\Delta E / k_B T$, where ΔE is the energy barrier and $k_B T$ is the thermal excitation energy. For a magnetic element with energy

$$\frac{E}{M_s^2 V} = \frac{D_x m_x^2}{2} + \frac{D_y m_y^2}{2} + \frac{D_z m_z^2}{2} - h_x m_x \tag{9.7}$$

as shown in Fig. 9.17, thermal stability energy is proportional to $\mathrm{M}_s^2 V(D_y - D_x)$ and critical switching current is proportional to:

$$\frac{\mathrm{M}_s^2 V \alpha}{\eta} \sqrt{(D_z - D_y)(D_z - D_x)} \tag{9.8}$$

where D_x, D_y, D_z are demagnetization factors including shape and crystal anisotropy, α is the damping parameter and η is the spin torque efficiency. For a thin film element as in Fig. 9.17, D_z ? $D_y > D_x$ due to strong demagnetization field out of plane. Formula (9.8) shows that spin torque critical switching current scales differently as thermal stability barrier. Spin torque switching current scales with the difference between out-of-plane demagnetization factor D_z and in-plane demagnetization factors D_x, D_y, while thermal stability energy scales as difference between in-plane demagnetization factors D_x, D_y.

The scaling of spin torque switching current with element demagnetization factor is quite different from that of the magnetic field induced magnetization switching. For magnetic field induced switching, the critical switching field scales with the difference between in-plane demagnetization factors D_x, D_y $\mathrm{M}_s(D_y - D_x)$, similar to that of the thermal stability barrier.

The scaling difference between spin torque switching and magnetic field switching is a result of different magnetization switching dynamics. As shown in Fig. 9.17, magnetization information 0 and 1 corresponds to equilibrium magnetization multi-stable states with a stable energy barrier between them. For magnetic field induced magnetization switching, external magnetic fields raise the energy of the initial magnetization state and lower the energy of the final magnetization state.

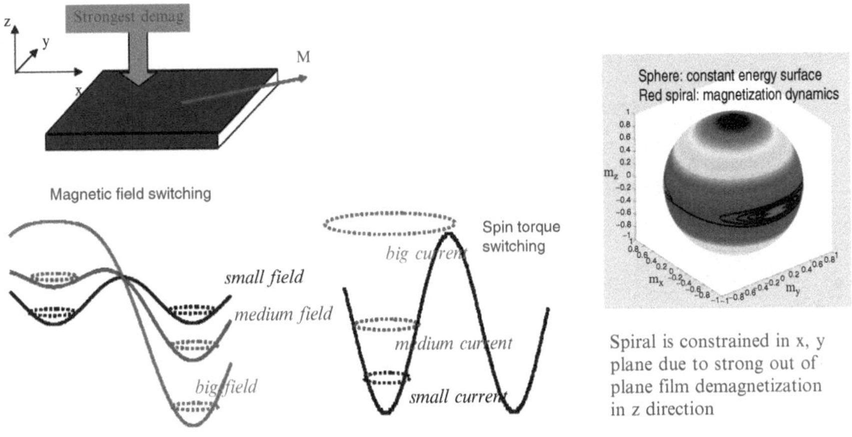

Fig. 9.17 Magnetization Switching through spin torque and magnetic field

When the magnetic field induced energy is bigger than energy barrier between multi-stable magnetization states, initial magnetization equilibrium condition is destroyed and magnetization evolves to final equilibrium condition through magnetization relaxation. Thus the field required to switch magnetization scales the same as the energy barrier between two stable magnetization sates. For spin torque induced magnetization switching, spin torque excites magnetization precession out of the initial equilibrium state. In this process, there are always two equilibrium magnetization states and damping (or magnetization relaxation) tries to pull magnetization back to initial equilibrium condition. Spin torque switching is a competition between spin torque excitation and magnetization damping. Critical switching current is proportional to damping over spin polarization efficiency as in Eq. (9.5). The important consequence of spin torque excitation competing damping for magnetization switching is that magnetization needs to precess out of the thin film plane. The strong out of plane demagnetization constrain for magnetization motion in thin film element brings an energy penalty that results critical switching current proportional to demagnetization factor out of the plane.

The solution based upon this scaling difference between thermal stability and spin torque switching current is to decrease out of plan demagnetization factor and maintain the difference in lateral demagnetization factor. One approach is to add perpendicular anisotropy in out of plane z direction [20]. The other approach is to decrease magnetization saturation and at the same time increase uniaxial anisotropy in x direction. One can also think about coupling magnetic elements to magnetic films with low magnetization saturation and high anisotropy to achieve out-of-plane demagnetization factor reduction. There could be many different methods to reduce out-of-plane demagnetization factor. The ultimate goal of these methods is to reach cylindrical symmetry in magnetization dynamics as illustrated in Fig. 9.18. When magnetization dynamics is cylindrically symmetric, spin torque critical switching current scales the same as thermal stability barrier. The challenge here is to keep thin

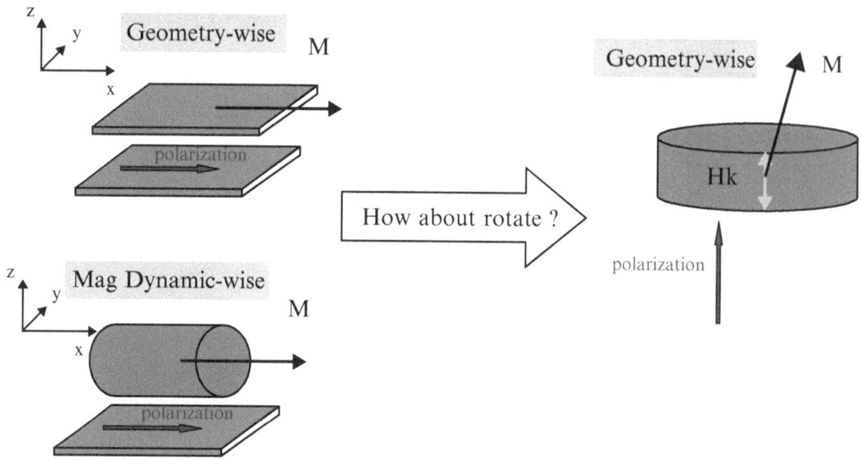

Fig. 9.18 Road leads to the same scaling of spin torque switching current and thermal stability barrier

film structure geometrically as illustrated in Fig. 9.18, because thin film structure is essential for TMR and other required properties of SPRAM device.

This ultimate structure of cylindrical symmetry in magnetization dynamics can be rotated as illustrated in Fig. 9.18. This rotated configuration can be achieved by thin film with a perpendicular crystal anisotropy field much bigger than out-of-plane demagnetization field. Indeed for a thin film with equilibrium magnetization in vertical direction (the so called "perpendicular stack"), critical spin torque switching current scales the same as thermal stability barrier.

9.4.2 *Current Reduction Through Decreasing Damping by Changing Interfacial Tunneling Properties*

Spin torque magnetization switching is achieved through competition between spin torque excitation and magnetization relaxation damping. Formula (9.8) shows critical switching current is proportional to damping parameter. The damping term $\alpha \vec{m} \times \vec{m} \times \vec{h}$ in Eqs. (9.1) and (9.8) for spin torque switching should include all the physical mechanism through which a magnetic element loses its magnetization moment. One approach to reduce switching current is to reduce the magnetization damping.

A spin current can exert a finite torque on ferromagnetic order parameter, and, vice versa, a moving magnetization vector loses torque by emitting a spin current [21]. This magnetization moment loss through "spin pumping" provides a different damping mechanism than that of the intrinsic ferromagnetic relaxation. For ferromagnetic/metal interface, the loss of magnetic moment in MTJ free layer due to spin pumping is derived to be in a Gilbert damping format through microscopic quantum pumping approach [22, 23]:

$$\vec{I}_{pump} \propto A\vec{m} \times \frac{d\vec{m}}{dt} \tag{9.9}$$

This format can also be derived based upon a phenomenological "radiant reaction spin interaction" approach [24]. In this approach a dissipative torque is introduced to the precessing magnetization dynamics through radiant spin interaction correspondence with the radiation reaction force acting on magnetization itself. It was shown in [24] that this dissipative torque is indeed in general can be written in a Gilbert damping format (9.9).

There are experimental evidences indicating the importance of spin pumping damping in magnetic tunneling junction [25]. In these experiments, dependence of magnetization switching voltage, electric resistance and tunneling magnetoresistance ratio on insulating barrier properties of magnetic tunneling junctions (MTJ) are studied. The magnetic properties and patterning geometric dimensions of these MTJ devices are the same. For the first set of MTJ devices, only MgO thickness between free layer and reference layer is changed through controlling deposition time. For the second set of MTJ devices, besides the MgO insulating layer between free layer and reference layer, an insulating moment conservation layer is inserted between ferromagnetic free layer and metallic cap layer. Critical switching voltage, RA (resistance multiplying area) and TMR (tunneling magnetoresistance) for these MTJ devices are measured to characterize magnetization switching and electronic transport behavior.

Figure 9.19 shows the measured critical switching voltage, RA and TMR for different MgO thickness. The figure shows both the mean and the error bar for a population of devices. Here TMR is defined as (high resistance-low resistance)/(low resistance) at zero biasing voltage and RA is measured for low resistance state at 50 mV. While TMR remains almost constant for different MgO thickness, both RA and switching voltage V_c increase as MgO thickness increases. However, the ratio of V_c divided by RA is not a constant for different MgO thickness. It decreases as MgO thickness increases. From RA = 37 to RA = 17, Vc/RA drops 0.57 times while TMR maintain 160%. TMR starts to drop for RA below 15. However the

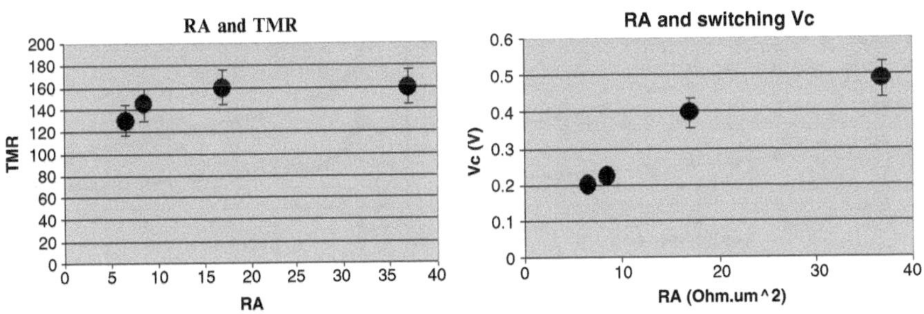

Fig. 9.19 Measured dependence of switching voltage Vc, RA and TMR on MgO thickness

polarization efficiency loss due to TMR dropping between RA = 5 and RA = 37 is smaller than the increased switching current density (for a simplest estimation, polarization is proportional to $\sqrt{TMR/(TMR+2)}$

This measured magnetization switching and electronic transport behaviors in MTJ cannot be explained by LLG model with ballistic spin torque term and intrinsic ferromagnetic relaxation. According to spin torque LLG Eq. (9.1), for free layer with the same material properties and geometric dimensions, the spin torque magnitude required to switch free layer magnetization should be the same for different MgO thickness. For constant TMR of the measured MTJs, the magnitude of spin torque should be more or less linearly proportional to current density. This has been shown through ballistic quantum spin transport model calculation on spin torque magnitude versus RA for varying MgO thickness. Thus LLG with spin torque term and intrinsic ferromagnetic relaxation would predict that the ratio of Vc over RA be the same for different MgO thickness. This is clearly inconsistent with the measured decreasing of Vc/RA as MgO thickness increases.

Figure 9.20a shows a comparison of switching current density versus pulse duration time between conventional MTJ and MTJ with moment conservation layer. Not only insulating specular layer decreases MTJ critical switching current density, but also it changes the slope of switching current versus pulse time.

Above experiments can be understood if we assume that there is spin pumping for MTJs due to free layer magnetization precessing. Spin pumping mechanism predicts spin momentum loss proportional to tunneling transmission probability. Increasing MgO thickness or inserting moment conservation layer decreases tunneling transmission probability and as a result decrease spin momentum loss of ferromagnetic free layer. This reduced effective magnetization damping could explain the decreased switching current density for above measurements. Figure 9.20b shows modeled switching behavior of a MTJ device with decreased effective magnetization damping. It gives the same trend as specular layer MTJ measurement.

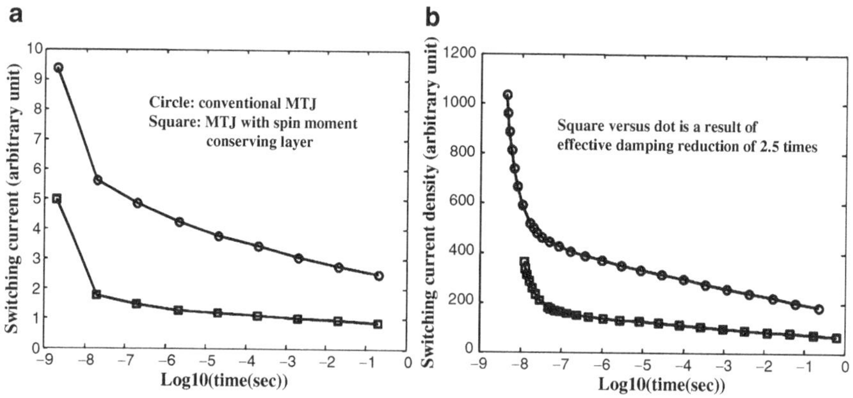

Fig. 9.20 Measured (**a**) and modeled (**b**) switching current density vs pulse duration for conventional MTJ and MTJ with spin moment conserving layer

Dependence of spin torque induced magnetization switching upon insulating interfacial layers can be used to lower switching current for SPRAM. However for practical application there is the trade-off between critical switching current and MTJ device resistance. Spin pumping induced magnetization moment loss depends upon insulating layers properties. When insulating layers become more opaque, MTJ free layer magnetization precessing losses less moment to surrounding metal layers. This results a reduction of effective damping for free layer magnetization and reduces the required critical spin torque switching current. However more opaque insulating layers at the same time means less electrons tunneling through MTJ insulating layers per unit area. This increases MTJ RA value. Thus there is the trades-off between critical switching current amplitude and RA magnitude.

Another path to decrease damping is to decrease intrinsic ferromagnetic relaxation rate of the material. This requires changing free layer material properties. One example is the reduction of intrinsic ferromagnetic relaxation rate in epitaxial iron alloy through vanadium (V) doping [26].

9.4.3 *Current Reduction Through Increasing Spin Torque Efficiency*

The carriers of the spin torque for magnetic element are the polarized electrons passing through magnetic element. In principle the magnitude of spin torque introduced to magnetic element is proportional to the number of electrons passing through the element, Spin torque switching current can be reduced by increase spin torque transfer efficiency. The spin torque efficiency is defined as the ratio of spin torque to electronic current density. The spin torque magnitude and its ratio over electronic current depend upon quantum electronic spin transport mechanism. Increasing spin torque efficiency here means increased spin torque magnitude for the same electronic current magnitude, or increasing the ratio of spin torque magnitude over current magnitude.

One mechanism discussed in Refs. [14, 27] was based upon resonance in quantum spin confinement structure. The quantum transport of electronic charge and spin were given in (9.3) and (9.4) through non-equilibrium Green's function. In the weak coupling regime, the Green's functions of the coupled system can be approximated with those of the uncoupled system: $G_R^{\sigma\sigma} \sim \left[E - E^{\sigma} + i\eta\right]$ is the retarded green function for spin channel σ and η is the dissipation due to system interaction with contact. Thus the spin transport can be approximated as:

$$T \sim \int \frac{\mathrm{dE}}{\left[\left(\left(E-E^{+}\right)^{2}+\eta^{2}\right)\left(\left(E-E^{-}\right)^{2}+\eta^{2}\right)\right]} \tag{9.10}$$

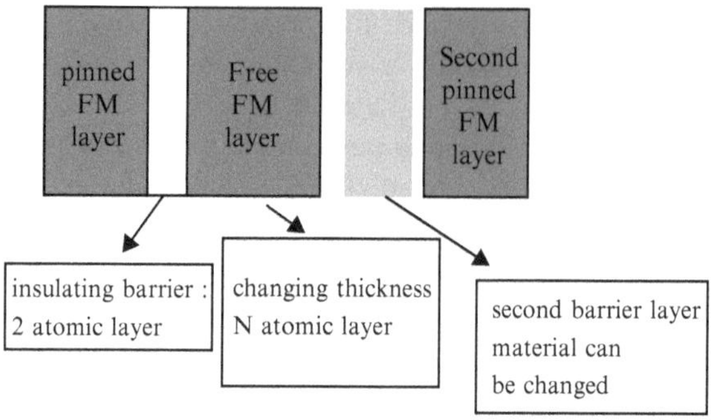

Fig. 9.21 An example of increasing spin torque efficiency through resonance in quantum confinement configuration

and the electronic transport is approximated as:

$$I \sim \int \frac{dE}{\left[\left(E-E^{+}\right)^{2}+\eta^{2}\right]}, \int \frac{dE}{\left[\left(E-E^{-}\right)^{2}+\eta^{2}\right]} \tag{9.11}$$

where + and – sign denote spin up and spin down channels, and the final spin and electronic transport are summations of above integral over all the available discrete energy bands. When $E^{+} \approx E^{-}$ for some discrete bands due to quantum confinement effects, the resonance condition at $E \approx E^{+}, E^{-}$ will enhance spin torque transfer much more than electric current transport.

Let's consider a tunneling structure in Fig. 9.21. Besides pinned ferromagnetic (FM) layer and free ferromagnetic (FM) layer sandwiching an insulating barrier as in conventional SPRAM MTJ stack, there is a second pinned ferromagnetic (FM) layer at the other side of free ferromagnetic layer. In the study here, the thickness of the free layer and the material properties of the tunneling barrier between free FM layer and second pinning FM layer are changed. All the other parameters and the configuration are fixed. The thickness of the films is in the unit of atomic layer number with atomic spacing 2 A. The band offset parameter E_b^{σ} and hopping parameter t_h^{σ} for insulating layer are chosen to be 1 and 6.5 eV. The band offset parameter E_b^{σ}, hopping parameter t_h^{σ} for ferromagnetic spin up and spin down channel are 0.4, 0.318eV and 0.4, 0.736eV. These parameters are used in Ref. [27] and consistent with the ab initio values for one dimension Co FM nanowires.

Figure 9.22 shows the current and spin torque magnitude at center of the ferromagnetic free layer for different FM free layer thickness and different second barrier material properties. For insulating second barrier layer, the offset energy parameter $E_b = 6.5eV$. For conducting second barrier layer, the offset energy parameter $E_b = 0.65eV$. The absolute number of this barrier thickness is not critical as long as they are chosen to represent insulating and conducting cases. It can be seen that for insulating second barrier layer, when the FM free layer thickness is 7 atomic layer spacing, the increasing of spin torque magnitude is much higher than the increasing of current magnitude (Fig. 9.22a, b). Thus, there is a dramatic

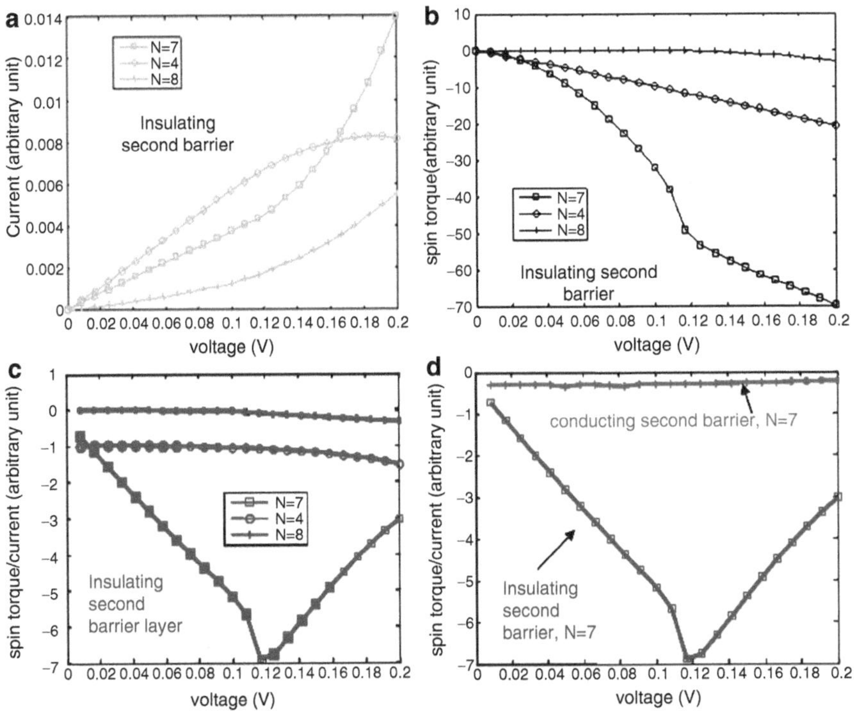

Fig. 9.22 Dramatic increasing of spin torque magnitude through resonance in quantum confinement configuration (**a**) current vs bias voltage for insulating second barrier layer with different FM free layer thickness. (**b**) Spin torque magnitude vs bias voltage for insulating second barrier layer with different FM free layer thickness. (**c**) Ratio of spin torque magnitude over current vs bias voltage for insulating second barrier layer with different FM free layer thickness. (**d**) Comparison of spin torque magnitude over current between insulating second barrier layer and conducting second barrier layer

increase of spin torque magnitude over current magnitude for resonance FM free layer thickness of 7 atomic spacing (Fig. 9.22c). Figure 9.22d compares spin torque magnitude over current between insulating second barrier layer case and conducting second barrier layer case. Dramatic increasing of spin torque magnitude over current magnitude happens only for insulating second barrier layer.

When second barrier layer is insulating, it forms a quantum confinement well between FM free layer and pinned FM layers. Resonant spin torque condition can be excited in this structure for particular resonant free layer thickness and spin torque efficiency can be greatly enhanced. On the other hand, when second barrier layer becomes conducting, the quantum confinement well is destroyed and the enhancement of spin torque through resonance is lost.

The electronic spin transports in realistic MTJ structure are usually more complex than the model here, especially for high TMR MgO barrier MTJ, where spin filtering effects in crystalline structure are important [28, 29]. However above modeling exercise shows the physical possibility of dramatically increasing spin torque efficiency through engineering features like double or multiple barrier tunneling. Interesting enough, recent experiments on double MgO barrier MTJ device confirms significant reduction of switching current compared to that of the conventional single MgO barrier MTJ device [30, 31]. In Ref. [30], for dual barrier MTJ, the switching current density is reduced from $2.7\ \mathrm{MA/cm^2}$ of single barrier to $0.62\ \mathrm{MA/cm^2}$. The thermal stability of the double barrier and single barrier device is about the same $43k_BT$. The reduction of switching current density by double barrier is more than double. Although the detailed physical mechanism requires further analysis (for example, one possible mechanism could be due to damping reduction of the second insulating layer, similar to moment conservation layer in previous section), the experiments show the possibility of switching current reduction though engineering new tunneling structures to enhance spin torque efficiency.

9.4.4 Currnt Reduction Through Time and Spatial Varying Polarized Current

The concept of current reduction through time varying polarized current comes from radio frequency (RF) assisted magnetization switching [32] and/or magnetization precessional switching [33–35]. Radio frequency assisted magnetization switching and precessional magnetization switching have been proposed to reduce the head field magnitude required for magnetic field induced magnetization switching in magnetic data storage. In order to illustrate the concept behind time varying polarization current, we first review basic ideas behind DC and RF magnetic field induced magnetization switching.

Figure 9.23a shows the directions of gyro-magnetic torque and damping torque in Landau–Lifshitz equation (9.1). For DC magnetic field induced magnetization switching, the switching is accomplished through magnetization relaxation term as shown in Fig. 9.23b. When a constant magnetic field is applied to the opposite direction of the initial magnetization, the relaxation torque pulls magnetization toward magnetic field direction (as shown in Fig. 9.23b). However, due to small damping parameter in ferromagnetic material, the magnitude of this relaxation torque usually is much smaller than the magnitude of gyro-magnetic torque for the same magnetic field amplitude. For a typical damping parameter around 0.005–0.01 in ferromagnetic thin film, the magnitude of relaxation torque is one to two hundred times smaller than the magnitude of gyro-magnetic torque. Thus if gyro-magnetic torque can be used directly to switch the magnetization, the switching magnetic field can be decreased significantly.

Figure 9.23c shows a configuration to switch magnetization through gyro-magnetic torque. A magnetic field perpendicular to the initial magnetization direction provides a gyro-magntic torque pulling magnetization away from the initial equilibrium condition. However in order to successfully switch the magnetization, the magnetic field direction needs to be changed during the magnetization precession period

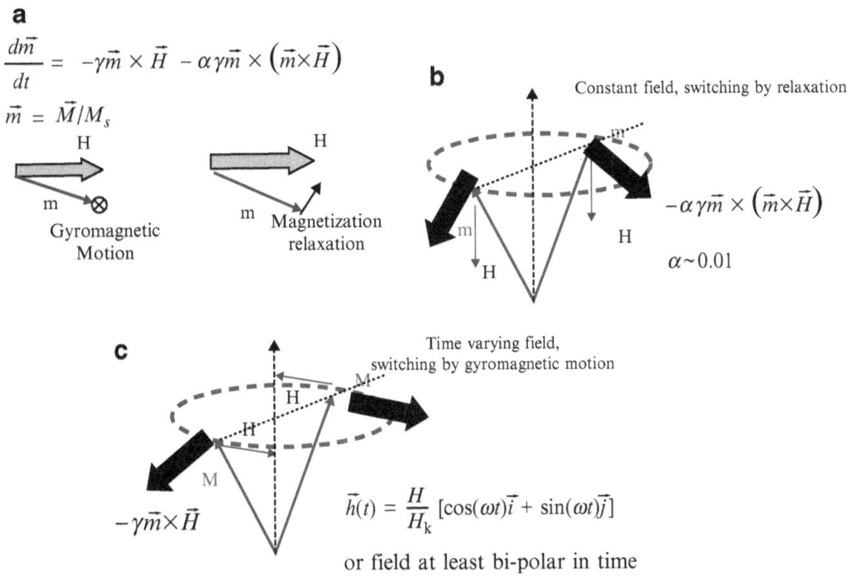

Fig. 9.23 Magnetization switching through DC and AC magnetic field

as shown in the Fig. 9.23c. If the magnetic field direction is constant as in DC switching case, the gyro-magnetic torque directions in the first half period and the second half period of the magnetization precession are exactly opposite. The averaged gyro-magnetic torque on magnetization vector will be zero during one period of magnetization precession. Because magnetization precession is in the gigahertz frequency range, the successful switching of magnetization through AC magnetic field requires a magnetic field direction changing in gigahertz radio frequency range.

The additional complexity of RF field assisted switching comes from the fact that magnetization precession is a nonlinear oscillation. The gyro-magnetic precession frequency depends upon the angle between magnetization direction and external magnetic field direction. For a fixed frequency RF magnetic field, the exact frequency matching between RF magnetic field and magnetization precession could only happen for certain magnetization direction during magnetization reversal. In order to achieve the frequency matching during the whole magnetization reversal process, the frequency of external RF field needs to be adjusted according to magnetization direction. If this indeed can be achieved, the switching coercivity reduction is dramatic. As shown in [36, 37], when the magnetization field frequency matches magnetization precession dynamics during the whole reversal process, the AC field switching coercivity can be reduced to a fraction of the DC magnetic field coercivity.

Above concept of RF magnetic field assisted switching can be extended to spin torque induced magnetization switching [38, 39]. Figure 9.24 illustrates the case of time varying spin torque magnetization reversal. Instead of a fixed spin polarization direction, the spin polarization direction changes with time. If the direction of spin

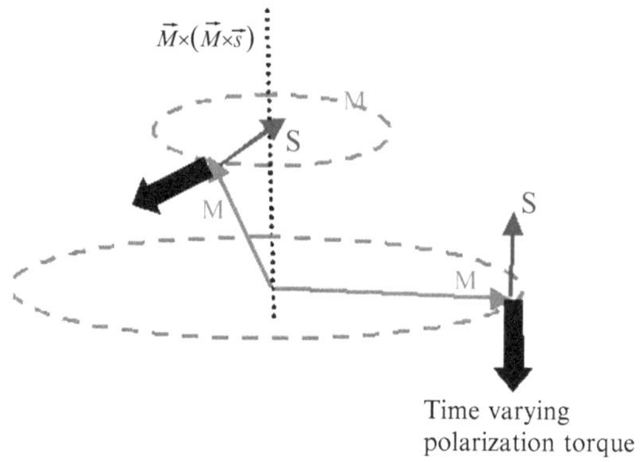

Fig. 9.24 Reduction of switching current magnitude for spin torque polarization direction changing with time, providing maximum spin torque during whole reversal process

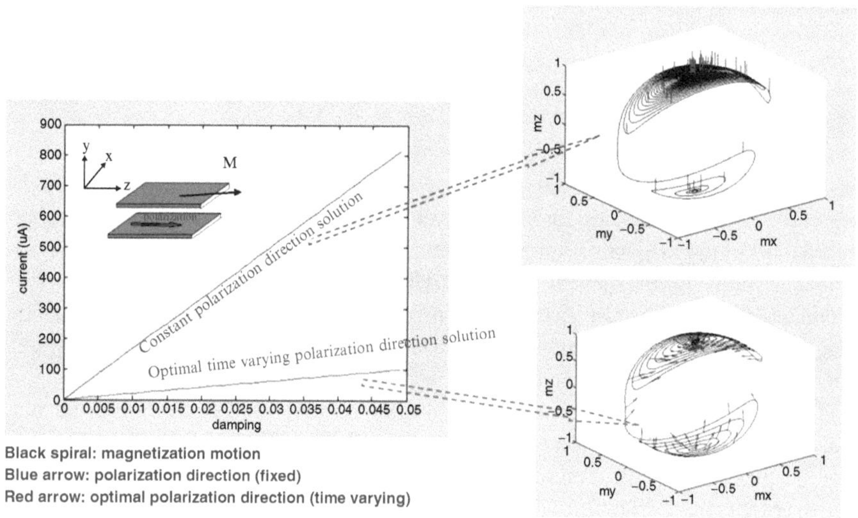

Fig. 9.25 Reduction of switching current by optimal time varying polarization direction matching magnetization direction during whole magnetization reversal, providing maximum spin torque

polarization can match magnetization precession to provide the maximum spin torque during whole magnetization reversal, significant switching current reduction can be achieved. Figure 9.25 compares switching current magnitude of the constant polarization direction to the switching current magnitude of the optimal time varying polarization direction for a thin film stack with energy in the form of (9.7). The optimal spin polarization direction solution is obtained by a time varying polarization direction that provides maximum spin torque during the whole magnetization reversal process. The magnetization trajectory and spin polarization direction evolutions are shown in the figure. Significant critical switching current reduction is obtained.

Exact matching of polarization direction to magnetization direction during whole spin torque magnetization reversal is difficult to achieve in practical application. However there could be intermediate solutions between constant polarization direction and optimal time varying polarization direction that may be easy to implement practically. The key requirement here is the positive feedback of magnetization direction information to spin polarization direction. We can imagine this could be achieved either by electronic circuit or coupled magnetic films.

Spatial varying spin torque can also reduce switching current. When spin polarization current is confined to local surface regions of the thin film element, the total switching current can be reduced [40]. Compared to the case of uniform current passing through the whole element surface area, the switching current density is increased for current confinement structures. However the total switching current for current confined structures could be smaller than that of the uniform current structure. This is because the reduction of surface area is bigger than the increasing of switching current density for current confined structures.

9.4.5 Current Reduction Through Coupled Magnetic Elements and Nonunifrom Magnetization Switching

Another approach to reduce spin torque switching current magnitude is through coupled magnetic elements and non-uniform magnetization switching. Coupled magnetic grain structures have been proposed to solve "writability versus thermal stability" dilemma in magnetic data storage [41–43]. The proposals are to reduce ratio of dynamic coercivity over thermal stability energy barrier through non-uniform magnetization reversal in composite grain structures. For media with reduced coercivity over thermal energy barrier ratio, the required head field to switch media grains is decreased for the same thermal stability criterion.

An example of composite grain structure is shown in Fig. 9.26a. The hard magnetic grain with crystal anisotropy 15,000 Oe in perpendicular direction is vertically coupled to the soft magnetic grain with zero anisotropy. The two grains have same thickness and the magnetization saturation is 500 emu/cc. By tuning the exchange coupling between two grains, magnetic field switching dynamic coercivity can be reduced to 0.3 times of the dynamic coercivity of the hard gain only case (15,000 Oe) using LLG simulation. For this composite grain structure, the thermal energy barrier is about the half of the hard grain only structure with the same volume. Thus the coercivity over energy barrier is reduced about 0.6 for this composite grain structure.

Switching coercivity reduction in composite structure is a complex dynamic process. As shown in Ref. [44], the magnitude of coercivity reduction depends upon many dynamic effects, such as head field rising time.

We can imagine switching current reduction for SPRAM through coupled magnetic elements. For spin torque magnetization switching in coupled magnetic elements, besides the complexity of magnetization dynamics under spin torque

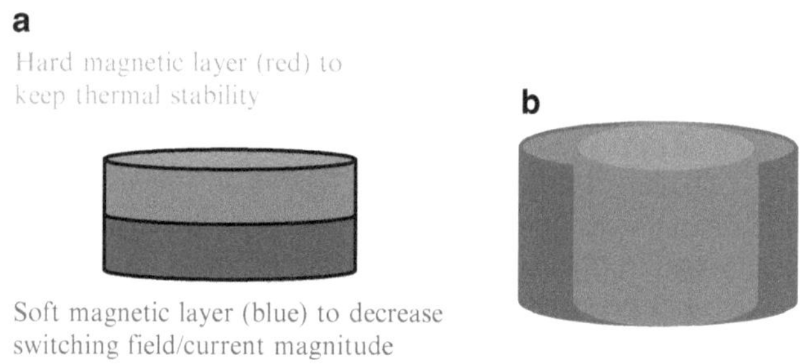

Fig. 9.26 Composite element structures to reduce switching magnetic field magnitude and switching current magnitude

excitation, an additional complexity comes from spin torque transport across inhomogeneous composite structure. Spin torque is an interfacial effect for regions with different magnetization orientations. Magnetization switching in the free layer of a composite structure in Fig. 9.26a not only depends upon the spin torque transfer between pinned reference layer and free layer, but also depends upon the spin torque transfer within different composite magnetic layers in the free layer. The composite structure benefit on switching current reduction depends on the detailed spin torque transfer within composite magnetic free layers.

In order to illustrate dynamics of composite effects on spin torque induced magnetization switching, we consider the example element in Fig. 26a. For the sake of simplicity, the spin torque at interface is averaged within the whole grain volume and detailed quantum spin torque transfer across the composite structures of the free layer is neglected. The purpose here is to illustrate composite effects on magnetization dynamics. Figure 9.27 shows the spin torque induced magnetization switching for this composite grain. When excited by spin torque, the soft magnetic element starts to reverse first and this helps the switching of hard magnetic element. It is this non-uniform dynamic magnetization switching that has the potential to reduce spin torque switching current. For the particular example here, the switching current can be reduced to half of the switching current for the element with hard anisotropy only. Of course detailed quantum spin torque transfer across composite free layer structure is required to fully characterize spin torque induced magnetization switching in composite grain structure. This will be an interesting topic for further exploration.

Another possible composite element structure for spin torque switching is shown in Fig. 9.26b. Instead of vertical magnetic inhomogeneous structure, lateral magnetic inhomogeneous structure could be designed to lower switching current and at the same time maintain thermal stability. For the polarized spin current injection in the vertical direction, as in Fig. 9.26a, b, lateral inhomogeneous structure could have benefit compared to vertical inhomogeneous when spin torque interfacial effects are considered. Another example of switching current reduction through non-uniform magnetization motion has been demonstrated both numerically and experimentally in Ref. [45].

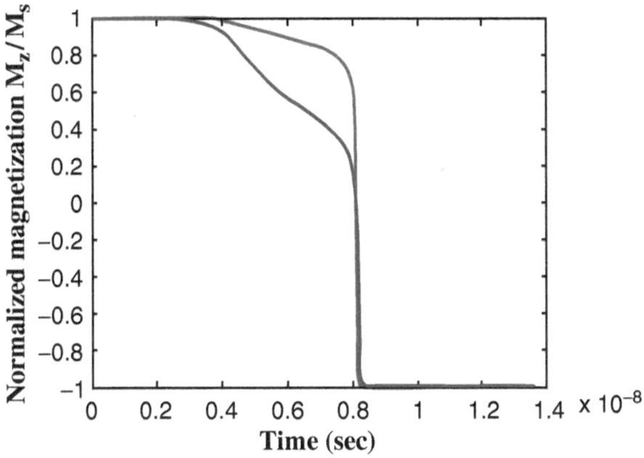

Fig. 9.27 Spin torque induced composite elements magnetization switching

9.4.6 Current Reduction Through Thermal Spin Torque Switching

As SPRAM scales down, in order to achieve maximum potential capacity supported by CMOS technology node, not only the current threshold is required to reduce, but also the MTJ physical dimension is required to shrink (Table 9.1 and Ref. [19]). This leads to another technology bottleneck for MTJ. The challenge here is to maintain thermal stability of MTJ with spherical or square magnetic element shape and at the same time the element can be switched at very low current.

One solution is to introduce surface anti-ferromagnetic coupled (AFC) magnetic layer with relatively low Curie temperature. At room temperature, the AFC coupling provides surface anisotropy to maintain thermal stability of the square magnetic element. During the writing process, the joule heating of spin torque current raises AFC surface magnetic layer temperature above Curie temperature and the AFC induced surface anisotropy disappears. The element can be switched at low threshold current.

Thermally assisted writing for magnetic field switching of CIP spin valve structure has been discussed in Ref. [46]. Thermal assisted writing for MTJ structure is studied in Ref. [47] and system trade-offs are discussed in Ref. [47, 48]. One result is the optimal MTJ element size due to a trade-off between heating current strength and MTJ resistance. For thermal switching of MTJ element, similar to conventional spin torque switching, writing current decreases with decreasing memory element dimension. However, for the same RA, the MTJ resistance increases with decreasing memory element dimension. Thus, for a given technology node with the same targeted RA, there is an optimal memory cell size due to this tradeoff between critical switching current and MTJ resistance.

9.4.7 Variability Control at Device Level

Variation control at device level can be realized by new writing or reading scheme based upon the understanding of device physics. Here we present several examples of reading scheme for variation control, based upon some unique SPRAM device characters.

Figure 9.28 depicts a traditional voltage sensing scheme for SPRAM design [49]. Read current I_R is sent to the SPRAM cell and generates the BL voltage as:

$$\begin{aligned} &\text{if MTJ is in low resistance state: } V_{BL,L} = I_R \cdot (R_L + R_{TR}) \text{ or} \\ &\text{if MTJ is in high resistance state: } V_{BL,H} = I_R \cdot (R_H + R_{TR}) \end{aligned} \tag{9.12}$$

Here R_L and R_H are the low and the high MTJ resistance at read current I_R, respectively. R_{TR} is the resistance of NMOS transistor. $V_{BL,L}$ and $V_{BL,H}$ are the BL voltage when the MTJ is at the low resistance state or the high resistance state, respectively. V_{REF} is a reference voltage which is between $V_{BL,L}$ and $V_{BL,H}$. By comparing the BL voltage to V_{REF}, the MTJ resistance state can be readout. If a V_{REF} is shared by multiple SPRAM bits, it needs to satisfy:

$$Max\left(V_{BL,L}\right) < V_{REF} < Min\left(V_{BL,H}\right) \tag{9.13}$$

Here $Max(V_{BL,L})$ and $Min(V_{BL,H})$ denote the maximal $V_{BL,L}$ and the minimal $V_{BL,H}$ generated by all involved SPRAM bits. However, $Max(V_{BL,L}) < Min(V_{BL,H})$ may not be always true when the bit-to-bit variation of MTJ resistance is large: When there is the distribution of low and high resistance due to processing variations, the sensing margin decreases. One example of dependence of detection error on resistance variation is shown in Fig. 9.14.

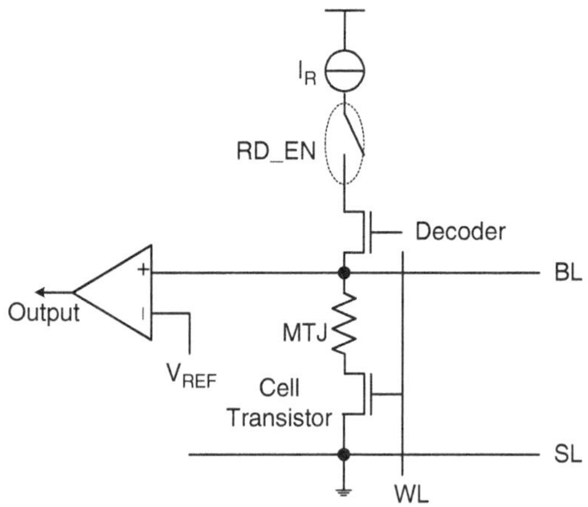

Fig. 9.28 Traditional SPRAM sensing scheme

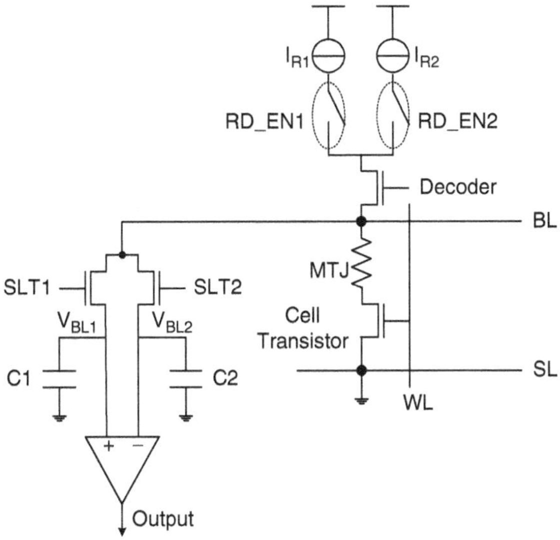

Fig. 9.29 Destructive self-reference SPRAM sensing scheme

To overcome the impact of bit-to-bit variation of MTJ resistance on the read operation of SPRAM, a so called "self-reference" sensing scheme was proposed, as shown in Fig. 9.29 [49]: The drains of two switch transistors SLT1 and SLT2 are connected to BL while the sources of them are connected to the corresponding voltage storage elements, i.e., capacitors C1 and C2, respectively. The top connect points of C1 and C2 are also connected to a voltage sense amplifier.
The operation of conventional self-reference scheme is follows:

1. *First read*: A read current I_{R1} is applied and incurs the corresponding BL voltage V_{BL1}, which is stored in C1. V_{BL1} can be either $V_{BL,L1}$ or $V_{BL,H1}$, which are the corresponding BL voltages when the MTJ is at the low resistance state or the high resistance state, respectively.
2. *Erase*: Value "0" is written into the same SPRAM bit.
3. *Second read*: Another read current I_{R2} ($>I_{R1}$) is applied and incurs BL voltage V_{BL2}, which is stored in C2. Here I_{R2} is carefully chosen to make sure:

$$V_{BL,L1} < V_{BL2} < V_{BL,H1} \tag{9.14}$$

   By comparing V_{BL1} and V_{BL2}, the original value of SPRAM bit (represented by V_{BL1}) can be readout.
4. *Write back*: Because the original value of SPRAM bit has been overwritten in step (2), the value readout in step (3) needs to be written back to SPRAM bit.

The information that the R–I curve of a MTJ provides is actually much richer than just two resistance states. Our observation shows that the current dependence of the high and the low resistance states of a MgO-based MTJ are quite different: the current roll-off slope of high resistance is much steeper than that of low resistance, as shown in Fig. 9.4. This special MTJ device property motivates us to design a novel self-reference scheme to tolerate the variation of MTJ resistance without erasing the original value.

The schematic of our new self-reference scheme is shown in Fig. 9.30. A switch transistor SLT1 is connected to BL as well as their corresponding voltage storage elements C1. The other switch transistor SLT2 is connected to a voltage divider. The top connect point of C1 and the output of the voltage divider (V_{BL2O}) are connected to a voltage sense amplifier.

The operation of our proposed self-reference scheme is follows:

1. *First read*: A read current I_{R1} is applied and incurs the corresponding BL voltage V_{BL1}, which is stored in C1. V_{BL1} can be either $V_{BL,L1}$ or $V_{BL,H1}$, which are the corresponding BL voltages when the MTJ is at the low resistance state or the high resistance state, respectively;
2. *Second read*: Another read current I_{R2} (usually I_{Rmax}) is applied and incurs BL voltage V_{BL2}. Here we have:

$$\frac{I_{R1}}{I_{R2}} = \alpha \tag{9.15}$$

   where $\alpha = V_{BL2O}/V_{BL2}$. V_{BL2O} is the output of voltage divider.
3. *Sensing*: V_{BL1} and V_{BL2O} are compared by the voltage sense amplifier. If V_{BL1} is *significantly larger* than V_{BL2O}, the original value of SPRAM bit is "1"

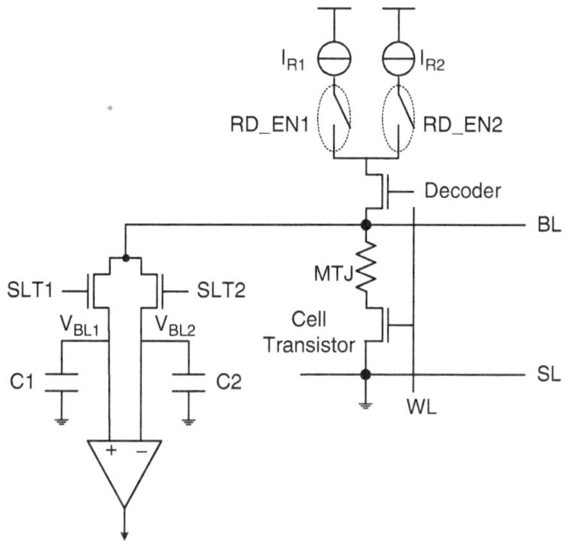

Fig. 9.30 Non-destructive new self-reference SPRAM sensing scheme

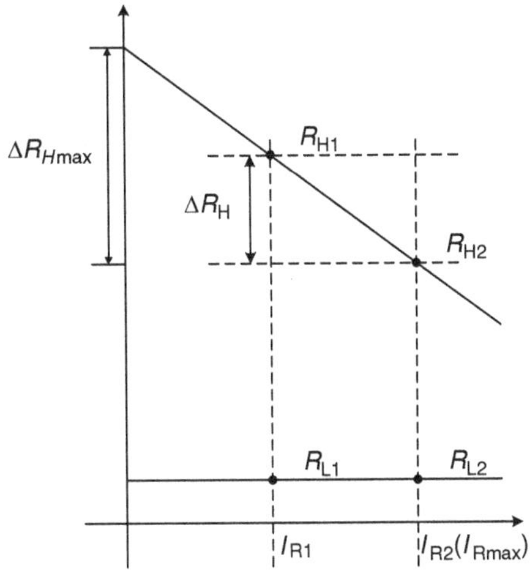

Fig. 9.31 Theory of new self-reference sensing scheme

(high resistance state). Otherwise, the original value of SPRAM bit is "0" (low resistance state).

The explanation is follows: Considering the ideal case that the MTJ resistance at low-resistance state does not change when read current varies, as shown in Fig. 9.31. If the original value of SPRAM bit is "1", we have:

$$V_{BL1} = V_{BL,H1} = I_{R1} \cdot (R_{H1} + R_{TR1}) \tag{9.16}$$

$$V_{BL2} = V_{BL,H2} = I_{R2} \cdot (R_{H2} + R_{TR2}) \tag{9.17}$$

and

$$V_{BL2O} = V_{BL,H2O} = \alpha \cdot V_{BL2} = \frac{I_{R1}}{I_{R2}} \cdot I_{R2} \cdot (R_{H2} + R_{TR2}) = I_{R1} \cdot (R_{H2} + R_{TR2}) \tag{9.18}$$

Here R_{H1} and R_{H2} are the resistances of MTJ at high resistance state, under read current I_{R1} and I_{R2}, respectively; and $V_{BL,H1}$ and $V_{BL1,H2}$ are the corresponding BL voltages, respectively. $V_{BL,H2O}$ is the output of voltage divider when MTJ is at high resistance state, under I_{R2}. If we assume $R_{TR1} = R_{TR2} = R_{TR}$, then:

$$V_{BL1} = V_{BL,H1} = I_{R1} \cdot (R_{H1} + R_{TR}) > V_{BL2O} = I_{R1} \cdot (R_{H2} + R_{TR}) \tag{9.19}$$

because R_{H1} is significantly larger than R_{H2}.
Similarly, if the original value of SPRAM bit is "0", we have:

$$V_{BL1} = V_{BL,L1} = I_{R1} \cdot (R_{L1} + R_{TR1}) \approx V_{BL2O} = V_{BL,L2O} = I_{R1} \cdot (R_{L2} + R_{TR2}) \tag{9.20}$$

because R_{L1} is the same as R_{L2}. Here $V_{BL,L1}$ and $V_{BL1,L2}$ are the BL voltages when the MTJ resistance equals R_{L1} or R_{L2}, respectively. $V_{BL,L2O}$ is the output of voltage divider when MTJ is at low resistance state, under I_{R2}.

Compared to conventional self-reference scheme, our new self-reference scheme is a *nondestructive* solution because it eliminates the "erase" and "write back" steps.

Here we ignored the leakage current along the voltage divider and the resistance variation of SLT2 in different scenarios.

9.4.8 Variability Control at System Level

Variability control can also be achieved at system level. Here we present an example of writing variation control through a system fault-tolerance scheme [50]. As the technology scales down, although the mean value of MTJ write threshold current reduces, its cell-to-cell variability will inevitably increase due to the process variability. As a result, if we follow the conventional design practice that sizes the nMOS transistor for the *worst-case* MTJ write current threshold (e.g., 6σ larger than the mean), the increasing process variability could significantly degrade the potential of storage density improvement as the technology continues to scale down.

We argue that jointly considering within-cell transistor sizing and memory defect tolerance holds a great promise to minimize the impact of significant MTJ write current threshold variability. This leads to a so-called better-than-worst-case transistor sizing design strategy, as illustrated in Fig. 9.32. The basic idea is simple: we intentionally use smaller-than-worst-case transistor sizing to reduce the memory cell size at the cost of more cell defects that are further compensated by system-level defect tolerance techniques. This joint design approach clearly involves a trade-off between the reduced individual memory cell size and increased memory defect tolerance redundancy. In spite of the simple basic idea, how to realize this design methodology to ensure a net increase of storage density may not be trivial.

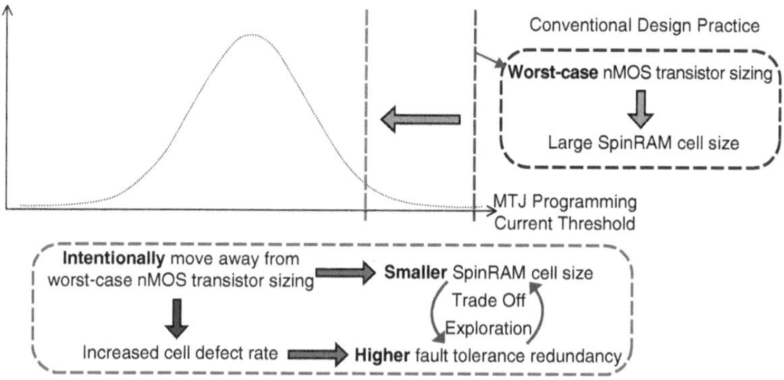

Fig. 9.32 Illustration of the proposed smaller-than-worst-case memory cell transistor sizing design methodology

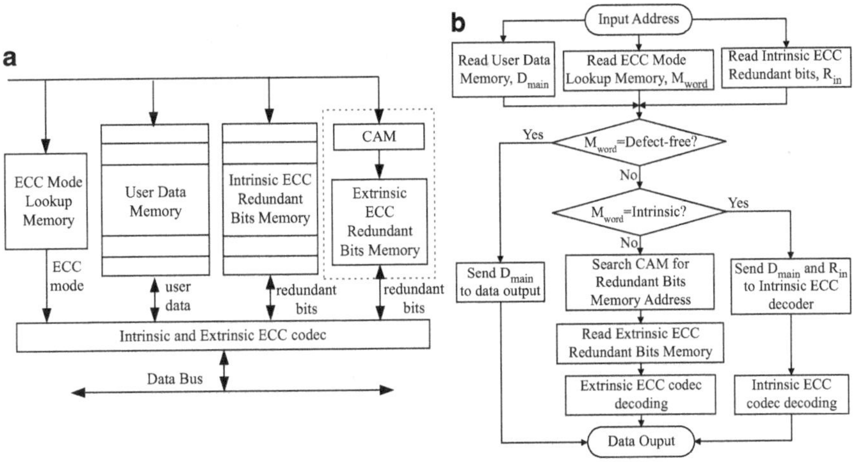

Fig. 9.33 (**a**) dual-ECC memory architecture with intrinsic and extrinsic ECCs, and (**b**) the read flow diagram for the proposed dual-ECC architecture

Moreover, memory defect tolerance may possibly result in memory read latency penalty that must be very carefully taken into account.

To address this design challenge, a *dual-ECC* memory defect tolerance architecture has been proposed in Ref. [50], as illustrated in Fig. 9.33a, which can minimize the memory defect tolerance implementation overhead and hence enable a large improvement of effective storage density. In this dual-ECC architecture, only two different ECCs (error correction codes) are used: one is relatively weak (hence incurs a small amount of coding redundancy) and called intrinsic ECC, another one is relatively strong (hence incurs a large amount of coding redundancy) and called extrinsic ECC. Its main features are: (a) The weak intrinsic ECC is *uniformly* applicable to all the memory words. (b) An ECC mode lookup memory is used to store a 2-bit information per memory word to indicate whether each memory word is being protected by the intrinsic ECC, extrinsic ECC, or defect-free. Accordingly, each memory read operation in the dual-ECC memory may fall into three scenarios, as illustrated in Fig. 9.33b, including (i) the memory word is defect-free, hence it does not invoke ECC decoding, leading to the minimal read access latency, (ii) the memory word is being protected by the intrinsic ECC, leading to a moderate latency overhead due to intrinsic ECC decoding, and (iii) the memory word is being protected by the extrinsic ECC, leading to the worst-case latency due to the sequential operations of CAM search, redundant bits memory access, and extrinsic ECC decoding.

The effectiveness of this design solution has been demonstrated using 256 Mb SPRAM design at 45 nm node as a test vehicle. Results show that, under a normalized write current threshold deviation of 20%, the overall memory chip size can be reduced by 20% while maintaining almost the same average memory read latency compared with the design using the worst-case transistor sizing. Finally, because several other emerging memory technologies such as phase-change and resistive memories share the same current-driven resistance-based storage nature as SPRAM, this design principle can be readily applied to these memories in a straightforward manner.

References

1. L. Berger, Emission of spin waves by a magnetic multilayer traversed by a current. Phys. Rev. B. **54**, 9353–9358 (1996)
2. J.C. Slonczewski, Current-driven excitation of magnetic multilayers. J. Magn. Magn. Mater. **159**, L1–L7 (1996)
3. H. Zhou, N.H. Bertram, Scaling of hysteresis and transition parameter with grain size in longitudinal thin film media. J. Appl. Phys. **85**, 4982–4984 (1999)
4. X. Wang, N.H. Bertram, Simple transition parameter expression including grain size and intergranular exchange. J. Appl. Phys. **93**, 7005–7007 (2003)
5. X. Wang, B. Valcu, N.H. Yeh, Transition width limit in magnetic recording. Appl. Phys. Lett. **94**, 202508 1–202508 3 (2009)
6. H.N. Bertram, X. Wang, V.L. Safonov, Dynamic-thermal effects in thin film media. IEEE Trans. Magn. **37**, 1521–1527 (2001)
7. X. Wang, W. Zhu, S. Markus, D. Dimitrov, Spin torque induced magnetization switching variations. IEEE Trans. Magn. **45**(4), 2038–2041 (2009)
8. W.F. Brown, Thermal fluctuations of a single-domain particle. Jr. Phys. Rev. **130**, 1677–1686 (1963); I.N. Krivorotov, N.C. Emley, A.G.F. Garcia, J.C. Sankey, S.I. Kiselev, D.C. Ralph, R.A. Buhrman, Temperature dependence of spin-transfer-induced switching of nanomagnets. Phys. Rev. Lett. **93**, 166603 1–166603 5 (2004); Z. Li, S. Zhang, Thermally assisted magnetization reversal in the presence of a spin-transfer torque. Phys. Rev. B **69**, 134416 1–134416 6 (2004); D.M. Apalkov, P.B. Visscher, Spin-torque switching: Fokker–Planck rate calculation. Phys. Rev. B **72**, 180405 1–180405 4 (2005); J.L. Garcia-Palacios, F.J. Lazaro, Langevin-dynamics study of the dynamical properties of small magnetic particles. Phys. Rev B **58**, 14937–14958 (1998)
9. I. Theodonis, N. Kioussis, A. Kalitsov, M. Chshiev, W.H. Butler, Anomalous bias dependence of spin torque in magnetic tunnel junctions. Phys. Rev. Lett. **97**, 237205 1–237205 4 (2006)
10. S. Salahuddin, D. Datta, P. Srivastava and S. Datta, IEEE Electronic Device Meeting, Washington DC, Dec 2007, pp. 121, 10
11. J. Xiao, G.E. Bauer, Spin-transfer torque in magnetic tunnel junctions: Scattering theory. Phys. Rev. B **97**, 224419 1–224419 9 (2008)
12. C. Heiliger, M.D. Stitles, Ab initio studies of the spin-transfer torque in magnetic tunnel junctions. Phys. Rev. Lett. **100**, 186805 1–186805 4 (2008)
13. A. Rebei, W.N.G. Hitchon, G.J. Parker, s-d–type exchange interactions in inhomogeneous ferromagnets. Phys. Rev. B **72**, 064408 1–064408 12 (2005)
14. A. Manchon, N. Ryzhanova, N. Strelkov, A. Vedyayev, M. Chshiev, B. Dieny, Description of current-driven torques in magnetic tunnel junctions. J. Phys. Condens. Matter **20**, 145208 1–145208 14 (2008)
15. P.M. Haney, R.A. Duine, A.S. Nunezc, A.H. MacDonald, Current-induced torques in magnetic metals: Beyond spin-transfer. J. Magn. Magn. Mater. **320**, 1300–1311 (2008)
16. X. Wang, Y. Zheng, H. Xi, D. Dimitar, Thermal fluctuation effects on spin torque induced switching: Mean and variations. J. Appl. Phys. **103**, 034507 1–034507 4 (2008)
17. X. Wang, W. Zhu, D. Dimitrov, Quantum transport and stochastic magnetization dynamics simulation on intrinsic spin torque switching. Phys. Rev. B **79**, 104408 1–104408 5 (2009)
18. Y. Chen, X. Wang, H. Li, H. Liu, D. Dimitar, presented at International Symposium on Quality Electronic Design, 2008

19. X. Wang, Y. Chen, H. Li, D. Dimitrov, H. Liu, Spin torque random access memory down to 22 nm technology. IEEE Trans. Magn. **44**(11), 2479–2482 (2008)
20. J.Z. Sun, R. Allenspach, S. Parkin, J.C. Slonczewski, B.D. Terris, U.S. Patent Application 20,050,104,101
21. Y. Tserkovnyak, A. Brataas, G.E.W. Bauer, B.I. Halperin, Nonlocal magnetization dynamics in ferromagnetic heterostructures. Rev. Mod. Phys. **77**, 1375–1421 (2005)
22. Y. Tserkovnyak, A. Brataas, G.E.W. Bauer, Enhanced Gilbert damping in thin ferromagnetic films. Phys. Rev. Lett. **88**, 117601 1–117601 4 (2002)
23. Y. Tserkovnyak, A. Brataas, Spin pumping and magnetization dynamics in metallic multilayers. Phys. Rev. B **66**, 224403 1–224403 10 (2002)
24. J. Ho, F.C. Khanna, B.C. Choi, Radiation-spin interaction, Gilbert Damping, and spin torque. Phys. Rev. Lett. **92**, 097601 1–097601 4 (2004)
25. X. Wang, W. Zhu, Y. Zheng, Z. Gao, H. Xi, presented at IEEE International Magnetics Conference 2009, to be published on IEEE Trans. Magn. **45**(10) (2009)
26. C. Scheck, L. Cheng, I. Barsukov, Z. Frait, W.E. Bailey, Low relaxation rate in epitaxial vanadium-Doped ultrathin iron films. Phys. Rev. Lett. **98**, 117601 1–117601 4 (2007)
27. I. Theodonis, A. Kalitsov, N. Kioussis, Enhancing spin-transfer torque through the proximity of quantum well states. Phys. Rev. B **76**, 224406 1–224406 6 (2007)
28. W.H. Butler, X.G. Zhang, T.C. Schulthess, J.M. MacLaren, Spin-dependent tunneling conductance of Fe|MgO|Fe sandwiches. Phys. Rev. B **63**, 054416 1–054416 12 (2001)
29. W.H. Butler, X.G. Zhang, S. Vutukuri, M. Chshiev, T.C. Schulthess, Theory of tunneling magnetoresistance for epitaxial systems. IEEE Trans. Magn. **41**, 2645–2648 (2005)
30. Y. Zheng, W. Zhu, X. Wang, D. Wang, Z. Gao, D. Dimitrov1, X. Feng, W. Jung, W. Tian, H. Xi1, E. Estrine, A. Abdurahaman, R. Kos, C. Hutchinson, M. Tang, Ultra-low switching current density of spin transfer memory with double junction barrier, unpublished, Seagate Internal Report.
31. Z. Diao, A. Panchula, Y. Ding, M. Pakala, S. Wang, Z. Li, D. Apalkov, H. Nagai, A. Driskill-Smith, L. Wang, E. Chen, Y. Huai, Spin transfer switching in dual MgO magnetic tunnel junctions. Appl. Phys. Lett. **90**, 132508 1–132508 3 (2007)
32. J. Zhu, MMM 2005 conference (unpublished), Paper No. CC-12; J. Zhu, X. Zhu, and Y. Tang, TMRC 2007 conference (unpublished), Paper No. B6; J. Zhu, X. Zhu, Y. Tang, IEEE. Trans. Magn. **44**, 125 (2008)
33. L. He, W.D. Doyle, IEEE Trans. Magn. **30**, 4086; W.K. Hiebert, A. Stankiewicz, M.R. Freeman, Phys. Rev. Lett. **79**, 1134 (1997); T.M. Crawford, T.J. Silva, C.W. Teplin, C.T. Rogers, Appl. Phys. Lett. **74**, 3386 (1999); Y. Acremann, C.H. Back, M. Buess, O. Portmann, A. Vaterlaus, D. Pescia, H. Melchior, Science **290**, 492 (2000); C. Thirion, W. Wernsdorfer, D. Mailly, Nat. Mater. **2**, 524 (2003); D. Xiao, M. Tsoi, Q. Niu, J. Appl. Phys. **99**, 013903 (2006)
34. H.W. Schumacher, C. Chappert, P. Crozat, R.C. Sousa, P.P. Freitas, J. Miltat, J. Fassbender, B. Hillebrands, Phys. Rev. Lett. **90**, 017201 (2003); H.W. Schumacher, C. Chappert, R.C. Sousa, P.P. Freitas, J. Miltat, Phys. Rev. Lett. **90**, 017204 (2003)
35. A.D. Kent, B. Zyilmaz, E. del Barco, Spin-transfer-induced precessional magnetization reversal. Appl. Phys. Lett. **84**, 3897–3899 (2004)
36. Z.Z. Sun, X.R. Wang, Theoretical limit of the minimal magnetization switching field and the optimal field pulse for stoner particles. Phys. Rev. Lett. **97**, 077205 1–077205 4 (2006)
37. K. Rivkin, J.B. Kettersen, Magnetization reversal in the anisotropy-dominated regime using time-dependent magnetic fields. Appl. Phys. Lett. **89**, 252507 1–252507 3 (2006)
38. X.R. Wang, Z.Z. Sun, Theoretical limit in the magnetization reversal of stoner particles. Phys. Rev. Lett. **98**, 077201 1–077201 4 (2007)
39. K. Rivkin, J.B. Kettersen, Switching spin valves using rf currents. Appl. Phys. Lett. **88**, 192515 1–192515 3 (2006)
40. H. Meng, J.-P. Wang, Composite free layer for high density magnetic random access memory with lower spin transfer current. Appl. Phys. Lett. **89**, 152509 1–152509 3 (2006)
41. R.H. Victora, X. Shen, Exchange coupled composite media for perpendicular magnetic recording. IEEE Trans. Magn. **41**, 2828–2833 (2005)

42. S. Li, K. Gao, L. Wang, W. Zhu, X. Wang, US patent application 11/235208, 2005
43. D. Suess, Multilayer exchange spring media for magnetic recording. Appl. Phys. Lett. **89**, 113105 1–113105 3 (2006)
44. B. Livshitz, R. Choi, A. Inomata, H.N. Bertram, V. Lomakin, Fast precessional reversal in perpendicular composite patterned media. J. Appl. Phys. **103**, 07C516 1–07C516 3 (2008)
45. P.M. Braganca, O. Ozatay, G.F. Garcia, J. Lee, D.C. Ralph, R.A. Buhrman, Enhancement in spin-torque efficiency by nonuniform spin current generated within a tapered nanopillar spin valve. Phys. Rev. B **77**, 144423 1–144423 6 (2008)
46. R. Beech, J. Anderson, A. Pohm, J. Daughton, Curie point written magnetoresistive memory. J. Appl. Phys. **87**, 6403–6405 (2000)
47. H. Xi, J. Stricklin, H. Li, Y. Chen, X. Wang, Y. Zheng, Z. Gao, M. Tang, T. Zhang, Spin transfer torque memory with thermal assist mechanism: A case study. IEEE Trans. On Magn. **46**, 860–865 (2010)
48. H. Li, H. Xi, Y. Chen, J. Stricklin, X. Wang and T. Zhang, Thermal Assisted Spin Transfer Torque Memory (STT-RAM) Cell Design, *IEEE Computer Society Annual Symposium on VLSI*, 2009, pp. 217–222
49. G. Jeong, W. Cho, S. Ahn, H. Jeong, G. Koh, Y. Hwang, K. Kim, A 0.24-μm 2.0-V 1T1MTJ 16-kb Nonvolatile Magnetoresistance RAM With Self-Reference Sensing Scheme, IEEE J Solid State Circuits **38**, 1906–1910 (2003)
50. W. Xu, Y. Chen, X. Wang, T. Zhang, Improving STT MRAM storage density through smaller-than-worst-case transistor sizing, to be presented at 46th Design Automation Conference, 2009

Chapter 10
High Performance Embedded Dynamic Random Access Memory in Nano-Scale Technologies

Toshiaki Kirihata

Abstract Described are the high performance embedded DRAMs in nano-scale technology. The chapter starts with a discussion of the evolution of high-performance embedded DRAMs for the previous 15 years. It will then look into the principles of the embedded DRAMs, which include technology, macro and array architectures, mode of operations, wordline and bitline architectures, and sensing schemes. The discussion will also address ideas unique to the high-performance embedded DRAM such as Direct Write, Negative Wordline, Concurrent Refresh, Dataline Redundancy, BIST and Self-Repair methodology. After covering these key technical attributes, the chapter will detail the IBM embedded DRAM macros starting from ASIC to the most recent SOI embedded DRAM and the cache prototype designs for microprocessors. To conclude the chapter research, and development for future embedded DRAM with floating body cell, gain cell, and 3-dimensional embedded DRAM approach will be explored.

Keywords Embedded DRAM • eDRAM • SOI • ASIC • Processor • Cache • Array • Sensing scheme • μSA • Direct write • Negative wordline • Redundancy • Concurrent refresh • BIST • 3D • Gain cell • Floating body cell • POWER7™ • IBM

10.1 Introduction

Evolutionary improvements in semiconductor technology have been responsible not only for high performance servers, but also for introducing consumers to personal computing and the information era. High performance microprocessors and high

T. Kirihata (✉)
IBM Systems and Technology Group, Semiconductor Research and Development Center, 2070 Route 52, Hopewell Junction, NY 12533, USA
e-mail: kirihata@us.ibm.com

K. Iniewski (ed.), *CMOS Processors and Memories*, Analog Circuits and Signal Processing, DOI 10.1007/978-90-481-9216-8_10,

density semiconductor memories are the most vital microelectronic components. They have been the fastest growing segments to drive the semiconductor industries for over three decades. A Dynamic Random Access Memory (DRAM) has been the predominant choice for high density and low cost semiconductor memory in computing systems. Since the announcement of the first 4 Kb DRAM in the market in 1973, the chip density has quadrupled every 3 years. Currently 1–2 Gb DRAMs [1] are in mass production, and a 4 Gb DRAM prototype [2, 3] has already been presented.

However, the performance of the DRAM has not kept pace with the gate speed in a logic technology, leading to a performance mismatch in a computing system. The difference in operation speed between microprocessor and DRAM does not allow for a system performance improvement, because of the many wait states for the command execution. As technology is scaled in a nano-meter generation, it is equally, if not more important to integrate more functions within a single die. DRAM integration with high performance logic process has long been desired, because the embedded DRAM (eDRAM) [4] not only reduces packaging cost, but also significantly increases the memory bandwidth while eliminating a power hungry and noisy IO communication in an electric system. Because of the smaller memory cell size, the eDRAM can be ~3–6 times denser than embedded SRAMs (eSRAM), and operates with low power dissipation and 1,000x better soft error rate. High performance logic based embedded DRAM macros are not only the key enablers for realizing a truly optimized embedded system for ASIC [5] and cellular super computing products [6], but also the key elements for boosting the performance of the multi-core microprocessor with high density CACHE memory in the next generation.

In this chapter, we discuss research and development for high performance embedded DRAM circuit design in nano-scale technology. The discussion focuses on the circuit technology for boosting the performance in embedded systems rather than density improvement, which can be found in many other DRAM articles [7, 8]. Section 10.2 starts with a history of high performance DRAM development. The discussion in Section 10.3 moves into the details of high-performance eDRAM designs from technology, circuits, and architecture. In Section 10.4, we explore the eDRAM designs while following IBM eDRAM development. The discussion in Section 10.5 introduces the high density cache memory prototype with the eDRAM for high-performance SOI microprocessors. To conclude this chapter, Section 10.6 looks into future work for high performance embedded DRAMs.

10.2 Evolution for High Performance Embedded DRAMs

Table 10.1 shows the evolution of eDRAMs for the previous 15 years. The earlier efforts in eDRAM development targeted the DRAM performance improvement with on-chip cache and graphic applications, both using a DRAM based process. 32 bank 256 Mb DRAM [9] incorporated a CACHE and TAG, demonstrating 23 ns random access latency when the cache hits with an additional 2% silicon area overhead. In 1996, the very wide I/O available from the eDRAMs provided a data

Table 10.1 Evolution for the embedded DRAMs

94~96	97~04		95~10		10~
0.8~0.5 μm	0.5~0.09 μm		0.09~0.045 μm		0.045 μm~
EARLY INTEGRATION	ARCHITECTURE INNOVATION	ASIC	HIGH. PERF. CELL	SOI eDRAM	FUTUTE
Graphics Multimedia ATM Switch 32 bit RISC	IRAM μCELL Fully Pipeline Dual Port Destructive Read	BIST Configurable Virtual Socket Access optimizer	2.2 nm GOX TR Concurrent Ref. Clock Multiplier Hierarchical BIST	μSA Orthogonal WL CACHE Prototype	FBC cell Gain cell 3Di

bandwidth of 12.4 Gb/s [10] for 8 Mb density, demonstrating a potential solution for ATM switch. A further demonstration integrated 16 Mb of DRAM with a 32 bit RISC microprocessor for multimedia [11]. Although these eDRAMs demonstrated some performance boost, they mostly followed a conventional commodity DRAM design and standards, and the advantage of the eDRAM solution was limited.

In 1997, the intelligent RAM (IRAM) [12] was proposed. It demonstrated the feasibility of 8GFLOPs by utilizing massive eDRAMs to configure a vector processor. In order to satisfy the IRAM like applications, it is very important to architect the embedded DRAM for significantly higher performance than the commodity DRAMs. This is because I/O width can be very large in embedded VLSI system, and page mode operation commonly used for commodity DRAMs is less effective. Instead, improvements in random access time (or latency) and cycle time (or address bandwidth) are the keys in boosting the system performance. Random access performance was first addressed by utilizing a short bitline and wordline array named micro-cell architecture [13]. This demonstrated a random access time of 6.8 ns and a random access cycle time of 9.1 ns for 64 Mb density. Dual port memory cell architecture [14] enabled an 8 ns random access cycle time, halving the DRAM array time constant by interleaving two ports alternatively. A logic technology-based embedded DRAM [15] employing a fully pipelined architecture was the first to demonstrate GHz operation with 3.7 ns random access time. This resulted in 1 Tb/s bandwidth delivered from 1,024 I/Os with a 1 GHz clock. Destructive read architecture [16] explored the on-chip SRAM cache to the cycle time improvement, demonstrating a random access speed as fast as 2.9 ns in 130 nm technology. These high performance eDRAMs opened the door to new applications such as CACAE and high speed network, which were typically designed with SRAMs.

The considerations of system level integration are also important requirements to enable a System-On-Chip (SOC) with eDRAMs. A configurable eDRAM macro [17] enabled 2,112 derivative organizations by using a software memory generator, or memory compiler. Virtual socket architecture [18] employed hardware design languages (HDL) for designing peripheral circuitry, as a software macro, realizing quick turn-around time. An access optimizer [19] overcomes bank contention problems by reordering the addressing sequence. Besides these ideas, ASIC-compatible DRAM macros [5, 20, 21] employed a processor based built-in self-test (BIST) with a two-dimensional redundancy analyzer, making the DRAMs testable and repairable using ASIC design, verification, and test methodology. The ASIC eDRAM macros are supported with the memory compiler such that the memory configuration can be customized for the application, making it easy to develop the embedded system – the key for the first-design right.

As technology is scaled in the 90 nm generation and beyond, it is difficult to improve eDRAM performance without reducing the gate oxide thickness and the threshold voltage of the access transistor. A 800 MHz eDRAM [22] with concurrent refresh mode employed a 2.2 nm gate oxide (G_{OX}) access transistor, resulting in a 312 MHz random access frequency, while improving memory availability for the short retention cell. A 65 nm SOI eDRAM [23] with µSA architecture demonstrated 1.5 ns access time, and 500 MHz random cycle operation. A one MB CACHE subsystem prototype [24] integrated a 45 nm SOI eDRAM, on-chip-wordline voltage generator, one time-programmable read-access-only-memory (OTPROM) [25], clock generator, and BIST, demonstrating the feasibility of high density eDRAM cache for high-performance microprocessors.

In a revolutionary step, SOI eDRAMs [26, 27] used a floating body in SOI technology as a storage element, and realized a DRAM cell without having a capacitor. Gain cell eDRAMs [28, 29] employed a 2-transistor-memory cell (2T) and a 2-transistor and 1-gated-diode Cell (2T1D) for high performance on-chip cache memory. A 3-dimensional memory stacking approach with Through-Silicon-Via (TSV) [30] is now a hot topic in research, and is expected to open the new era for VLSI applications, technology, designs, and test.

10.3 Principles of High Performance Embedded DRAM Technology, Architecture, and Designs

The DRAM array employs a one-transistor and one-capacitor (1T1C) as a memory cell, where the capacitor stores a charge as a data bit. The charge in the capacitor is read or written through the access transistor when it is switched. The capacitor maintains the data bit (i.e. 1 for VDD, and 0 for 0 V) when the access transistor is off. Because of this simplicity, the DRAM cell area is 1/6th to 1/10th the size of the memory cell of static random access memory (SRAM), which typically requires six transistors. It also enables 1,000x soft error rate. However, unlike SRAM, DRAM requires not only a special process for creating a memory cell with three-dimensional capacitor structure, but also unique circuits to access and maintain the data bits stored in the memory cell.

10.3.1 Technology

The reduction of memory cell size and the prevention of cell leakage are the primary focuses for traditional DRAM development. However, these are not as important for embedded DRAM technology. Instead, the embedded DRAM technology needs to be compatible to the logic technology. As discussed in the introduction, some System-On-Chip (SOC) solutions [9, 11] used the DRAM base technology to integrate the high density DRAM memory cell. However, the lessons learned from development is that DRAM base technology is not necessarily a low cost solution from the embedded system design point of view, because this increases the complexity of the logic technology. The logic gate performance with the DRAM base technology is too slow to support high performance logic, making the embedded DRAM solution less attractive. In addition, developing the logic library only for the embedded DRAM is not a cost effective approach from the development point of view. The goal of the high performance embedded DRAM technology is to enable 100% compatibility to the existing logic transistor in high performance logic technology. This will allow the use of the existing logic library, and support embedded DRAM IP integration with few additional masks.

Integration of the DRAM cell in the high performance logic technology must not degrade logic transistor performance. The use of a planar DRAM storage capacitor makes it easy to integrate the DRAM to the logic technology, which was somewhat accepted in early generations [10]. However, as the CMOS technology is scaled in the nano-scale generation, the area required for the planar cell is too big. In order to reduce the cell size, the three-dimensional capacitor is therefore necessary, which can be created by either stacked or trench technology.

Figure 10.1 shows the cross-section view for (a) stacked cell and (b) trench cell. Historically, a stacked cell, Fig. 10.1a, approach [31] was commonly used for commodity DRAMs. However, it is necessary to build the capacitor over the device after the device has been built. This makes it difficult to keep the logic transistor performance. In addition, it is difficult to planarize the surface of the silicon due to the stacked capacitor. Keeping tight wiring pitch can therefore be difficult. If the wiring rule is changed, most logic library elements need to be redeveloped, which can be expensive. This may cause performance degradation due to wiring capacitance and resistance delays. The trench capacitor, Fig. 10.1b, [32] can be fabricated

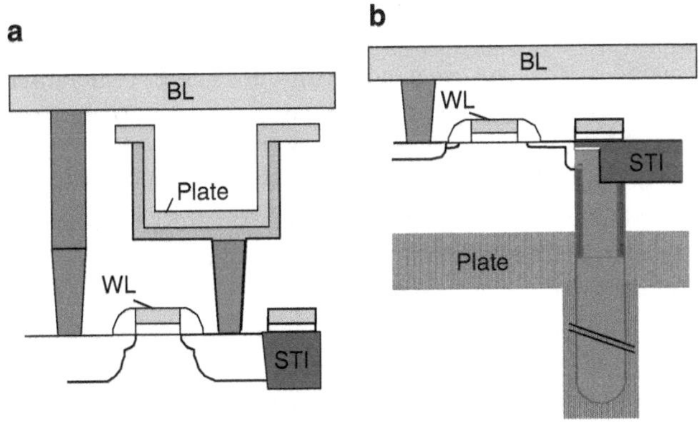

Fig. 10.1 Cells: (**a**) stacked capacitor and (**b**) trench capacitor

by digging the hole in the silicon before the device has been built. This method will allow for a fully compatible process to logic technology and will not degrade transistor performance, giving an ideal technology solution for embedded DRAM integration in a logic chip.

The logic compatible DRAM process also creates a significant difference in the cell designs between commodity DRAM and the embedded DRAM. Commodity DRAM employs a self-aligned borderless contact to the bitline, and buried strap to the capacitor, resulting in $8F^2$ cell [33], where F is the minimum lithography feature. The access transistors use a gate oxide (G_{OX}) of >5 nm and the transistor design is optimized to provide long retention. Although these innovations are attractive for reducing cell size and improving data retention, they require complex and expensive process steps. In a chip with an embedded DRAM, these techniques that reduce DRAM cell size do not necessarily pay off to reduce overall chip cost because chip size cannot be reduced more than the logic chip integration will allow.

The embedded DRAM cell [34] uses a longer WL pitch with tighter BL pitch. In order to minimize the additional process step for embedded DRAMs, the access transistor of the cell should be preferably designed by the devices available for logic technology. A logic thin oxide device of <2 nm G_{OX} achieves a significantly high ION current, but because of the data retention and wordline boost requirement, it is not a good choice for the DRAM transfer gate. Therefore, IO or analog type device with thicker oxides should be used for designing the embedded DRAM. The challenge in embedded DRAM cell design is to find an optimum gate length with additional threshold adjustment, using the IO or analog devices available to the existing logic technology. Because of the logic compatibility and the performance requirement, the cell size of the embedded DRAM is ~2x of the commodity DRAM cell from the same generation, which is acceptable for most embedded applications.

10.3.2 Macro Architecture

Figure10.2 shows a high level block diagram for the embedded DRAM macro. It consists of memory arrays, each having memory cells (CELL) arranged in two-dimensional matrices. They are accessed by wordlines (WLs) and bitlines (BLs) for row and column, respectively. The memory arrays are stacked in the Y-direction, and are supported by a peripheral circuit block, typically located at the bottom of the macro. The peripheral circuit block consists of command and address receivers and decoders, and the macro IO circuitries, which control the memory arrays with the given command. Because the boundary of each memory array creates a natural break, the number of the stacked arrays can be programmed. This allows for memory size customization with the granularity of the memory array size.

Each memory array is organized with 2^N rows and 2^M columns, allowing for 2^{N+M} bit memory cells in a two-dimensional array, where N and M are chosen by the

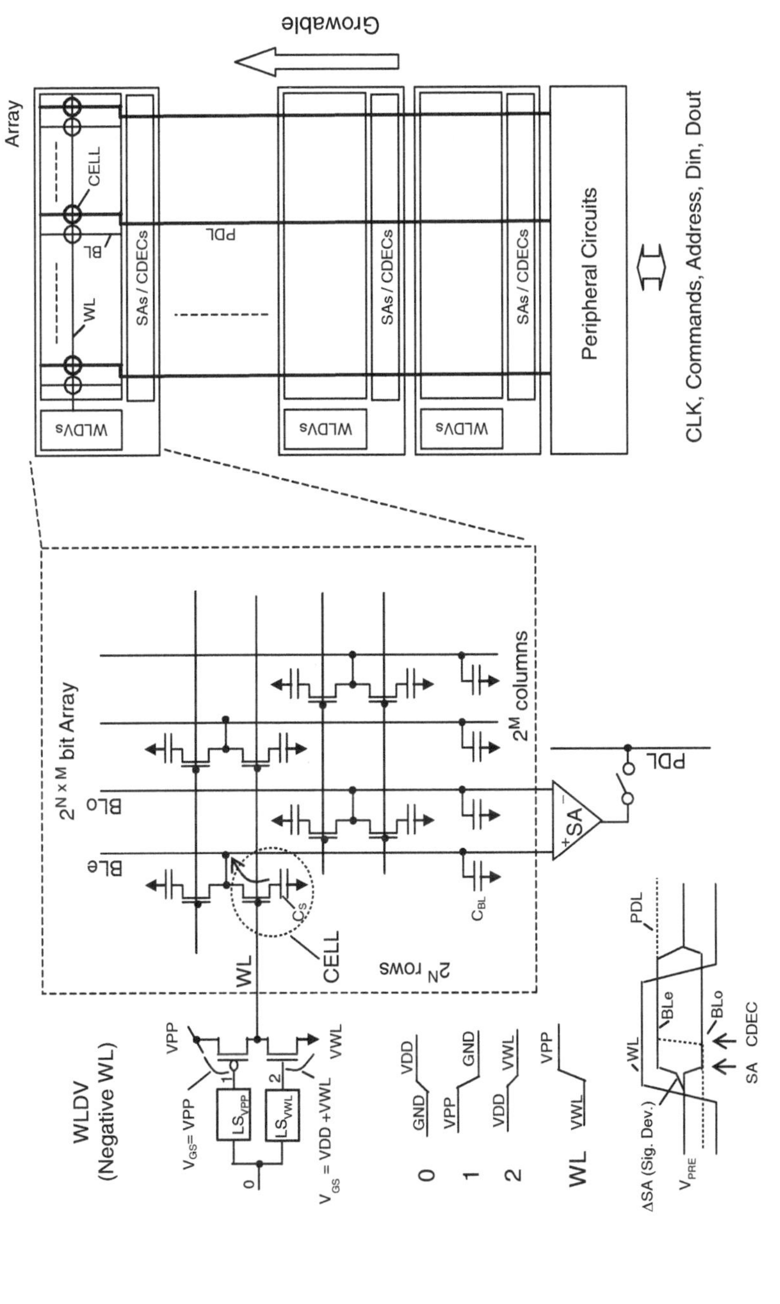

Fig. 10.2 Macro architecture

trade-off with respect to the performance, power, IO, and multi-bank requirement. For example, to configure a 1 Mb eDRAM macro, the macro can contain four 256 Kb arrays, each consisting of 256 rows (256 WLs N = 8) and 1,024 columns (1,024 BL pairs M = 10), or sixteen 64 Kb arrays, each consisting of 256 rows (256 WLs N = 8) and 256 columns (256 BL pairs M = 8). Employing a smaller array will improve the performance, but the overhead of WL drivers (WLDVs), column decoders (CDECs), and the sense amplifiers (SAs) will also be increased [13]. Because of the high performance requirement, the array size of the embedded DRAM macro is much smaller than the array size for the cost sensitive commodity DRAMs.

Each WL supports the 2^M cells coupling to either the even BL (BLe's) or the odd BL (BLo's). This allows the creation of a BL pair using BLe's and BLo's for a differential sensing scheme. Prior to memory access, all BLs are precharged to BL precharge voltage (V_{PRE}). When a WL rises, 2^M cells coupled to the corresponding WL are simultaneously activated. This couples the storage nodes in the selected cells to the corresponding BLs (i.e. BLe's). This results in charge sharing between the cell capacitor (C_S) and the BL capacitor (C_{BL}). The other BLs (i.e. BLo's) keep a V_{PRE} level, creating a differential voltage between each BL pair. Time for the charge sharing is called "signal development time", and the differential voltage is called the "sensing signal" (ΔSA). Because of this charge sharing operation, the data bits in the selected storage nodes are destroyed (destructive read). The ΔSA on the BLs is then amplified to a full CMOS level and latched by the sense amplifiers (SAs). At the same time, the amplified CMOS level data bits on the BLs will be written back to the selected cells. For a read operation, the 2^M bits held in the sense amplifiers are multiplexed by the column decoders (CDECs), and the selected subset bits are transferred to the primary datalines (PDLs). Because of the wide IO requirement (typically 128 bits or more), the number of bits transferred to the PDLs in the embedded DRAM macro is significantly more than the commodity DRAM (typically less than 32 IOs). The PDLs are arranged over the arrays and communicate to the peripheral circuits using fourth metal or above. Once the data bits have been latched in sense amplifiers, the data bits can be read from the sense amplifiers in an operation known as page mode. However, because of the wide IO organization of the eDRAM macro, page mode is less effective, and therefore recent eDRAM macros do not support it. Instead, eDRAM supports a fast random cycle and a multibank operation to boost memory bandwidth.

Because of the dynamic cell feature using a capacitor, the charges stored in the memory cells leak as time goes on, and therefore they should be periodically read and written back. This is a unique but important requirement for DRAMs, and is known as refresh operation. As long as this refresh is executed before the storage node voltage has leaked lower than the voltage detectable by the sense amplifier, the data bits are maintained. The time interval to maintain the storage node is called retention time, and the time between refresh operations is called refresh cycle. Embedded DRAM requires operation at higher temperatures while using a logic compatible device, and therefore the retention time and refresh cycle are significantly shorter than that of commodity DRAMs.

10.3.3 Modes of Operation

Commodity DRAMs need to follow an industry standard of address multiplexing and common input and output pins. Because of the packaging constraints, the number of IOs is limited (≤ 32). Therefore, in order to improve the memory bandwidth, recent high performance DRAMs employ Double-Data-Rate 2–3 (DDR2 [35] / DDR3 [36]) or Rambus-signalling-level (RSL [37]) interfaces. Typical embedded DRAMs employ a Single-Data-Rate (SDR [38]) interface with a single clock. This is because there are significantly more IOs in the embedded system than in the memory system with commodity DRAMs. The commands are given by signals: bREAD and bWRITE with broad-side addressing: bank addresses BA<0:3> (pre-decoded), row addresses RA<0:7> and column addresses CA<0:2>. To increase the flexibility of the bank, the BA<0:3> are assigned to the corresponding bank<0:3>, respectively. Data input pins DI<0:127> and output pins DO<0:127> are not shared, and all signals including the DIs and DOs swing with CMOS level without using expensive IO interface. The detailed operation of single bank and multibank modes are described next.

10.3.3.1 Single Bank Fast Random Access Cycle Mode

Single bank mode follows a typical SRAM interface protocol. The goal of this design is to emulate the SRAM interface and improve the random access cycle and latency. Figure 10.3a shows the timing diagram for write and read operations. The eDRAM accepts either the bREAD or bWRITE command at the falling edge of the CLK, which starts a memory access operation. When both signals bREAD and bWRITE are high at the CLK edge, the eDRAM is in a stand-by state. For a read mode, the bREAD signal is low at the CLK edge. The addresses BA<0:3>, RA<0:7>, and CA<0:2> are given at the same CLK edge with the bREAD command, which immediately starts row and column decoding operations. Unlike conventional commodity DRAMs, these row and column operations are executed completely in parallel. This activates the corresponding WL, SA, and column switch. The SA is internally timed to develop a sufficient sensing signal on the bitline before the activation. This allows the read data bit to be transferred to the PDL, while also allowing for the restoration of the data bits to the cells. The PDL data bits are then transferred to the DO<0:127>. For a faster CLK cycle, the data transfer to the DO can fall behind the subsequent CLK edge, resulting in a DO error. In this case, the DO transfer is delayed in the data output circuits until the subsequent CLK edge is given. This operation is called a 1 stage-pipeline mode. The WL is immediately turned off and BLs are precharged to the stand-by state after the restore operation has been completed. This completes a read cycle in the array within a cycle, improving random access cycle performance.

For a write mode, the bWRITE signal should be low at the CLK edge. The addresses BA<0:3>, RA<0:7>, and CA<0:2> are given at the same CLK edge with bWRITE, which immediately starts a decoding operation to activate the corresponding

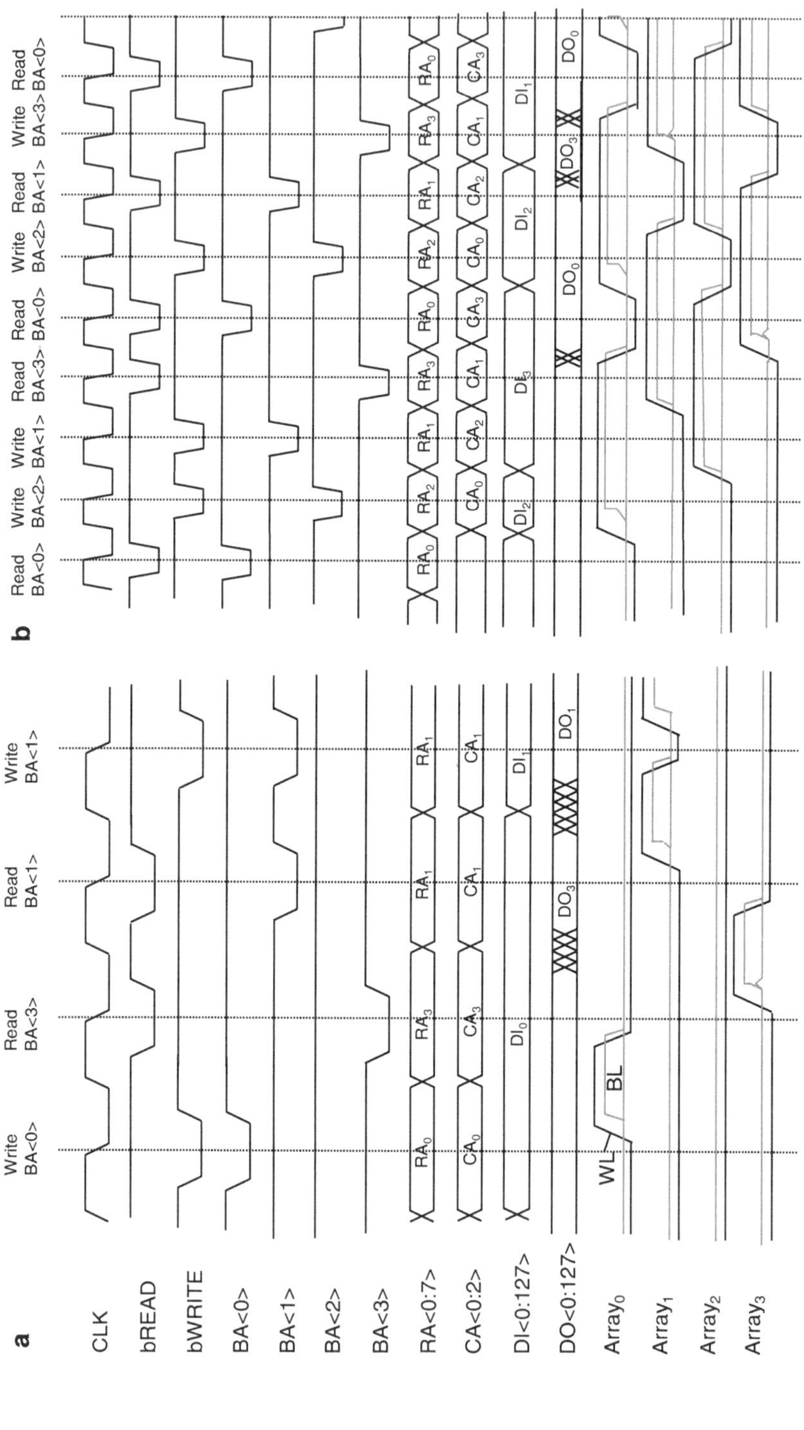

Fig. 10.3 Modes of operations: **(a)** single bank and **(b)** multi-bank

WL. At the same time, the data inputs DI<0:127> are also accepted, transferring the data inputs to the PDL and also to the corresponding BLs in the array. For a conventional DRAM, BL swing for the write mode is typically enabled after the sensing has been finished (late write). For embedded DRAM, the write data bit transfer to the BLs should be earlier than the WL activation, which is known as a direct write scheme [5]. The BLs not selected by the column decoders operate as a read mode.

The single bank fast random cycle mode is particularly advantageous for network applications, because it accepts any command at any clock cycle, improving data through-put in a random addressing network. However, because the bandwidth is not improved more than the array cycle, it is not preferable to employ this mode for a large density macro.

10.3.3.2 Multi Bank Pipeline Mode

Multi-bank mode [5] optimizes the operation to improve the memory bandwidth. Unlike single bank fast random cycle mode, the DRAM employs a pipeline architecture operates with a faster clock cycle, and allows the activation of a new array at each clock cycle while working on the current array operation.

Figure 10.3b shows the timing diagram for write and read operations in multi-bank mode with 4 stage pipeline. Similar to a single bank access mode, the bREAD or bWRITE command is given at the time when CLK falls. This enables the first bank row addressing to the WL selection and sense amplifier activation. At the second CLK falling edge, the data bits are read from the sense amplifier to the PDL. At the same time, the eDRAM starts the second bank row addressing to the WL selection and sense amplifier activation. The third clock edge allows the data bits from first bank to be output to the data-output circuits, while transferring the data bits in the second array to the PDL. At the same time, the WL activation starts in the third bank. The fourth clock edge allows for the first bank memory to be precharged so that it is ready for the subsequent command, while also transferring the data bits to the DOs. At the same time, the data output circuits receive the data bits from the second array, while transferring the data from the third array to the PDL and starting to decode the fourth bank to activate WL and the sense amplifier. Multi-bank operation with four stage pipeline allows the execution of up to four memory arrays at the same time, improving the memory bandwidth by 4x the single bank fast random cycle mode. However, because of these additional pipeline control circuitries, actual random cycle time (defined by the command interval for the same bank) and the latency (defined by the command to the output) are slower than the single bank fast random cycle mode.

The multi-bank mode is preferred for applications with predictable addressing such as graphical applications. When this mode is used for random access application, the macro should contain a sufficient number of banks to minimize a bank contention. As an alternative solution, it is always possible to architect a memory system by creating banks by using multiple single bank fast random cycle macros and employing additional logic external to the DRAM to realize interleaving, scheduling, and pipelining.

10.3.4 Wordline Architectures

Improving the WL transition speed is an important consideration for the DRAM design. Arranging the WL drivers at the center shortens the WL length by half, resulting in ¼ the delay. In order to further improve speed, the array should be divided into several small array segments, however, this will increase the array area. Figure 10.4 shows a sampling of different wordline architectures.

A stitched WL architecture, Fig. 10.4a, [38] arranges second metal layer (M2) over each polysilicon WL. The M2 is used to reduce the WL resistance by stitching the WL periodically in an array. The number of cells between the M2 stitches in a segment is determined by the local polysilicon WL resistance and the capacitance within the segment. The number of segments coupling to the wordline driver (WLDV) is determined by the M2 resistance and the WL capacitance in all segments. This hierarchical approach results in a faster WL transition, and therefore has been commonly used since the 1 Mb generation. However, as technology is scaled to a nano-scale generation, it is becoming more difficult to manufacture the M2 layers coupling to the pitched limited WL. A larger WL pitch in embedded DRAMs allows for M2 WL stitching with less significant yield concern. Because of the simplicity and a small design overhead for the stitch, the stitched WL architecture makes it easy to develop a small and high speed array, such as for cache application, which typically requires <2 ns latency [23].

A segmented WL architecture, Fig. 10.4b, [35] distributes the local WL drivers (LWLDVs) into each array segment. The master WL drivers (MWLDVs) are located at the side (i.e. left) of the array to support multiple segments. The LWLDVs are driven by the master WLs (MWLs) over the arrays using the second metal layer (M2). The LWLDV includes a final decoding stage (i.e. one out of four selections with two address bits). This allows for the increase of the M2 MWL pitch as large as 4x that of the stitched WL architecture. Although the design overhead of the LWLV is larger than the M2 stitching, it allows for the support of more array segments, typically 2–4x that of the stitched WL architecture. Configuring a larger array using segmented WL architecture compensates for the LWLDV area overhead. In addition, because of the natural array break by the LWLDV, it allows the number of segments coupling to the MWLDV to be programmable. This is particularly advantageous for eDRAM design, because the macro can be grown as a two-dimensional configurable macro by tiling the array segment bounded LWLDV and SA [17]. This gives significant flexibility to customize the memory organization for ASIC applications.

The stitched WL, Fig. 10.4a, and segmented WL, Fig. 10.4b, architectures discussed above are still commonly used for eDRAM designs, however, they do not enjoy the advantage of the embedded systems. The use of additional metal layers is a free process in an embedded system, resulting in no increase in the process cost. For example, the ASIC eDRAM [5] employs the segmented WL architecture, and within the segment, the WLs are further stitched using stitched WL architecture. This segmented stitched WL architecture, Fig. 10.4c, allows for the increase in array density to configure the large macro similar to the segmented WL architecture, while reducing the design overhead to support a smaller macro similar to the stitched WL architecture.

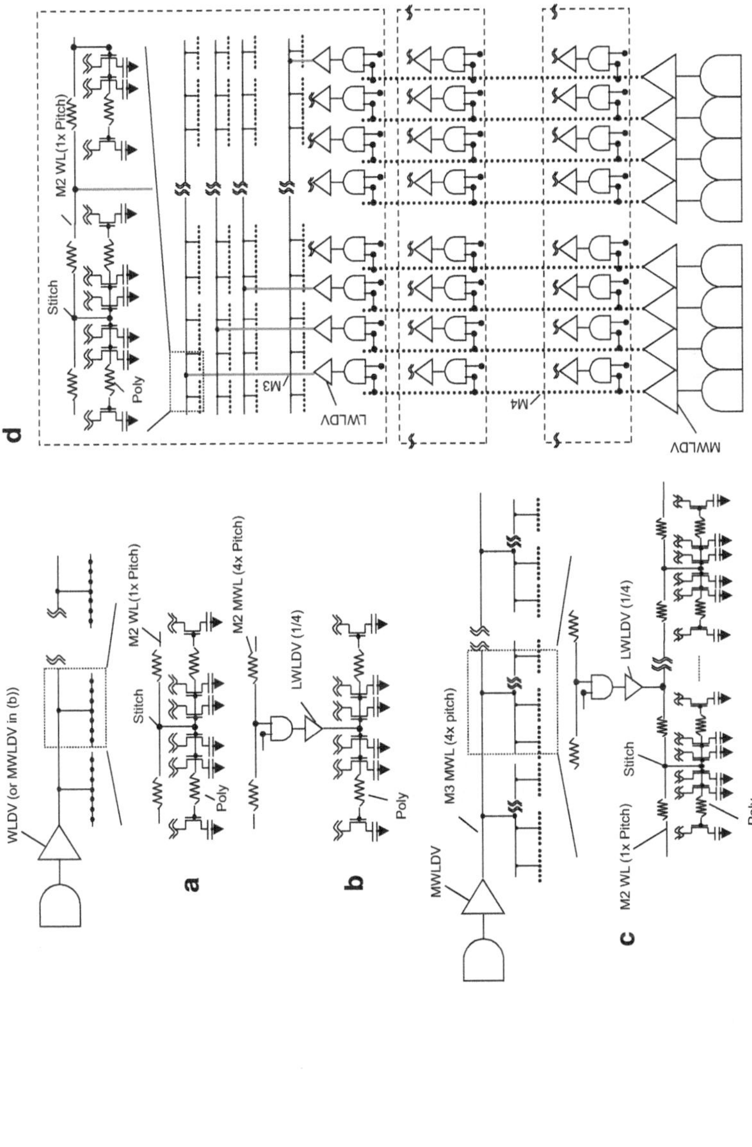

Fig. 10.4 Wordline architectures: (**a**) stitched WL, (**b**) segmented WL, (**c**) segmented stitched WL, and (**d**) orthogonal WL

The revolutionary approach employs an orthogonal segmented WL architecture, Fig. 10.4d, [23]. In this architecture, the LWLDVs are arranged in the sense amplifier area, and drive the WL vertically using third metal layer (M3). The vertically arranged M3 metal layers are coupled to the horizontal M2 metal layers for the WL stitch to support the array. The MWLDVs are located on the bottom of the macro and drive the LWLDVs using fourth metal layer (M4), typically the same layer for the global datalines (PDLs). The orthogonal WL architecture eliminates the pitch limited layout constraints for the LWLDV, while reducing the macro width. It thus allows for the arrangement of the macro more horizontally, making it easy to communicate the data bits to other IPs located on the bottom of the multiple eDRAM IPs.

10.3.5 Bitline Architectures

The steady state sensing signal (ΔSA) is mathematically calculated by:

$$\Delta SA = (V_S - V_{PRE}) \times C_S / (C_{BL} + C_S)$$

where V_S is the cell voltage (0 for 0 V and 1 for VDD), V_{PRE} is the BL precharge voltage (i.e. ½ VDD), and C_S and C_{BL} are the cell and BL capacitances, respectively. A larger ΔSA results in faster sensing speed, which can be realized by either increasing the cell capacitance (C_S) or reducing bitline capacitance (C_{BL}). However, a larger C_S increases the time constant for the write operation, which should be avoided. Therefore, a strategy for producing a high performance BL architecture is to reduce the BL capacitance (C_{BL}) with small silicon overhead.

Figure 10.5 shows a sampling of bitline architectures for eDRAMs. The conventional approach, Fig. 10.5a, employs a folded BL architecture to create a differential pair (BLe and BLo). The reduction of the C_{BL} can be realized by reducing the number of cells coupled to the BL. This approach is well known as short BL architecture, or micro cell array architecture [13]. However, reducing the BL length requires more sense amplifiers in an embedded DRAM macro. The overhead of the short BL architecture is significantly increased, as the BL length becomes shorter than 128 cells per BL.

The hierarchical BL (HBL) architecture approach, Fig. 10.5b [39], employs a main BL (MBL) and a local BL (LBL). The MBL couples to the LBL with NMOS switch and the LBL supports a fewer number of cells. This thus reduces the C_{BL} with smaller overhead than conventional short BL architecture. The HBL architecture does, however, have several design disadvantages. First, the NMOS switch (Switch) should be integrated into the array in order to not break the array. This is because breaking the array increases the silicon overhead significantly, which makes the HBL architecture less attractive. Second, the gate of the NMOS switches needs to be boosted by at least the threshold voltage of the NMOS switch over VDD during the write. An insufficient voltage boost not only reduces the write performance, but also reduces the storage node voltage. In addition, boosting the gate voltage requires a thicker oxide device for the NMOS, which is expensive. Third, local LBLs should be precharged through the NMOS switches. This requires all NMOS switches to be turned on and off, except for the selected LBL segment during the cycle, resulting in a large power dissipation.

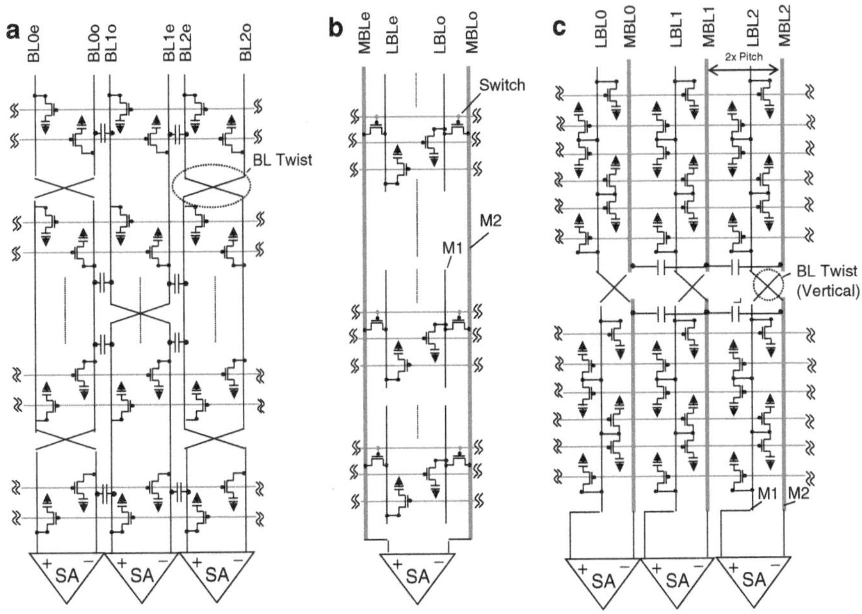

Fig. 10.5 Bitline Architectures: (**a**) folded BL, (**b**) hierarchical BL, and (**c**) vertical twisted BL

In addition to the C_{BL} reduction consideration, it is important to architect the BLs to be immune to the noise during signal development and sensing, generated both internal to the DRAM and from other signals arranged over the array. In a logic-process based embedded DRAM, the metal wires used as bitlines are relatively thick and thus create a large coupling noise between bitlines during signal development and sensing. To improve the signal margin, every other bitline is transposed with its neighbor at various points along its length, providing some cancellation of coupling noise in the differential sensing scheme.

As shown in Fig. 10.5a in the conventional folded BL architecture, the twisted BL [40] creates common noise for the BL pair, thus keeping the ΔSA. In some cases, it may be better to consider creating the BL twist vertically using additional layers. In this scheme, a BL pair for each column is configured with a local BL (LBL: M1) and a master BL (MBL: M2) arranged over the LBL. This vertical twisted BL architecture approach, Fig. 10.5c [35, 41], allows one to increase the BL pitch by as large as 2x, reducing the BL capacitance, and increasing the sensing signal (ΔSA).

10.3.6 Sensing Schemes

Figure 10.6 shows the BL sensing schematic (a) and the waveforms in read modes (b) and (c), and in write modes (d)–(f). This sense circuit employs a conventional cross coupled NMOS and PMOS sense amplifiers (NSA and PSA), precharge

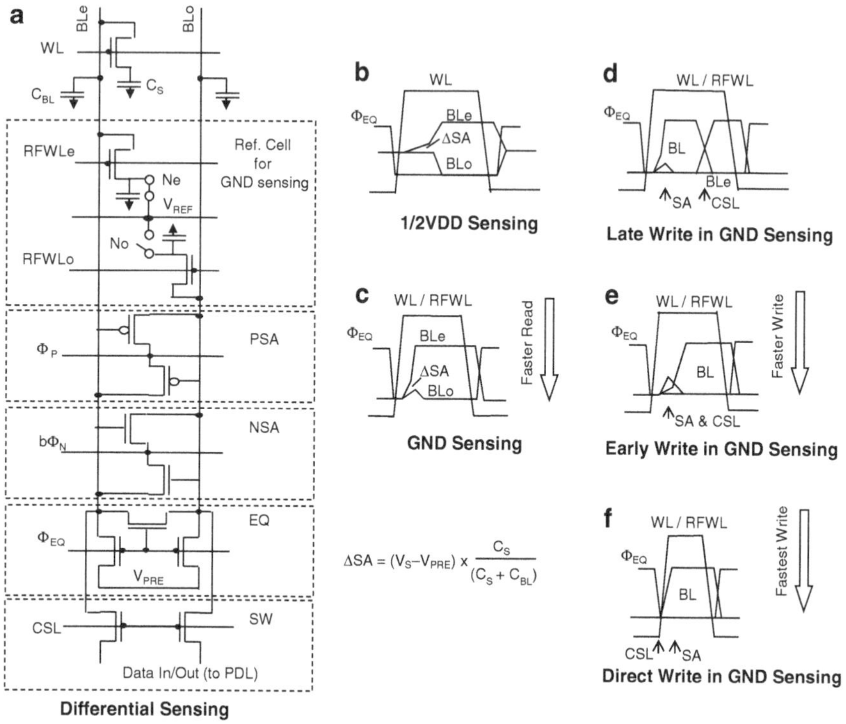

Fig. 10.6 Sensing schemes: (**a**) differential cross-coupled sense amplifier, (**b**) ½ VDD sensing, (**c**) GND sensing, (**d**) late write, (**e**) early write, and (**f**) direct write

devices (EQ), and column switch devices (SW), where bitline precharge voltage (V_{PRE}) plays an important role. Precharging the BLs at ½ VDD, Fig. 10.6b, allows the BL precharge level itself to serve as an ideal reference voltage, and is commonly used for commodity DRAMs (1/2 VDD sensing scheme [42]). In a standby state, the BLe and BLo are precharged at a 1/2VDD level through the precharge devices (EQ). Prior to the WL activation, the precharge devices (EQ) are disabled. Activation of the WL then creates the sensing signal (ΔSA) on the BL pair. The signal $b\Phi_N$ and Φ_P then go low and high, respectively, which amplifies the BL voltage to CMOS level (VDD and 0 V). This naturally starts a write back operation to the cells. Activation of a column-select-signal, CSL, opens a column switch (SW), transferring the bit from the BL to the PDL for a read mode, or vice-versa for a write mode. After the transfer of the data bits and the restoration of the data bits to the cells, the WL is turned off and the precharge device (EQ) is turned on. This naturally generates the BL precharge voltage of 1/2 VDD by short-circuiting the BL pairs. This results in a low power sensing scheme because the sensing current is required only for 1/2 VDD BL swing during the sense operation. The 1/2 VDD sensing scheme does, however, lose gate overdrive significantly as VDD is reduced, causing the sensing operation to stall at low voltages. The signal development time for reading "1" data is also a concern, because it does not start until the WL voltage reaches 1/2 VDD + array threshold voltage (Vta).

In order to overcome these two fundamental disadvantages, it is preferred to precharge the BLs at GND level, Fig. 10.6c, for embedded DRAMs. To enable proper discrimination of a stored "1" or "0", the GND sensing scheme [21] requires a reference cell, which stores a reference voltage (VREF). The VREF is supplied directly from the VREF voltage generator through the switches (Ne and No shown as switches for simplicity), or by simply short circuiting a two reference cell pair (storing VDD and GND after the SA has been activated) similar to the BL precharge operation. The gates of the Ne and No are controlled by the inverted RFWLe and RFWLo, respectively. For a read operation, the WL and the corresponding RFWL (i.e. RFWLo) are activated at the same time. This allows charge sharing between BL (i.e. BLe) and the selected cells, and the other BL (i.e. BLo) with the reference cell. Because of the reference cell, the reference BL voltage will be between the "0" and "1" level of the BL after signal development. The GND sensing scheme allows sufficient gate overdrive during amplification even if the VDD voltage is dropped to 0.6 V. The GND level BL provides speedy signal development, since the signal transfer from the cell to the bitline occurs immediately after the wordline rises over the threshold of the memory access gate.

10.3.7 Late Write, Early Write, and Direct Write

The random access cycle time is determined by either (1) read mode, which requires the writing back of the data to the original cell, or (2) write mode, which may require the flipping of a data pattern to the cell. Conventional design activates the column-select-signal (CSL) after the sense amplifier (SA) has been activated. This late write, Fig. 10.6d, is required when ΔSA is small, because the early activation of the column may reduce, or in the worst case, destroy the ΔSA during signal development because of the BL coupling. However, delaying the write operation to the cell results in write limited random access cycle time. In order to minimize this delay, an early write approach, Fig. 10.6e [43], is introduced, which activates the BL swing for the write and SA activation at the same time. Recent high performance eDRAM employs a direct write, Fig. 10.6f [5], which allows for the swinging of the write BLs earlier than the SA activation. The detailed operation of the direct write will be discussed in the section of the Embedded DRAMs for ASIC.

10.3.8 Negative Wordline Architecture

As VDD is scaled to 1 V or below, the device threshold (Vt) is a key concern, because it cannot be scaled due to device leakage. The Vt of the access transistor (Vta) is a more severe concern, because the transistor should satisfy the retention requirement. Non-scaling Vta requires more boosting of the WL, which can cause a device reliability problem. In order to overcome this fundamental problem, negative WL (or offset WL) architecture [44] is introduced, where a WL swings from a negative voltage (VWL) to a boost voltage (VPP). This allows the Vta and the VPP to be reduced by the amount of VWL, because the device leakage problem with lower Vta will be overcome by the negative WL. Lowering the Vta of the access transistor also contributes to reducing the Vta fluctuation, giving more margin for embedded DRAMs. The device stress can also be reduced because the gate-to-source voltage (V_{GS}) applied to the access transistor is determined by the WL high voltage of VPP and BL low voltage of GND. Optimum VWL and VPP voltages should be determined on the basis of the device off current, GIDL current, device performance, Vta fluctuation, and reliability.

The negative WL architecture requires a unique design for the WL drivers. A conventional inverter, which allows swinging from VWL-to-VPP, is not allowed for the driver, because the V_{GS} of the transistors will be VWL + VPP, which is too large a voltage across the thin gate oxide (G_{OX}) of the transistors. Using a transistor with thicker G_{OX} than the array G_{OX} can overcome the problem, but is expensive. A dual level-shifter approach using LS_{VPP} and LS_{VWL}, shown in Fig. 10.2, is used to reduce the stress of the NMOS and PMOS devices in the WLDV. In this scheme, the LS_{VPP} converts the GND-to-VDD input (node 0) to VPP-to-GND output (node 1), reducing the V_{GS} of the PMOS to VPP. Similarly, the LS_{VWL} converts the GND-to-VDD input (node 0) to VDD-to-VWL output (node 2), reducing the V_{GS} of the NMOS to VDD + VWL. These result in reducing the reliability concern for the WL driver devices.

10.3.9 Concurrent Refresh Mode

Concurrent refresh mode [22, 45] is a technique that improves the memory availability. Fig. 10.7 (Copyright © IEEE 2005) shows the architecture, which supports concurrent refresh mode. The macro includes two bank select ports: BSELs and RBSELs. Each BSEL activates the corresponding array BANK for the memory access mode (read or write). RBSEL activates the corresponding array for the concurrent refresh independently from the memory access operation. This allows a memory access operation for the array (i.e. $BANK_i$) indicated by bank select signal (i.e. $BSEL_i$), while simultaneously allowing a refresh operation for a different array (i.e. $BANK_j$) as selected by the refresh bank select signal (i.e. $RBSEL_j$). Each array includes a row address counter (RAC) to identify the word address to be refreshed. For memory access, a row address (RADDs) issued with the bank select signal (BSEL) is supplied to the row decoder. On the other hand, when the bank refresh command (RBSEL) is issued, the RAC counter associated with the selected bank is connected to the row decoder, allowing the corresponding wordline to be refreshed. This distributed RAC counter arrangement eliminates complexity in the management of refresh addresses in each bank independently in a concurrent refresh mode. Assuming that each array

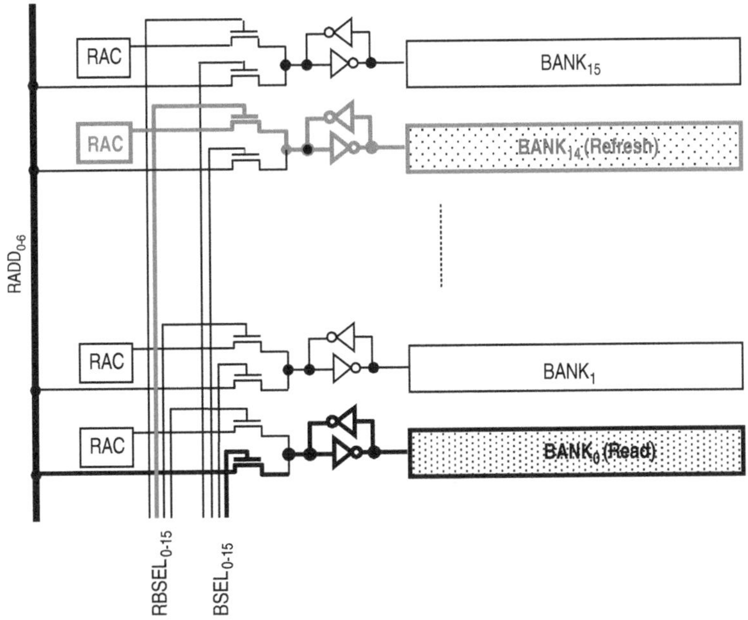

Fig. 10.7 Concurrent refresh architecture (Copyright © IEEE 2005) [22]

contains 128 wordlines, as long as 128 refresh commands are issued for each bank within a retention time, the data are maintained. This results in 100% memory utilization as long as the bank select signal and the refresh bank select signals are correctly managed to avoid a bank contention. Multiple logical memory banks may be refreshed simultaneously while enabling a memory access at each clock cycle. This is realized by activating the corresponding refresh bank control signals ($RBSEL_{0-15}$) simultaneously. By refreshing multiple arrays in a concurrent refresh mode, a larger logical bank can be configured for a large density memory.

10.3.10 Redundancy

System integration requires special care for redundancy design and its test methodology. This is because yield loss due to unfixable eDRAM is much more expensive than the stand-alone commodity DRAMs. Figure 10.8 shows redundancy and self-repair architecture with array-built-in-self-test (ABIST) circuitry [20, 21]. Similar to the commodity DRAMs, a distributed row redundancy replacement is commonly used for eDRAM, where a defect in the array is repaired with a row redundancy element (RWL) within the array. However, because of the wide IO organization, column redundancy replacement with bitline replacement is not easy to implement. Therefore, eDRAM employs dataline redundancy.

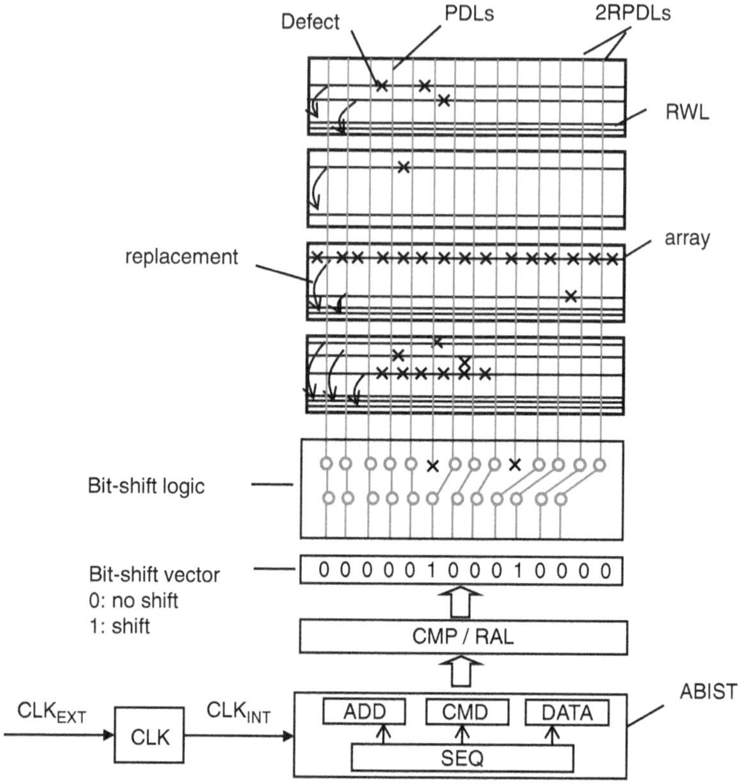

Fig. 10.8 Self repair with redundancy and ABIST

As discussed in the architecture, after the sensing operation, the data bits are read or written through the primary datalines (PDL) arranged over the array. Each PDL typically supports eight adjacent columns (or BLs). For dataline redundancies, the eDRAM includes additional two redundant PDLs (RPDLs), each coupling to eight redundant BLs (RBLs). The repair of the dataline is achieved by the bit-shift steering operation. Up to two PDLs can be repaired within a column domain. The bit-shift location is controlled by the bit shift vector, where the bit "1" identifies the location of the bit-shift. The bit-shift steering can be changed in each predetermined row domain (i.e. array). This allows the use of the dataline redundancy elements more effectively, improving the bit yield, particularly for a larger macro.

In addition to these conventional redundancy solutions, it is important to improve the yield at a system level. For example, including additional macros allows the replacement of unfixable macros as a block redundancy. For truly random fails, ECC solutions are sometimes less expensive approaches. Alternatively, it may be advantageous to employ a small SRAM as a single bit replacement. Without using a repair solution, managing the addressing to not use the defective domain is also an interesting idea to be considered.

10.3.11 Test Methodology

Another challenge for the eDRAM is to establish the test methodology in an embedded system. The embedded DRAM requires the boundary test and the array test. The boundary test can be realized easily by using a logic tester and the flow based on the application netlist. However, testing the eDRAM array requires functions typically available only in a memory tester. This results in two-pass testing with logic tester and memory tester, which is expensive. Even with this two-pass testing approach, testing eDRAM is not a simple task. This is because the eDRAM has significantly more IOs than the commodity DRAM, and also the performance test of the eDRAM requires >1 GHz tester speed. Therefore, it is important to implement a single test flow using logic tester while following logic test methodology. Array-Built-In-Test (ABIST) [20, 21] is introduced to provide the unique array tests.

An ABIST engine, shown in Fig. 10.8, includes a Command Generator (CMD), an Address Generator (ADD), and a Data Generator (DATA) to control eDRAM. They are controlled by a command sequencer (SEQ). The output of the eDRAM is verified by the read-compare (CMP) function. The results are analyzed by redundancy allocation logic (RAL), which determines if the eDRAM is perfect or repairable with the available redundancy element. A typical RAL includes a function to update the redundancy programming (such as in bit-shift register), enabling self-repair. It is important to handle the allocation while considering row and column redundancy availability to improve the yield. Recent eDRAM also supports a bit map capability, which is the key requirement to characterize the process technology not only for development, but also for manufacturing.

Besides this pattern generation and detection capability with ABIST, it is also important to enable a performance test, which requires a high speed clock. A clock multiplier (CLK) [22] boosts the external tester clock (CLK_{EXT}) to the high speed clock (CLK_{INT}), allowing the ABIST to generate the command for speed test. In addition to the high speed clock generation, the trend of the recent BIST is to support multiple macros as a remote BIST engine. The hierarchical ABIST approach [45] generates one seed command globally with ¼ frequency. A command multiplier is located near each eDRAM macro, which generates four commands by using one global seed command. The hierarchical ABIST approach improves the test speed while reducing the overhead of the ABIST in an embedded system.

10.4 IBM Embedded DRAM Macros

IBM has a long history in the development of commodity DRAMs [1, 32, 33, 35, 38, 41] using deep trench storage capacitors. The trench capacitors have a fundamental advantage in the embedded environment, in that they can be built entirely below the silicon surface at a very early stage in wafer processing. All subsequent process steps after the trench capacitor formation can be identical to those of a high performance logic process. In this section, we will discuss the innovative design features in three different areas, which were not discussed in Section 10.3.

10.4.1 Embedded DRAMs for ASIC

To ensure that the first pass design is correct with minimum design resources and schedule, it is important to follow a classic ASIC library approach. The ASIC environment requires significant design robustness to guarantee eDRAM operation over voltage ranges that can fluctuate by as much as 25%, temperature ranges up to 115°, and large logic noise conditions.

Figure 10.9a shows a chip microphotograph of a general purpose embedded DRAM [5] for the IBM Blue Logic ASIC system (Cu-08). Segmented stitched WL architecture allows the macro to be constructed from Tileable 512 Kb (or 584 Kb) arrays, each having 512 rows by 8 columns by 128 (or 146) IOs. The Tileable arrays can be arranged into up to 8 and 16 for row direction and column direction respectively, resulting in a maximum 64 Mb with 1,024 IOs (or 72 Mb macro with 1,168 IOs).

Figure 10.1b, c show the circuit and the timing diagrams for this eDRAM macro. It incorporates a PMOS cross-coupled sense amplifier (PSA), a distributed PSA set driver (PSET), a PMOS column switch (PSW), and a local buffer (LBUF). In a standby state, the signals Φ_{EQ} is high, precharging the BLs (BLe and BLo) and the LDLs (LDLe and LDLo) at a GND level (0 V) by the equalizers. The reference cell (REF) is precharged at VREF by the equalizer (Φ_{RFEQ}). To improve the signal margin for 1 data, VREF voltage is set at 1/3 VDD or ~0.35 V. Prior to the WL and reference WL (RFWL) activation, the signals Φ_{EQ} and Φ_{RFEQ} become low, disabling the precharging operation. When WL and RFWL are activated, the charge-sharing starts between BLs and the cells, creating a sensing signal (ΔSA) on the BL pair. Because of the GND BL precharge, a signal development starts immediately after WL reaches a Vta level. RFWL is immediately turned off after signal development. As a result, the reference cells start restoring the initial VREF voltage for the next command cycle. The signal $b\Phi_{PSET}$ then becomes low, which makes Φ_{PSA} become high. This allows the BLs of the pair to become high and low (or low and high), amplifying ΔSA on the BLs to CMOS level (VDD and 0 V). Because of the GND BL precharge, PSA sees a gate overdrive of VDD, which is significantly larger than 1/2 VDD sensing scheme. This results in high speed sensing while improving a low voltage margin.

Unlike commodity DRAM, PMOS column switch (PSW) is turned on immediately by the corresponding column select (i.e. bCSL0) after the command is given. The threshold of PSW (Vpsw) protects the ΔSA on the BLs during signal development. When BL is amplified over the Vpsw, the BL data are naturally transferred to the local datalines (LDL pair) through the PSW. For a read mode, read primary datalines (RPDLe and RPDLo) and Φ_{RD} goes high prior to the $b\Phi_{PSET}$ activation. The LDLs are coupled to the read NMOS (RD), transferring the data bit on the BL pair to the RPDL pair without having any additional trigger when one of the LDL of the pair (i.e. LDLe) becomes high.

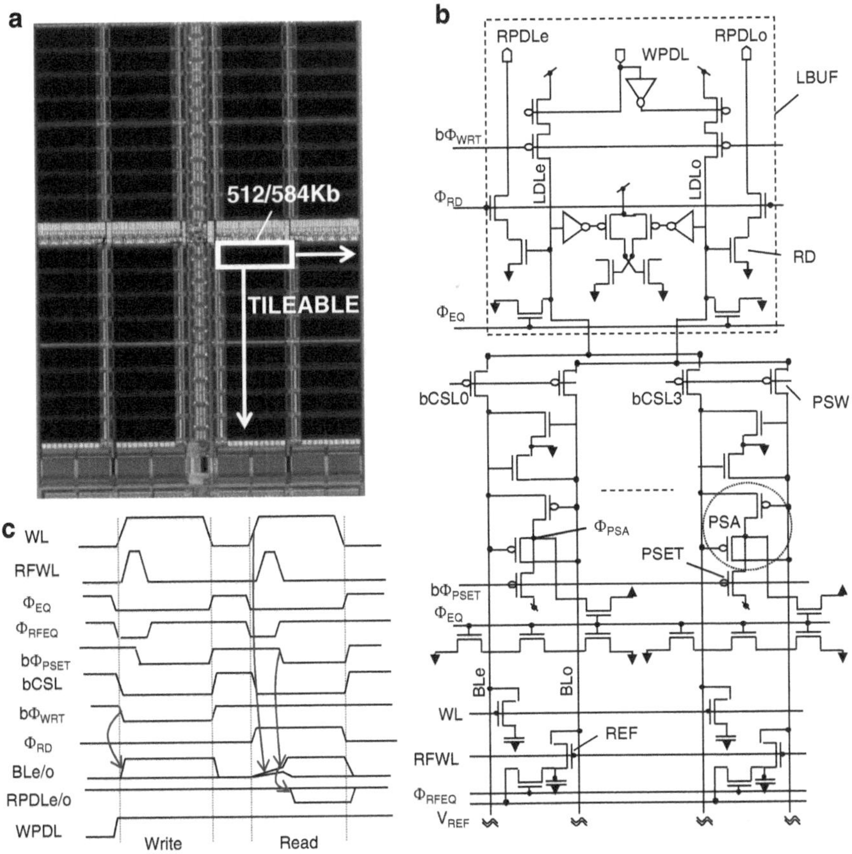

Fig. 10.9 500 MHz multibank embedded DRAM macro for ASIC: (**a**) chip microphotograph, (**b**) circuit diagram, and (**c**) timing diagram

For a write mode, a write-enable-signal ($b\Phi_{WRT}$) goes low, which drives one of the LDL of the pair to become high depending on the state of write primary dataline (WPDL). As a result, one of the corresponding BL of the pair is driven through the PSW as a direct write mode. The direct write mode is enabled equal or earlier than the WL activation. A BL-to-BL coupling concern is overcome by the BL twisting scheme. Because of the distributed PSET for each BL, BLs in the signal development phase are protected from pre-amplification due to the adjacent BL swing in a direct write mode. These techniques result in fast access time (latency) and fast random cycle time.

The macro is designed with 1, 2, and 4 stage pipelines to support various ASIC applications. A testchip was fabricated with a variety of memory configurations, which were constructed using a 2-dimensional memory compiler. The hardware data demonstrated a 500 MHz multi-bank operation at 1.05 V and 105°. The detailed study confirmed that the random cycle time is faster than 8 ns, which is sufficient to support 100% memory utilization by utilizing four or more memory banks. The macro contains row and column redundancies, an array-built-in-self-test-engine (ABIST), and a 2-dimensional redundancy analyzer, resulting in full compatibility with the ASIC design and test flow.

10.4.2 Embedded DRAM with Destructive Read Architecture

Rapid growth of the worldwide information network has created urgent demand for increasing transmission capacity in wired networks. These high speed backbone systems require large amounts of high-speed memory and logic. Because of the nature of the network, the most important feature of the network memory is a fast random access cycle time. Destructive read architecture [16] reduces the random access cycle time of the eDRAM effectively by half by disintegrating read and write-back operations.

Figure 10.10 (Copyright © IEEE 2002) shows (a) the chip micro-photograph, the conceptual waveforms of (b) conventional DRAM array, (c) destructive read DRAM array, and (d) internal pipeline diagram for the embedded DRAM with destructive read architecture. Conventional array cycle, Fig. 10.10b, consists of three steps: (1) signal development and sense, (2) write, and (3) precharge. Instead, the

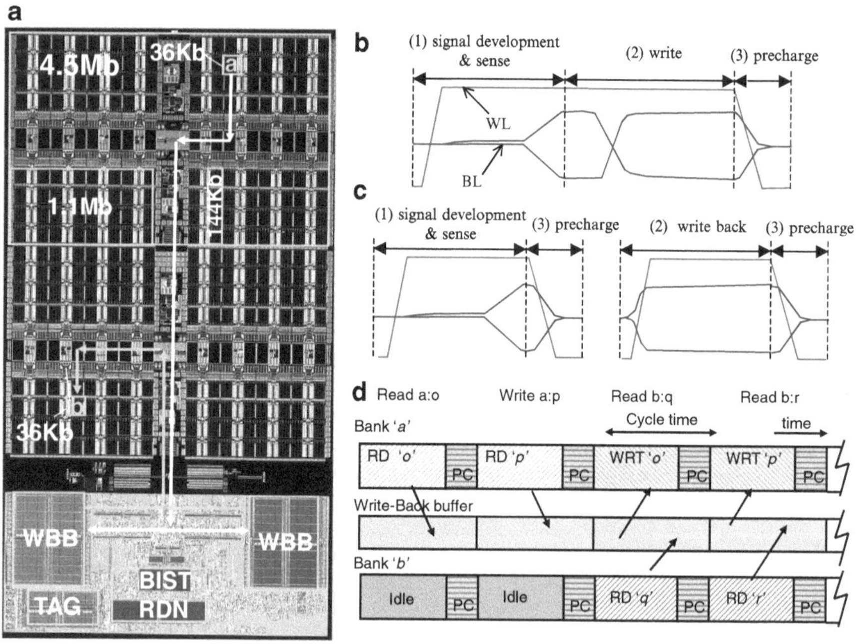

Fig. 10.10 3.2 ns random cycle embedded DRAM with destructive read architecture: (**a**) chip microphotograph, (**b**) conventional approach, (**c**) destructive read, and (**d**) pipeline for destructive architecture (Copyright © IEEE 2002) [16]

access cycle of the destructive-read array, Fig. 10.10c, consists of either (1) read or (2) write, and (3) precharge, resulting in a fast array cycle.

The 9 Mb proto-type macro is constructed in a four level design hierarchy. This 9 Mb prototype consists of two 4.5 Mb units, however, it is increasable from 4.5 to 18 Mb in 4.5 Mb increments (first hierarchy). Each 4.5 Mb unit contains four 1.1 Mb quadrants arranged in a 2 x 2 matrix (second hierarchy). The 1.1 Mb quadrant contains eight 144 Kb banks (third hierarchy). Any of the 144 Kb banks may be accessed for read or write at any given clock cycle. The 144 Kb bank contains four 36 Kb micro cell arrays, each having 256 wordlines and 144 bitlines (fourth hierarchy).

The peripheral block consists of a write-back buffer (WBB), TAG memory, redundancy circuitry (RDN) and BIST engine. The WBB is organized with the same array size as the 36 Kb micro cell DRAM array, and stores the 36 Kb data from any of the destructive 36 Kb bank. A tag RAM (TAG) stores an address of the word entries in the WBB. The WBB, which works like a direct-mapped cache, is implemented with a two-port SRAM with a cycle time that is the same or faster than that of the DRAM core. When a command is issued to the memory macro, the control circuitry will check the tag RAM to see if the data of the corresponding address are stored in the WBB. If it is a hit, only the WBB will be accessed. Otherwise, the data will be fetched from the DRAM core, transferred to the output port if required, and stored in the WBB. If the corresponding address in the WBB originally contains data from a different bank, these data will be written back to the DRAM core at the same time. In order to avoid the cycle time penalty in a write-back operation, the capability of simultaneously performing a read from a different bank is implemented. The DRAM may accept a new set of commands and data at each cycle. 100% write-back to DRAM cells is guaranteed without any data loss. The internal pipeline diagram, Fig. 10.10d, shows the consecutive read commands from bank a, which store the data bits 'o' and 'p' to the WBB, respectively, The data bits 'o' and 'p' are written back to bank a when bank b is addressed for reading the data bits 'q' and 'r' to the WBB.

Hardware data for this prototype demonstrate <4 ns random access cycle time at 1.5 VDD for 130 nm bulk technology, which is 2x faster than the eDRAM [21] using the same technology. The detailed SHMOO tests were performed separately on the destructive read DRAM and writeback buffer components of the 9 Mb macro. At 1.5 V power-supply voltage, the writeback buffer cycles at 3.8 ns, while the destructive-read DRAM cycles at 3.2 ns. A custom designed write-back buffer or faster SRAM could be used so that the overall cycle time would be limited only by the destructive read DRAM array.

10.4.3 Embedded DRAM for High-performance SOI Microprocessor

Embedded DRAMs [5, 21] have been shown to be the key enabler not only for a high-capacity L3 cache memory [46], but also for super computing [6]. However, they use bulk technology, and therefore it is not possible to integrate it with main stream

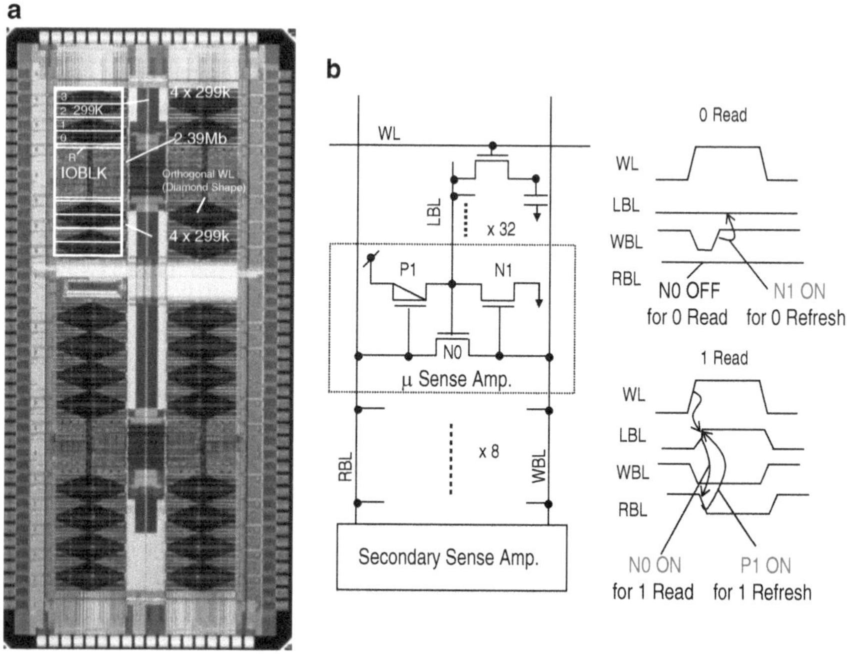

Fig. 10.11 500 MHz random cycle embedded DRAM in 45 nm SOI CMOS: (**a**) chip microphotograph, and (**b**) μSA circuit and timing diagrams

high performance microprocessor with SOI technology. The design challenge for this application was to realize 500 MHz random access cycle time while managing a SOI floating body effect. The breakthroughs of this macro include orthogonal WL architecture and ultra-short BL architecture with micro sense amplifier (μSA).

2.39 Mb eDRAM macros [23], Fig. 10.11a, are organized by upper and lower memory units, each consisting of four 299 Kb sub-arrays (0 through 3) and one 16 row redundancy array (R). They are supported by the Input and Output circuitry (IOBLK) located between two memory units. When the macro decoder receives a valid macro-select-signal and a valid read or write command, the decoder evaluates the address, activating either the top or bottom memory units. The activation of the memory unit enables one of the four arrays, and sends 1 of 256 valid addresses. In order to support a CACHE associatively with minimum speed penalty, one of the eight late-select-addresses (LSELs) is activated later than the memory command. Once a LSEL command has been received and a corresponding column-select-signal (CSL) has been issued, the corresponding data are driven onto the data bus for a read operation. For a write operation, the CSL selection is enabled immediately with the memory command.

The 299 Kb sub-array consists of 256 WLs by 1,168 BLs, which includes ECC bits for IBM servers. An evolutionary approach in this eDRAM locates the segmented WL drivers in the sense amplifier area, which couples to the horizontally arranged WLs through the vertically arranged wires. They are controlled by the MWL vertically from the peripheral circuits located at the bottom of the embedded DRAM macro. In order to minimize the skew between WLs, the connection points are arranged with a pyramid-like pattern as shown in the diamond shape in the chip micro-photograph. This "Orthogonal WL architecture" reduces the eDRAM macro width, allowing for more IOs in limited space. The architecture also eliminates the pitch limited layout constraints for the segmented WL driver, overcoming the bit yield concern of the conventional non-orthogonal wordline architecture.

One of the keys in fast read operation is bit-line segmentation using a micro-sense amplifier architecture (μSA) as shown in Fig. 10.11b. In this approach, only a small number of cells (32) are connected to the local bit line, so the Local Bit Line (LBL) has a relatively small capacitive load, resulting in faster charge and discharge times. The LBLs are coupled to the gate of the NMOS transistor (N0) in order to detect the LBL transition as a sense amplifier. They are supported by additional PMOS (P1) and NMOS (N1), resulting in 3 transistor design. Eight LBLs in each column are coupled to the global read bitline (RBL) and write bitline (WBL) through the 3T μSA in each segment. The RBL and WBL are supported by the main sense amplifier, located at the edge of the 299 Kb array.

The 0.127 μm^2 eDRAM cell is constructed from the IO device with 2.35 nm gate oxide (G_{OX}), and fabricated on 65 nm SOI with floating-body. 20 fF capacitor is built by digging through the SOI and buried oxide (BOX) to the silicon substrate, where the BOX acts as isolation between SOI and the substrate, eliminating the oxide collar process for bulk technology. The detailed operations are as follows.

During pre-charge, WL is low, turning all cell transfer devices off. WBL and RBL are pre-charged high, which turns on the N1 and turns off P1. The LBL is therefore precharged to GND though the N1. Prior to WL activation, WBL becomes low, which floats the LBL at GND level. When WL rises to 1.7 V, the signal development starts on the LBL. When the cell stores "0" data, the LBL remains at low voltage, keeping the RBL at high level. The secondary sense amplifier senses a "1" at RBL and ground at WBL, interpreting this to be a "0" value. Detection of "0" allows the WBL to become high, forcing the LBL to become low. This thus allows the writing back of 0 data to the memory cell.

When the cell stores "1", the LBL becomes high because of the charge sharing between high storage node and GNDed BL. The 32 cells/LBL architecture allows the large LBL swing, which allows the turning on of the NMOS (N0). This results in coupling the RBL to the WBL, discharging the RBL to GND though the NMOS (N0). The secondary sense amplifier senses a "0" at RBL and ground at WBL, interpreting this to be a "1" value. When the RBL becomes low, the PMOS (P1) turns on, which makes the LBL become high. The LBL that becomes high gives a positive feedback to the N0, further accelerating the sensing speed. The cells are naturally restored to the full VDD.

When a "1" is to be written, RBL is pulled to ground by the secondary sense amplifier. This allows the LBL to be high, writing the high voltage to the cell as a direct write. For writing "0" to the cell, the WBL stays high, which keeps the LBL at low voltage during the WL activation, allowing the writing of the 0 data to the corresponding cells.

The hardware results show 1.5 ns access latency with 99.99% bit yield at 1 V and 85°. The cycle time study demonstrates 500 MHz random cycle at 1 V over 32 K address space (limited by high frequency tester). Typical active power for 500 MHz operation is 45 mW, which is approximately 1/5 of the embedded SRAM power in the same 65 nm technology. Precharging the BL at GND during μSA operation causes the SOI body to float closer to GND level, preventing sub-threshold leakage and a bipolar current problem. No significant SOI history effect has been observed, realizing 40 μs data retention. The eDRAM is functional with as low as 0.6 V, demonstrating 4.0 ns access latency.

10.5 High Performance Cache with Embedded DRAM Macros

Recently, high performance microprocessors have begun to integrate multiple CPU cores onto a single chip [47]. To keep the system balanced and productive, each core must have a steady supply of data, and so the multi-core trend has greatly increased the total number of cache bits (typically SRAM) needed per chip. SRAM leakage power tends to increase as technology dimensions decrease and is proportional to the number of transistors used. In addition, as CMOS technology has been scaled, intrinsic device fluctuations have significantly reduced SRAM cell stability and static noise margin, in particular at low voltage operation.

In this section, we explore the low overhead on-chip cache designs for high performance SOI microprocessor with embedded DRAMs. A 1 MB cache prototype [24] provides flexible solutions for word line voltage generation, testing, and self-repair, as well as programmable timing interfaces, to allow the eDRAM subsystem to work in multiple contents. An ABIST engine allows the eDRAM to be tested, and redundancy repair may be invoked in conjunction with the on-chip OTPROM.

10.5.1 Architecture

The cache prototype, shown in Fig. 10.12 (Copyright © IEEE 2009), contains four 2.39 Mb eDRAM macros. These are supported by four sets of charge pumps [48], which generate word line high (V_{PP}) and low (V_{WL}) voltages, as well as an OTPROM programming voltage (V_{PRG}). An external tester clock (CLK_{EXT}) is received by the internal clock generation circuit and may be used directly or boosted to an internal clock (CLK_{INT}) by either a clock multiplier or an All-Digital Phase-Locked Loop (ADPLL) [49]. The CLK_{INT} is distributed through the clock distribution system throughout the prototype. User logic may be used to control address and command generation, simulating cache behavior. A complete cache subsystem would additionally include a directory structure, to store cache line tags, valid bits, etc. To accelerate the design cycle and improve debug speed and yield, these were not

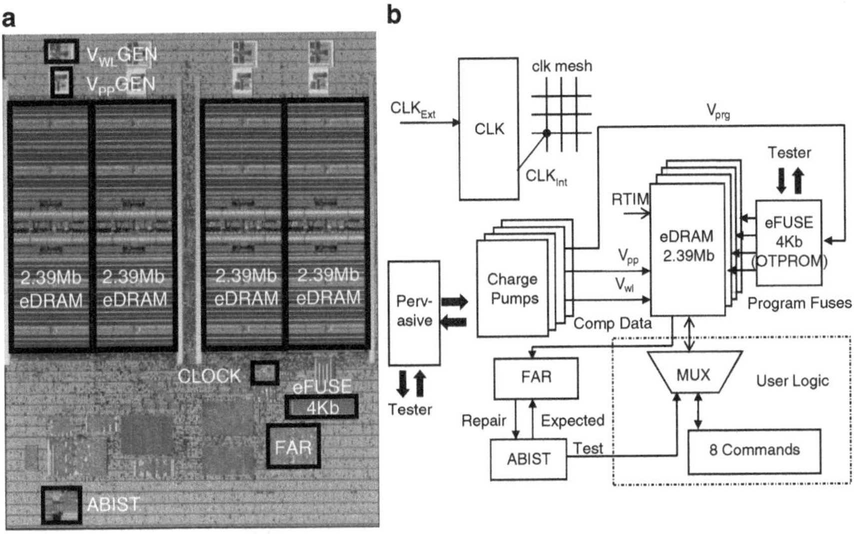

Fig. 10.12 1 MB L3 cache prototype with embedded DRAM: (**a**) chip microphotograph, and (**b**) architecture (Copyright © IEEE 2009) [24]

included, but a programmable timing interface is provided at the eDRAM macro for receiving late select addresses (L_{SEL}) 1, 2, or 3 clock cycles after eDRAM access is initiated (P_{LS}).

The Array-Built-In-Self-Test (ABIST) logic and Fail Accumulation Register (FAR) communicate at test time with individual eDRAM macros through a multiplexer (MUX). The ABIST and FAR support detailed array diagnostics through a full-chip bitmap capability. Redundancy allocation logic is integrated within each eDRAM macro to realize a self-repairing system with the ABIST engine and the FAR. Calculated redundancy allocation results are stored in the 4 KB One-Time Programmable Read-Only-Memory (OTPROM) [25]. Pervasive logic initializes and sets up all components of the prototype. It regulates the flow of control signals, data, and scan-information into and out of the prototype and can also be set to extend the internal pipeline depths of the eDRAM macro and internal support logic.

10.5.2 ABIST and FAR

The Programmable Pattern Instruction Memory in the ABIST block is serially loaded by the tester through the pervasive logic. The ABIST engine formats the pattern into usable commands, addresses and data. Write, refresh and read transactions are sent to the eDRAM through the user logic MUX and expected data are sent to the FAR Comparison and Control Logic. The synchronization of expected data and eDRAM data out at the FAR, for all eDRAM latency modes, is enabled by the programmable depth of the expected data pipe.

When mismatches between eDRAM data out and expected data are detected, the FAR compares the failing address to those already stored in the FAR. Upon completion of the fail accumulation for the repair test pass, the FAR delivers their contents to the ABIST, where repair information is formatted and sent to the eDRAM as special repair commands, through the normal system interface bus. The eDRAM captures the incoming repair information in its redundancy registers, completing the autonomous repair operation. At this point the eDRAM repair solution is "soft-set" in the eDRAM redundancy registers, the array has been repaired and post-repair testing that avoids the defective addresses can be executed.

To capture the repair solution in a non-volatile format the eDRAM redundancy register contents are unloaded to the tester. The easiest way to associate the repair solution with the chip is to program the eFuses in the OTPROM. Configurable paths enable data stored in the OTPROM to be sensed and shifted in several directions, to either the eDRAM row redundancy registers, column redundancy registers, or off-chip. This configuration enables the storage of electronic chip identification data, as well as repair solutions that can be updated with new repairs from multiple temperature test passes. The integration of ABIST, FAR and eDRAM has been done in a way that delivers an autonomic repair capability, enabling in-system repair.

10.5.3 Refresh Management

To avoid blocking all memory accesses when refresh is performed, the cache prototype includes a concurrent refresh engine [22]. This engine may issue a refresh to a sub-array concurrent with a user read or write access to a different sub-array in the same bank. An external clock (RTIM) is brought onto the cache prototype to represent the refresh time-base. For a 40 μs retention time, and a 256 word line sub-array, the RTIM cycle time (t_{RTIM}) is set to 156 ns (40 μs/256 rows per sub-array). The intent is to refresh 1 address in each of the four sub-arrays every 156 ns (256 addresses per sub-array). Because three out of four sub-arrays are guaranteed to be refreshed within 156 ns, the probability that a bit would NOT be refreshed during truly random accesses with a 2.5 ns cycle time is given by $\frac{1}{4}^{(156\text{ ns}/2.5\text{ ns} - 3)}$. Although this is mathematically unlikely, the cache system must handle the case to avoid data loss. To facilitate refresh management, a refresh request protocol was developed.

Figure 10.13a (Copyright © IEEE 2009) explains the details of refresh management in a system. When the RTIM goes low, the concurrent refresh engine checks if all four arrays within a bank have already been refreshed within the t_{RTIM} interval (156 ns). If not, a refresh request signal (RREQ) goes high, indicating that the eDRAM needs refresh. For example, a R/W command to sub-array 1 is in progress. Concurrently, the address N, set by the Refresh Address Counter (RAC) is refreshed in sub-array 0. During the next cycle, sub-array 2 receives a command and performs an access cycle. Simultaneously address N is refreshed in sub-array 1. During the third command cycle, sub-array 2 is accessed again. The concurrent refresh engine picks address N in

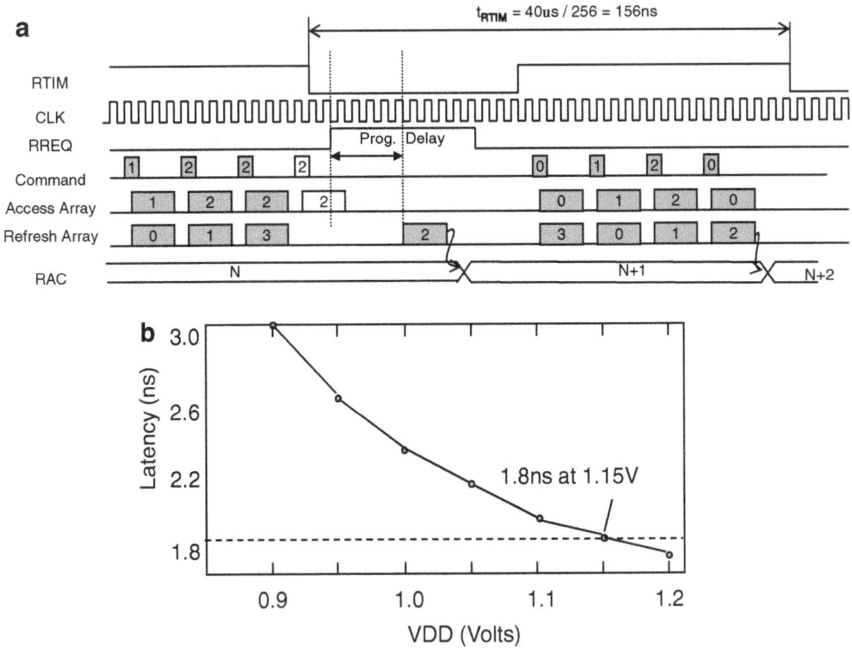

Fig. 10.13 (**a**) Refresh management with concurrent refresh mode (Copyright © IEEE 2009) [24], and (**b**) measured access time with ABIST (Copyright © IEEE 2009) [24]

sub-array 3 to be refreshed. In the final cycle before RTIM goes low, sub-array 2 is accessed again. Address N in sub-arrays 0, 1, and 3 has been refreshed. Sub-array 2 cannot be refreshed, because an access is in progress. As soon as the bank senses RTIM going low, it issues a refresh request, RREQ. When the system recognizes the RREQ generation, no new memory access commands should be issued until RREQ has been disabled. In order to avoid a data loss problem during the RREQ event, the sub-array 2 access operation with the already given command is continued until the data bit communication has been completed. The time interval between RREQ generation to the internal refresh operation is programmable. After the preprogrammed delay, refresh to address N in sub-array 2 is enforced by the concurrent refresh engine. When all sub-arrays have been refreshed with address N, the RAC is incremented to address N + 1 and RREQ goes inactive. It is important to note that during the pre-programmed delay, external commands can still be accepted by the eDRAM, and concurrent refreshes will continue to be attempted during this time. Thus, one can avoid internally-enforced "auto-refresh" if desired.

The 1 MB eDRAM cache prototype has been fabricated in 45 nm partially depleted silicon-on-insulator (SOI) CMOS. The 0.438 mm^2 2.39 Mb macro (with ECC 2392064 bits) employs 0.0672 μm^2 deep trench cell, resulting in 37% cell efficiency. There are three additional masks required when eDRAM is included on a logic chip – one critical mask for the deep trench, one non-critical mask for the buried plate and one non-critical threshold tailor implant for the cell transfer device. The inclusion of deep trench technology has no effect on the logic transistors, while providing a significantly larger decoupling capacitor using the deep trenches. The eDRAM employs an orthogonal WL and μSA architecture discussed in 65 nm SOI embedded DRAM [23]. The self-test system's ability to identify failing addresses and autonomously invoke repair solutions facilitates the evaluation of array performance, pattern sensitivities and guardband optimization. The access time measured with BIST, Fig. 10.13b (Copyright © IEEE 2009), shows 1.8 ns latency at 1.15 V. The prototype has been successfully redeveloped for the IBM POWER7™ microprocessor [50] to construct a large 32 MB L3 on-chip cache memory.

10.6 Future Work

In this chapter, we have seen advanced circuit technologies for high performance embedded DRAM designs and their implementation in nano-scale technology. Most ideas discussed in this chapter have already been well studied, or demonstrated in actual hardware for product use. To conclude this chapter, it is useful to look into additional ideas which may have the potential to boost the embedded DRAM performance using alternative cells and 3-dimensional memory integration.

10.6.1 Embedded DRAM with Floating Body Cell

A revolutionary approach employs a floating body transistor cell (FBC) [26] in Partially Depleted Silicon-On-Insulator technology (PD-SOI). In this scheme, the floating body effect is used as a storage mechanism. Unlike a conventional 1T1C cell, the FBC cell does not have a separate storage capacitor, and therefore is also called "zero capacitor" [51] or "capacitor-less cell" [27] DRAM. Fig. 10.14 (Copyright © IEEE 2002) shows the cross section of the FBC cell. The source,

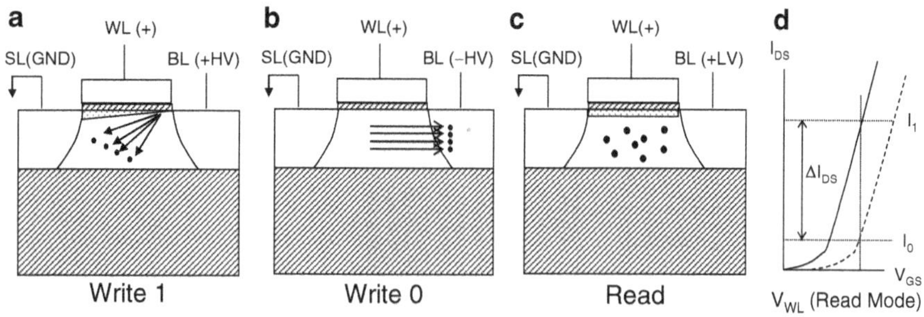

Fig. 10.14 Principles of FBC cell: (**a**) write 1, (**b**) write 0, (**c**) read, and (**d**) I_{DS} in read mode (Copyright © IEEE 2002) [26]

drain, and gate of the FBC NMOS transistor are coupled to Source Line (SL) biased at GND (0 V), BL, and the WL, respectively. Similar to the conventional 1T1C DRAM, BL runs perpendicular to the WL. The SL runs along the WL. For writing "1", the BL is raised after the WL activation. As a result, holes are injected into the floating body by operating the NMOS transistor in saturation, leading to impact ionization. For writing "0", the negative BL voltage is applied, which allows for the discharge of the hole completely from the floating body because of the forward biased pn junction. The number of holes stored in the body determines the threshold voltage (Vt) of the NMOS transistor. This threshold variation is used to sense the amount of holes stored and hence the state of the memory cell. To read the cell, the BL is raised so that the NMOS operates in linear mode. This protects a "0" data from the impact ionization, theoretically allowing for non destructive read. A recent study [52] finds that in use, a few holes are eliminated by each WL cycle. This results in quasi-nondestructive read for an actual memory operation. The NMOS Vt difference between "0" and "1" states results in a different drain current (I_0 for 0 or I_1 for 1) depending on the data pattern (0 or 1) that is detected by a current mode sense-amplifier. Similar to a conventional 1T1C cell, because of the pn junction, the holes in the FBC cell will leak as time goes by, requiring a refresh operation.

The FBC cell may ultimately achieve $4F^2$ cross point cell density with a borderless contact. A $4F^2$ cell with zero capacitor may make the FBC DRAM a main stream memory in the future. A technical challenge of the FBC cell DRAM is to enable the cell scalability without reducing the cell gain. From the design point of view, it is important to develop a robust sense amplifier to detect the ΔI_{DS} reliably, while improving sensing speed. It may also be interesting to develop a unique memory to enjoy a quasi-nondestructive read advantage.

10.6.2 Embedded DRAM with Gain Cell

An evolutionary approach employs a gain cell targeting for high performance cache applications with density of 2x of that of SRAM. Fig. 10.15a shows the 3 transistor (3T) gain cell. A write switching transistor (NMOS1) is used to couple the write bitline (WBL) to the storage node when the write wordline (WWL) is on. The data bit written in the storage node is maintained by the NMOS2 device capacitor, as long as WWL is low. The storage node is coupled to the gate of the read gain transistor (NMOS2). A read switching transistor (NMOS3) is used to couple the read bitline (RBL) to the drain of the read gain transistor (NMOS2). This allows for the discharge of the RBL to GND for the "1" data when the read wordline (RWL) goes high. Because the read pass structure is similar to the conventional SRAM design and isolated from the storage node, it ultimately allows for a truly SRAM-like read performance with the feature of non-destructive read. In addition to this read access performance advantage, the completely independent read and write WLs and BLs allow the 3T gain cell to perform read and write operations simultaneously. A technical challenge of the 3T gain cell is cell size reduction, where the device designs for NMOS1-3 are important. The NMOS1 is the key device to determine the write performance and the data retention. Therefore, it requires a high threshold voltage and thicker oxide than the other two devices because the WWL should be boosted to allow for a full VDD write operation to the cell. The NMOSs 2–3 may use a logic-like device with thin oxide, however, the threshold of these two FETs should be optimized to maximize the read gain. Lowering the VT of the NMOS2 improves the gain, however, if this is too low, the gain is reduced because the read current is determined by the device conductance of the NMOS3. In addition, it is preferable to integrate a capacitor (such as trench as shown) in the storage node to improve data retention, which may be realized by the existing 3-dimensional capacitor structure. To reduce cell area, the RBL and WBL may be shared if simultaneous read and write functions are not needed.

Cell size reduction can be achieved by employing a two transistor (2T) gain cell [28] shown in Fig. 10.15b. Similar to the 3T gain cell, a write switching transistor (NMOS1) is used to couple the WBL to the storage node when the WWL is on. The written data bits in a storage node are maintained by the NMOS2 device capacitor, as long as WWL is low. The storage node is coupled to the gate of the read gain transistor (NMOS2). The source and drain of the NMOS 2 are coupled to the RWL and RBL, respectively. The RWL runs along with WWL, and RBL runs perpendicular to the RWL and WWL, or along with WBL. Prior to the low-going RWL, the RBL is precharged at the VDD. When a read command is accepted, the RWL goes low, which allows the discharge of the RBL when the storage node is high, or stays high when the storage node is low. The low-going RBL for reading "1" is clamped to VDD – VT of the NMOS2 due to the NMOS2's coupling to the unselected RWLs. Similar to the 3T gain cell, the 2T gain cell features a non-destructive read, and may allow SRAM like performance with 1/3 of the 6T SRAM area. However, it requires a RWL driving device (NMOS3), which is an additional requirement to employ the 2T gain cell. From the array point of view, 2T gain cell is a shared read switching NMOS approach (NMOS3 in Fig. 10.15a) used for 3T gain cell to reduce the cell area. The 2T PMOS gain cell memory prototype [28] employed a fully pipelined architecture, demonstrating eight cycle successive read access to the same bank at 2 GHz clock rate for 2 Mb density in 65 nm node. The 2T gain cell allows full compatibility to the existing embedded SRAM, which potentially replaces the SRAM TAG for high performance on-chip cache memory.

Another gain cell approach [29], Fig. 10.15c, uses a gain cell consisting of two transistor and 1 gated diode (2T1D), where the gate voltage of the read transistor is amplified by the gated diode. Similar to the 3T gain cell, a write switching transistor (NMOS1) is used to couple the WBL to the storage node when the WWL is on. The storage node is coupled to the gate of the read gain transistor (NMOS2). The storage node is additionally coupled to the NMOS gated diode (GDNMOS), which can be formed using an open-drain NMOS or alternatively by connecting the source and drain. During the read operation, the source of the diode devices is raised by the RWL. When reading a "0", the gated diode is off, resulting in less coupling to the read transistor. When reading a "1", the gated diode is on, resulting in a large coupling to the read access transistor, amplifying the read signal. Because of the self and selective coupling

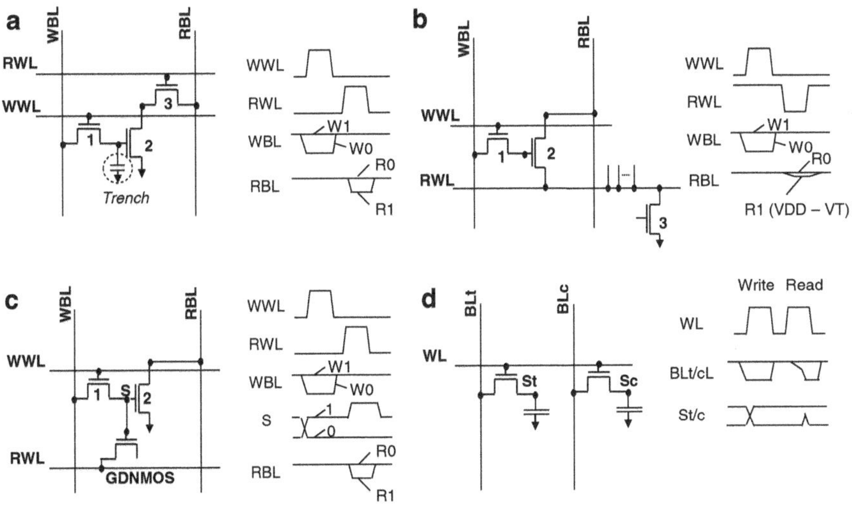

Fig. 10.15 Alternative cells: (**a**) 3T gain cell, (**b**) 2T gain cell, (**c**) 2T1D gain cell, and (**d**) twin cell

to the gate of the read transistor, the eDRAM with 2T1D gain cell allows the read gain to be doubled. The storage voltage amplified by the GDNMOS is detected by the RBL transition (going 0 for the data "1" or maintained at "VDD" for the data "0") similar to other gain cells. A technical challenge for the 2T1D cell is the reduction of cell size while improving a capacitive coupling to the storage node by the GDNOMS. Similar to the 3T gain cell, the WBL and RBL may be shared to reduce cell size.

10.6.3 Embedded DRAM with Twin Cell

Although FBC and gain cell approaches have several potential advantages for future embedded DRAMs, there are still many technical challenges to overcome. Instead, a twin cell approach, Fig. 10.15d, is a simple extension of the existing known technology, and may boost the performance by as much as 2x the existing eDRAM with 1T1C cell. Unlike the conventional 1T1C cell, the twin cell stores "0" and "1" data as a pair. A VDD sensing scheme with a cross-coupled sense amplifier is preferably used. This allows the discharge of either the true or complement BL during a signal development, which is then amplified by the NMOS cross-coupled sense amplifier. No complicated reference or control is required for the twin cell sensing scheme. In addition, there is no need to write a full voltage of "1" data to the cell, because the sensing signal is determined by only the "0" data. Because of the VDD precharge, "1" data is protected from the leakage. Although "0" data may leak to high as the storage node goes high, the WL voltage with respect to the BL will be negative, self-limiting the "0" to "1" leakage. This thus allows for the use of lower wordline boost voltage (VPP), and ultimately a non-boosting approach may be possible, while reducing the array Vt. Although the twin cell size is 2x the existing 1T1C cell, because of the simplification of the sensing scheme or word system, the area of the twin cell embedded DRAM is less than 2x the conventional 1T1C cell macro, which may make it a main stream approach for high performance embedded DRAMs sooner than any other approaches.

10.6.4 Embedded DRAM for 3 Dimensional Integration

As an alternative approach to the aggressive scaling with main stream embedded DRAM, it is possible to consider stacking one or more memories to boost the memory density. Traditionally, stacking memory chips has been realized by wire-bond in peripheral pads, and used for consumer applications. This wire-bond 3D approach reduces the circuit board area, or allows the keeping of the existing memory module standard by creating a stacked memory module using two or more dies. Although these known 3D integration approaches still have several technical advantages, they are not suitable for general purpose application, particularly for high performance application. This is because the key for the high performance system is the reduction of latency and increase in bandwidth, which cannot be realized by the wire-bond technology using peripheral connection. Through-Silicon-Via (TSV) technology [30, 53], Fig. 10.16 (Copyright © IEEE 2009), creates holes in a silicon wafer, and allows

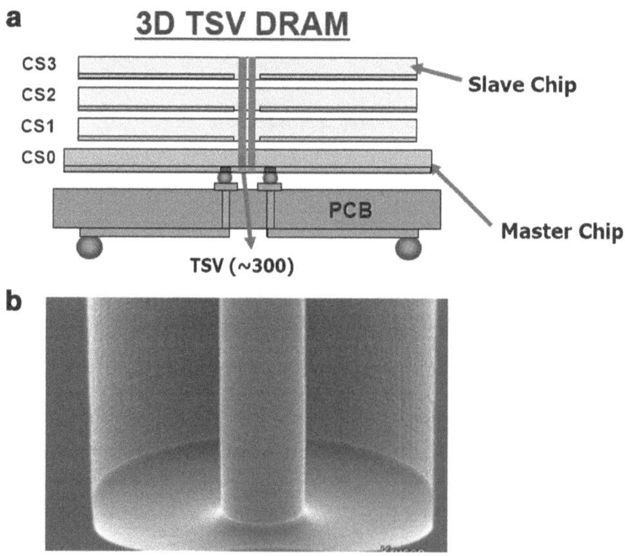

Fig. 10.16 3D DRAM integration with Through-Silicon-Via (TSV): (**a**) a master chip, CS0, controls 3 eDRAM memory chips, CS1-3, through the TSVs and (**b**) Cu TSV (Copyright © IEEE 2009) [30, 53]

1Mbit DRAM Macro

- 32 Macro per Core = 4MByte
- 8 Core per Chip = 32MByte

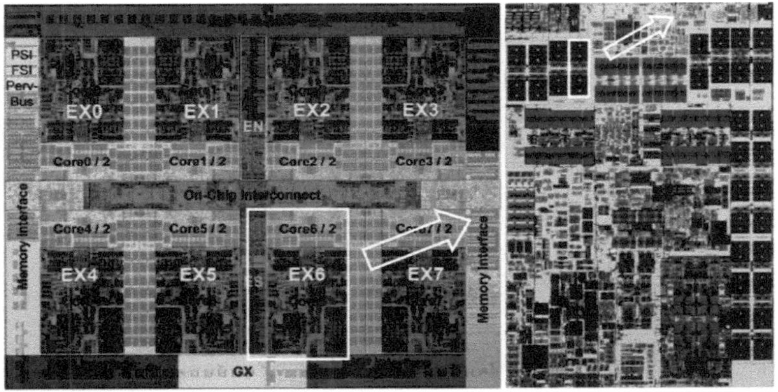

Technology	45nm PD SOI
Cell Size	152nm x 221nm(0.0672μm^2)
Macro Size	377μm x 634μm (0.239mm^2)
Organization	1k row x 8 col x 146 (1Mbit)
Sub-Array	Hierarchical μSense 264 word-line x 1200 bit-line
Supply	1.05V
Performance	1.7ns/1.35ns (cycle/access)

Fig. 10.17 POWER7™ Die and Core Microphotographs and embedded DRAM features (Copyright © IEEE 2010) [50, 54]

for the communication of the data and control signals in two or more chips, which are vertically stacked. Because of the silicon technology process, the number and the speed of the TSV communication are significantly more than the existing wired-bond approach, eliminating pin and speed bottom neck.

The 3D integration with TSV is expected to enable a new era for technology, VLSI chips, and applications. Several 3D chip prototypes have been proposed, but the killer application has not yet been identified. The 3D high density cache memory with embedded DRAM is one of the most significant candidates. This is because the 3D integration allows for the boost of the VLSI density significantly more than the silicon scaling rate, which allows the integration of high density cache memory for boosting both memory bandwidth and latency. This new cache hierarchy will boost the system performance significantly for a multi-core microprocessor, which cannot be satisfied by using simple commodity DRAM or SRAM stacking.

10.6.5 Summary

As discussed in this chapter, the performance of embedded DRAM is closing the gap with SRAM performance while offering more than a 2x density improvement. Figure 10.17 (Copyright © IEEE 2010) shows the chip microphotographs and the embedded DRAM features of the POWER7™ microprocessor [50, 54]. This highly parallel, scalable, next generation microprocessor employs one 4 MB embedded DRAM L3 cache per core, which is constructed from thirty-two 1 Mb embedded DRAM macros having 1.35 ns random access and 1.7 ns random cycle performance. The embedded DRAM provides approximately threefold density improvement while reducing power consumption by more than one half that of equivalent SRAM L3 designs. These performance benefits are achieved by using μSA and orthogonal WL architectures discussed in this chapter, resulting in an eight core per chip, 567 mm^2 die using 1.2 Billion transistors in 45 nm SOI CMOS. This is the most advanced VLSI design using embedded DRAM so far reported. However, the integration of high density high performance eDRAM in microprocessors has just begun, and new applications are expected within a couple of years.

In order to further improve performance with nano-scale technology, we need to develop new memory cells and 3D integration technology. While these challenges will drive novel memory architectures for the next generation ecosystem, consideration must still be given to technology, VLSI design, and system requirements. It is also crucial to develop new methodologies to test the array, repair failures, use partial good memories, and perhaps ultimately to manage the cache memory allocation to accommodate defective memory elements, and to enable autonomic solutions for optimizing the system for each chip. The high performance embedded DRAM with 3D integration is the most promising emerging technology for high performance microprocessors with low power consumption, and will open a new era for VLSI designs in both consumer and high-end applications.

References

1. T. Kirihata, G. Mueller, B. Ji, G. Frankowsky, J. Ross, H. Terletzki, D. Netis, O. Weinfurtner, D. Hanson, G. Daniel, L. Hsu, D. Storaska, A. Reith, M. Hug, K. Guay, M. Selz, P. Poechmueller, H. Hoenigschmid, M. Wordeman, A 390 mm^2, 16-Bank, 1 Gb DDR SDRAM with Hybrid Bitline Architecture. JSCC **34**, 1580–1588 (1999).
2. T. Murotani, I. Naritake, T. Matano, T. Ohtsuki, N. Kasai, H. Koga, K. Koyama, K. Nakajima, H. Yamaguchi, H. Watanabe, T. Okuda, A 4-level storage 4Gb DRAM, ISSCC Dig. Tech. Papers, Feb 1997, pp. 74–75.
3. H.S. Jeong, W.S. Yang, T.S. Hwang, C.H. Cho, S. Park, S.J. Ahn, Y.S. Chun, S.H. Shin, S.H. Song, J.Y. Lee, S.M. Jang, C.H. Lee, J.H. Jeong, M.H. Cho, J.K. Lee, H.S. Kinam Kim, Highly manufactuable 4Gb DRAM using 0.11μm DRAM technology, IEDM Dig. Tech. Papers, Dec 2000, pp. 353–356.
4. T. Kirihata, Embedded dynamic random access memory, VLSI technology, systems, and applications, Dig. Tech. Papers, Oct 2003, pp. 155–158.
5. J. Barth, D. Anand, J. Dreibelbis, J. Fifield, K. Gorman, M. Nelms, G. Pomichter, D. Pontius, A 500-MHz multi-banked compliable DRAM macro with direct write and programmable pipleing. JSCC **40**(1), 213–222 (Apr 2005).
6. S.S. Iyer, J. Barth, P. Parries, J. Norum, J. Rice, L. Logan, D. Hoyniak, Embedded DRAM the technology platform for the BlueGene/L chip. IBM J. Res. Dev. **49**(2, 3), 333–349 (2005).
7. B. Keeth, R. J. Baker, B. Johnson, and F. Lin, *DRAM Circuit Design – Fundermental and High-Speed Topics*. IEEE Press, Wiley-Interscience.
8. I. Kiyoo, The history of DRAM circuit designs –At the forefront of DRAM development. JSSC **13**(1), 27–31 (Winter 2008).
9. S. Tanoi, Y. Tanaka, T. Tanabe, A. Kita, T. Inada, R. Hamazaki, Y. Ohtsuki, M. Uesugi, A 32-bank 256Mb DRAM with cache and TAG, ISSCC Dig. Tech. Papers, Feb 1994, pp.144–145.
10. P. Gillingham, B. Hold, I. Mes, C. O'Connell, P. Schofield, K. Skjaveland, R. Torrance, T. Wojcicki, and H. Chow, A. 768k embedded DRAM for 1.244Gb/s ATM switch in 0.8μm logic process, ISSCC Dig. Tech. Papers, Feb 1996, pp. 262–263.
11. T. Shimizu, J. Korematu, M. Satou, H. Kondo, S. Iwata, K. Sawai, N. Okumura, K. Ishimi, Y. Nakamoto, M. Kumanoya, K. Dosaka, A. Yamazaki, Y. Ajioka, H. Tsubota, Y. Nunomura, T. Urabe, J. Hinata, and K. Saitoh, A multimedia 32b RISC microprocessor with 16Mb DRAM, ISSCC Dig. Tech. Papers, Feb 1996, pp.216–217.
12. D. Patterson, T. Anderson, N. Cardwell, R. Fromm, K. Keeton, C. Kozyrakis, R. Thomas, K. Yelick, Intelligent RAM(IRAM): chips that remember and compute, ISSCC Dig. Tech. Papers, Feb 1997, pp. 224–225.
13. T. Kimuta, K. Takeda, Y. Aimoto, N. Nakamura, T. Iwasaki, Y. Nakazawa, H. Toyoshima, M. Hamada, M. Togo, H. Nobusawa, T. Tanigawa, 64 Mb 6.8 ns random ROW access DRAM macro for ASICs, ISSCC, Dig. Tech. Papers, Feb 1999, pp. 416–417.
14. Y. Agata, K. Motomochi, Y. Fukushima, M. Shirahama, M. Kurumada, N. Kuroda, H. Sadakata, K. Hayashi, T. Yamada, K. Takahashi, T. Fujita, An 8-ns Random Cycle Embedded RAM Macro With Dual-Port Interleaved DRAM Architecture (D^2RAM). JSSC **44**(11), 1668–1672 (Nov 2000).
15. O. Takahashi, S.H. Dhong, M. Ohkubo, S. Onishi, R.H. Dennard, R. Hannon, S. Crowder, S.S. Iyer, M.R. Wordeman, B. Davari, W.B. Weinberger, N. Aoki, 1 GHz fully pipelined 3.7 ns address access time 8 k x 1024 embedded DRAM macro. JSSC **35**(11), 1673–1679 (Nov. 2000).
16. C. Hwang, T. Kirihata, M. Wordernan, J. Fifield, D. Storaska, D. Pontius, G. Fredernan, B. Ji, S. Tomashot, S.H. Dhong, A 2.9ns random access cycle embedded DRAM with a destructive-read architecture, In Symp. VLSI Circuits, Dig. Tech. Papers, Jun 2002, pp. 174–175.

17. T. Yabe, S. Miyano, K. Sato, M. Wada, R. Haga, O. Wada, M. Enkaku, T. Hojyo, K. Mimoto, M. Tazawa, T. Ohkubo, K. Numata, A configurable DRAM macro design for 2112 derivative organizations to be synthesized using a memory generator, ISSCC, Dig. Tech. Papers, Feb 1998, pp. 72–73.
18. T. Yamauchi, M. Kinoshita, T. Amano, K. Dosaka, K. Arimoto, H. Ozaki, M. Yamada, T. Yoshihara, Design methodology of embedded DRAM with virtual-socket architecture. JSSC **36**(1), 46–54 (Jan 2001).
19. T. Watanabe, K. Ayukawa, S. Miura, M. Toda, T. Iwamura, K. Hoshi, J. Sato, K. Yanagisawa, Access optimizer to overcome the "future walls of embedded DRAMs" in the era of systems on silicon, ISSCC, Dig. Tech. Papers, Feb 1999, pp. 370–371.
20. J. Dreibelbis, J. Barth, H. Kalter, R. Kho, Processor-based built-in self-test for embedded DRAM. JSSC **33**(11), 173–1740 (Nov 1998).
21. J. Barth, D. Anand, J. Dreibelbis, E. Nelson, A 300 MHz multi-banked eDRAM macro featuring GND sense, bit-line twisting and direct reference cell write, ISSCC Dig. Tech. Papers, Feb 2002, pp. 156–157.
22. T. Kirihata, P. Parries, D.R. Hanson, H. Kim, J. Golz, G. Fredeman, R. Rajeevakumar, J. Griesemer, N. Robson, A. Cestero, B.A. Khan, G. Wang, M. Wordeman, S.S. Iyer, An 800MHz embedded DRAM with a concurrent refresh mode. JSSC **40**(6), 1377–1387 (Jun 2005).
23. J. Barth, W.R. Reohr, P. Parries, G. Fredeman, J. Golz, S.E. Schuster, R.E. Matick, H. Hunter, C.C. Tanner, J. Harig, H. Kim, B.A. Khan, J. Griesemer, R.P. Havreluk, K. Yanagisawa, T. Kirihata, S.S. Iyer, A 500MHz random cycle, 1.5ns latency, SOI embedded DRAM macro featuring a three transistor micro sense amplifier. JSSC **43**(1), 86–95 (Jan 2008).
24. P. Klim, J. Barth, W.R. Reohr, P.D. Dick, G. Fredeman, G. Koch, H.M. Le, A. Khargonekar, P. Wilcox, J. Golz, J.B. Kuang, A. Mathews, J.C. Law, T. Luong, H.C. Ngo, R. Freese, H.C. Hunter, E. Nelson, P. Parries, T. Kirihata, S.S. Iyer, A 1MB cache subsystem prototype with 1.8ns embedded DRAMs in 45nm SOI CMOS. JSSC **43**(1), 86–95 (Apr 2009).
25. G. Uhlmann, T. Aipperspach, T. Kirihata, K. Chandrasekharan, Y.Z. Li, C. Paone, B. Reed, N. Robson, J. Safran, D. Schmitt, S.S. Iyer, A commercial field-programmable dense eFUSE array memory within 99.999% sense yield for 45nm SOI CMOS, ISSCC Dig. Tech. Papers, Feb 2008, pp.406–407.
26. T. Ohsawa, K. Fujita, T. Higashi, Y. Iwata, T. Kajiyama, Y. Asao, K. Sunouchi, Memory design using a one-transistor gain cell on SOI. JSSC **37**(11), 1510–1522 (Nov. 2002).
27. P. C. Fazan, S. Okhonin, M. Nagoga, and J-M. Sallese, A simple 1-transistor capacitor-less memory cell for high performance embedded DRAMs, IEEE CICC Dig. Tech. Papers, 2002, pp. 99–102.
28. D. Somasekhar, Y. Yibin, P. Aseron, S.-L. Lu, M.M. Khellah, J. Howard, G. Ruhl, T. Karnik, S. Borkar, V.K. De, A. Keshavarzi, 2GHz 2Mb 2T gain cell memory macro with 128GBytes/sec bandwidth in a 65nm logic process technology. JSSC **44**(1), 174–185 (Nov 2009).
29. W. Luk, R.H. Dennard, A novel dynamic memory cell with internal voltage gain. JSSC **40**(4), 884–894 (Nov 2005).
30. S.S. Iyer, T. Kirihata, M.R. Wordeman, J. Barth, R.H. Hannon, R. Malik, Process-design considerations for three dimensional integration, in Symp. VLSI Technology, Dig. Tech. Papers, Jun 2009, pp. 60–63.
31. M. Takeuchi, K. Inoue, M. Sakao, T. Sakoh, T. Kitamura, S. Arai, T. Iizuka, T. Yamamoto, H. Shirai, Y. Aoki, M. Hamada, R. Kubota, S. Kishi, A 0.15μm logic based embedded DRAM technology featuring 0.425μm^2 stacked cell using MIM (metal–insulator–metal) capacitor, in Symp. VLSI Tech., Dig. Tech. Papers, Jun 2001, pp.29–30.
32. G. Bronner, H. Aochi, M. Gall, J. Gambino, S. Gernhardt, E. Hammerl, H. Ho, J. Iba, H. Ishiuchi, M. Jaso, R. Kleinhenz, T. Mii, M. Narita, L. Nesbit, W. Neumueller, A. Nitayama, T. Ohiwa, S. Parke, J. Ryan, T. Sato, H. Takato, S. Yoshikawa, A fully planarized 0.25μm CMOS technology for 256Mb DRAM and beyond, in Symp. VLSI Tech., Dig. Tech. Papers, Jun 1995, pp. 15–16.
33. H. Akatsu, R. Weis, K. Cheng, M. Seitz, M.-S. Kim, R. Ramachandran, T. Dyer, B. Kim, D.-K. Kim, R. Malik, J. Strane, T. Goebel, O.-J. Kwon, C, Y. Sung, P. Parkinson, K. Wilson, I. McStay,

M. Chudzik, D. Dobuzinsky, M. Jacunski, C. Ransom, K. Settlemyer, L. Economikos, A. Simpson, A. Knorr, M. Naeem, G. Stojakovic, W. Robl, O. Gluschenkov, B. Liegl, C.-H. Wu, Q. Wu, W.-K. Li, C.J. Choi, N. Arnold, T. Joseph, K. Varn, M. Weybright, K. McStay, W.-T. Kang, Y. Li, S. Bukofsky, R. Jammy, R. Schutz, A. Gutmann, W. Bergner, R. Divakaruni, D. Back, E. Crabbe, W. Mueller, G. Bronner, A highly ,manufacturable 110 nm DRAM technology with $8F^2$ vertical transistor cell for 1Gb and beyond, in Symp. VLSI Technology, Dig. Tech. Papers, Jun 2002, pp. 52–53.

34. G. Wang, P. Parries, B. Khan, J. Liu, Y. Otani, J. Norum, N. Robson, T. Kirihata, S.S. Iyer, A $0.168\mu m^2/0.11\mu m^2$ highly scalable high performance embedded DRAM cell for 90/65nm logic applications, in Symp. VLSI Technology, Systems, and Applications. Dig. Tech. Papers, Apr 2005, pp. 31–32.
35. T. Kirihata, G. Mueller, M. Clinton, S. Loeffler, B. Ji, H. Terletzki, D. Hanson, C. Hwang, G. Lehmann, D. Storaska, G. Daniel, L. Hsu, O. Weinfurtner, T. Boehler, J. Schnell, G. Frankowsky, D. Netis, J. Ross, A. Reith, O. Kiehl, M. Wordeman, A $113mm^2$ 600Mb/s/pin 512Mb DDR2 SDRAM with vertically-folded bitline architecture, ISSCC, Feb 2001, pp.382–383.
36. H. Fujisawa, S. Kubouchi, K. Kuroki, N. Nishioka, Y. Riho, H. Noda, I. Fujii, H. Yoko, R. Takishita, T. Ito, H. Tanaka, M. Nakamura, An 8.1-ns column-access 1.6-Gb/s/pin DDR3 SDRAM with an 8:4 multiplexed data-transfer scheme. JSSC **42**(1), 201–209 (Jan 2007).
37. S. Takase, N. Kushiyama, A 1.6GB/s DRAM with flexible mapping redundancy technique and additional refresh scheme, ISSCC Dig. Tech. Papers, Feb 1999, pp. 410–411.
38. T. Kirihata, M. Gall, K. Hosokawa, J.-M. Dortu, H. Wong, K.-P. Pfefferl, B. Ji, O. Weinfurtner, J. DeBrosse, H. Terletzki, M. Selz, W. Ellis, M. Wordeman, O. Kiehl, A $220mm^2$, four and eight Bank, 256Mb SDRAM with single sided stitched WL architecture, JSSC, **33**(11), 1711–1719 (Nov 1998).
39. M. Asakura, T. Oishi, S. Tomishima, H. Hidaka, K. Arimoto, K. Fujishima, A hierarchical bitline architecture with flexible redundancy and block compare test for 256Mb DRAM, in Symp. VLSI Circuits, Dig. Tech. Papers, May 1993, pp. 93–94.
40. H. Hidaka, K. Fujishima, Y. Matsuda, M. Asakura, T. Yoshihara, Twisted bit-line architecture for mult-megabit DRAM's. JSSC **24**(1), 21–27 (Feb 1989).
41. H. Hoenigschmid, A. Frey, J.K. DeBrosse, T. Kirihata, G. Mueller, D.W. Storaska, G. Daniel, G. Frankowsky, K.P. Guay, D.R. Hanson, L.L.-C. Hsu, B. Ji, D.G. Netis, S. Panaroni, C. Radens, A.M. Reith, H. Terletzki, O. Weinfurtner, J. Alsmeier, W. Weber, M.R. Wordeman, $7F^2$ cell and bitline architecture featuring tilted array devices and penalty-free vertical BL twists for 4-Gb DRAMs. JSSC **35**(5), 713–718 (May 2000).
42. N. Lu, H.H. Chao, Half-VDD bitline sensing scheme in CMOS DRAM's. JSSC, sc-19 **4**(4), 451–454 (Nov 1984).
43. H. Pilo, D. Anand, J. Barth, S. Burns, P. Corson, J. Covino, S. Lamphier, A 5.6ns Random Cycle 144Mb DRAM with 1.4Gb/s/pin and DDR3-SRAM interface. JSSC **38**(11), 1974–1980 (Nov 2003).
44. R. Scheuerlein, J.D. Meindl, Offset wordline architecture for scaling DRAM's to the gigabit level. JSSC, sc-19 **23**(1), 41–47 (Feb 1988).
45. D. Anand, J. Covino, J. Dreibelbis, J. Fifield, K. Gorman, M. Jacunski, J. Paparelli, G. Pomichter, D. Pontius, M. Roberge, S. Sliva, A 1.0GHz multi-banked embedded DRAM is 65nm CMOS featuring concurrent refresh and hieratical BIST, IEEE 2007 CICC, Dig. Tech. Papers, pp 795–797.
46. J. Clabes, J. Friedrich, M. Sweet, J. DiLullo, S. Chu, D. Plass, J. Dawson, P. Muench, L. Powell, M. Floyd, B. Sinharoy, M. Lee, M. Goulet, J. Wagoner, N. Schwartz, S. Runyon, G. Gorman, P. Restle, R. Kalla, J. McGill, S. Dodson, Design and implementation of the POWER5™ microprocessor, ISSCC Dig. Tech. Papers, Feb 2004, pp. 56–57.
47. U.M. Nawathe, M. Hassan, L. Warriner, K. Yen, B. Upputuri, D. Greenhill, A. Kumar, H. Park, An 8-core, 64-thread, 64-bit, power efficient SPARC SoC (Niagara2), ISSCC Dig. Tech. Papers, Feb 2007, pp.108–109.
48. J.B. Kuang, A. Mathews, J. Barth, F. Gebara, T. Nguyen, J. Schaub, K. Nowka, G. Carpenter, D. Plass, E. Nelson, I. Vo, W. Reohr, T. Kirihata, An on chip dual supply charge pump system for 45nm eDRAM, ESSCIRC 2008, Sept 2008, pp. 66–69.

49. A. Rylyakov, J. Tierno, G. English, M. Sperling, D. Friedman, A wide power-supply range (0.5V-to-1.3V) wide tuning range (500 MHz-to-GHz) all-static CMOS ADPLL in 65nm SOI, ISSCC Dig. Tech Papers, Feb 2007, pp. 172–173.
50. J. Barth, D. Plass, E. Nelson, C. Hwang, G. Fredeman, M. Sperling, A. Mathews, W. Reohr, K. Nair, N. Cao, A 45nm SOI Embedded DRAM Macro for POWER7™ 32MB On-Chip L3 Cache, ISSCC Dig. Tech Papers, Feb 2010, pp. 342–343.
51. S. Okhonin, P. Fazan, M.-E. Jones, Zero capacitor embedded memory technology for system on chip, Memory Technology, Design and Testing, Aug 2005.
52. T. Ohsawa, K. Fujita, K. Hatsuda, T. Higashi, T. Shino, Y. Minami, H. Nakajima, M. Morikado, K. Inoh, T. Hamamoto, S. Watanabe, S. Fujii, T. Furuyama, Design of a 128Mb SOI DRAM using the Floating Body Cell (FBC). JSSC **41**(1), 135–145 (Jan 2006).
53. U. Kang, H.-J. Chung, S. Heo, S.-H. Ahn, H.-Lee, S.-H. Cha, J. Ahn, D. Kwon, J. H. Kim, J.W. Lee, H.-S. Joo, W.-S. Kim, H.-K. Kim, E.-M. Lee, S.-R. Kim, K.-H. Ma, D.-H. Jang, N.-S. Kim, M.-S. Choi, S.-J. Oh, J.-B. Lee, T.-K. Jung, J.-H. Yoo, C. Kim, 8Gb 3D DDR3 SDRAM using through-silicon-via technology, ISSCC Dig. Tech. Papers, Feb 2009, pp. 130–131.
54. D. Wendel, R. Kalla, R. Cargoni, J. Clables, J. Friedrich, R. Frech, J. Kahle, B. Sinharoy, W. Starke, S. Taylor, S. Weitzer, S. G. Chu, S. Islam, V. Zyuban, The implementation of Power7™: A highly parallel and scalable multi-core high-end server processor, ISSCC Dig. Tech. Papers, Feb 2010, pp. 102–103.

Chapter 11
Timing Circuit Design in High Performance DRAM

Feng (Dan) Lin

Abstract The need for high performance dynamic random access memory (DRAM) becomes a pressing concern with processor speeds racing into the multi-gigahertz range. For a high performance computing system, such as high-end graphics cards or game consoles, memory bandwidth is one of the key limitations. One way to address this ever increasing bandwidth requirement is by increasing the data transfer rate. As the I/O speed jumps up, precise timing control becomes critical and challenge with variations and limitations in nano-scaled DRAM processes. In this chapter, we will explore two most important timing circuits used in high performance DRAM: clock distribution network (CDN) and clock synchronization circuits (CSC).

Keywords Memory interface • DRAM • Memory bandwidth • Clock distribution network (CDN) • Clock synchronization circuit (CSC) • Source-synchronous • Analog phase generator (APG) • De-skewing • Duty-cycle correction (DCC) Delay-locked loop (DLL) • 3D integration

11.1 Introduction

11.1.1 Memory Interface

A typical memory system, shown in Fig. 11.1, generally includes a memory controller, DRAM devices and memory channels between them. As part of the DRAM device, memory interface serves as a gateway between the external world and internal periphery logics (i.e. array interface and command and address logic). The state-of-art memory interface may include data transceivers, SERDES (serializer/deserializer),

F. Lin (✉)
Senior Member of Technical Staff, Micron Technology, Inc.
e-mail: flin@micron.com

K. Iniewski (ed.), *CMOS Processors and Memories*,
Analog Circuits and Signal Processing, DOI 10.1007/978-90-481-9216-8_11,

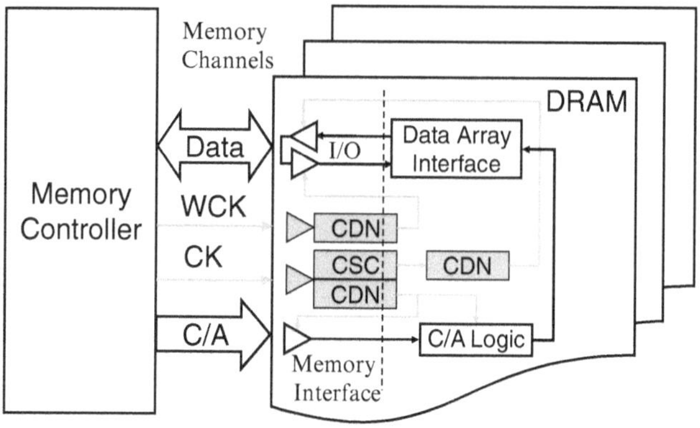

Fig. 11.1 High-performance Memory Interfaces (*clock paths* and *timing circuits* in *Gray colors*)

clock distribution networks (CDNs), and clock synchronization circuits (CSC). Besides bidirectional data I/Os, which move data in and out of the memory interface, there are also clocks and command and address (C/A) inputs, which are unidirectional. The data and C/A are retimed within the memory interface using the distributed clocks. External to the DRAM interface, the memory controller communicates to one or multiple DRAMs via memory channels and buses. Various bussing topologies and different characteristics of memory channels will introduce latency and timing skews, and may profoundly impact the overall system timing performance.

In some literature, the data channel is also referred to as a *link*, which includes transceivers in both the memory controller and the DRAM, and the connection between them. Total memory bandwidth can be increased either by increasing the number of links or by increasing the per link data transfer rate. The bandwidth limitation in both the channel and the DRAM interface put an upper boundary for the data rate. The package size, the number of DRAMs, and metal routing layers on the PCB board also put some constrains on the number of links. Power efficiency (usually specified in mW/Gb/s) and cost are the other two considerations. Although each pieces of the link and associate clocking scheme worth a detail analysis when pursuing higher system performance, with limited space, we now turn our focus to the timing circuits design within the DRAM interface.

11.1.2 Evolution of the DRAM Interface and Timing Specifications

Timing circuit is generally related to a clock. Following the advent of synchronous operation, DRAM device specifications and performance began a slow migration towards metrics related to clock frequency [1]. From EDO (extended data out) asynchronous DRAM to DDR (double data rate) synchronous DRAM, the memory interface also evolved from clock-less asynchronous operation to synchronous data operation on both rising and falling clock edges. The various generations of DDR and its faster graphics cousin GDDR, such as DDR2, DDR3, GDDR3, GDDR4, and GDDR5, encompass evolutionary advances in technology to achieve higher overall data transfer rates. The fastest data transfer rate reported for GDDR5 [4] is around 6 Giga bits per second per pin, compared to only 0.133 Giga bits per second for a low-end DDR system. To enable such speedup, changes in burst length (BL), improvements in signaling technology and packaging, and significant advances in circuit design are imperative.

Design a timing circuitry may start with specifications. Two important timing parameters define how fast the DRAM device can cycle the data (clock cycle time or t_{CK}) and how many clock cycles the system must wait for data following a Read request (CAS latency or CL). To achieve higher performance, it is desire to have DRAM devices with the lowest CAS latency possible at any given operating frequency. Unfortunately, timing delays and timing variations associated with data and clock path make it tough to meet with higher overall data rates. Such timing delays and variations are generally referred to as clock skew and jitter.

When a signal (e.g. *CK*) propagating through a memory interface, like in Fig. 11.1, a timing delay is introduced, also known as clock skew. Besides static timing delay, high-speed signal distribution is also susceptible to duty-cycle distortion and timing mismatch. Under process, voltage and temperature (PVT) variations, the propagation delay may change dynamically (i.e. jitter) and further degrade timing performance. Both skew and jitter make it difficult to predict and track latency [14] (e.g. CL) as speeds increase.

11.1.3 Source-Synchronous Interface and Matched Routing

There are several ways to mitigate the timing variations caused by the on-die signal distribution. Traditionally, memory interfaces favor *source-synchronous* relationship between data and clock (or strobe) to alleviate PVT sensitivities. In some literatures, the clock bundled with data is also referred as a *forwarded* clock. Shown in Fig. 11.1, the additional write clocks (*WCK*) can facilitate high-speed data capture by placing the clock close to the destination (e.g. transceivers), which in turn, shorten the clock distribution. Differential write clock can further reduce duty-cycle distortion at the cost of additional pins. A forwarded read clock (not shown in Fig. 11.1) synchronized with returned data may facilitate data capture at the memory controller side as well.

To maintain source-synchronicity, a technique called *matched routing* [1–2] has been utilized to match the internal clocks and data all the way from the device pins to the capture latches. *Logical effort* [3] matching and *point-to-point* matching are both used in this method to address various topologies of the clock and data paths. With this technique, the clock distribution delay is modeled and replicated in each data path. Any timing change caused by PVT variations can be tracked and zeroed out as long as the matched routing is applied. Although the gate and wire delay may be scaled linearly with DRAM processes, matched routing adds extra circuits for each data path and increases power and area consumption. With single-ended signaling for the data transceivers and the lengthened data path, duty-cycle distortion is a major concern for this approach. At higher speeds and for increased data bus width (pin count or number of links), matched routing may not be practical.

An alternative scheme employs adjustable timing for the clock distribution network (CDN). In this case, data capture latches are located close to the data input pads. The latches are generally built from sense-amp style circuits to maximize their gain and input sensitivity and minimize setup and hold. The CDN delay can be backed out through training by the memory controller. The goal of training is to optimize capture timing at the input data latches. The memory controller accomplishes this by delaying or advancing the clock until the latches operate in the center of the data eye. Although process variation and static timing offsets can be tuned out by initial training, delay variations across the CDN due to voltage and temperature (VT) drift may still exist. Retraining may be necessary if these delay variations are too great, further complicating the link interface design.

To mitigate VT sensitivity for high-speed memory operating, a novel multi-phase, voltage and temperature insensitive CDN [6] will be introduced in Section 11.2. Design consideration for different CDN topologies will be analyzed with a focus on power and performance. Simulation data using a 50 nm DRAM process will be presented for evaluation.

11.1.4 Timing Adjust Circuitry

Besides training, another way to compensate the timing skew is by using a clock synchronization circuit (CSC), such as *phase-locked loop* (PLL) or *delay-locked loop* (DLL). Both the PLL and DLL are closed-loop feedback systems with adjustable delays and a timing comparator (i.e. phase frequency detector). A feedback signal is compared against a reference signal and the delay gets adjusted until the phase relationship between the feedback and reference signals approaches a predetermined condition. For the PLL, however, the feedback signal generated by an oscillator must achieve both phase and frequency locks to the reference.

PLLs are usually found in communication systems and microprocessors for clock data recovery (CDR) and frequency synthesis.

DLLs, on the other hand, are widely used in memory interfaces for clock synchronization and de-skewing. For instance, a DLL plays an important role for the DRAM output timing, in which the output data are synchronized with the incoming system clock (e.g. *CK*). The overall timing path monitored by the DLL not only includes input data capture and clock distribution network, but also the output data path. With the closed-loop nature, the DLL can provide a dynamic on-die voltage and temperature tracking, which is essential to meet DRAM timing specs (e.g. t_{AC} and CL).

Based on various design choices, the timing adjust circuitry can be implemented using either digital [7–8] or analog [5, 9, 11] techniques. An analog approach generally includes a voltage-controlled delay line (VCDL) and a charge-pump phase detector (CPPD). An analog signal resulted from the phase difference of the CPPD and filtered by a loop filter is ported to adjust the delay of the VCDL. With a fixed number of delay stages of the VCDL, analog DLL is a good candidate for precise phase alignment and multi-phase generation.

On the other hand, a digital DLL is usually made up with digital-controlled delay line, phase detector and digital loop filter (i.e. counter). Delay gets adjusted by adding or removing a logic gate, which in turn resulting in a larger area and poorer resolution. However, the digital solution is great for scalability and portability across DRAM process generations. Since logic gate delay stays fairly constant as feature size and supply voltage scaled, an all-digital design can be reused with little modification and is also easy for test and debug, an issue which cannot be over emphasized for modern DRAM design.

A case study with the combination of both digital and analog DLL will be analyzed in Section 11.3 using a three-metal, 95 nm CMOS process. As part of the clock distribution network, the analog DLL serves as a multi-phase generator. Wide lock range and fast locking are achieved by using a modified digital DLL. By mixing the two different circuit techniques together, measured data show great improvements on the timing performance over a GDDR3/4 combined memory system.

11.2 Clock Distribution Network

11.2.1 CML Versus CMOS

In a memory system, a clock signal must be buffered and treed out to multiple locations for high-speed parallel data operations. A circuitry which is organized to deliver the clock equally to different destinations is referred to as a clock distribution network (CDN), or a clock tree. Conventionally, a CDN is made up from simple CMOS inverters. The inverters are sized to drive certain loads over a specific distance at predetermined rise and fall times. Despite its elegance and simplicity, a CMOS-based CDN injects switching noise into power supplies and is more susceptible to supply voltage variation.

Current-mode logic (CML), on the other hand, appears superb because of high supply noise immunity. A CML buffer is basically a differential pair with resistive loads and bias current. If bias current and voltage swing are kept constant over PVT corners, a constant delay can be achieved, which keeps PVT sensitivity at a minimum. The CML-based CDN is, however, more complicated and consumes more static power than its CMOS counterpart.

A comparison for delay of a two-stage CDN is shown in Fig. 11.2. Over process (slow – SS, typical – TT, and fast – FF) and voltage corners, the CMOS delay variation is six times greater than that of the CML CDN. However, the CML tree consumes three times more current than the CMOS tree. Similar 0.26 ps/°C temperature sensitivity is recorded for a given process.

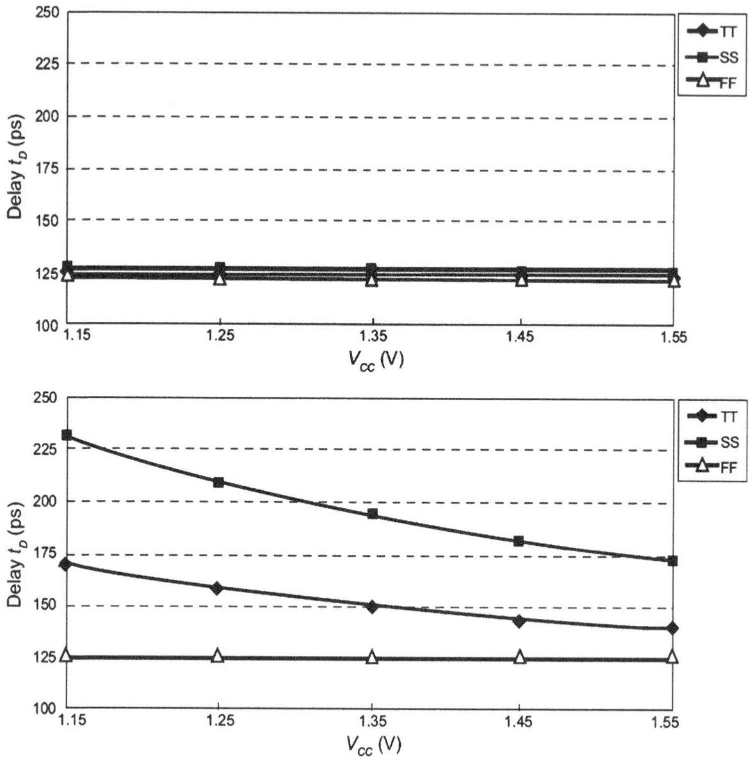

Fig. 11.2 Propogation delay comparison across process and voltage corners at 95°C, 1 GHz for two-stage CML (*top*) and CMOS (*bottom*) CDNs

Voltage sensitivity factor (α) is a good metric for comparison of competing CDN designs. The α factor is defined as

$$\alpha = (\Delta T / T) / (\Delta V / V) \tag{11.1}$$

where T is the total propagation delay at a given supply voltage (V) and changes in delay and voltage are depicted as ΔT and ΔV respectively. The bigger the number (α), the worse the performance is to supply noise. For example, if α is calculated using the data from Fig. 11.2 (across a VCC range of 1.25–1.35 V at the typical corner) we find that α_{CML} is approximately 0.1, while α_{CMOS} approaches 1. Conversion circuits that translate the small-swing CML signals to full CMOS signals may increase the total α_{CML} up to 0.5. While a CML design exhibits superior voltage sensitivity, it comes at the cost of higher power. A mixture of CML and CMOS elements may actually produce a CDN design with balanced power and timing performance.

11.2.2 Clock Division and Multiphase Clocking

In order to improve signal quality and reduce jitter and ISI for high-speed clock distribution, a clock divider is generally applied to slow down the internal clock speed. To maintain double-data rate operation, both rising and falling edges of the full speed clock must be counted for multiphase clocking after division, which results in two pairs of differential half-speed clocks, one originated from the *CK*, the other from the *CK#*. Depicted in Fig. 11.3a, total 4 bits of data are clocked at each rising edge of the four 90-degree phase-shifted clocks, *C0*, *C90*, *C180* and *C270*. The clocks generated by two clock dividers run at a quarter of the data rate. For 4-Gb/s data rate, the internal divided clocks operate at only 1 GHz. This technique greatly improves the achievable speed for an existing DRAM process. Use of multiphase ¼ rate clocks subsequently simplifies design of the capture latches and deserializers due to the relaxed clock speed.

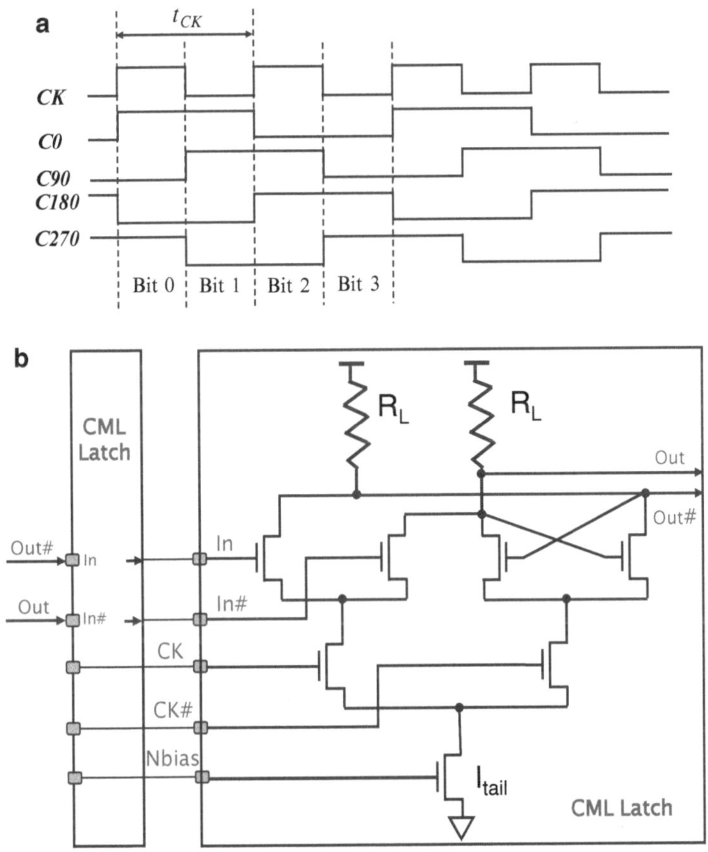

Fig. 11.3 (**a**) Multiphase clocking diagram. (**b**) CML latch-based clock divider

One drawback to including a CMOS divider in the critical timing path is increased voltage sensitivity (remember $\alpha_{CMOS} \approx 1$ from the previous example). Therefore, a CML latch based divider, shown in Fig. 11.3b, is instead used in this design. The *CK* and *CK#* are the differential full-speed clocks, and *Nbias* is the CML bias voltage. The half speed differential outputs, *Out* and *Out#* are fed back to the inputs of the divider. Two CML latches are configured as a master and slave for each divider. While the added CML divider helps to reduce voltage sensitivity for the CDN, it still produces higher temperature sensitivity, which must be mitigated.

11.2.3 Voltage and Temperature Insensitive CDN

By combining both CML and CMOS circuits, a mixed-mode, voltage and temperature insensitive CDN topology is depicted in Fig. 11.4. The clock receiver (Rx) is a CML based differential receiver with hooks for duty-cycle correction (DCC). The outputs of the Rx are fed into a CML divider (Fig. 11.3) for 4-phase ¼ rate clock generation. After conversion from CML to CMOS (via c2c), the 4-phase clocks are distributed through CMOS inverters to data capture latches and deserializers. Four copies of this CMOS tree (only one is shown in Fig. 11.4) are matched to deliver the 90-degree phase shifted ¼ rate clocks to 16 different locations.

With a conventional 2-phase CML CDN (*CK* and *CK#* running full speed all the way through the CDN), there is no divider in the data capture path (most critical timing path) and the global routes are driven by CML buffers. Only two copies of the local CMOS trees running full speed are needed to drive the data capture latches. However, after data capturing, there is still a need for a clock divider and an additional 4-phase clock tree for subsequent deserialization (data stream is deserialized from 2-bit into 4-bit). Simulations were run to measure average current for

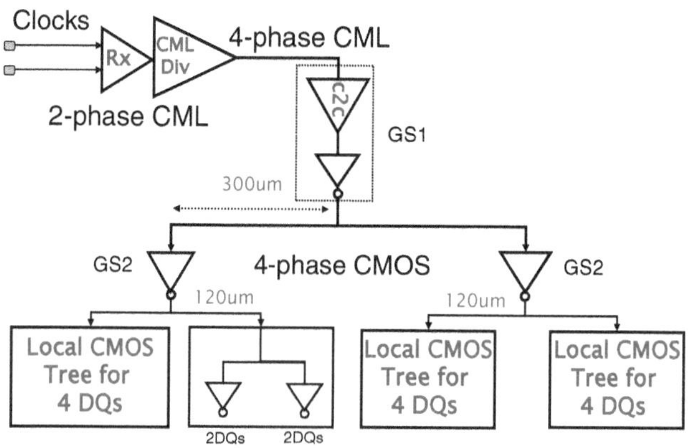

Fig. 11.4 A mixed-mode 4-phase CDN with 16 DQs

both of the described configurations at 4-Gb/s, 1.35 V, 85°C and a typical process corner. The 2-phase CML CDN produced the shortest delay at 252 ps versus 372 ps for the 4-phase mixed-mode CDN. Average current is 30 and 10 mA for the 2-phase CML and the proposed 4-phase CDN, respectively.

As mentioned before, we try to minimize VT sensitivity in the CDN to avoid the need for either retraining or on-die timing circuits, i.e., DLL. By mixing CML and CMOS circuits in the proposed CDN, we reduce supply sensitivity while minimizing power consumption. However, temperature behavior is degraded because of the CML divider. Since CML and CMOS buffers have similar positive temperature coefficients (0.26 ps/°C from the previous example), what opportunity is there to improve the temperature sensitivity?

Let's first look at how we generate the bias voltage (e.g., *Nbias* in Fig. 11.3b) for the CML circuits. To achieve a near constant delay, *Nbias* must be adjusted to provide constant current (I_{tail}) within the CML elements. With a constant current flowing through the load resistor (R_L), as shown in Fig. 11.3b, the output swing of the CML also remains constant. The propagation delay (τ) of a CML buffer can be defined as

$$\tau = C * R_L = C * \Delta V / I_{tail} \tag{11.2}$$

where the C is the load capacitance and ΔV is the output swing. Notice the delay has no direct relationship to the supply voltage.

Generation of a constant current reference across different voltages and temperatures is best accomplished using Bandgap Reference (BGR) based circuitry. An example configuration is shown in Fig. 11.5 that employs a simple current mirror and diode load referenced to a BGR. The resulting *Vinit* exhibits a negative temperature coefficient that tracks threshold voltage behavior over temperature. To cancel the positive temperature coefficient evident in the CMOS buffers, we explored using the fixed bias voltage scenario depicted in Fig. 11.5. A resistor (R_L) serves as the current mirror load for fixed bias generation. For a given process, a fixed bias voltage (*VA*) can be selected and applied to the CML circuits. The fixed biasing scheme makes the CML circuit experiencing a negative temperature coefficient (the τ is reduced as the temperature goes up), which, combined with a positive temperature response from the CMOS tree, resulting a net flat temperature response for the overall design. A plot of temperature sensitivity across three

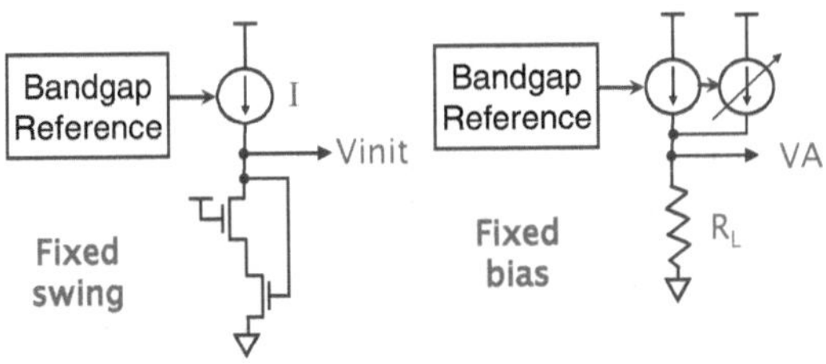

Fig. 11.5 CML bias generation: fixed-swing versus fixed-bias

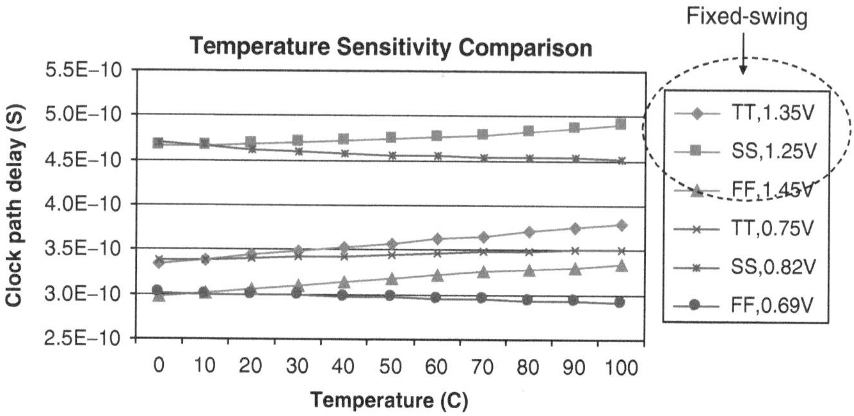

Fig. 11.6 Temperature sensitivity of the CDN across processes and voltages with two difference bias schemes: fixed-swing or fixed-bias

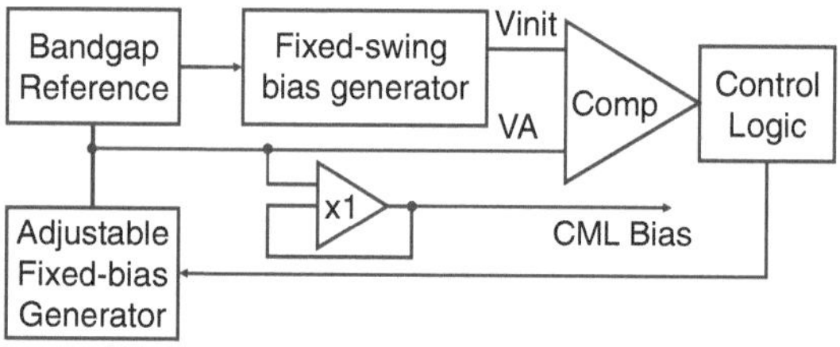

Fig. 11.7 Block diagram of a self-adaptive bias generator

different processes (SS, TT and FF) for the proposed 4-phase CDN is shown in Fig. 11.6. Overall, a 70% reduction in temperature sensitivity was achieved through the use of fixed bias voltages. However, there is only a merely 6% increase in power at 100C and typical corner with the new bias scheme.

11.2.4 Self-adaptive Bias Generator

The remaining question is how to select the fixed bias voltage across different process corners? A straightforward solution is exploiting the fixed swing bias circuit initially as a reference. A block diagram for a proposed self-adaptive bias generator is shown in Fig. 11.7. The self-adaptive process starts with the BGR initialization. After the BGR is started and settled, the temperature-compensated current is applied to both fixed-swing and fixed-bias generators. The process information is captured by the fixed-swing generator and represented by the *Vinit*.

One implementation of the control logic sets the fixed bias value *VA* to its minimum initially. The range of the *VA* can be determined by sweeping the process corners. After comparing *Vinit* and *VA* via a comparator, the results are fed into the control logic to determine if the *VA* needs to be adjusted for the process or not. When the *VA* is greater than the *Vinit*, the calibrating process is completed. A unit gain buffer is inserted between the *VA* and the final CML bias voltage. The fixed-swing generator, comparator, and control logic can be idled or shut off to save power after the calibration.

For a given device on a given process, the calibration only needs to be run once during the power-up initialization. Three-bit binary weighted current sources are used for adjustment (Fig. 11.5). A 3-bit UP counter selects the value of the current. The step size is around 20 mV with a 10 kΩ resistor. The calibration takes a couple hundred cycles, depending on the process corner and clock frequency. Calibration can be part of the power-up sequence.

11.2.5 Simulation Results

Simulation results based upon the proposed CDN are summarized in Tables 11.1 and 11.2. Both results for the fixed-swing and fixed-bias generators are listed for comparison. The mixed-mode CDN shown in Fig. 11.4 was targeted with different biasing schemes. Using typical models (TT) from a 3-metal 50 nm DRAM process, the fixed bias is set around 0.77 V, where the CML bias for the fixed-swing generator can vary between 0.71 and 0.78 V. The timing improvement over a 100 mV and 85°C VT change is about 40%. For a temperature change only at 1.45 V, from Table 11.2, a 68% improvement is achieved. Compared to the 2-phase CML CDN described before, the proposed 4-phase CDN approaches a similar timing performance over the voltage and temperature variations while at the same time, consumes far less power and simplifies the timing path design (with only one CDN and consolidated data capture and deserialization).

For low-voltage operations, the improvement using the proposed VT insensitive CDN is not as significant as that at higher voltages. At 1.35 V, only 45% improvement is recorded over 85°C temperature change, as compared to 68% at 1.45 V. For low-voltage high-speed memory applications, voltage sensitivity is dominant factor for the timing budget. A clock distribution design with more CML components may be more appropriate in this case.

Table 11.1 Simulation results across voltage and temperature

	Propagation delay of the clock distribution network		
TT, 4 Gpbs	**Voltage, Temperature**	**Fixed-swing (ps)**	**Fixed-bias (ps)**
1	1.45 V, 0°C	378	383
2	1.45 V, 85°C	409	373
3	1.35 V, 0°C	394	402
4	1.35 V, 85°C	425	385

Table 11.2 Simulation results across temperature

	Propagation delay of the clock distribution network		
TT, 4 Gpbs	**Voltage, Temperature**	**Fixed-swing (ps)**	**Fixed-bias (ps)**
1	1.45 V, 0°C	378	383
2	1.45 V, 85°C	409	373
Delta	0 V, 85°C	31	10

11.3 Clock Synchronization Circuits

Clock synchronization circuit (CSC) is another critical timing circuitry related to high performance DRAM. As the clock speed for I/O interfaces push well into the gigahertz range, designing CSC, or de-skewing circuitry, becomes a challenge. The on-die sync circuit must cover a wide operation range and, at the same time, achieve good jitter performance. Inclusion of on-die clock synchronization is necessary for control of output timing skew between the system clock and the output data from the DRAM. However, there are side effects from the use of sync circuits when the system clock period becomes comparable with the I/O delay of the DRAM. Variations in process, voltage, and temperature (PVT) can easily cross cycle boundaries and make read latency (i.e. CL) prediction difficult. Besides latency control, high-speed clock distribution is also susceptible to duty-cycle distortion, power supply noise, and timing mismatch. Power and jitter for timing circuits must be carefully gauged to achieve overall better performance.

From all-digital approach [7–8] to pure analog implementation, from open-loop synchronous mirror delay (SMD) [10] to close-loop architecture, there are various flavors in designing a sync circuit based on different applications [4–15]. Traditionally, a delay-locked loop (DLL) is selected for memory interfaces due to its close-loop architecture and relatively simple implementations. In this section, we will explore a mixed-mode DLL (MDLL) used in a combination 512 Mb 2.0 Gb/s/pin GDDR3 and 2.5 Gb/s/pin GDDR4 synchronous DRAM. The chip was fabricated in a 1.5 V 95 nm triple-metal CMOS process. The measured maximum achievable clock frequency is 2 GHz at 2.12 V supply.

11.3.1 MDLL Clocking Architecture

The proposed mixed-mode DLL (MDLL) [2] clocking system is shown in Fig. 11.8. A conventional digital DLL with fast-lock and fine resolution (via phase interpolation) is adopted for clock synchronization and de-skewing. A clock divide-by-2 circuit is inserted into the input path to slow down the internal clock speed for high-speed GDDR4 operation. To extend the lock range of the MDLL, a MUX is either used to select full-speed (i.e. GDDR3) or half-speed (i.e. GDDR4) clock. As

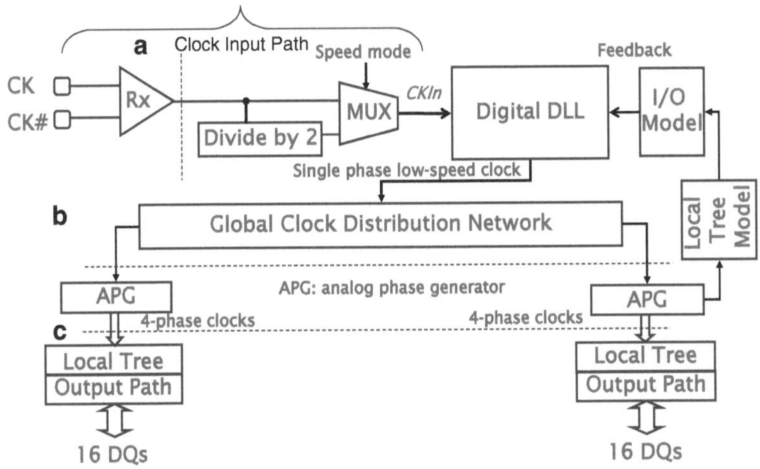

Fig. 11.8 Mixed-mode DLL clocking architecture

mentioned in the CDN design, the clock divider effectively doubles the I/O operating speed with minimum impact on internal timing circuits. When half-speed clock is used for power saving and better signal integrity for high-speed clock distribution, an edge recovery circuit is required for a DDR interface, where falling edges of the clock are also used for data operations. An analog DLL using a voltage-controlled delay line (VCDL) is picked for the edge recovery. During power-up and initialization, the analog phase generator (APG) is locked first, followed by the digital DLL synchronization. During digital synchronization, the phase detector of the APG is disabled for a short period of time when the output of the DLL is switched for quick initialization. Digital delay interpolation is the last step of the lock process for tight synchronization (more details follow).

The MDLL-based CDN can be grouped into two parts: global clock tree and local clock tree (as shown in Fig. 11.8). With 32 DQs placed in four quadrants of the die for a graphic DRAM, clock trees can consume significant power. Matching is another concern when distributing multiphase clocks over a long distance. Instead, in this design, only a single-phase clock out of the divider and the digital DLL is distributed globally across the die. Two analog phase generators (APG) are placed at each end of the global tree to generate multiphase clocks. The outputs of the APG will drive the local clock tree, which eventually clocks the output data path. For low-speed operation, a clock divider, like the one shown in Fig. 11.3, can be used instead to generate 4-phase clocks to save power and lock time. The duty-cycle for each clock phase is not critical as long as each clock distribution path is matched within the local CDNs. Except for the APG, which using CML-typed delay elements, all the clock distributions are based on CMOS gates primarily for power saving purpose.

In order to evaluate the proposed clocking architecture, a comparison between the digital DLL (DDLL) and the proposed MDLL is summarized in Table 11.3. An 800

Table 11.3 Comparison between conventional digital DLL (DDLL) and proposed mixed-mode DLL (MDLL) at 1.5 V, 85°C, and 800 MHz external clock

	DLL current (mA)	Global clock tree + phase generator (mA)	Local clock trees (mA)	Lock time (ns)
DDLL	14	13.3	50	56
MDLL	8.2	13.1	50	146

MHz external clock is fed into both DLLs. The clock is running at full speed for the DDLL and half speed for the MDLL. Four-phase clocks are generated by a clock divider for the DDLL and an APG for the MDLL, respectively. Simulation is performed at 1.5 V, typical process and 85°C. Duty-cycle correction (DCC) for the DDLL is not included in this simulation, which requires additional power, area, and lock time for the conventional approach. From Table 11.3, the proposed MDLL consumed less power for both DLL and global clock tree (including two APGs) due to half-speed clocking. However, the lock time did increase for analog phase generation.

11.3.2 Fast-Lock Digital DLL

The digital DLL used in the MDLL can achieve fast lock with a measure-controlled delay (MCD) line, shown in Fig. 11.9a. The time-to-digital conversion (TDC) and digital-to-time conversion (DTC) will start during measure initialization. The input delay (t_{in}) and output delay (t_{out}) are modeled with an I/O model ($t_{in} + t_{out}$). Delay for clock distribution is t_{Tree} in Fig. 11.9a. The intrinsic loop delay (t_{ILD}) is defined as

$$t_{ILD} = t_{Tree} + t_{in} + t_{out} \tag{11.3}$$

The t_{ILD} is measured by the measure delay array (MDA) and the difference ($N^*t_{CK} - t_{ILD}$) is stored in the forward delay array (FDA). Parameter t_{CK} is the clock period and N is an integer number. When $t_{CK} > t_{ILD}$, $N = 1$.

During initial measurement, the FDA is bypassed by the two-input MUX. A *Start* signal triggered by the rising edge of the internal clock (*CKIn*) initializes the measurement. As shown in Fig. 11.9, the *Start* signal propagates through the clock tree and I/O model and becomes the *Stop* signal at the input of the MDA. The time difference between the two signals (*Start* and *Stop*) is t_{ILD}. The outputs of the MDA are sampled by the *CKIn* and the output digital codes are stored in the Measure Control block. After the difference between N^*t_{CK} and t_{ILD} is captured and reflected in the FDA, the initialization is completed and the loop is locked (total delay from external to output clock is N^*t_{CK}). An example timing diagram with $N = 2$ is depicted in Fig. 11.9b. The delay gets adjusted by adding or removing delay stages. Assuming the delay per stage is t_D, the number (n) of delay stages picked after the MCD initialization is approximately

$$n = (N * t_{CK} - t_{ILD}) / t_D \pm 1 \tag{11.4}$$

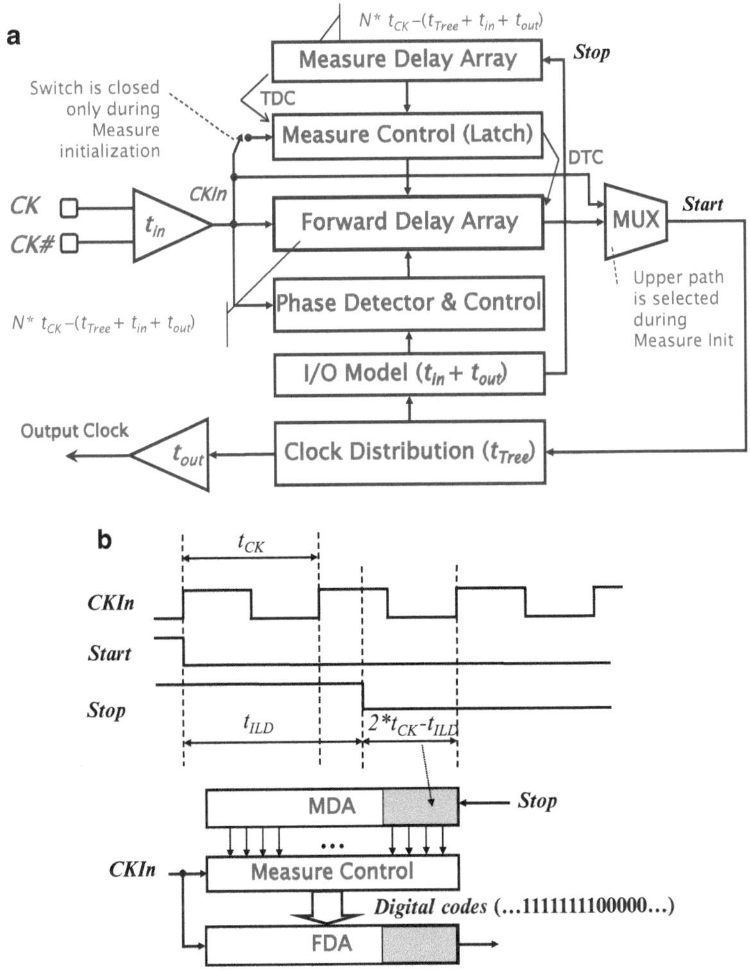

Fig. 11.9 (**a**) Fast-lock, measure-controlled delay (MCD) based digital DLL. (**b**) Timing diagram of the MCD operation ($N = 2$)

After the measurement, the output of the FDA is selected and normal DLL loop is resumed thereafter for voltage and temperature tracking. Because MDA and FDA are identical and operating mutual exclusively, the MDA and FDA can be combined to save area. Typical digital delay for two logic gates (t_D) is around 100–150 ps. Typical lock time for the MCD-based DLL is around ten clock cycles.

A digital delay interpolation (DDI) [15] or mixing out of one FDA element is followed to achieve better resolution, generally in 5–15 ps range. Typical lock time for the entire DDLL is less than 60 cycles, depending on clock frequency, intrinsic loop delay, adjustment rate, and final resolution. Compared to the two-cycle lock of a conventional synchronous mirror delay (SMD) [10] or clock-synchronized delay (CSD), the lock time is longer for the MCD because of switching the output via the

MUX. Longer delay through the digital loop, clock path and I/O model with regarding to a shorter t_{CK} also requires extra time for the feedback signal to settle. Finer resolution achieved by subsequent delay interpolation further increases the lock time. The data quoted in Table 11.3 shows a lock time about 45 cycles for the DDLL, which is significantly better than the 200-cycle spec requirement. Even with the analog DLL, the lock time is still within the spec. Although the finer adjustment circuitry can be designed using an analog approach, an all-digital implementation is still adopted in the clock architecture for its simplicity, scalability, and fast lock time.

11.3.3 Analog Phase Generator (APG)

Based on the proposed clocking architecture, a multiphase generator is required to recover the edges lost due to clock division. For this design an analog DLL was selected. Diagrams for a conventional APG and the proposed DCC-enabled APG are shown in Fig. 11.10a, b, respectively. Both APGs are locked to 180-degree phase of the input reference instead of 360-degree phase for fast-locking, reduced power and area, and better linearity. Not like the digital delay line, the delay gets adjusted by varying the delay per stage of the voltage-controlled delay line (VCDL). If an *N* number of delay stages are connected in series in the VCDL, when locked, the delay per stage for 180-degree phase locking can be defined as

$$t_D = 0.5 * t_{CK} / N \tag{11.5}$$

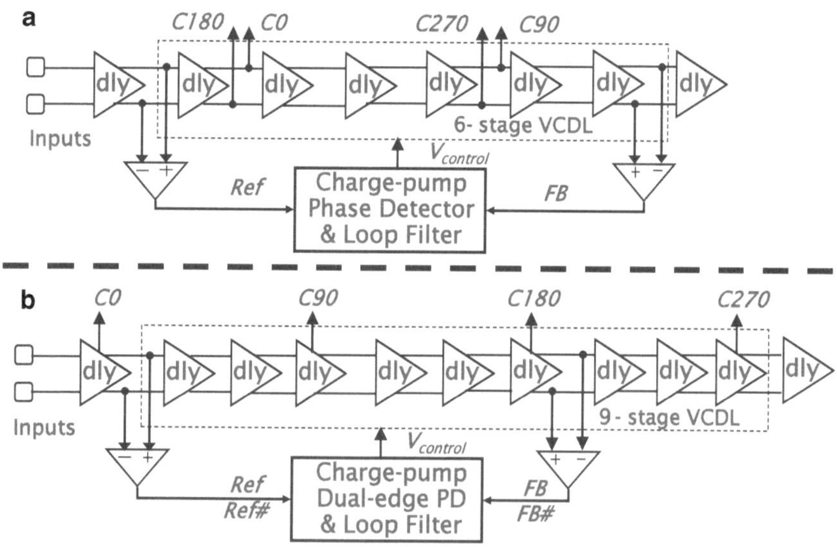

Fig. 11.10 (**a**) Conventional APG with 6-stage VCDL; (**b**) proposed APG with built-in DCC

For 800 MHz clock and 6-stage VCDL shown in Fig. 11.10a, the analog delay per stage is around 104 ps when the APG is locked. Compared to digital adjustment, analog VCDL is quite different with a fixed number of delay stages and generally smaller layout area to cover the full operating range. Built with differential CML logic, the analog delay line has much less intrinsic delay and better supply noise rejection capability (α_{VCDL} close to 0.5 with a CML to CMOS converter). To maintain similar characteristics of the analog delay, the length (*N*) of the VCDL needs to be doubled to 12 when a 360-degree phase locking is selected.

One drawback for 180-degree locking is sensitivity to input duty-cycle distortion (DCD). Because the falling edge of the input signal is used to generate feedback for the phase detector (PD), any input DCD will cause a misalignment for the multiphase outputs. For the proposed MDLL-based CDN, the DCD is likely generated from the DDLL and global clock distribution network. A separated duty-cycle correction circuit is generally required in the conventional APG for high-speed operation.

Instead of using complementary phases, such as clock 180 and 270 in Fig. 11.10a, the proposed APG adds extra three delay stages to form a 9-stage VCDL. Multiple phases are tapped off from the equal length delay stages of the VCDL, as shown in Fig. 11.10b. The dependency of output DCD due to input duty-cycle is removed with the proposed configuration. However, output DCD dependency still needs to be solved for the feedback signal (*FB*), which is generated by using the falling clock edges.

A fully differential dual-edge phase detector (DDPD) can be used to perform automatic DCC, as is shown in Fig. 11.11a. Inputs *Ref* and *Ref#* are differential references, while *FB* and *FB#* are differential feedback signals. Based on transitions of both rising and falling edges, pulses are generated to trigger the SR latches and create *UP* and *DN* signals, respectively. Timing diagrams for this DDPD are shown in Fig. 11.11b with only *UP* and *DN* signals for clarity. Whenever both *UP* and *DN* are in the same state, either '1' or '0', there is no net charge dumped into the loop filter, and the control voltage ($V_{control}$) remains unchanged. During normal phase locking, $V_{control}$ gets adjusted twice – one at rising edges and one at falling edges. With input duty-cycle distortion (DCD) present, the relationship is plotted in Fig. 11.11b with equal intervals for pumping up and down when locked. The average impact on $V_{control}$ is zero and the loop is locked. With 1 GHz external clock and 46% duty-cycle internal clock (2 ns period), simulation data shows that the proposed APG maintains the desired phase alignment (500 ps apart from each other) while the conventional APG produced a worst-case bit time of 403 ps (40% duty cycle relative to the external clock as, measured between the four phases).

A self-bias VCDL [17] is used for the proposed APG. The VCDL is clamped to its minimum delay initially to avoid false locking. A fully differential current steering charge pump phase detector (CPPD) [18] is used to generate the control voltage for the VCDL with no dead zone. The typical lock time for the APG is in the order of 100 cycles. The analog implementation including all decoupling capacitors occupies roughly 133 × 317um², which is one third of the size of the DDLL. Average current drawn from the APG is around 3–5 mA depending on the frequency and operating conditions.

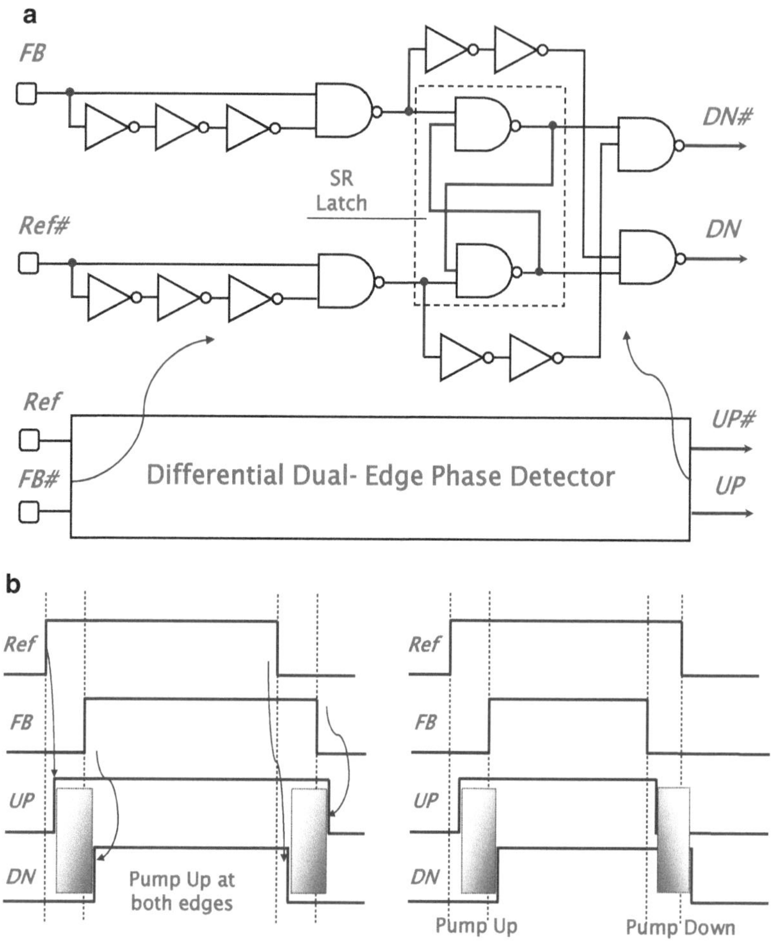

Fig. 11.11 (**a**) Fully differential dual-edge phase detector (DDPD). (**b**) Timing diagrams of a dual-edge phase detector (locking w/o DCD in the *left* and locked w/ DCD in the *right*)

11.3.4 Measurement Results

Using the proposed mixed-mode DLL, an 88 mm^2 512 Mb × 32 GDDR3/GDDR4 device, is fabricated in a 1.5 V 3-metal 95 nm CMOS process. Figure 11.12a is a measured t_{CK}-*VDD* Shmoo data showing device performance reaching beyond 2 Gbps/pin in GDDR3 operation for page fast write and read vectors that produce 2 ns column cycle times (data failure limited). The CAS latency (CL) is 12 with a burst length (BL) of 4 for the GDDR3 operation. Figure 11.12b is a measured jitter histogram of read strobe (*RDQS*), where the part is running at 1.5V, 1 GHz in GDDR4 mode. The RMS jitter and the peak-to-peak jitter are 4.63 and 38 ps, respectively. Performance results are summarized in Table 11.4.

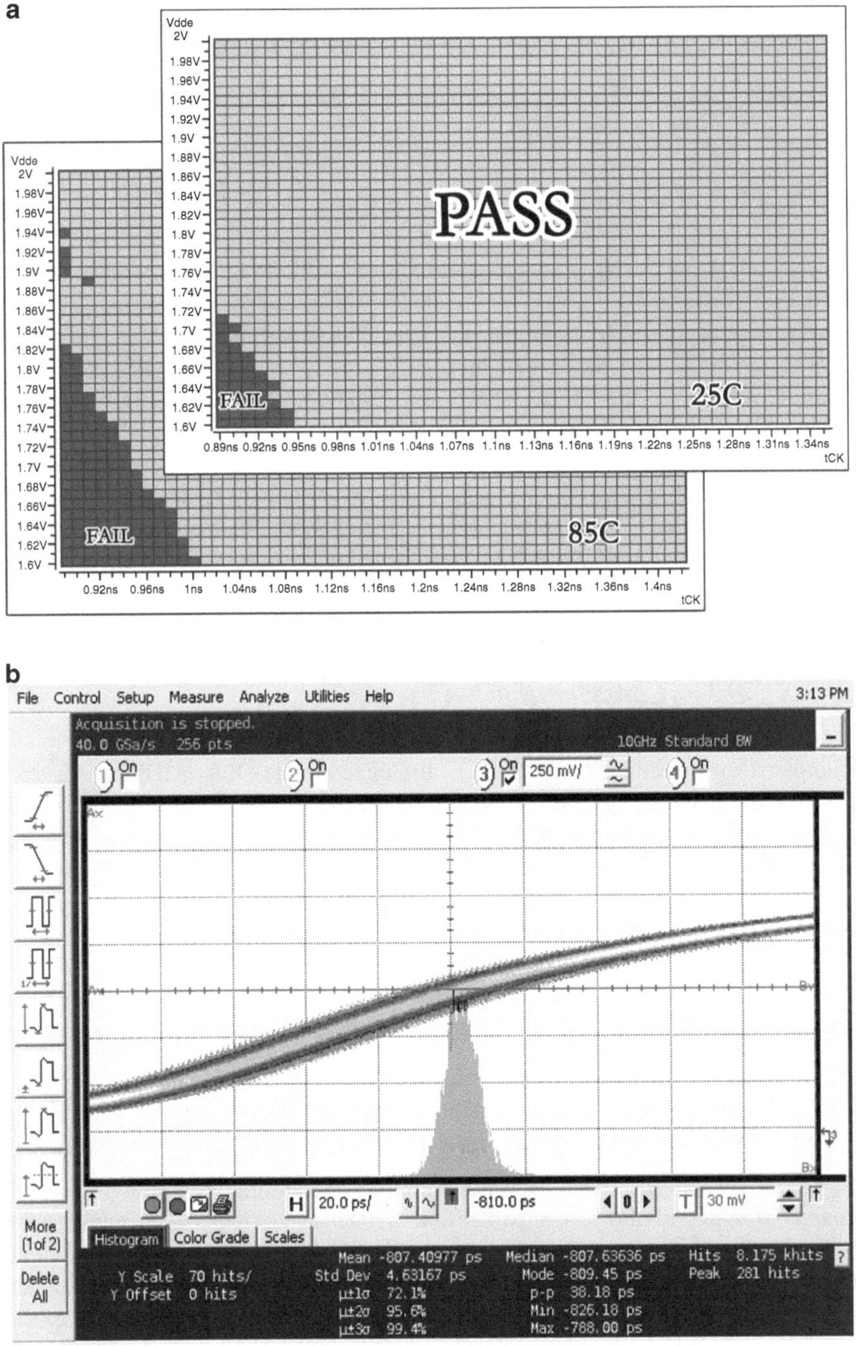

Fig. 11.12 (**a**) *tCK* vs. *VDD* Schmoo GDDR3, BL = 4, temperature = 25/85°C, CL = 12. (**b**) Measured jitter histogram at 1 GHz, 1.5 V and GDDR4

Table 11.4 Performance summary of the proposed MDLL

Technology	3-metal 95 nm CMOS process
Supply voltage	1.5–1.8 V nominal
Operating frequency	400–1,250 MHz
Jitter at 1.5 V, 1 GHz, GDDR4	38 ps (peak-to-peak), 4.63 ps (RMS)
Lock time	<200 clock cycles
Maximum speed	4 Gb/s/pin at 2.12 V supply

The *RDQS* is treated as a special data (*DQ*s) with a '0101' toggling pattern. A Read strobe is bundled with 8-bit of data and forwarded back to memory controllers for data capture. The timing spec for the *RDQS* and *DQ*s (i.e. t_{DQSQ}) is much tighter than the one for the system clock (e.g. *CK*) and *DQ*s (i.e. t_{AC}). A typical t_{DQSQ} is limited to only 80 ps for a GDDR4 operation and the skew is mainly due to mismatches from the CDN and channel links. The t_{AC} spec, on the other hand, is proportional to t_{CK} (e.g. $0.25*t_{CK}$ or 200 ps for 2.5 Gb/s data rate) and dictated by I/O modeling and sync circuit performance. The forwarded strobes keep the memory interface source synchronous and facilitate high-speed data operation. The data rate can be extended well beyond 2.5 Gbps by using the proposed half-speed clocking and mixed-mode DLL.

11.3.5 Other Consideration

Compared to a traditional digital DLL, the mixed-mode DLL (MDLL) is a step forward in both operating speed and lock range. However, the MDLL may not be suitable for the input timing adjust due to its overall longer delay path, which is more susceptible to voltage and temperature (VT) variations. If on-die VT tracking and deskewing via a sync circuit is desire for the input path, like the one shown in Fig. 11.4, circuit modifications are necessary to further improve the timing performance.

Let's first revisit the clocking architecture shown in Fig. 11.8. Assuming a differential CML receiver (Rx) for the system clock, there are at least three places (dotted lines) in Fig. 11.8 where CML to CMOS or CMOS to CML conversion takes place:

(a) from the CML Rx to the CMOS input (*CKIn*)
(b) from single-phase CMOS output of the global CDN to CML inputs of the APG
(c) from the VCDL of the APG to 4-phase CMOS clocks fed to local trees

Each logic conversion adds delay and impacts overall timing. A straightforward upgrade is to get rid of the logic conversion as much as you can and change circuits between (a) and (b) into CML-based logic. A CML-typed clock divider (shown in Fig. 11.3b) and CML-based MUX can be used instead for the clock input path. An analog DLL (ADLL), like the one shown in Fig. 11.10a can be adopted for clock synchronization. Duty-cycle distortion may not be an issue due to fully differential CML operation through the clock input path. The 6-stage VCDL in Fig. 11.10a can provide 12 phases with a 30° phase shift for analog phase interpolation (API) [9].

An exemplary 15-bit thermometer code can be applied to interpolate between any two adjacent phases. The final resolution after interpolation is about two degree (or about 7–10 ps for an internal speed of 800 MHz).

The locking procedure for the ADLL can be divided into three steps. The first step is to lock the delay of the VCDL to half of the clock period (i.e. 0.5 * t_{CK}). As shown in Fig. 11.10a, the charge-pump phase detector (CPPD) and the loop filter will adjust the control voltage ($V_{control}$) until the feedback (*FB*) and reference (*Ref*) are in sync. The initial $V_{control}$ can be clamped to achieve a minimum delay in the VCDL. If a similar analog delay is replicated in the analog phase generator (APG), the control voltage ($V_{control}$) from the ADLL can be ported to bias the 9-stage VCDL of the APGs. Therefore, even without enabling its own CPPD, the APG can acquire lock at the same time when the ADLL is locked. Matching the various VCDLs (Fig. 11.10a, b) in both loading and interconnects makes it easy to consolidate the initial locking steps and reduce overall lock time.

After the ADLL is locked, the dual-edge CPPD (Fig. 11.11) of the APG is enabled for duty-cycle adjustment and further polish for its tight locking. The final step is analog phase interpolation (API) [19] between two adjacent phases out of the VCDL of the ADLL. Instead of using a charge-pump circuitry, a separate phase detector (PD) with window detection and digital filtering can be applied for phase selection and interpolation. Just like digital DLL, the added window detection and digital filtering can further improve the jitter performance of the API. Any jitter coupled from the I/O model and presented at the input of the PD via the feedback signal can be tracked and filtered before any adjustment. The digital codes won't be changed until the phase error is beyond a predetermined window (e.g. 20 ps). The resulted digital codes for the API can be stored in a shift register or an up-down counter. Similar interpolation control logic makes the transition from digital DLL to analog DLL a lot easier for design, test and debug. The learning curve is not as steep when similar circuit blocks (e.g. the APG and the digital logic) have been scrutinized and investigated before in the MDLL.

Besides the sync circuits and clock input path, the global clock tree can also be upgraded to CML based CDN (2-phase ¼ rate), which eliminating the interface (b) between the global CDN and APGs. Although the ADLL and APG can be further consolidated into one with multiple APIs, the power consumption over a 4-phase global CML CDN may not justify the modification.

The last piece of the clock path stays the same with a local CMOS clock tree to deliver the 4-phase ¼ rate CMOS clocks to the output path. The techniques discussed in the Section 11.2 for a voltage and temperature insensitive CDN can also be applied here. The CML-based clock input path and the global CDN are biased by a fixed voltage, with process information calibrated by a self-adaptive bias generator (Fig. 11.7). The power out of the global CDN can be further scaled down according to the operating frequency, which can be monitored by the ADLL.

Instead of using current-mode logic, another way to mitigate the power supply sensitivity is by supply regulation. The internal power supplies for the critical timing circuits can be regulated and isolated from the rest of the periphery circuits. Based on the Eq. 11.1, the α factor gets bigger at lower voltages, therefore, supply regulation may not be a good candidate to solve timing problem as the feature size and supply voltage keep scaling down. For example, the typical supply voltage (VDD) for the 95 nm DRAM process is 1.5 V (GDDR4) or 1.8 V (GDDR3), but the VDD for the 50 nm process is only 1.35 V. There are not enough margins left for voltage regulation as the process scaled into deep submicron region. Dedicated power supply pins and good on-die power delivery for the timing circuits are a few keys to keep timing specs on check.

11.4 Future Directions for Nanoscaled DRAM Interface

High-performance computing system, such as multi-core processor or high-end game console, continues to advance and demands higher bandwidth and lower power from the DRAM. These demands directly impact the DRAM interface protocol and hence, the timing circuit design. At the same time, cutting throat competition in the commodity DRAM market forces DRAM manufacturers pushing down the cost and chasing the endless process scaling. The bleeding edge technologies create a slew of challenges, not necessary new but in a different scale, facing to today's DRAM designer, such as density, I/O transfer rate, power consumption and device variability. Given the historical trends, the future for the nanoscaled DRAM interface should continue to grow in speed, burn more power, and operate at lower supply voltages. But these trends may not be sustainable due to cost and power concerns. Instead of a skyrocket clock frequency, like the race in the microprocessor world a few years back, a course change (just like from single-core to multi-core change in the microprocessor) is necessary to fuel a higher performance DRAM with a reasonable power and cost structure. It is hard to predict what those changes are, but hopefully from the selective topics in this chapter, the reader will appreciate of the difficulties we are encountering and maybe get an idea for potential solutions to problems related to DRAM interfaces.

Moving forward, there are other areas of consideration for high-performance DRAM that may impact timing circuit design. One area only briefly mentioned in this chapter is link initialization for high-speed data operations. Link initialization conducted by memory controller via extensive training may be beneficial to offload some of the burden of the timing circuit design from the DRAM. Per-pin deskewing [16] looks promising but adds a lot of complexity and overhead in the memory controller. The link initialization may also add a significant test and debug time to validate a memory interface.

Another possibility is moving the timing circuits and high-speed I/O into an intermediate chip, like the advanced memory buffer (AMB) used in fully-buffered dual in-line memory module (FB-DIMM). The AMB, fabricated with logic processes, serves as an intermediate buffer between memory controller and DRAM devices. A narrow point-to-point channel supported by the AMB enables high-speed communication between the host controller and AMB. At the same time, a wide channel between the AMB and memory devices provides higher bandwidth.

The additional latency and cost resulted from the intermediate chip may or may not be justified by the improved performance.

A radical change recently for high density DRAM devices is by 3-dimensional integration (3-DI), which stacking memory devices vertically using *through-silicon vias* (TSVs). Aside from the existing chip-stacking technologies, such as wire-bonding, flip-chip, multichip packaging (MCP) and system-in-package (SiP), the TSV technique explores the chip-stacking and interconnecting vertically by deploying TSVs. An 8-gigabit 3D DDR3 DRAM based on 50 nm technology [20] has been reported with a single master device and three slave chips connected using 300 TSVs. The pitch of the TSVs is only 60-μm. The master device acts as a buffer that isolates the channel and the slave chips. The timing circuits are located only in the master device, which in turn, saves power compared to conventional quad-die package structures. The prototype device also utilizes a separate datapath (a bidirectional local datapath and unidirectional global datapath – read/write separated), a modified repair scheme and a power-noise reduction technology.

The TSV technology opens a door for continuous circuit innovations in array architecture, signaling, timing control, error detection and correction, redundant and repair algorithm, and test and debug. Although there is still a lot challenges ahead with the new technology, by shortening the interconnections between the chips, it's not hard to envision a next generation memory system emerged with reduced power and die size, improved timing and elevated bandwidths.

References

1. B. Keeth, R.J. Baker, B. Johnson, F. Lin, *DRAM Circuit Design – Fundamental and High-Speed Topics* (Wiley-IEEE Press, Piscataway, NJ, 2007)
2. F. Lin, R. Royer, B. Johnson, B. Keeth, A wide-range mixed-mode DLL for a combination 512 Mb 2.0 Gb/s/pin GDDR3 and 2.5 Gb/s/pin GDDR4 SDRAM. IEEE J. Solid-St. Circ. **43**(3), 631–641 (2008)
3. I.E. Sutherland, R.F. Sproull, D. Harris, *Logical Effort: Designing Fast CMOS Circuits* (Academic, San Diego, CA, 1999)
4. S. Bae, Y. Sohn, K. Park, K. Kim, D. Chung, J. Kim, S. Kim, M. Park, J. Lee, S. Bang, H. Lee, I. Park, J. Kim, D. Kim, H. Kim, Y. Shin, C. Park, G. Moon, K. Yeom, K. Kim, J. Lee, H. Yang, S. Jang, J. Choi, Y. Jun, K. Kim, A 60nm 6Gb/s/pin GDDR5 graphics DRAM with multifaceted clocking and ISI/SSN-reduction techniques, in *IEEE ISSCC Dig. Tech. Papers*, 278–279 (2008)
5. K. Lee, J. Cho, B. Choi, G. Lee, H. Jung, W. Lee, K. Park, Y. Joo, J. Cha, Y. Choi, P.B. Moran, J. Ahn, A 1.5-V 3.2 Gb/s/pin Graphic DDR4 SDRAM with dual-clock system, four-phase input strobing, and low-jitter fully analog DLL. IEEE J. Solid-St. Circ. **42**(11), 2369–2377 (Nov. 2007)
6. F. Lin, B. Keeth, *A Self-Adaptive and PVT Insensitive Clock Distribution Network Design for High-Speed Memory Interfaces*. Proceedings of the IEEE/EDS Workshop on Microelectronics and Electron Devices (WMED) (2009), pp. 55–58, Boise, ID
7. F. Lin, J. Miller, A. Schoenfeld, M. Ma, R.J. Baker, A register-controlled symmetrical DLL for double-data-rate DRAM. IEEE J. Solid-St. Circ. **34**(4), 565–568 (1999)
8. T. Matano, Y. Takai, T. Takahashi, Y. Sakito, I. Fujii, Y. Takaishi, H. Fujisawa, S. Kubouchi, S. Narui, K. Arai, M. Morino, M. Nakamura, S. Miyatake, T. Sekiguchi, K. Koyama, A

1-Gb/s/pin 512-Mb DDRII SDRAM using a digital DLL and a slew-rate-controlled output buffer. IEEE J. Solid-St. Circ. **38**(5), 762–768 (2003)
9. S. Sidiropoulos, M.A. Horowitz, A semi-digital dual delay-locked loop. IEEE J. of Solid-St. Circ. **32**(11), 1683–1692 (1997)
10. T. Saeki, Y. Nakaoka, M. Fujita, A. Tanaka, K. Nagata, K. Sakakibara, T. Matano, Y. Hoshino, K. Miyano, S. Isa, S. Nakazawa, E. Kakehashi, J.M. Drynan, M. Komuro, T. Fukase, H. Iwasaki, M. Takenaka, J. Sekine, M. Igeta, N. Nakanishi, T. Itani, K. Yoshida, H. Yoshino, S. Hashimoto, T. Yoshii, M. Ichinose, T. Imura, M. Uziie, S. Kikuchi, K. Koyama, Y. Fukuzo, T. Okuda, A 2.5-ns clock access, 250-MHz, 256-Mb SDRAM with synchronous mirror delay. IEEE J. Solid-St. Circ. **31**, 1656–1665 (1996)
11. Y. Jung, S. Lee, D. Shim, W. Kim, C. Kim, S. Cho, A dual-loop delay-locked loop using multiple voltage-controlled delay lines. IEEE J. Solid-St. Circ. **36**(5), 784–791 (2001)
12. J. Kim, S. Lee, T. Jung, C. Kim, S. Cho, B. Kim, A low-jitter mixed-mode DLL for high-speed DRAM applications. IEEE J. Solid-St. Circ. **35**(10), 1430–1436 (2000)
13. G. Dehng, J. Lin, S. Liu, A fast-lock mixed-mode DLL using a 2-b SAR algorithm. IEEE J. of Solid-St. Circ. **35**(10), 1464–1471 (2001)
14. B. Johnson, B. Keeth, F. Lin, H. Zheng, Phase-tolerant latency control for a combination 512Mb 2.0gb/s/pin GDDR3 and 2.5gb/s/pin GDDR4 SDRAM. *ISSCC 2007 Digest of Tech. Papers*, 494–495 (2007)
15. B. Garlepp, K.S. Donnelly, K. Jun, P.S. Chau, J.L. Zerbe, C. Huang, C.V. Tran, C.L. Portmann, D. Stark, Y.F. Chan, T.H. Lee, M.A. Horowitz, A portable digital DLL for high-speed CMOS interface circuits. IEEE J. Solid-St. Circ. **34**(5), 632–642 (May 1999)
16. E. Yeung, M.A. Horowitz, A 2.4 Gb/s/pin Simultaneous Bidirectional Parallel Link with Per-Pin Skew Compensation, *IEEE J. Solid-St. Circ.* **35**(11), 1619–1628 (2000)
17. J. Maneatis, Low-jitter process-independent DLL and PLL based on self-biased techniques. IEEE J. of Solid-St. Circ. **31**(11), 1723–1732 (1996)
18. K. Wong, E. Fayneh, E. Knoll, R. Law, C. Lim, R. Parker, F. Wang, C. Zhao, Cascaded PLL design for a 90nm CMOS high-performance microprocessor. *ISSCC Dig. Tech. Papers*, 422–424 (2003)
19. B. Kim, L.S. Kim, K. Park, Y.H. Jun, S.I. Cho, A DLL with jitter reduction techniques and quadrature phase generation for DRAM interface. IEEE J. of Solid-St. Circ. **44**(5), 1522–1530 (2009)
20. Uksong Kang, Hoe-Ju Chung, Seongmoo Heo, Soon-Hong Ahn, Hoon Lee, Soo-Ho Cha, Jaesung Ahn, DukMin Kwon, Jin Ho Kim, Jae-Wook Lee, Han-Sung Joo, Woo-Seop Kim, Hyun-Kyung Kim, Eun-Mi Lee, So-Ra Kim, Keum-Hee Ma, Dong-Hyun Jang, Nam-Seog Kim, Man-Sik Choi, Sae-Jang Oh, Jung-Bae Lee, Tae-Kyung Jung, Jei-Hwan Yoo, Changhyun Kim, 8Gb 3D DDR3 DRAM using through-silicon-via technology. *ISSCC Dig. Tech. Papers*, 130–131 (2009)

Chapter 12
Overview and Scaling Prospect of Ferroelectric Memories

Daisaburo Takashima

Abstract In this chapter, the overview and scaling prospect of ferroelectric random access memory (FeRAM) are presented. First, the memory cell structure, material, operating principle, and current status of FeRAMs and chain FeRAMs are introduced. Second, several key techniques to achieve stable FeRAM operation and realize FeRAM scaling are described: (1) the scaling techniques to reduce bitline capacitance to obtain sufficient cell signal in scaled FeRAMs; (2) the dummy cell techniques to generate adequate reference voltage; and (3) the cell signal enhancement techniques to squeeze cell charge from capacitor and maximize cell signal. Third, the reliability of FeRAM and future cell directions using 3-dimensional capacitor and new approaches are discussed. Finally, the application of FeRAM to nonvolatile RAM cache in solid-state drive (SSD) / hard-disk drive (HDD) and performance improvement are demonstrated.

Keywords Ferroelectric • FeRAM • Chain FeRAM • Memory • RAM

12.1 Introduction

A ferroelectric random access memory (FeRAM) using a mechanism of polarization switching of ferroelectric film is one of the most promising candidates for future high-performance nonvolatile RAM, because the FeRAM provides excellent advantages such as no-volatility, low power consumption, fast read/write operation and high endurance of over 10^{13} read/write cycles. The FeRAM is applicable to many kinds of portable electronic equipments such as smart card, cellar phone and nonvolatile cache memory in memory systems. In the late 1980s, early papers on 512-b

D. Takashima (✉)
Center for Semiconductor Research & Development, Semiconductor Company, Toshiba Corporation, Yokohama, Japan
e-mail: daisaburo.takashima@toshiba.co.jp

K. Iniewski (ed.), *CMOS Processors and Memories*, Analog Circuits and Signal Processing, DOI 10.1007/978-90-481-9216-8_12,

and 16-kb FeRAMs [1, 2] with 2-transistor and 2-capacitor cell (2T/2C) were reported. The first 256-kb FeRAM using 1-transistor and 1-capacitor cell (1T/1C) was demonstrated in 1994 [3]. Since then, many efforts to shrink memory cell and die sizes as well as fast operation have been made, and many new architectures and circuits/device technologies have been developed. Currently, high-density 64 Mb [4–6] and 128 Mb FeRAM [7] have been developed. The high read/write bandwidth of 1.6 GB/s with DDR2 SDRAM interface [7] is the highest in all non-volatile memories and the same as that of the current volatile DRAM products. This enables DRAM replacement with non-volatile FeRAM.

One of the most promising new FeRAM architecture is a chain FeRAM™, which was reported in 1997 [8, 9]. The chain FeRAM realizes small die with fast operation and achieves large cell signal even in small cell. In this chapter, the overview of significant key FeRAM techniques developed for last two decades, the-state-of-the-art FeRAM technologies, and scaling prospect of FeRAM are introduced. Most of them are technologies installed in 8–128 Mb chain FeRAMs, and others are cited from other company's FeRAM products and papers.

In Section 12.2, the principle of FeRAM cell structure and material, and basic read/write operation are presented. In Section 12.3, an array configuration and current status of the conventional FeRAM are presented. In Section 12.4, the concept of chain FeRAM is presented, and the development history of chain FeRAM technologies is also demonstrated. In Section 12.5, two array architectures to overcome signal degradation in scaled FeRAMs and obtain sufficient cell signal are demonstrated. And practical sense amplifier configuration and its sensing scheme are also shown. In Section 12.6, the dummy cell techniques, which match FeRAM operation and generate intermediate voltage between "1" and "0" signals, are introduced. In Section 12.7, two cell signal enhancement techniques to maximize cell signal are introduced. These techniques are significant to realize low voltage high-density FeRAMs. In Section 12.8, a variety of reliability issues for FeRAM are discussed. In Section 12.9, the future prospect and concern of FeRAM scaling are discussed, and some challenging approaches including three-dimensional cells for the conventional FeRAM and chain FeRAM, a cross-point cell and FET type cells are also introduced. In Section 12.10, a new application to enhance memory system performance and the measured performance of the prototype solid-sate drive (SSD) with nonvolatile FeRAM cache are demonstrated. Section 12.11 summaries the results of this review.

12.2 FeRAM Principle and Read/Write Mechanism

A ferroelectric thin film using $PbZr_xTi_{1-x}O_3$ (PZT) and $SrBi_2Ta_2O_9$ (SBT) materials as shown in Fig. 12.1a has perovskite crystalline structure. When an electric field is applied to this film, the Ti or Zr atom moves. And it stays there even after disappearance of electric field. This ferroelectric film has the hysteresis curve shown in Fig. 12.1b. The X-axis shows voltage applied to two electrodes. The Y-axis shows

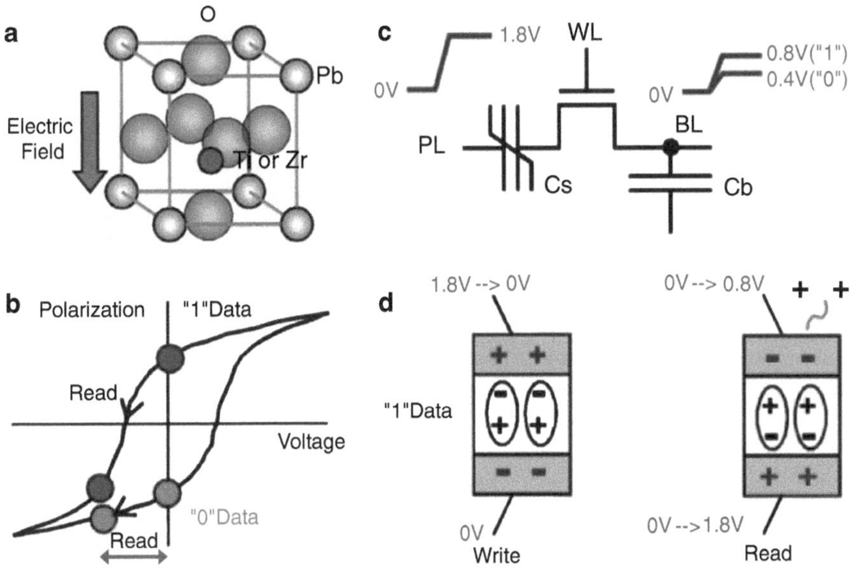

Fig. 12.1 Structure and operating principle of ferroelectric memory: (**a**) perovskite structure; (**b**) hysteresis curve and loci of read operation; (**c**) memory cell configuration and read operation; and (**d**) read/write mechanism

polarization. This material has two stable points of "1" and "0" data at no bias condition. When bias is applied, a polarization switching occurs. The basic cell configuration of the conventional 1-transistor and 1-ferroelectric capacitor (1T/1C) FeRAM is shown in Fig. 12.1c. A cell transistor is connected to a wordline (WL) to select memory cell. One terminal of memory cell is connected to a bitline (BL) to read/write cell data. Other terminal is connected to a plateline (PL) to induce polarization switching.

The read/write mechanism is explained in Figs. 12.1c, d. In the "1" data write operation, the PL is pulled down to 0 V and the BL is pulled up to 1.8 V. In the "0" data write operation, the PL and BL are set to opposite state biases. When a bias is applied to ferroelectric capacitor, the hole or electron is collected on each side of electrode to negate the electric field caused by polarization switching. In the read operation, as shown in Fig.12.1c, when PL is pulled up to internal array operation voltage of Vint, the read bias (Vread), which is determined by capacitance ratio of Cb/(Cb + Cs) × Vint, is applied to the selected ferroelectric capacitor, where Cs is cell capacitance and Cb is total BL capacitance. Therefore, when the memory cell has "1" data, the polarization switching occurs, and the collected charge on electrode is released as shown in Fig. 12.1d, and the released charge is read out to BL. All readout cell data are amplified by sense amplifiers, and these amplified data are rewritten in memory cell again.

12.3 Conventional 1T/1C FeRAM and Current Memory Cell Structures

The conventional 1T/1C FeRAM cell configuration [3] is shown in Fig. 12.2a. A memory cell consists of one transistor and one ferroelectric capacitor connected in series. It is the same cell configuration as that of DRAM, except ferroelectric materials such as PZT and SBT are utilized for FeRAM capacitor, and cell data are held using polarization switch mechanism. This 1T/1C FeRAM cell configuration is simple. The memory cell is placed at a half of intersections of wordline (WL) and bitline (BL) in order to adopt the folded BL scheme [10] to minimize cell array noise. The cell data is readout to one bitline (/BL), and other bitline (BL) is used as reference bitline which is designed to be the intermediate value between “1” and “0” data. The voltage difference between BL pair is sensed and amplified in a sense amplifier. The plateline (PL) is divided into each row. The selected PL is driven to switch polarization of memory cell capacitor.

The current smallest 64 Mb memory cell structures in the conventional FeRAM are shown in Fig. 12.2b-1, b-2. The memory cell in Fig. 12.2b-1 is the capacitor-under-bitline (CUB)-type cell [4]. The bitline is formed after ferroelectric capacitor fabrication. This type fits to FeRAM macro embedded in logic LSI, because a metal BL such Al and Cu can be also used as wiring metal in logic LSI. The memory cell size of 64 Mb cell is 0.54 μm^2, and the capacitor size is 0.25 μm^2 in 130 nm logic process. The other memory cell in Fig. 12.2b-2 is the capacitor-over-bitline (COB)-type cell [5]. The bitline is formed before ferroelectric capacitor fabrication. This type fits to standalone FeRAM, because tungsten (W) is used for bitline (BL) to endure thermal process of PZT capacitor crystallization around

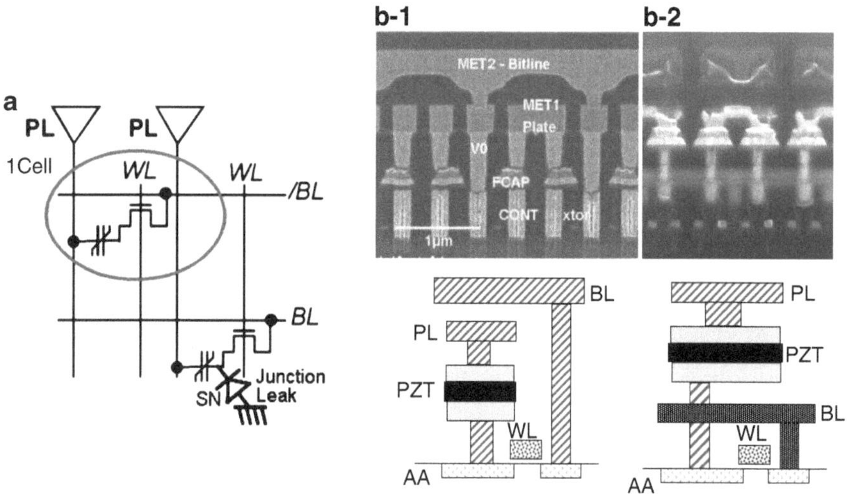

Fig. 12.2 Conventional FeRAM cell configuration: (**a**) schematic diagram; (**b-1**) capacitor-under-bitline (CUB) structure; and (**b-2**) capacitor-over-bitline (COB) structure

600–650°C. On the other hand, this COB type memory cell has an advantage of large capacitor area thanks to no via to pass through in gap between capacitors. The memory cell size of 64 Mb cell is 0.25 μm², and the capacitor size is 0.12 μm² in 130 nm DRAM process. These small memory cell is realized by introducing two key technologies; a capacitor plug technology [11], which directly connects an active area with a bottom electrode of capacitor, and one-mask capacitor etching technology [12], which realizes small capacitor by minimizing taper angle of capacitor fabrication.

12.4 Chain FeRAM Architecture and Development History

The conventional FeRAM has achieved small cell size as discussed in previous session. However, there are some problems to be solved. First, the memory cell size is limited to 8F² like DRAM by adopting the folded bitline (BL) scheme, where F means the minimum feature size. Second, the plateline (PL) cannot be shared with neighboring cells, because the repeating plateline-up of unselected cells destroys cell data due to junction leakage. Therefore, the PL and its driver should be divided into each row. This causes large PL delay and large PL driver size due to large capacitive load of PL, which is the order of 10 pF. Third, cell signal is degraded in scaled FeRAMs, due to large BL capacitance because of many BL contact attachments to BL.

A chain FeRAM proposed in 1997 [8, 9] overcomes these problems. The chain FeRAM realizes smallest 4F² size memory cell in ideal case using a planar cell transistor thanks to cross-point-type cell, and achieves fast random access. Figure 12.3a shows the circuit diagram of the chain cell block. One memory cell

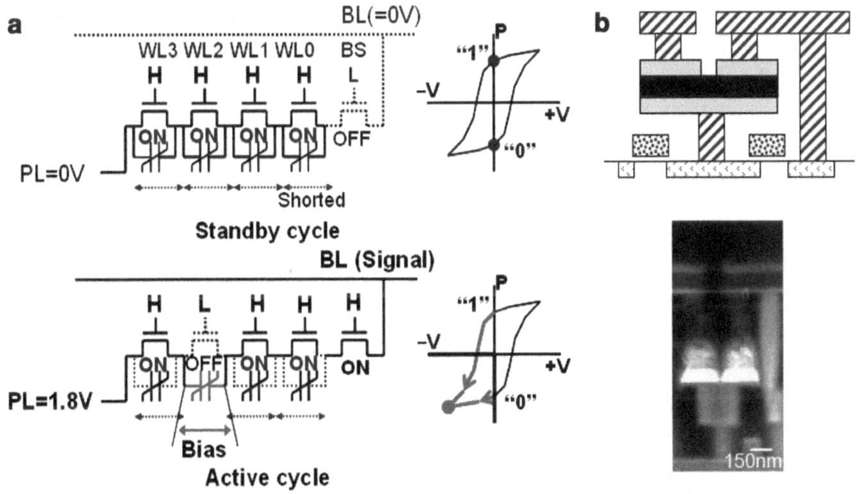

Fig. 12.3 Chain FeRAM cell configuration: (**a**) chain cell string and basic operation; and (**b**) memory cell structure

consists of one transistor and one ferroelectric capacitor connected in parallel, and one chain cell string consists of plural memory cells connected in series. One terminal of the chain cell string is connected to BL via the block selecting transistor, and the other terminal is connected to PL. One PL is shared with plural memory cells. This enables larger PL driver with small PL driver area penalty and fast random access. In the practical design, 1 PL is shared with 16 cells of 2 chain cell strings. The BL contact to memory cell is also shared with plural memory cells. This reduces BL capacitance per cell to 1/2–1/3 of that of the conventional FeRAM, and achieves large readout cell signal due to small total BL capacitance.

Figure 12.3a also shows basic operation of chain FeRAM. During standby cycle, all wordlines (WL0–WL3) are boosted to high voltage, the block selecting signal (BS) and PL are set at ground, and BL is precharged at ground. Therefore all ferroelectric capacitors are short-circuited by turned-on cell transistors. As shown in the hysteresis loop, both "1" and "0" data are held steadily at zero bias points without noise from capacitor electrodes. During active cycle, when accessing cell data of one cell of chain cell string, the selected wordline (WL2) is pulled down, and the block selecting signal (BS) is pulled up. After that, the PL is pulled up to array operating voltage of Vint of 1.8 V. Therefore, PL bias is applied only to the selected ferroelectric capacitor, and the cell data is read out to BL, as shown in the loci of the hysteresis loop. When "1" data is read out, a large charge is read out to BL by polarization switching, and when "0" data is read out, a small charge is read out to BL by non-polarization switching. A sense amplifier amplifies the readout cell signal. On the other hand, cell data of unselected cells are protected by short-circuiting capacitors. Thus, a complete random access is realized.

The smallest 0.25 μm^2 memory cell with 0.084 μm^2 capacitor [13] of 128 Mb chain FeRAM using 130 nm dual oxide 4-metal CMOS process is shown in Fig. 12.3b. The bottom electrode and capacitor plug are shared with twin cells. This minimizes memory cell size. Ninety nm PZT film is used for 1.5 V low voltage array operation. The development history of chain FeRAM chips are summarized in Fig. 12.4. The first prototype of 16 Kb chain FeRAM [14] has been developed in 1999. Since then, 8 Mb [15] in 2001, 32 Mb [16] in 2003, 64 Mb [6] in 2006 and 128 Mb [7] in 2009 have been developed so far. The average cell size including block selector region has been shrunk from 13 to 0.32 μm^2. A random cycle time for each chip is around 70 ns, whereas the page/burst cycle time has been drastically reduced from 80 ns to 1.25 ns.

The latest 128 Mb chain FeRAM has the highest memory capacity in all nonvolatile RAMs. The die size is 87.7 mm^2. This chip adopts DDR2 SDRAM interface of 1.6 GB/s (= 16 b IO/1.25 ns) read/write bandwidth at 400 MHz clock and 1.8 V Vdd, which is the highest in all nonvolatile memories and equivalent to the current DDR2 DRAM product. Features of 4-bank interleave operation, /CAS latency of 6 and burst length of 4 and 8 meet to JEDEC standard. The read/write burst of 1×10^{13} cycles fits to nonvolatile cache application in memory systems.

The development history of chain memory cells is also shown in Fig. 12.4. Until 8 Mb, the offset cell was used. In this type cell, a bottom electrode was connected to an active area via M1 metal. This did not need capacitor plug technology. But

Design Rule	0.5um	0.25um	0.2um	130nm	130nm +DualOx
Capacity	16Kb	8Mb	32Mb	64Mb	128Mb
Chip	 ISSCC1999	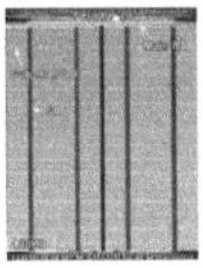 ISSCC2001	 ISSCC2003	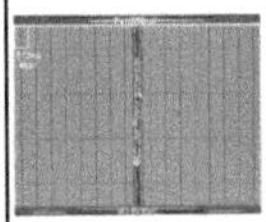 ISSCC2006	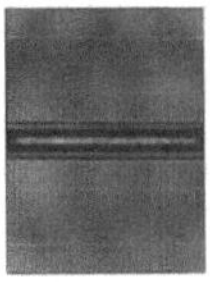 ISSCC2009
Cell Structure	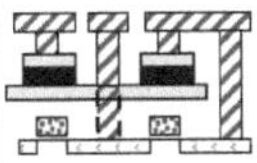 Offset Cell	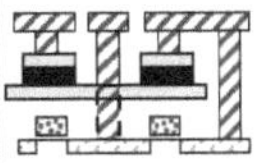 Offset Cell	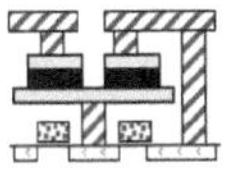 Stacked Cell	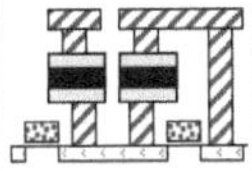 1-Mask Cell	 Non-separated PZT film
Cell Size(Ave.)	13.26um^2	5.2um^2	1.875um^2	0.7191um^2	0.32um^2
Chip Size	1.9mm^2	76mm^2	96mm^2	87.5mm^2	87.7mm^2
R/W Cycle	80ns	70ns	75ns	60ns	75ns
Page/Burst	80ns	40ns	25ns	10ns	1.25ns

Fig. 12.4 Development history of chain FeRAMs

the cell size was relatively large. Since 32 Mb, the stacked chain cell technology has been introduced to connect the bottom electrode directly to the active area by capacitor plug. And one bottom electrode is shared with neighboring twin cells to reduce cell size [17]. In 64 Mb, one mask cell was introduced to reduce process cost and minimize taper of capacitor [18]. In 128 Mb, A design rule of front-end-of-line (FEOL) process limits memory cell size. Therefore, the bottom electrode and capacitor plug are shared with twin cells again. Furthermore, the ferroelectric films in twin cells are connected to each other without etching, as shown in black region of Fig. 12.4. This structure enables perfect step coverage, less PZT RIE damage and large capacitor size. The signal improvement of 18% has been obtained [13]. The memory cell size is reduced to 14F^2. In 128 Mb, $IrO_2/SrRuO_3$/MOCVD-PZT/Ir/TiAlN/W-plug is adopted as capacitor material layers. This Ir/TiAlN/W-plug structure prevents plug-oxidation even at 650°C 1 h. This is enough for PZT annealing for crystallization [13, 17, 18].

12.5 Scaling Techniques to Reduce Bitline Capacitance

A serious problem of FeRAM scaling is the cell signal degradation due to ferroelectric capacitor scaling. The difficulty of 3-dimensional ferroelectric capacity causes capacitor area reduction by FeRAM cell scaling. This is because a cell signal is proportional to ratio of cell polarization P_{cell} to bitline (BL) capacitance Cb; (P_{cell}/Cb), and the Cb can not be scaled sufficiently according to ferroelectric capacitor area shrink.

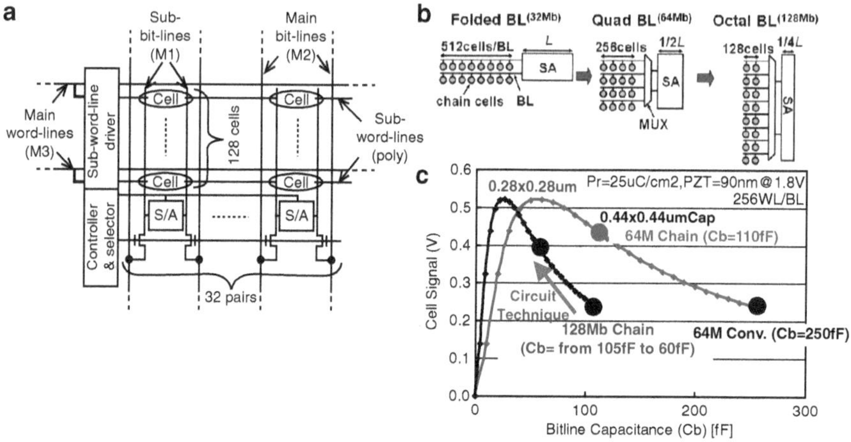

Fig. 12.5 Memory array architectures to obtain enough cell signal: (**a**) hierarchical bitline architecture; (**b**) quad and octal bitline architectures for chain FeRAM; and (**c**) bitline capacitance vs. cell signal

One solution to reduce BL capacitance Cb is to introduce hierarchical BL scheme [19] to FeRAM as shown in Fig. 12.5a. A BL is divided into plural sub-BLs, and a local sense amplifier (S/A) is equipped to each sub-BL. And the data amplified in the local sense amplifier is transferred to a main sense amplifier. This technique reduces total BL capacitance. However, the area overhead for local sense amplifier is required. Another similar approach to equip a gain transistor for each sub-BL or for each chain cell string is also proposed [20]. The issue is how to realize small and simple local amplifier to reduce area penalty.

In the chain FeRAM, the total BL capacitance is essentially smaller than that of the conventional FeRAM, and this is suitable for memory cell scaling. However, an additional technique to reduce BL capacitance further is required to obtain enough cell signal in 64 Mb and beyond. Therefore, a quad BL architecture [6] and an octal BL architecture [7] have been introduced in 64 and 128 Mb chain FeRAMs, respectively, as shown in Fig. 12.5b. These techniques reduce total BL capacitance to about 1/2, 1/4 without increasing sense amplifier area overhead. In the 32 Mb chain FeRAM [16], the folded BL scheme was adopted. The cell data is read out to one to two BLs, and the other is used as reference BL. In the quad BL of 64 Mb, the cell data is read out to one of four BLs, another is used as reference BL, two others are used as shield BL to reduce BL-to-BL coupling noise. Therefore, the number of sense amplifier (SA) is reduced to a half of that of the folded BL scheme, because a sense amplifier is shared with four BLs. This merit can be used to reduce BL length in array area to a half without sense amplifier area penalty. This effect is enhanced in the octal BL of 128 Mb chain FeRAM. One BL is used for reading cell data, another is used as reference BL, and six others are used as shield BLs. BL length is reduced to 1/4 of that of the folded BL scheme without sense amplifier area penalty.

Figure 12.5c shows the signal comparison between the conventional 64 Mb FeRAM, 64 Mb chain FeRAM and 128 Mb chain FeRAM. The x-axis shows total

BL capacitance. The y-axis shows the readout cell signal ("1"data-"0"data). The 64 Mb chain FeRAM has larger cell signal than that of the conventional FeRAM due to smaller BL capacitance. However, the simple cell shrink reduces cell signal from 0.44 V to 0.23 V because of insufficient BL capacitance reduction. However, the introduction of octal BL architecture reduces BL capacitance from 105 to 60 fF. This achieves enough cell signal of 0.4 V.

Figure 12.6a shows the practical octal BL design in 128 Mb chain FeRAM. Each chain cell is depicted as circle. In order to select one out of eight chain cell strings, eight kinds of plateline (PL) are introduced. When the PL connected to the selected chain cell string is pulled up, the cell data is read out to one of eight BLs. This BL and the reference bitline are connected to a sense amplifier via 8:2 multiplexer. The area penalty to arrange eight PLs is eliminated by placing eight PLs over area of two chain cell strings. A problem in the octal BL architecture is difficulty of connecting eight kinds of PLs to ends of chain cell strings, because total width of eight PLs spreads on 30% of two chain cell string area. The octal BL architecture using four kinds of string lengths of 6, 7, 9 and 10 solves this problem. The different contact positions, which are obtained by different chain string lengths, enable easy contacts between PLs and ends of chain cell strings, even at straight platelines placed in parallel.

A concern for chain cell strings with a variety of lengths is complicated design of block-selector selection. This needs to calculate which side of chain cell string is

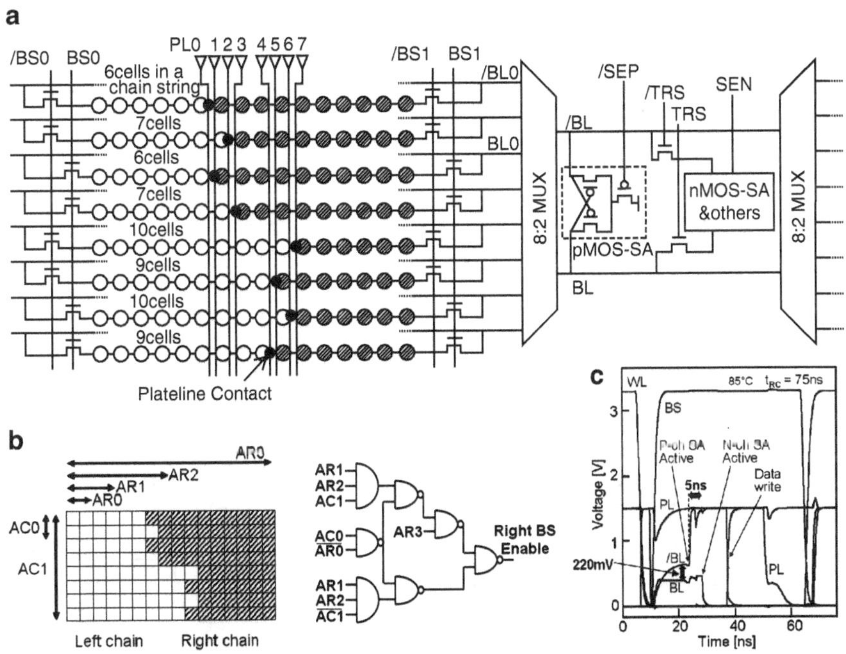

Fig. 12.6 Octal bitline architecture for chain FeRAM: (**a**) array configuration; (**b**) address mapping and control scheme for activating one of both side block selectors; and (**c**) simulated waveform of cell array

selected. However, this block-selector selection is easily realized by introducing the selection circuit as shown in Fig. 12.6b. The time penalty for decoding is only 0.9 ns.

Another problem of FeRAM scaling is difficulty of parasitic capacitance reduction of sense amplifier (SA) nodes due to complicated SA circuits. A low parasitic capacitance sensing scheme shown in the right side of Fig. 12.6a reduces parasitic capacitance at sensing. In initial sensing, the readout cell data is transferred to only Pch-SA nodes, and is amplified, whereas other nodes of SA circuits are separated from Pch-SA. After that, /TRS is pulled up, the amplified data is transferred to other SA circuits including Nch-SA. The time penalty from Pch-SA to Nch-SA activations is 5 ns, as shown in the simulated waveforms of Fig. 12.6c. As a result, the proposed two schemes reduce total BL capacitance (Cb) to 60 fF, which is a half of that of 64 Mb, and obtain sufficient cell signal of ±200 mV as shown in Fig. 12.6c.

12.6 Dummy Cell Design Techniques

Another concern inherent to FeRAM is how to make the reference BL voltage of intermediate between "1" and "0" readout cell data. This is because the bitline (BL) voltage of not only "1" data but also "0" data rises from precharged grounded bitline level as shown in read operation of FeRAM of Fig. 12.7a. The readout signal voltage of "1" data is derived from polarization switching shown as Fig. 12.1b. The readout signal voltage of "0" data is also derived from paradielectric capacitance in case of non-switching. Therefore, the relatively high reference BL voltage

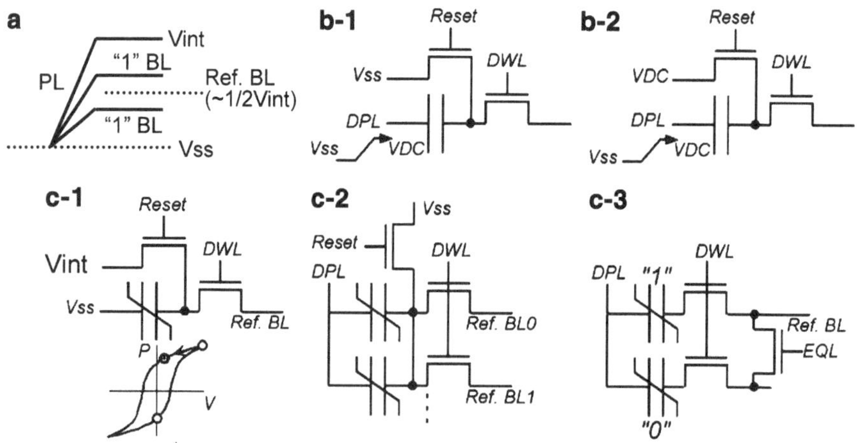

Fig. 12.7 (a) Reference bitline voltage design, (b) dummy cell designs using MOS capacitor and (c) dummy cell designs using ferroelectric capacitor

must be generated from precharged grounded BL level within thigh read-to-sense period. There are two kinds of dummy cells to generate this reference BL. One type of dummy cell uses MOS capacitor, and the other uses ferroelectric capacitor. The MOS capacitor type has better than the ferroelectric capacitor type with regard to stability, capacitance fluctuation and reliability degradation. On the other hand, the ferroelectric capacitor type occupies smaller area than the MOS capacitor type, thanks to high-dielectric constant.

In the standard dummy cell using MOS capacitor in Fig. 12.7b-1 [21], the reference BL voltage is generated by pulling up MOS capacitor. This type requires large capacitor equivalent to a total bitline capacitance in order to make a half of array operating voltage (Vint) [15]. The dummy cell of Fig. 12.7b-2 [15] overcomes this problem. The dummy cell capacitor node is precharged to VDC level during standby, and precharged charge is transferred to BL at read operation. After that, the dummy cell capacitor is pulled up to VDC level. This doubles the reference BL voltage. In other words, the dummy cell capacitor size is reduced to 1/3 to obtain the necessary reference BL voltage [15].

One type of dummy cell using a ferroelectric capacitor is shown in Fig. 12.7c-1 [22]. The paradielectric part of hysteresis curve is used to make a reference BL level. This can avoid the influence of depolarization phenomena, in which stored polarization is degraded during data retention, as well as fatigue degradation. Another type of dummy cell using a ferroelectric capacitor is shown in Fig. 12.7c-2 [23]. In this type, many reference BLs connected to dummy ferroelectric capacitors are short-circuited during reading operation. This configuration makes an average of them to minimize fluctuation of ferroelectric capacitors. The other type is also shown in Fig. 12.7c-3 [24]. This dummy cell makes an average of "1" and "0" data stored in two ferroelectric capacitors to obtain the intermediate value.

12.7 Cell Signal Enhancement Techniques

A severe problem in FeRAM scaling is how to obtain sufficient signal even in small ferroelectric capacitor. In this session, two kinds of cell signal enhancement techniques are introduced. One problem in read FeRAM operation is that read bias (Vread) which is smaller than array operation voltage Vint is applied to the selected ferroelectric capacitor. The Vread is determined by capacitance ratio of $Cb/(Cb + Cs) \times Vint$ when the plateline (PL) is pulled up to Vint, where Cs is cell capacitance and Cb is total bitline (BL) capacitance. This insufficient bias to ferroelectric capacitor in read operation causes insufficient polarization switching especially in low voltage operation.

A bitline-ground-sensing scheme [25] of Fig. 12.8a is one solution to overcome this problem. In this scheme, a BL voltage is kept at ground level even when cell data is read out to BL, while the readout charge is transferred to a sense amplifier. The similar approach called the charge-transfer-technique is sometime used for

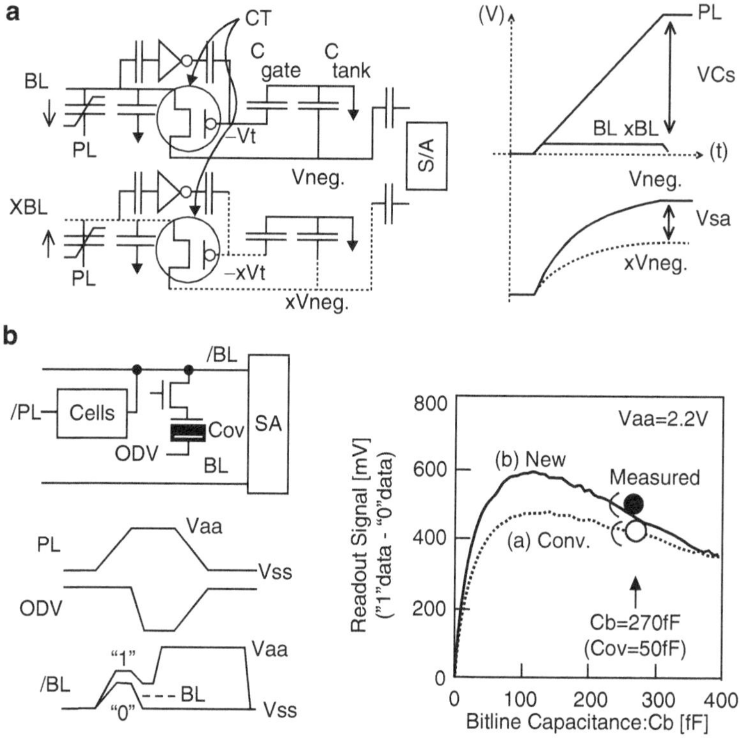

Fig. 12.8 Cell signal enhancement schemes: (**a**) bitline-ground-sensing scheme; and (**b**) ferroelectric capacitor overdrive scheme

kinds of memories. When the BL level is pulled up by readout charge from memory cell, the increased charge is fully transferred to a sense amplifier by utilizing threshold voltage drop of transfer transistor inserted between BL and sense amplifier. The node voltage of sense amplifier changes largely by charge derived from BL, because parasitic capacitance of sense amplifier node is small. Though this technique applies full Vint bias to the selected ferroelectric capacitor, this technique requires somewhat large sense amplifier area by introducing complex circuits and the negative bias generator as shown in Fig. 12.8a.

A simple signal enhancement technique called the ferroelectric capacitor overdrive scheme [15] is shown in Fig. 12.8b. In this scheme, the newly introduced overdrive capacitor (Cov) is pulled down on the way to reading out cell data to BL after activating PL. This increases read bias (Vread) to the selected ferroelectric capacitor without increasing stress bias applied for BL and memory cell. The signal enhancement about 100 mV is observed in 8 Mb chain FeRAM as shown in Fig. 12.8b. This scheme is simple, but extra capacitor is required. The advanced overdrive technique without area penalty was also reported [26].

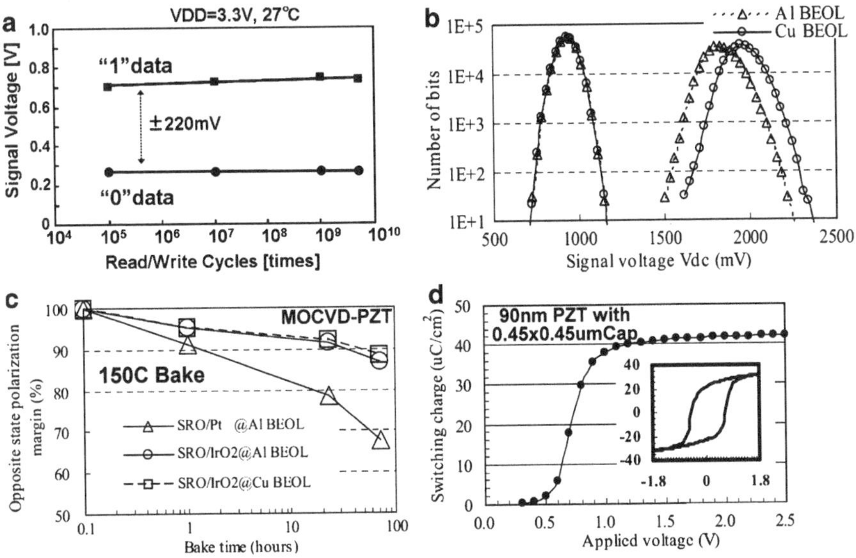

Fig. 12.9 Reliability of 64 Mb chain FeRAM: (**a**) fatigue property; (**b**) signal distribution; (**c**) data retention characteristics; and (**d**) low voltage saturation

12.8 Reliability Issues

In this section, several kinds of reliability issues are discussed. In FeRAM, many reliability items such as (1) fatigue, (2) cell signal distribution, (3) data retention after static imprint, and (4) polarization saturation especially in low voltage operation should be considered. Figure 12.9 shows some results of 64 Mb chain FeRAM. The fatigue property up to 1×10^{10} write-cycles [6] is shown in Fig. 12.9a. Introduction of $SrRuO_3/IrO_2$ top electrode has suppressed signal degradation due to fatigue cycles, whereas Pt top electrode degrades cell signal after fatigue. Figure 12.9b shows the cell signal distribution comparison between back-end-of-line (BEOL) processes using Al and Cu [27]. The tail-to-tail signals in both BEOL processes are sufficient for reliability. Moreover, the Cu process has less damage than Al process. This result is significant to realize FeRAM embedded in logic LSI. Figure 12.9c shows the data retention after static imprint [27]. The worst case of FeRAM data retention is the data retention of opposite-data-write after long data retention. For example, when "1" data is held for 10 years at 85°C, this causes imprint of ferroelectric capacitor and hysteresis curve shift. After that, when the opposite "0" data is written into a ferroelectric capacitor, the initial polarization value is degraded due to the imprint effect. This worsens "0" data retention. In the test in Fig. 12.9c, the retention-after-imprint test is accelerated by 150°C bake. 100 h at 150°C is equivalent to 10 years at 85°C. A

MOCVD-PZT with $SrRuO_3/IrO_2$ top electrode maintains data polarization after 10 years at 85°C. Figure 12.9d shows the polarization switching property [27]. By introducing thin 90 nm MOCVD-PZT film, the switching saturation is observed at low voltage of less than 1.5 V. This low voltage operation is mandatory to apply FeRAM for logic LSI process.

12.9 Future Prospect of FeRAMs – 3D Capacitor and New FeRAM

In this session, the future prospect and strategy to realize gigabit scale FeRAMs are discussed. One approach to increase an amount of memory cell polarization and realize enough cell signal even in small memory cell is to introduce the cell with three-dimensional (3D) ferroelectric capacitor.

Figure 12.10a shows 3D-capacitor adopted for the conventional COB-type cell [28]. A challenging issue to make FeRAM with 3D-capacitor is how to control the orientation of polarization over non-flat bottom electrode, because ferroelectric film has highly orientated polycrystalline structure. Another problem of the 3D-ferroelectric capacitor in the conventional FeRAM is that memory capacitor size is limited to 6F × 6F size, as shown in Fig. 12.10a, where F means each material

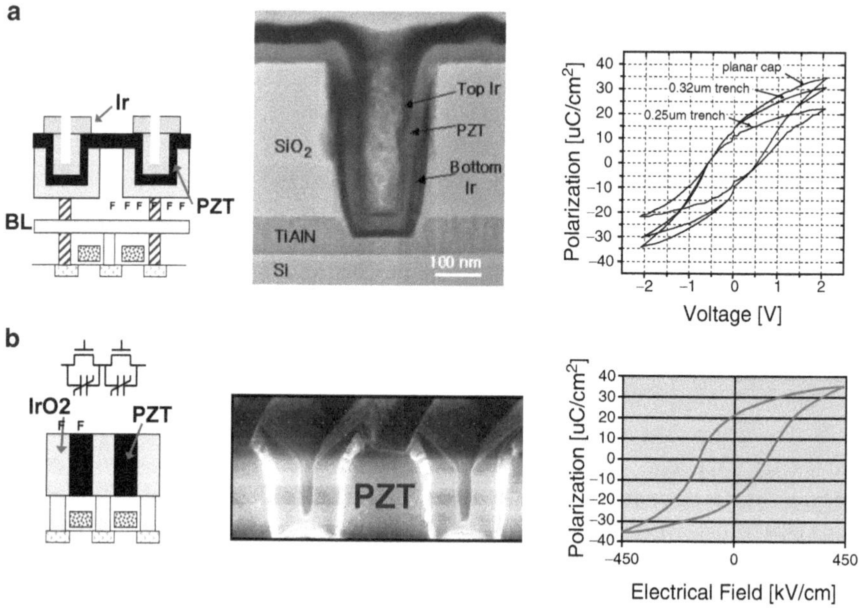

Fig. 12.10 Three dimensional FeRAM capacitor approaches: (**a**) 3D-PZT capacitor for the conventional FeRAM; and (**b**) vertical capacitor for chain FeRAM

thickness. For example, if F_{min} is determined by thickness of PZT film of 60 nm, the capacitor size is limited to 0.36 μm × 0.36 μm.

The vertical capacitor [29] applicable to chain FeRAM of Fig. 12.10b overcomes this problem and achieves 2F × 2F size. This vertical capacitor utilizes the polarization orientation only in horizontal direction thanks to simple chain FeRAM structure. Moreover, the fabricated ferroelectric capacitor has good symmetry property as shown in the hysteresis curve in Fig. 12.10b, because both electrodes are symmetrical and fabricated at a time.

Other new approaches for future high-density FeRAMs are a cross-point ferroelectric films stacked in multi-layers of Fig. 12.11a and a ferroelectric FET cell of Fig. 12.11b. In Fig. 12.11a [30], the ferroelectric films are sandwiched between plateline (PL) metal and bitline (BL) metal in cross-point type cell. This configuration does not need a cell transistor used for selecting memory cell, and enables the minimum cell size of $4F^2$ and multi-layer stacking. The problem of this type cell is

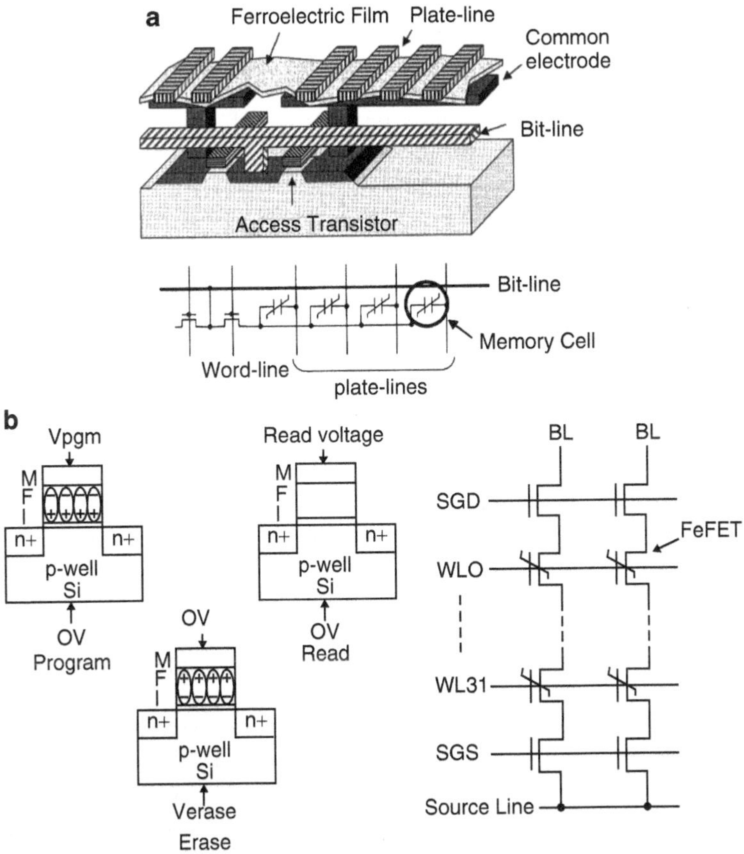

Fig. 12.11 New FeRAM approaches: (a) quasi-matrix ferroelectric memory; and (b) Fe(Ferroelectric)-NAND flash memory

the read/write-disturb to unselected ferroelectric capacitor, because of lacking a selecting transistor. The memory cell should endure against disturb of 1/3 Vint. In the prototype of Fig. 12.11a, the hierarchical BL architecture is adopted to minimize the number of disturb cycles and reduce total BL capacitance.

The concept of ferroelectric FET cell is reported in [31]. The first memory array fabrication using ferroelectric FET cells is reported in [32]. The advantage of FET type cell is scalability. The enough cell signal can be obtained even in small cell size, because small polarization is enough to change the driving current of FET. On the other hand, this type cell has many problems such as (1) device technology to realize enough data retention, (2) read/write disturb and (3) lack of stable operation method. The data retention is a severe problem in this type cell. If the electrons pass through the dielectric film and are collected at the interface between ferroelectric film and dielectric film after polarization switching, this negates the effect of polarization switching, and the stored data is destroyed. This fact needs high quality dielectric film equivalent to 8 nm tunnel oxide used for flash memory cell in order to achieve 10 year data retention. The read/write disturb problem also causes unwanted ferroelectric film switching. Good quality film should be developed to endure 1/3 Vint or 1/2 Vint disturb [32]. Last problem is the lack of stable operation method. The write data must be written only by voltage difference between gate and drain in the reported cell [32]. A new memory cell shown in Fig. 12.11b [33] overcomes this lack of stable operation. The structure and operation method is similar to NAND flash memory. As the data is written by the same sequence as NAND flash memory, data is written by voltage difference between gate and all others of source, drain and pwell. However, this type cell still has other problems discussed above and needs erase operation whose unit size is larger than program unit size.

12.10 Application as Nonvolatile FeRAM Cache

In this section, a new application to maximize potential of FeRAM is introduced. One good application of FeRAM is to use it as nonvolatile FeRAM cache in memory systems such as solid-state drive (SSD) and hard-disk drive (HDD). A FeRAM has good characteristics of low power dissipation because FeRAM data is written into memory cell with small 4–13 μA write current per cell [35] by switching polarization around several tens of femtocoulomb, whereas other nonvolatile RAMs such PRAM and MRAM requires relative large write current per cell of 50 μA to 2 mA due to current-drive-type cell. FeRAM requires one less digit of write energy per bit than others. This excellent property enables mass data read/write and achieves high read/write bandwidth with long page length. This enables DRAM replacement with non-volatile FeRAM. In order to take full advantage of low power dissipation, 128 Mb chain FeRAM with 1.6 GB/s DDR2 SDRAM interface has been developed [7].

Figure 12.12 shows the concept of SSD with a nonvolatile FeRAM cache and the measured SSD benchmark performance improvement by nonvolatile RAM cache [34]. The windows OS such as Windows XP, Vista and 7 installed in personal computer (PC) issues the flush cache commands at short intervals. This flush cache command requests SSD to move DRAM cache data into NAND flash to avoid sudden power failure, because DRAM data is destroyed at power-off. Due to this command, as shown in Fig. 12.12a, an average of valid write data in DRAM cache is only 180 KB when business user used PC for 22 days [34]. Therefore, a few valid data in DRAM cache not only degrades cache hit rate, but also causes a great deal of program access with fragmented data to NAND flash memory. On the other hand, in the nonvolatile FeRAM cache, the flush cache command can be ignored. Therefore, the 16 MB FeRAM cache can be filled with valid-data. This not only improves cache hit rate, but also achieves efficient write-back. A lot of data can be written back into NAND flash memory at a time.

Figure 12.12b shows a photograph of the prototype SSD with 16 MB chain FeRAM cache and the measured score of PC Mark05, which is popular PC bench mark test.

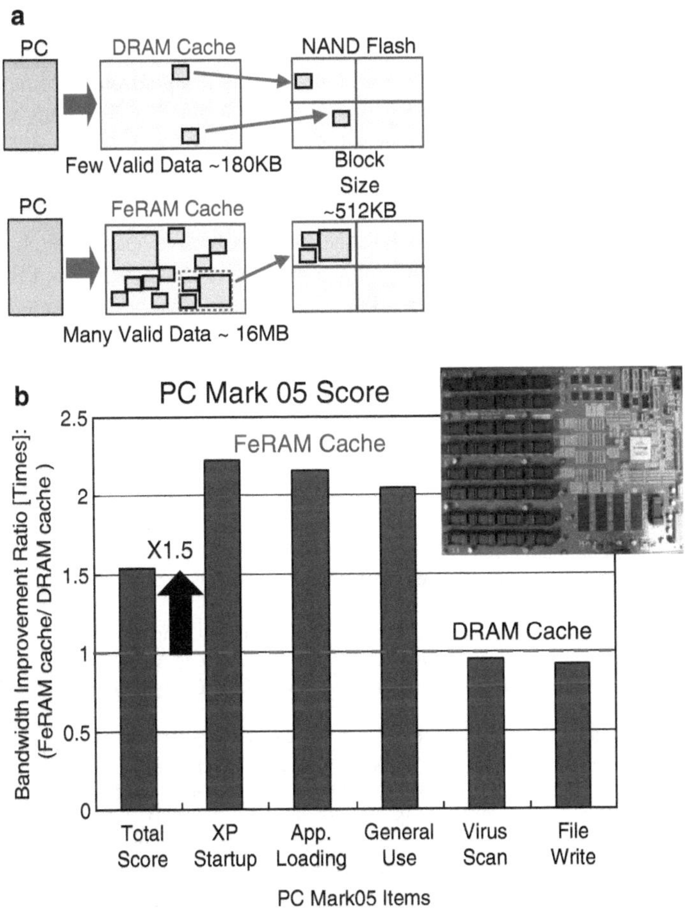

Fig. 12.12 (**a**) Concept of solid-state drive (SSD) with nonvolatile FeRAM cache and (**b**) PC mark 05 score improved by FeRAM cache

The X-axis shows the test items. The Y-axis shows the effective read/write bandwidth improvement ratio when using FeRAM cache. The FeRAM cache doubles read/write bandwidth in the XP startup, application loading, and general usage treating relatively small files. The total score has been improved by 1.5 times of that of DRAM cache. This concept is also applicable to HDD. The measured performance improvement of HDD by nonvolatile FeRAM cache also has been demonstrated in [35].

12.11 Conclusions

In this chapter, the overview and scaling prospect of ferroelectric random access memories (FeRAM) was presented. In the introduction, the summary of FeRAM development history was presented. In Section 12.2, the principle of FeRAM and read/write operation was presented. In Sections 12.3 and 12.4, the array configuration and current status of the conventional FeRAM and chain FeRAM were presented. In Sections 12.5–12.7, three key techniques to realize high-density FeRAMs: scalable array architectures, dummy cell schemes and cell signal enhancements were presented. In Section 12.8, a variety of reliability issues for FeRAMs were discussed. In Section 12.9, the concern and prospect of FeRAM scaling are discussed, and new challenging approaches for future FeRAMs were introduced. In final Section 12.10, the promising application to enhance memory system performance such as SSD was demonstrated. A FeRAM has excellent properties such as lower write-energy per cell compared with other nonvolatile RAMs such as PRAM and MRAM. This implies that FeRAM market will continue to grow if shipping products taking full advantage of FeRAM merits.

References

1. J.T. Evans, R. Womack, An experimental 512-bit nonvolatile memory with ferroelectric storage cell. IEEE J. Solid State Circuits **23**(5), 1171–1175 (Oct 1989)
2. R. Womack, D. Tolsch, A 16kb ferroelectric nonvolatile memory with a bit parallel architecture, in ISSCC Dig. Tech. Papers, Feb 1989, pp. 242–243
3. T. Sumi, N. Moriwaki, G. Nakane, T. Nakakuma, Y. Judai, Y. Uemoto, Y. Nagano, S. Hayashi, M. Azuma, E. Fujii, S. I. Katsu, T. Otsuki, L. McMillan, C. Paz de Araujo G. G. Kano, A 256kb nonvolatile ferroelectric memory at 3V and 100ns, in ISSCC Dig. Tech. Papers, Feb 1994, pp. 268–269
4. H.P. McAdams, R. Acklin, T. Blake, X.H. Du, J. Eliason, J. Fong, W. F. Kraus, D. Liu, S. Madan, T. Moise, S. Natarajan, N. Qian, Y. Qiu, K.A. Remack, J. Rodriguez, J. Roscher, A. Seshadri, S.R. Summerfelt. A 64-Mb embedded FRAM utilizing a 130-nm 5LM Cu/FSG logic process. IEEE J. Solid Sate Circuits **39**(4), 1625–1634, 667–677 (Apr 2004)
5. Y.K. Hong, D.J. Jung, S.K. Kang, H.S. Kim, J.Y. Jung, H.K. Koh, J.H. Park, D.Y. Choi, S.E. Kim, W.S. Ann, Y.M. Kang, H.H. Kim, J.-H. Kim, W.U. Jung, E.S. Lee, S.Y. Lee, H.S. Jeong, K. Kim, 130 nm-technology, 0.25 um^2, 1T1C FRAM cell for SoC (system-on-a-chip)-friendly applications, in Symposium on VLSI Technology Dig. Tech. Papers, June 2007, pp. 230–231

6. K. Hoya, D. Takashima, S. Shiratake, R. Ogiwara, T. Miyakawa, H. Shiga, S.M. Doumae, S. Ohtsuki, Y. Kumura, S. Syuto, T. Ozaki, K. Yamakawa, I. Kunishima, A. Nitayama, S. Fujii, A 64Mb chain FeRAM with quad BL architecture and 200MB/s burst mode, in ISSCC Dig. Tech. Papers, Feb 2006, pp. 134–135
7. H. Shiga, D. Takashima, S. Shiratake, K. Hoya, T. Miyakawa, R. Ogiwara, R. Fukuda, R. Takizawa, K. Hatsuda, F. Matsuoka, Y. Nagadomi, D. Hashimoto, H. Nishimura, T. Hioka, S. Doumae, S. Shimizu, M. Kawano, T. Taguchi, Y. Watanabe, S. Fujii, T. Ozaki, H. Kanaya, Y. Kumura, Y. Shimojo, Y. Yamada, Y. Minami, S. Shuto, K. Yamakawa, S. Yamazaki, I. Kunishima, T. Hamamoto, A. Nitayama, T. Furuyama, A 1.6GB/s DDR2 128Mb chain FeRAM with scalable octal bitline and sensing schemes. IEEE J. Solid State Circuits **45**(5), 141–152 (Jan 2010)
8. D. Takashima, I. Kunishima, M. Noguchi, S. Takagi, High-density chain ferroelectric random-access memory (CFRAM), in Symposium on VLSI Circuits Dig. Tech. Papers, June 1997, pp. 83–84
9. D. Takashima, I. Kunishima, High-density chain ferroelectric random access memory (chain FRAM). IEEE J. Solid State Circuits **33**(5), 787–792 (May 1998)
10. H. Masuda, R. Hori, R. Kamigaki, K. Itoh, H. Kawamoto, H. Hatto, A 5V-only 64 dynamic RAM based on high S/N design. IEEE J. Solid State Circuits **15**(5), 846–854 (Oct 1980)
11. S. Onishi, K. Hamada, K Ishihara, Y. Ito, S. Yokoyama, J. Kudo, K. Sakiyama, A half-micron ferroelectric memory cell technology with stacked capacitor structure, in IEDM Dig. Tech. Papers, Dec 1994, pp. 843–846
12. K. Shoji, M. Moniwa, H. Yamashita, T. Kisu, K. Torri, T. Kumihashi, T. Morimoto, H. Kawakami, Y. Gotoh, T. Itoga, T. Tanaka, N. Yokoyama, T. Kure, M. Ohkura, Y. Fujisaki, K. Sakata, K. Kimura, A 7.03-μm^2 Vcc/2-plate nonvolatile DRAM cell with Pt/PZT/Pt/TiN capacitor patterned by one-mask dry etching, in Symposium on VLSI Technology Dig. Tech. Papers, June 1996, pp. 28–29
13. Y. Shimojo, A. Konno, J. Nishimura, T. Okada, Y. Yamada, S. Kitazaki, H. Furuhashi, S. Yamazaki, K. Yahashi, K. Tomioka, Y. Minami, H. Kanaya, S. Shuto, K. Yamakawa, T. Ozaki, H. Shiga, T. Miyakawa, S. Shiratake, D. Takashima, I. Kunishima, T. Hamamoto, A. Nitayama. High-density and high-speed 128Mb chain FeRAM with SDRAM-Compatible DDR2 Interface, in Symposium on VLSI Technology Dig. Tech. Papers, June 2009, pp. 218–219
14. D. Takashima, S. Shuto, I. Kunishima, H. Takenaka, Y. Oowaki, S. Tanaka, A sub-40ns chain FRAM architecture with 7ns cell-plateline drive. IEEE J. Solid State Circuits **34**(11), 1557–1563 (Nov 1999)
15. D. Takashima, Y. Takeuchi, T. Miyakawa, Y. Itoh, R. Ogiwara, M. Kamoshida, K. Hoya, S.M. Doumae, T. Ozaki, H. Kanaya, K. Yamakawa, I. Kunishima, Y. Oowaki, A 76-mm2 8-Mb chain ferroelectric memory. IEEE J. Solid State Circuits **36**(11), 1713–1720 (Nov 2001)
16. S. Shiratake, T. Miyakawa, Y. Takeuchi, R. Ogiwara, M. Kamoshida, K. Hoya, K. Oikawa, T. Ozaki, I. Kunishima, K. Yamakawa, S. Sugimoto, D. Takashima, H.O. Joachim, N. Rehm, J. Wohlfahrt, N. Nagel, G. Beitel, M. Jacob, T. Roehr, A 32-Mb chain FeRAM with segment/stitch array architecture. IEEE J. Solid State Circuits **38**(11), 1911–1919 (Nov 2003)
17. T. Ozaki, N. Nagel, Y. Kumura, J. Lian, A. Hilliger, T. Tsuchiya, R. Bruchhaus, H. Kanaya, H. Koyama, U. Wellhausen, O. Hidaka, S. Shuto, S. Gemhardt, Y. Shimojo, B. K. Moon, H. Itokawa, U, Egger, H. Zhuang, K. Tomioka, M. Fukushima, K. Yamakawa, D. Takashima, I. Kunishima, Y. Oowaki, G. Beitel, Key technologies of first chain – 32Mbit ferroelectric RAM, in Extended Abstract of SSDM, Sept 2003, pp. 646–647
18. H. Kanaya, K. Homioka, T. Matsushita, M. Omura, T. Ozaki, Y. Kumura, Y. Shimojo, T. Morimoto, O. Hidaka, S. Shuto, H. Yoyama, Y. Yamada, K. Osari, N. Toko, F. Fujisaki, N. Iwabuchi, N. Yamaguchi, T. Watanabe, M. Yabuki, H. Shinomiya, N. Watanabe, E. Itoh, T. Tsuchiya, K. Yamakawa, K. Natori, S. Yamazaki, K. Nakazawa, D. Takashima, S. Shiratake, S. Ohtsuki, Y. Oowaki, I. Kunishima, A. Nitayama. A 0.602um^2 nestled chain cell structure formed by one mask etching process for 64Mbit FeRAM, in Symposium on VLSI Technology Dig. Tech. Papers, June 2004, pp. 150–151

19. J. Yamada, T. Miwa, H. Koike, H. Toyoshima, K. Amanuma, S. Kobayashi, T. Tatsumi, Y. Maejima, H. Hada, H. Mori, S. Takahashi, H. Takeuchi, T. Kunio, A 128-kb FeRAM macro for contact/contactless smart-card microcontrollers. IEEE J. Solid State Circuits **37**(8), 1073–1079 (Aug 2002)
20. D. Takashima, Y. Oowaki, I. Kunishima, Gain cell block architecture for gigabit-scale chain ferroelectric RAM, in Symposium on VLSI Circuits Dig. Tech. Papers, June 1999, pp. 103–104
21. R. Ogiwara, S. Tanaka, Y. Itoh, T. Miyakawa, Y. Takeuchi, S. Doumae, H. Takenaka, I. Kunishima, S. Shuto, O. Hidaka, S. Ohtsuki, A 0.5 μm 3V 1T1C FRAM with a variable reference bitline voltage scheme using a fatigue free reference capacitor. IEEE J. Solid State Circuits **34**(4), 545–551 (Apr 2000)
22. W. Kraus, L. Lehman, D. Wilson, T. Yamazaki, C. Ohno, E. Nagai, H. Yamazaki, H. Suzuki, A 42.5 mm^2 1Mb nonvolatile ferroelectric memory utilizing advanced architecture for enhanced reliability, in Symposium on VLSI Circuits Dig. Tech. Papers, June 1998, pp. 242–243
23. H.B. Kang, H.W. Kye, D.J. Kim, G.I. Lee, J.H. Park, J.K. Wee, S.S. Lee, S.K. Hong, N.S. Kang, J.Y. Chung, A pulse-tuned charge controlling scheme for uniform main and reference bitline voltage generation on 1T1C FeRAM, in Symposium on VLSI Circuits Dig. Tech. Papers, June 2001, pp. 125–126
24. H. Koike, T. Otsuki, T. Kimura, M. Fukuma, Y. Hayashi, Y. Maejima, K. Amantuma, N. Tanabe, T. Masuki, S. Saito, T. Takeuchi, S. Kobayashi, T. Kunio, T. Hase, Y. Miyasaka, N. Shohata, M. Takada, A 60 ns 1Mb nonvolatile ferroelectric memory with non-driven cell plate line write/read scheme, in ISSCC Dig. Tech. Papers, Feb 1996, pp. 268–269
25. S. Kawashima, T. Endo, A. Yamamoto, K. Nakabayashi, M. Nakazawa, K. Morita, M. Aoki, Bitline GND sensing technique for low-voltage operation FeRAM. IEEE J. Solid State Circuits **37**(5), 591–598 (May 2002)
26. D. Takashima, H. Shiga, T. Miyakawa, S. Shiratake, K. Hoya, R. Ogiwara, R. Takizawa, S. Doumae, R. Fukuda, Y. Watanabe, S. Fujii, T. Ozaki, H. Kanaya, S. Shuto, K. Yamakawa, I. Kunishima, T. Hamamoto, A. Nitayama, A scalable shield-bitline-overdrive technique for 1.3 V chain FeRAM, in ISSCC Dig. Tech. Papers, Feb 2010, pp. 262–263
27. Y. Kumura, T. Ozaki, H. Kanaya, O. Hidaka, Y. Shimojo, S. Shuto, Y. Yamada, K. Tomioka, K. Yamakawa, S. Yamazaki, D. Takashima, T. Miyakawa, S. Shiratake, S. Ohtsuki, I. Kunishima, A. Nitakaya, A SrRuO3/IrO2 top electrode FeRAM with Cu BEOL process for embedded memory of 130 nm generation and beyond. Solid State Electron. **50**(4), 606–612 (Apr 2006)
28. J.M. Koo, B.S. Seo, S. Kim, S. Shin, J.H. Lee, H. Baik, J.H. Lee, J.H. Lee, B.J. Bae, J.E. Lim, D.C. Yoo, S.O. Park, H.S. Kim, H. Han, S. Baik, J.Y. Choi, Y.J. Park, Y. Park, Fabrication of 3D trench PZT capacitors for 256Mbit FRAM device application, in IEDM Dig. Tech. Papers, Dec 2005, pp. 340–343
29. N. Nagel, R. Bruchhaus, K. Homik, U. Egger, H. Zhuang, H.-O. Joachim, T. Rohr, G Beitel T. Ozaki, I. Kunishima, New highly scalable 3 dimensional chain FeRAM cell with vertical capacitor, in Symposium on VLSI Technology Dig. Tech. Papers, June 2004, pp. 146–147
30. T. Nishihara, Y. Ito, A Quasi-matrix ferroelectric memory for future silicon storage. IEEE J. Solid State Circuits **37**(11), 1479–1484 (Nov 2002)
31. J.L. Moll, Y. Tarui, A new solid state memory resistor. IEEE Electron Dev. **ED-10**, 338–339 (Sept 1963)
32. T. Nakamura, Y. Nakao, A. Kamisawa, H.Takasu, A single-transistor ferroelectric memory cell, in ISSCC Dig. Tech. Papers, Feb 1995, pp. 68–69
33. S. Sakai, M. Takahashi, K. Takeuchi, Q.H. Li, T. Horiuchi, S. Wang, K.Y. Yun, M. Takamiya, T. Sakurai, Highly scalable Fe(ferroelectric)-NAND cell with MFIS (metal-ferroelectric-insulator-semiconductor) structure for sub-10nm Tera-bit capacity NAND flash memories, in NVSMW Dig. Tech. Papers, May 2008, pp. 103–105
34. D. Takashima, NAND flash memory and system designs for SSD application, in Forum of SSD Memory Subsystem Innovation, ISSCC, Feb 2009
35. D. Takashima, Y. Nagadomi, K. Hatsuda, Y. Watanabe, S. Fujii, A 128Mb chain FeRAM and system designs for HDD application and Enhanced HDD performance, in A-SSCC Dig. Tech. Papers, Nov 2009, pp. 13–16

FSC
www.fsc.org
MIX
Papier | Fördert gute Waldnutzung
FSC® C083411

Kolibri 360 GmbH
Ferdinand-Jühlke-Straße 7
99095 Erfurt, Deutschland
produktsicherheit@kolibri360.de